Discrete-Event System Simulation
FIFTH EDITION

Jerry Banks
Technológico de Monterrey, Campus Monterrey

John S. Carson II
Independent Simulation Consultant

Barry L. Nelson
Northwestern University

David M. Nicol
University of Illinois, Urbana-Champaign

Prentice Hall
Upper Saddle River · Boston · Columbus · San Francisco · New York
Indianapolis · London · Toronto · Sydney · Singapore · Tokyo · Montreal
Dubai · Madrid · Hong Kong · Mexico City · Munich · Paris · Amsterdam · Cape Town

Vice President and Editorial Director, ECS: *Marcia J. Horton*
Senior Editor: *Holly Stark*
Editorial Assistant: *William Opaluch*
Director of Team-Based Project Management: *Vince O'Brien*
Senior Managing Editor: *Scott Disanno*
Production Editor: *Paul Mailhot, PreTeX, Inc.*
Art Director: *Jane Conte*
Cover Designer: *Bruce Kenseelar*
Art Editor: *Gregory Dulles*
Manufacturing Manager: *Alan Fischer*
Manufacturing Buyer: *Lisa McDowell*
Director of Marketing: *Margaret Waples*
Senior Marketing Manager: *Tim Galligan*

To
Susie, Jay, Danny, and Zack
Jonna, Jennifer and Jonathan
Sharon and LeRoy
Elizabeth, Caitrin, Thomas, Galen, and John Patrick

The author and publisher of this book have used their best efforts in preparing this book. These efforts include the development, research, and testing of the theories and programs to determine their effectiveness. The author and publisher make no warranty of any kind, expressed or implied, with regard to these programs or the documentation contained in this book. The author and publisher shall not be liable in any event for incidental or consequential damages in connection with, or arising out of, the furnishing, performance, or use of these programs.

Mathematica and the Mathematica logo are registered trademarks of Wolfram Research, Inc., 100 Trade Center Drive, Champaign, IL 61820-7237.

Excel, Access, Visual Basic, Windows, Windows, 200, Windows XP, and Windows Vista are registered trademarks of Microsoft Corporation in the United States and other countries.

Library of Congress Cataloging-in-Publication Data

CIP data available.

Prentice Hall
is an imprint of

www.pearsonhighered.com

10 9 8 7 6 5 4 3 2
ISBN-13: 978-0-13-606212-7
ISBN-10:   0-13-606212-1

Contents

III Random Numbers   275

IV Analysis of Simulation Data   333

Preface

Our objective in this text is to provide a basic treatment of the important aspects of discrete-event simulation, with particular emphasis on examples to illustrate simulation principles and applications in manufacturing, services, and computing. The fifth edition, like all earlier editions, is meant for an upper-level-undergraduate or master's-level introduction to simulation, or for a second course with applications. We have updated the material extensively and revised some chapters completely. The associated website www.bcnn.net is an essential companion to the text. Although the book remains independent of any particular simulation language, we have continued the trend begun in the fourth edition of providing Excel-based examples and simulation support tools.

The first part of the book, entitled **Introduction to Discrete-Event System Simulation**, encompasses Chapters 1 through 4. Chapter 1, *Introduction to Simulation*, addresses the questions: What is simulation? What is it for? When is it the right tool? What are its advantages and disadvantages? What are the types of simulations? How do you carry out a simulation project? In this edition abstracts of many real simulation cases have been added.

Chapter 2, *Simulation Examples in a Spreadsheet*, introduces simulation with a series of demonstrations in Excel. To introduce random numbers, the chapter starts with simple examples to simulate coin tossing, and to simulate random service and arrival times using simple discrete distributions. From there, the instructor may choose any of the later extended examples for queueing, inventory, and other types of systems. All examples are based on a common framework, emphasizing model definition, the state variables needed for the Excel implementation, input specification, outputs, and performance measures—all in a simplified manner accessible to the student before covering the statistical review in Chapter 5. The Excel implementations, available on the website, are useful for hands-on experimentation to illustrate simulation concepts and as a starting point for some of the exercises at the end of the chapter.

While Excel is useful for teaching simulation concepts, for demonstrating the effect of statistical variability, and especially for analyzing and presenting results, it has severe limitations as a basis for simulation itself. Therefore Chapter 3, *General Principles*, discusses a generic framework centered on the key concepts of events and processes that is used as the basis for the implementation of nearly all discrete-event simulation software. Chapter 4, *Simulation Software*, provides a history of simulation and examples in GPSS and Java. We also include an updated discussion of the features and capabilities of currently available simulation software. Simulation software evolves rapidly, therefore, follow the Web links given in this chapter to access the latest vendor information.

The second part of the book, entitled **Mathematical and Statistical Models** and comprising Chapters 5 and 6, provides background material in statistics and queueing that is useful in simulation. Chapter 5, *Statistical Models in Simulation*, collects, in one location, the statistical background needed for the remainder of the book. In Chapter 6, *Queueing Models*, we introduce waiting-line concepts, a few simple steady-state queueing models, and performance measures for queueing systems. In this edition we have added a case study on rough-cut modeling of queueing systems prior to simulation. Some calculations are illustrated in MATLAB code. The spreadsheet `QueueingTools.xls`, available on the book's web site, computes queueing performance measures for all of the models in the chapter.

The third part of the book, entitled **Random Numbers**, discusses random-number generation in Chapter 7, and random-variate generation in Chapter 8. Its purpose is to introduce the concepts and some of the algorithms for generating simulation inputs, not so that a student can implement state-of-the-art methods, but so that they can use the tools intelligently and critically. The spreadsheet `RandomNumberTools.xls` contains implementations in Visual Basic for Applications (VBA) of a long-period random-number generator, such as those described in Chapter 7, *Random-Number Generation*; as well as implementations of the random-variate generation algorithms in Chapter 8, *Random-Variate Generation*.

The fourth part of the book, entitled **Analysis of Simulation Data**, encompasses Chapter 9 on input modeling, Chapter 10 on verification and validation, and Chapters 11 and 12 on output analysis.

Chapter 9, *Input Modeling*, focuses on the use of data to drive a simulation, specifically, choosing the statistical distributions to represent a model's random input variables. In this edition, the core example has been replaced with a series of brief mini-cases that illustrate the difficulties that can occur in input modeling, particularly violations of the usual "i.i.d." assumption. Input modeling when no data are available is also described.

Chapter 10, *Verification and Validation of Simulation Models*, addresses the questions: How do we know our model is "correct"? How accurate is it? What techniques can we use to assess the model's accuracy and correctness?

In this edition, the chapters on the analysis of simulation output have been renamed: Chapter 11 is *Estimation of Absolute Performance* and Chapter 12 is *Estimation of Relative Performance*. This reflects our opinion that the key issue is not how many systems designs are being simulated, but on whether performance in isolation, or relative to something else, is of interest. All examples in both chapters have been replaced with more timely applications. The spreadsheet `SimulationTools.xls`, available on the book's web site, implements many of the statistical procedures in these two chapters as menu-driven applications. In addition, the material in Chapter 12 on metamodeling now emphasizes issues that are special to simulation experiments as opposed to regression analysis in general.

The fifth and last part of the book, entitled **Applications**, has chapters on manufacturing and material handling systems, and on networked computer systems.

Chapter 13, *Simulation of Manufacturing and Material-Handling Systems*, discusses some of the issues and performance measures specific to manufacturing and material handling, and includes an extended example and analysis of a small job shop.

Chapter 14, *Simulation of Networked Computer Systems*, combines elements of 4/e Chapters 14 and 15, removing the material on simulation of CPUs and memory, and adding new material relevant to the simulation of wirelessly networked systems. In particular we describe common models of user mobility, and illustrate the range of complexities in models of radio propagation. The website has examples of simulations discussed in this chapter and provides extensive links to supporting material.

Discrete-Event System Simulation can serve as a textbook in the following types of courses:

1. An introductory simulation course in engineering, computer science, or management (Chapters 1 through 9 and selected parts of Chapters 10 through 12 when no companion language text is used; if a companion language text is used, skip Chapter 4, and use the application Chapters 13 and 14, as appropriate);
2. A second course in simulation (all of Chapters 10 through 12, a companion language text, and an outside project; add Chapter 13 or 14, as appropriate)

We gratefully acknowledge General Motors R&D for permission to use software the third author created for them as the basis for `SimulationTool.xls`, and the assistance of Feng Yang and Jun Xu for modifying the software for use with this book. Ira Gerhardt translated examples written in Maple for previous editions of the book into the MATLAB examples in this edition.

JERRY BANKS
JOHN S. CARSON II
BARRY L. NELSON
DAVID M. NICOL

What's New in the Fifth Edition

Chapter 1, *Introduction to Simulation*, adds abstracts of many real cases.

Chapter 2, *Simulation Examples*, begins with 3 simple spreadsheet simulations that cover the basics – how to obtain random numbers and generate a random variable from a simple discrete distribution, plus a few key concepts such as activities and system state, after which the instructor can choose among any of 9 examples in coin tossing, queueing, inventory policies, reliability and project activity networks to illustrate the basics of simulation modeling as well as experimentation with a simulation model.

Chapter 4, *Simulation Software,* updates the material on simulation software.

Chapter 6, *Queueing Models,* adds a case study on rough-cut modeling of queueing systems prior to simulation and replaces the Maple examples with Matlab. The Excel spreadsheet, `Queueing-Tools.xls`, available on the web site, computes queueing performance measures.

Chapter 7, *Random-Number Generation*, and Chapter 8, *Random-Variate Generation* now come with a spreadsheet, `RandomNumberTools.xls`, that contains implementations in Visual Basic for Applications (VBA) of a long-period random-number generator and random-variate generators for all the statistical distributions in Chapter 8.

Chapter 9, *Input Modeling* replaces the core example with a series of brief examples that illustrate the difficulties that can occur in input modeling. Examples using Maple were replaced by Matlab code.

Chapters 11 and 12 have been renamed to *Estimation of Absolute Performance* and *Estimation of Relative Performance*, respectively. The chapters come with updated examples and a spreadsheet `SimulationTools.xls`, available on the book web site, that implements many of the statistical procedures in these two chapters. The material in Chapter 12 on metamodeling now emphasizes issues that are special to simulation experiments as opposed to regression analysis in general.

Chapter 14 integrates the discussion of computer systems and networking, and provides new material on wirelessly networked systems. We describe models of how users move about in a domain and highlight a pitfall many have suffered using the random waypoint model. We also include material showing how radio propagation is modeled (so as to determine when broadcast messages are actually received), and point out the range of complexities : from the very simple free-space model, to the computationally intensive ray-tracing model.

List of Materials Available on www.bcnn.net

Chapter 2:

Example2.1CoinToss.xls - Tossing a Coin

Example2.2ServiceTimes.xls - Service Times

Example2.3ArrivalTimes.xls - Arrival Times

Example2.4CoinTossGame.xls - A Coin Tossing Question

Example2.5SingleServer.xls - Grocery Store (single-server queue)

Example2.6AbleBaker.xls - Call Center (two-channel queue)

Example2.7NewsDealer.xls - News Dealer's Inverntory Problem

Example2.8RefrigInventory.xls - Refrigerator Sales (inventory ordering)

Example2.9CurrentBearings.xls - Current Bearing Replacement Policy

Example2.9ProposedBearings.xls - Proposed Bearing Replacement Policy

Example2.10Target.xls - A Bombing Mission

Example2.11LeadTimeDemand - Lead-Time Demand Distribution

Example2.12Project.xls - Project Activity Network

BCNNvba.xla - Excel add-in (VBA code) to run examples

Chap2Excel5ed.zip - a zip file containing all Excel examples

Chapter 4: Base classes for programming-language based simulations

Chapter 6: QueueingTools.xls

Chapters 7 and 8: RandomNumberTools.xls

Chapters 11 and 12: SimulationTools.xls

Chapters 14 : Java and C++ code supporting the topics covered

About the Authors

Jerry Banks retired in 1999 as a professor in the School of Industrial and Systems Engineering, Georgia Institute of Technology, after which he worked as senior simulation technology advisor for Brooks Automation; he is currently a professor at Technoógico de Monterrey, México. He is the author, coauthor, editor, or coeditor of twelve books, one set of proceedings, several chapters in texts, and numerous technical papers. His most recent book is *RIFD Applied*, co-authored with three others, and published by John Wiley in 2007. He is the editor of the *Handbook of Simulation*, published in 1998 by John Wiley, which won the award for Excellence in Engineering Handbooks from the Professional Scholarly Publishing Division of the Association of American Publishers, Inc. He is also author or coauthor of *Getting Started with AutoMod*, Second Edition, *Introduction to SIMAN V and CINEMA V*, *Getting Started with GPSS/H*, Second Edition, *Forecasting and Management of Technology*, Second Edition (in preparation) and *Principles of Quality Control*. He was a founding partner in the simulation-consulting firm Carson/Banks & Associates, Inc., which was purchased by AutoSimulations, Inc. He is a full member of many technical societies, among them the Institute of Industrial Engineers (IIE); he served eight years as that organization's representative to the Board of the Winter Simulation Conference, including two years as board chair. He is the recipient of the INFORMS College on Simulation Distinguished Service Award for 1999 and was named a Fellow of IIE in 2002.

John S. Carson II is an independent simulation consultant. Formerly, he held management and consulting positions in the simulation services and software industry, including positions at AutoSimulations and the AutoMod Group at Brooks Automation. He was the co-founder and president of the simulation services firm Carson/Banks & Associates. He has over 30 years experience in simulation in a wide range of application areas, including manufacturing, distribution, warehousing and material handling, order fulfillment systems, postal systems, transportation and rapid transit systems, port operations (container terminals and bulk handling), and health-care systems. He has taught simulation and operations research at the Georgia Institute of Technology and the University of Florida.

Barry L. Nelson is the Charles Deering McCormick Professor and Chair of the Department of Industrial Engineering and Management Sciences at Northwestern University. His research centers on the design and analysis of computer simulation experiments on models of stochastic systems, concentrating on multivariate input modeling and output analysis, optimization via simulation and metamodeling. Application areas include financial engineering, computer performance modeling, quality control, manufacturing and transportation systems. He is the Editor in Chief of *Naval Research Logistics*, a Fellow of INFORMS, and was simulation area editor of *Operations Research*, president of the INFORMS (then TIMS) College on Simulation, and Chair of the Board of Directors of the Winter Simulation Conference.

David M. Nicol is professor of electrical and computer engineering at the University of Illinois at Urbana-Champaign. He is a long-time contributor in the field of parallel and distributed discrete-event simulations, having written one of the early Ph.D. dissertations on the topic. He has also worked in parallel algorithms, algorithms for mapping workload in parallel architectures, performance analysis, and reliability modeling and analysis. His research contributions extend to 180 articles in leading computer-science journals and conferences. His research is driven largely by problems encountered in industry and government—he has worked closely with researchers at NASA, IBM, AT&T, Bellcore, Motorola, and the Los Alamos, Sandia, and Oak Ridge National Laboratories, as well as a number of aerospace and communication companies. His current interests lie in modeling and simulation of very large systems, particularly communications and other infrastructure, with applications in evaluating system security. From 1997 to 2003 he was the editor-in-chief of the *ACM Transactions on Modeling and Computer Simulation*. Professor Nicol is a Fellow of the IEEE, a Fellow of the ACM, and the inaugural awardee of the ACM SIGSIM Distinguished Contributions award.

Part I

Introduction to Discrete-Event System Simulation

1

Introduction to Simulation

A *simulation* is the imitation of the operation of a real-world process or system over time. Whether done by hand or on a computer, simulation involves the generation of an artificial history of a system and the observation of that artificial history to draw inferences concerning the operating characteristics of the real system.

The behavior of a system as it evolves over time is studied by developing a simulation *model*. This model usually takes the form of a set of assumptions concerning the operation of the system. These assumptions are expressed in mathematical, logical, and symbolic relationships between the *entities*, or objects of interest, of the system. Once developed and validated, a model can be used to investigate a wide variety of "what if" questions about the real-world system. Potential changes to the system can first be simulated, in order to predict their impact on system performance. Simulation can also be used to study systems in the design stage, before such systems are built. Thus, simulation modeling can be used both as an analysis tool for predicting the effect of changes to existing systems and as a design tool to predict the performance of new systems under varying sets of circumstances.

In some instances, a model can be developed which is simple enough to be "solved" by mathematical methods. Such solutions might be found by the use of differential calculus, probability theory, algebraic methods, or other mathematical techniques. The solution usually consists of one or more numerical parameters, which are called measures of performance of the system. However, many real-world systems are so complex that models of these systems are virtually impossible to solve mathematically. In these instances, numerical, computer-based simulation can be used to imitate the behavior of the system over time. From the simulation, data are collected as if a real system were being observed. This simulation-generated data is used to estimate the measures of performance of the system.

This book provides an introductory treatment of the concepts and methods of one form of simulation modeling—discrete-event simulation modeling. The first chapter initially discusses when to use simulation, its advantages and disadvantages, and actual areas of its application. Then the concepts of system and model are explored. Finally, an outline is given of the steps in building and using a simulation model of a system.

1.1   When Simulation Is the Appropriate Tool

The availability of special-purpose simulation languages, of massive computing capabilities at a decreasing cost per operation, and of advances in simulation methodologies have made simulation one of the most widely used and accepted tools in operations research and systems analysis. Circumstances under which simulation is the appropriate tool to use have been discussed by many authors, from Naylor *et al.* [1966] to Shannon [1998]. Simulation can be used for the following purposes:

1. Simulation enables the study of, and experimentation with, the internal interactions of a complex system or of a subsystem within a complex system.
2. Informational, organizational, and environmental changes can be simulated, and the effect of these alterations on the model's behavior can be observed.
3. The knowledge gained during the designing of a simulation model could be of great value toward suggesting improvement in the system under investigation.
4. Changing simulation inputs and observing the resulting outputs can produce valuable insights about which variables are the most important and how variables interact.
5. Simulation can be used as a pedagogical device to reinforce analytic solution methodologies.
6. Simulation can be used to experiment with new designs or policies before implementation, so as to prepare for what might happen.
7. Simulation can be used to verify analytic solutions.
8. Simulating different capabilities for a machine can help determine its requirements.
9. Simulation models designed for training make learning possible, without the cost and disruption of on-the-job instruction.
10. Animation can show a system in simulated operation so that the plan can be visualized.
11. A modern system (factory, wafer fabrication plant, service organization, etc.) is so complex that its internal interactions can be treated only through simulation.

1.2   When Simulation Is Not Appropriate

This section is based on an article by Banks and Gibson [1997], which gives ten rules for evaluating when simulation is not appropriate. The first rule indicates that simulation should not be used when the problem can be solved by common sense. An example is given of an automobile tag facility serving customers who arrive randomly at an average rate of 100/hour and are served at a mean rate of 12/hour. To determine the minimum number of servers needed, simulation is not necessary. Just compute $100/12 = 8.33$, which indicates that nine or more servers are needed.

The second rule says that simulation should not be used if the problem can be solved analytically. For example, under certain conditions, the average waiting time in the example above can be found using the tools described in Chapter 6 and available at www.bcnn.net.

The next rule says that simulation should not be used if it is less expensive to perform direct experiments. The example of a fast-food drive-in restaurant is given, where it was less expensive to stage a person taking orders using a hand-held terminal and voice communication to determine the effect of adding another order station on customer waiting time.

The fourth rule says not to use simulation if the costs exceed the savings. There are many steps in completing a simulation, as will be discussed in Section 1.12, and these must be done thoroughly. If a simulation study costs $20,000 and the savings might be $10,000, simulation would not be appropriate.

Rules five and six indicate that simulation should not be performed if the resources or time are not available. If the simulation is estimated to cost $20,000 and there is only $10,000 available, the suggestion is not to venture into a simulation study. Similarly, if a decision is needed in two weeks and a simulation would take a month, the simulation study is not advised.

Simulation takes data, sometimes lots of data. If no data is available, not even estimates, simulation is not advised. The next rule concerns the ability to verify and validate the model. If there is not enough time or if the personnel are not available, simulation is not appropriate.

If managers have unreasonable expectations, if they ask for too much too soon, or if the power of simulation is overestimated, simulation might not be appropriate.

Last, if system behavior is too complex or cannot be defined, simulation is not appropriate. Human behavior is sometimes extremely complex to model.

1.3 Advantages and Disadvantages of Simulation

Simulation is intuitively appealing to a client because it mimics what happens in a real system or what is perceived for a system that is in the design stage. The output data from a simulation should directly correspond to the outputs that could be recorded from the real system. Additionally, it is possible to develop a simulation model of a system without dubious assumptions (such as the same statistical distribution for every random variable) of mathematically solvable models. For these and other reasons, simulation is frequently the technique of choice in problem solving.

In contrast to optimization models, simulation models are "run" rather than solved. Given a particular set of input and model characteristics, the model is run and the simulated behavior is observed. This process of changing inputs and model characteristics results in a set of scenarios that are evaluated. A good solution, either in the analysis of an existing system or in the design of a new system, is then recommended for implementation.

Simulation has many advantages, but some disadvantages. These are listed by Pegden, Shannon, and Sadowski [1995]. Some advantages are these:

1. New policies, operating procedures, decision rules, information flows, organizational procedures, and so on can be explored without disrupting ongoing operations of the real system.

2. New hardware designs, physical layouts, transportation systems, and so on can be tested without committing resources for their acquisition.
3. Hypotheses about how or why certain phenomena occur can be tested for feasibility.
4. Time can be compressed or expanded to allow for a speed-up or slow-down of the phenomena under investigation.
5. Insight can be obtained about the interaction of variables.
6. Insight can be obtained about the importance of variables to the performance of the system.
7. Bottleneck analysis can be performed to discover where work in process, information, materials, and so on are being delayed excessively.
8. A simulation study can help in understanding how the system operates rather than how individuals think the system operates.
9. "What if" questions can be answered. This is particularly useful in the design of new systems.

Some disadvantages are these:

1. Model building requires special training. It is an art that is learned over time and through experience. Furthermore, if two models are constructed by different competent individuals, they might have similarities, but it is highly unlikely that they will be the same.
2. Simulation results can be difficult to interpret. Most simulation outputs are essentially random variables (they are usually based on random inputs), so it can be hard to distinguish whether an observation is the result of system interrelationships or of randomness.
3. Simulation modeling and analysis can be time consuming and expensive. Skimping on resources for modeling and analysis could result in a simulation model or analysis that is not sufficient to the task.
4. Simulation is used in some cases when an analytical solution is possible, or even preferable, as was discussed in Section 1.2. This might be particularly true in the simulation of some waiting lines where closed-form queueing models are available.

In defense of simulation, these four disadvantages, respectively, can be offset as follows:

1. Vendors of simulation software have been actively developing packages that contain models that need only input data for their operation. Such models have the generic tag "simulator" or "template."
2. Many simulation software vendors have developed output-analysis capabilities within their packages for performing very thorough analyses.
3. Simulation can be performed faster today than yesterday and will be even faster tomorrow, because of advances in hardware that permit rapid running of scenarios and because of advances in many simulation packages. For example, some simulation software contains constructs for modeling material handling that uses such transporters as fork-lift trucks, conveyors, and automated guided vehicles.
4. Closed-form models are not able to analyze most of the complex systems that are encountered in practice. During the many years of consulting practice by two of the authors, not one problem was encountered that could have been solved by a closed-form solution.

1.4   Areas of Application

The applications of simulation are vast. The Winter Simulation Conference (WSC) is an excellent way to learn more about the latest in simulation applications and theory. There are also numerous tutorials at both the beginning and advanced levels. WSC is sponsored by six technical societies and the National Institute of Standards and Technology (NIST). The technical societies are the American Statistical Association (ASA), the Association for Computing Machinery/Special Interest Group on Simulation (ACM/SIGSIM), the Institute of Electrical and Electronics Engineers: Systems, Man and Cybernetics Society (IEEE/SMCS), the Institute of Industrial Engineers (IIE), the Institute for Operations Research and the Management Sciences: Simulation Society (INFORMS-SIM), and the Society for Modeling and Simulation International (SCS). Information about the upcoming WSC can be obtained from www.wintersim.org. WSC programs with full papers are available from www.informs-cs.org/wscpapers.html. Some presentations, by area, from a recent WSC are listed next:

Manufacturing Applications
Methodology for Selecting the Most Suitable Bottleneck Detection Method
Automating the Development of Shipyard Manufacturing Models
Emulation in Manufacturing Engineering Processes
Optimized Maintenance Design for Manufacturing Performance Improvement
Productivity Management in an Automotive-Parts Industry
Manufacturing Line Designs in Japanese Automobile Manufacturing Plants

Wafer Fabrication
A Paradigm Shift in Assigning Lots to Tools
Scheduling a Multi-Chip Package Assembly Line with Reentrant Processes
Execution Level Capacity Allocation Decisions for Assembly—Test Facilities
Managing WIP and Cycle Time with the Help of Loop Control

Business Processing
A New Policy for the Service Request Assignment Problem
Process Execution Monitoring and Adjustment Schemes
In-Store Merchandizing of Retail Stores
Sales Forecasting for Retail Small Stores

Construction Engineering and Project Management
Scheduling of Limited Bar-Benders over Multiple Building Sites
Constructing Repetitive Projects
Traffic Operations for Improved Planning of Road Construction Projects
Template for Modeling Tunnel Shaft Construction
Decision Support Tool for Planning Tunnel Construction

Logistics, Transportation, and Distribution
Operating Policies for a Barge Transportation System
Dispensing Plan for Emergency Medical Supplies in the Event of Bioterrorism

Analysis of a Complex Mail Transportation Network
Improving the Performance of Container Terminals
Yard Crane Dispatching Based on Real Time Data
Unit Loading Device Inventory in Airline Operations
Inventory Systems with Forecast Based Policy Updating
Dock Allocation in a Food Distribution Center
Operating Policies for a Barge Transportation System

Military Applications
Multinational Intra-Theatre Logistics Distribution
Examining Future Sustainability of Canadian Forces Operations
Feasibility Study for Replacing the MK19 Automatic Grenade Launching System
Training Joint Forces for Asymmetric Operations
Multi-Objective Unmanned Aerial Vehicle Mission Planning
Development of Operational Requirements Driven Federations

Health Care
Interventions to Reduce Appointment Lead-Time and Patient No-Show Rate
Supporting Smart Thinking to Improve Hospital Performance
Verification of Lean Improvement for Emergency Room Process
Reducing Emergency Department Overcrowding
Inventory Modeling of Perishable Pharmaceuticals
Implementation of an Outpatient Procedure Center
Infectious Disease Control Policy
Balancing Operating Room and Post-Anesthesia Resources
Cost Effectiveness of Colorectal Cancer Screening Tests

Additional Applications
Managing Workforce Resource Actions with Multiple Feedback Control
Analyzing the Impact of Hole-Size on Putting in Golf
Application of Particle Filters in Wildfire Spread Simulation
Predator-Prey Relationship in a Closed Habitat
Intensive Piglet Production Systems
Real-Time Delay Estimation in Call Centers
Pandemic Influenza Preparedness Plans for a Public University

For an article on the future of simulation appearing in the *ICS Newsletter* (Banks, 2008), sixteen simulationists, including well-known experts in the field, provided their response to the following question: "What remarkable results will we observe in simulation software in the long term, say, after three years?" The responses that are shown below appeared in the referenced article:

- After new fundamental ideas and methodologies such as agent-based modeling have been adopted by the software vendors (which is going to happen within a few years), progress in simulation modeling will be driven by gains in computing power. For example, at a certain point we will be able to simulate the detailed operation of very large supply chains and manufacturing facilities.

- Simulation modeling will be more of an "assembly" activity than a "build-from-scratch" activity. Intelligent, parameterized components will be used to piece together models rather than defining lots of detailed logic.
- Progress will be made in solving hard problems. To provide tools for really hard problems, simulation software developers will have to go back to the drawing board and seriously reconsider the fundamental question: "Who provides the power, and who provides the vision?" Getting this mix right is essential to further progress. When a software developer tries to provide both power and vision, end users are saddled with the developer's paradigms. This works well for easy problems, but poorly for hard problems. For all truly hard problems, users know more about the problems than software developers do. The obvious solution is for developers to provide well-thought-out collections of true primitives, along with ways of combining and packaging these primitives.
- Simulation software will be integrated more closely with control software.
- Modelers will have a single model that is shared across applications within the organization.
- Simulation applications will not be restricted to design applications but will also be used to make day-to-day operational decisions within an organization.
- Simulation modeling on powerful servers will be accessed with web-based interfaces.
- Better and easier modeling of human activities (e.g. embedding agent-based models into discrete-event models) will be available.
- More collaborative simulation project development can be expected.
- Interface standards and incorporation of web services allowing simulation software to not only work with each other as an integrated federation, but also standardize and simplify the way other applications interface with simulation software will be adopted.
- Analytical solving techniques (such as linear programming) will be integrated with simulation capabilities.
- The "remarkable" results will be derived from advances in other areas of computing technology and software engineering. But nothing will surpass the object-oriented paradigm that was ushered into mainstream software development by SIMULA 67, emanating from the simulation software community.

1.5   Some Recent Applications of Simulation

In this section, we present some recent applications of simulation. These have appeared in the literature indicated so you can find these cases and learn more details about them.

TITLE: "The Turkish Army Uses Simulation to Model and Optimize Its Fuel-Supply System"

AUTHOR(S): I. Sabuncuoglu, A. Hatip

REPORTED: November–December 2005 in *Interfaces*

CHALLENGE: Analysis of the Turkish army's fuel-supply system.

TECHNIQUE(S): (1) Measured performance of the existing and proposed systems under various scenarios, (2) developed a simulation optimization model based on a genetic algorithm to optimize system performance, (3) conducted extensive simulation experiments.

SAVINGS: Millions of US$.

TITLE: "PLATO Helps Athens Win Gold: Olympic Games Knowledge Modeling for Organizational Change and Resource Management"

AUTHOR(S): D.A. Beis, P. Poucopoulos, Y. Pyrgiotis, K.G. Zografos

REPORTED: January-February 2006 in *Interfaces*

CHALLENGE: Develop a systematic process for planning and designing venue operations. Develop a rich library of models that is directly transferable to future Olympic organizing committees and other sports-oriented events.

TECHNIQUE(S): Knowledge modeling and resource-management techniques and tools based on simulation and other decision analysis methodologies.

SAVINGS: Over US$69.7 million.

TITLE: "Schlumberger Uses Simulation in Bidding and Executing Land Seismic Surveys"

AUTHOR(S): P.W. Mullarkey, G. Butler, S. Gavirneni, D.J. Morrice

REPORTED: March-April 2007 in *Interfaces*

CHALLENGE: Quickly and accurately measure the cost of seismic surveys.

TECHNIQUE(S): Developed a simulation tool to evaluate the impact of crew sizes, survey area, geographical region, and weather conditions on survey costs and durations.

SAVINGS: US$1.5 to US$3.0 million annually.

TITLE: "Operations Research Helps Reshape Operations Strategy at Standard Register Company"

AUTHOR(S): S.L. Ahire, M.F. Gorman, D. Dwiggins, O. Mudry

REPORTED: November-December 2007 in *Interfaces*

CHALLENGE: Minimize the total costs to offer competitive pricing in the highly competitive traditional print market.

TECHNIQUE(S): (1) Regression to estimate costs and time attributes, (2) optimization modeling to determine the order-routing strategy, (3) simulation modeling of the production-distribution network

SAVINGS: Over US$10 million annually.

TITLE: "Simulation Implements Demand-Driven Workforce Scheduler for Service Industry"

AUTHOR(S): M. Zottolo, O.M. Ülgen, E. Williams

REPORTED: *Proceedings of the 2007 Winter Simulation Conference*, eds. S.G. Henderson, B. Biller, M.-H. Hsieh, J. Shortle, J.D. Tew, and R.R. Barton.

CONDUCTED BY: PMC (www.pmcorp.com)

CLIENT: Major US appliance company.

CHALLENGE: As a result of a time-consuming, manual, and inefficient scheduling process the client was experiencing over- and underscheduling of workgroups, inconsistent service levels being provided, and overwhelmed site managers.

TECHNIQUE(S): (1) Built simulation model to schedule different workgroups according to changes in customer demand during the day and servicing times, (2) determined work standards, (3) developed interface for input data as well as storing and publishing of schedules, (4) implemented and trained employees on the use of the scheduling tool.

SAVINGS: Estimated at US$80 million for client's US facilities.

TITLE: "Simulation Improves End-of-Line Sortation and Material Handling Pickup Scheduling at Appliance Manufacturer"

AUTHOR(S): N. Kale, M. Zottolo, O.M. Ülgen, E. Williams

CONDUCTED BY: PMC (www.pmcorp.com)

CLIENT: Major US car rental company.

REPORTED: *Proceedings of the 2007 Winter Simulation Conference*, eds. S.G. Henderson, B. Biller, M.-H. Hsieh, J. Shortle, J.D. Tew, and R.R. Barton.

CHALLENGE: To determine the most efficient method to distribute several types of finished appliances (SKUs) into a 12-lane sortation system in order to minimize the material handling and the number of forklifts required to pick them. This sortation task and its associated material-handling tasks were of a complexity meriting discrete-event simulation analysis due to volatile product mix, the large number of SKUs, and high overhead costs.

TECHNIQUE(S): Built simulation model and performed experimentation with respect to (1) assignment of different SKUs to the different lanes, (2) picking strategy and quantity, (3) SKU batch sizes.

SAVINGS: Estimated US$100,000 annually for a US$50,000 one-time capital investment.

1.6   Systems and System Environment

To model a system, it is necessary to understand the concept of a system and the system boundary. A *system* is defined as a group of objects that are joined together in some regular interaction or interdependence toward the accomplishment of some purpose. An example is a production system manufacturing automobiles. The machines, component parts, and workers operate jointly along an assembly line to produce a high-quality vehicle.

A system is often affected by changes occurring outside the system. Such changes are said to occur in the *system environment* [Gordon, 1978]. In modeling systems, it is necessary to decide on the *boundary* between the system and its environment. This decision may depend on the purpose of the study.

In the case of the factory system, for example, the factors controlling the arrival of orders may be considered to be outside the influence of the factory and therefore part of the environment. However, if the effect of supply on demand is to be considered, there will be a relationship between factory output and arrival of orders, and this relationship must be considered an activity of the system. Similarly, in the case of a bank system, there could be a limit on the maximum interest rate that can be paid. For the study of a single bank, this would be regarded as a constraint imposed by the environment. In a study of the effects of monetary laws on the banking industry, however, the setting of the limit would be an activity of the system [Gordon, 1978].

1.7   Components of a System

In order to understand and analyze a system, a number of terms need to be defined. An *entity* is an object of interest in the system. An *attribute* is a property of an entity. An *activity* represents a time period of specified length. If a bank is being studied, customers might be one of the entities, the balance in their checking accounts might be an attribute, and making deposits might be an activity.

The collection of entities that compose a system for one study might only be a subset of the overall system for another study [Law, 2007]. For example, if the aforementioned bank is being studied to determine the number of tellers needed to provide for paying and receiving, the system can be defined as that portion of the bank consisting of the regular tellers and the customers waiting in line. If the purpose of the study is expanded to determine the number of special tellers needed (to prepare cashier's checks, to conduct commercial transactions, etc.), the definition of the system must be expanded.

The *state* of a system is defined to be that collection of variables necessary to describe the system at any time, relative to the objectives of the study. In the study of a bank, possible state variables are the number of busy tellers, the number of customers waiting in line or being served, and the arrival time of the next customer. An *event* is defined as an instantaneous occurrence that might change the state of the system. The term *endogenous* is used to describe activities and events occurring within a system, and the term *exogenous* is used to describe activities and events in the environment that affect the system. In the bank study, the arrival of a customer is an exogenous event, and the completion of service of a customer is an endogenous event.

Table 1.1 lists examples of entities, attributes, activities, events, and state variables for several systems. Only a partial listing of the system components is shown. A complete list cannot be

Table 1.1 Examples of Systems and Their Components

System	Entities	Attributes	Activities	Events	State Variables
Banking	Customers	Checking-account balance	Making deposits	Arrival; departure	Number of busy tellers; number of customers waiting
Rapid rail	Riders	Origin; destination	Traveling	Arrival at station; arrival at destination	Number of riders waiting at each station; number of riders in transit
Production	Machines	Speed; capacity; breakdown rate	Welding; stamping	Breakdown	Status of machines (busy, idle, or down)
Communications	Messages	Length; destination	Transmitting	Arrival at destination	Number waiting to be transmitted
Inventory	Warehouse	Capacity	Withdrawing	Demand	Levels of inventory; backlogged demands

13

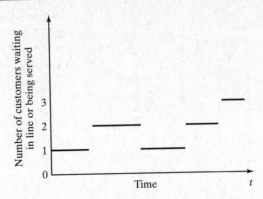

Figure 1.1   Discrete-system state variable.

developed unless the purpose of the study is known. Depending on the purpose, various aspects of the system will be of interest, and then the listing of components can be completed.

1.8   Discrete and Continuous Systems

Systems can be categorized as discrete or continuous. "Few systems in practice are wholly discrete or continuous, but since one type of change predominates for most systems, it will usually be possible to classify a system as being either discrete or continuous" [Law, 2007]. A *discrete system* is one in which the state variable(s) change only at a discrete set of points in time. The bank is an example of a discrete system: The state variable, the number of customers in the bank, changes only when a customer arrives or when the service provided a customer is completed. Figure 1.1 shows how the number of customers changes only at discrete points in time.

A *continuous system* is one in which the state variable(s) change continuously over time. An example is the head of water behind a dam. During and for some time after a rain storm, water flows into the lake behind the dam. Water is drawn from the dam for flood control and to make electricity. Evaporation also decreases the water level. Figure 1.2 shows how the state variable *head of water behind the dam* changes for this continuous system.

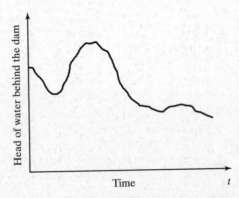

Figure 1.2   Continuous-system state variable.

1.9 Model of a System

Sometimes it is of interest to study a system to understand the relationships between its components or to predict how the system will operate under a new policy. To study the system, it is sometimes possible to experiment with the system itself. However, this is not always an option. A new system might not yet exist; it could be in only hypothetical form or at the design stage. Even if the system exists, it might be impractical to experiment with it. In the case of a bank, reducing the numbers of tellers to study the effect on the length of waiting lines might infuriate the customers so greatly that they will move their accounts to a competitor. Consequently, studies of systems are often accomplished with a model of a system.

We had a consulting job for the simulation of a redesigned port in Western Australia. At US$300 million for a loading/unloading berth, it is not advisable to invest that amount only to find that the berth is inadequate for the task.

A *model* is defined as a representation of a system for the purpose of studying that system. For most studies, it is only necessary to consider those aspects of the system that affect the problem under investigation. These aspects are represented in a model of the system; the model, by definition, is a simplification of the system. On the other hand, the model should be sufficiently detailed to permit valid conclusions to be drawn about the real system. Different models of the same system could be required as the purpose of investigation changes.

Just as the components of a system include entities, attributes, and activities, so too models are represented. However, the model contains only those components that are relevant to the study. The components of a model are discussed more extensively in Chapter 3.

1.10 Types of Models

Models can be classified as being mathematical or physical. A mathematical model uses symbolic notation and mathematical equations to represent a system. A simulation model is a particular type of mathematical model of a system. A physical model is a larger or smaller version of an object such as the enlargement of an atom or a scaled-down version of the solar system.

Simulation models may be further classified as being static or dynamic, deterministic or stochastic, and discrete or continuous. A *static* simulation model, sometimes called a Monte Carlo simulation, represents a system at a particular point in time. *Dynamic* simulation models represent systems as they change over time. The simulation of a bank from 9:00 A.M. to 4:00 P.M. is an example of a dynamic simulation.

Simulation models that contain no random variables are classified as *deterministic*. Deterministic models have a known set of inputs, that will result in a unique set of outputs. Deterministic arrivals would occur at a dentist's office if all patients arrived at their scheduled appointment times. A *stochastic* simulation model has one or more random variables as inputs. Random inputs lead to random outputs. Since the outputs are random, they can be considered only as estimates of the true characteristics of a model. The simulation of a bank would usually involve random interarrival times and random service times. Thus, in a stochastic simulation, the output measures—the average number of people waiting, the average waiting time of a customer—must be treated as statistical estimates of the true characteristics of the system.

Discrete and continuous systems were defined in Section 1.7. Discrete and continuous models are defined in an analogous manner. However, a discrete simulation model is not always used to model a discrete system, nor is a continuous simulation model always used to model a continuous system. Tanks and pipes might be modeled discretely, even though we know that fluid flow is continuous. In addition, simulation models may be mixed, both discrete and continuous. The choice of whether to use a discrete or continuous (or both discrete and continuous) simulation model is a function of the characteristics of the system and the objective of the study. Thus, a communication channel could be modeled discretely if the characteristics and movement of each message were deemed important. Conversely, if the flow of messages in aggregate over the channel were of importance, modeling the system via continuous simulation could be more appropriate. The models emphasized in this text are dynamic, stochastic, and discrete.

1.11   Discrete-Event System Simulation

This is a textbook about discrete-event system simulation. Discrete-event system simulation is the modeling of systems in which the state variable changes only at a discrete set of points in time. The simulation models are analyzed by numerical rather than analytical methods. *Analytical* methods employ the deductive reasoning of mathematics to "solve" the model. For example, differential calculus can be used to compute the minimum-cost policy for some inventory models. *Numerical* methods employ computational procedures to "solve" mathematical models. In the case of simulation models, which employ numerical methods, models are "run" rather than solved—that is, an artificial history of the system is generated from the model assumptions, and observations are collected to be analyzed and to estimate the true system performance measures. Real-world simulation models are rather large, and the amount of data stored and manipulated is vast, so such runs are usually conducted with the aid of a computer.

In summary, this textbook is about discrete-event system simulation in which the models of interest are analyzed numerically, usually with the aid of a computer.

1.12   Steps in a Simulation Study

Figure 1.3 shows a set of steps to guide a model builder in a thorough and sound simulation study. Similar figures and discussion of steps can be found in other sources [Shannon, 1975; Gordon, 1978; Law, 2007]. The number beside each symbol in Figure 1.3 refers to the more detailed discussion in the text. The steps in a simulation study are as follows:

1. Problem formulation   Every study should begin with a statement of the problem. If the statement is provided by the policymakers or those that have the problem, the analyst must ensure that the problem being described is clearly understood. If a problem statement is being developed by the analyst, it is important that the policymakers understand and agree with the formulation. Although not shown in Figure 1.3, there are occasions where the problem must be reformulated as the study progresses. In many instances, policymakers and analysts are aware that there is a problem long before the nature of the problem is known.

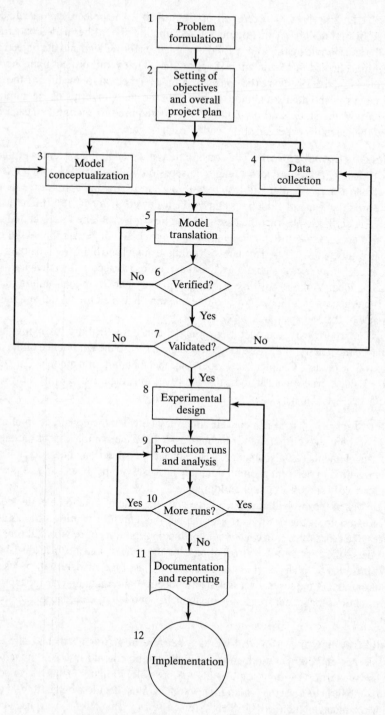

Figure 1.3   Steps in a simulation study.

2. Setting of objectives and overall project plan   The objectives indicate the questions to be answered by simulation. At this point, a determination should be made concerning whether simulation is the appropriate methodology for the problem as formulated and the objectives as stated. Assuming that it is decided that simulation is appropriate, the overall project plan should include a statement of the alternative systems to be considered and of a method for evaluating the effectiveness of these alternatives. It should also include the plans for the study in terms of the number of people involved, the cost of the study, and the number of days required to accomplish each phase of the work, along with the results expected at the end of each stage.

3. Model conceptualization   The construction of a model of a system is probably as much art as science. Pritsker [1998] provides a lengthy discussion of this step. "Although it is not possible to provide a set of instructions that will lead to building successful and appropriate models in every instance, there are some general guidelines that can be followed" [Morris, 1967]. The art of modeling is enhanced by an ability to abstract the essential features of a problem, to select and modify basic assumptions that characterize the system, and then to enrich and elaborate the model until a useful approximation results. Thus, it is best to start with a simple model and build toward greater complexity. However, the model complexity need not exceed that required to accomplish the purposes for which the model is intended. Violation of this principle will only add to model-building and computer expenses. It is not necessary to have a one-to-one mapping between the model and the real system. Only the essence of the real system is needed.

It is advisable to involve the model user in model conceptualization. Involving the model user will both enhance the quality of the resulting model and increase the confidence of the model user in the application of the model. (Chapter 2 describes a number of simulation models. Chapter 6 describes queueing models that can be solved analytically. However, only experience with real systems—versus textbook problems—can "teach" the art of model building.)

4. Data collection   There is a constant interplay between the construction of the model and the collection of the needed input data [Shannon, 1975]. As the complexity of the model changes, the required data elements can also change. Also, since data collection takes such a large portion of the total time required to perform a simulation, it is necessary to begin as early as possible, usually together with the early stages of model building.

The objectives of the study dictate, in a large way, the kind of data to be collected. In the study of a bank, for example, if the desire is to learn about the length of waiting lines as the number of tellers changes, the types of data needed would be the distributions of interarrival times (at different times of the day), the service-time distributions for the tellers, and historic distributions on the lengths of waiting lines under varying conditions. This last type of data will be used to validate the simulation model. (Chapter 9 discusses data collection and data analysis; Chapter 5 discusses statistical distributions that occur frequently in simulation modeling. See also an excellent discussion by Henderson [2003].)

5. Model translation   Most real-world systems result in models that require a great deal of information storage and computation, so the model must be entered into a computer-recognizable format. We use the term *program* even though it is possible, in many instances, to accomplish the desired result with little or no actual coding. The modeler must decide whether to program the model in a simulation language, such as GPSS/H^{TM} (discussed in Chapter 4), or to use special-purpose simulation software. For manufacturing and material handling, Chapter 4 discusses such software as AnyLogic®, Arena®, AutoMod™, Enterprise Dynamics®, Extend™, Flexsim, ProModel®, and SIMUL8®. Simulation languages are powerful and flexible. However, if the problem is amenable to solution with the simulation software, the model development time is greatly reduced. Furthermore, most simulation software packages have added features that enhance their flexibility, although the amount of flexibility varies greatly.

6. Verified? Verification pertains to the computer program that has been prepared for the simulation model. Is the computer program performing properly? With complex models, it is difficult, if not impossible, to translate a model successfully in its entirety without a good deal of debugging; if the input parameters and logical structure of the model are correctly represented in the computer, verification has been completed. For the most part, common sense is used in completing this step. (Chapter 10 discusses verification of simulation models, and Sargent [2007] also discusses this topic.)

7. Validated? Validation usually is achieved through the calibration of the model, an iterative process of comparing the model against actual system behavior and using the discrepancies between the two, and the insights gained, to improve the model. This process is repeated until model accuracy is judged acceptable. In the previously mentioned example of a bank, data was collected concerning the length of waiting lines under current conditions. Does the simulation model replicate this system measure? This is one means of validation. (Chapter 10 discusses the validation of simulation models, and Sargent [2007] also discusses this topic.)

8. Experimental design The alternatives that are to be simulated must be determined. Often, the decision concerning which alternatives to simulate will be a function of runs that have been completed and analyzed. For each system design that is simulated, decisions need to be made concerning the length of the initialization period, the length of simulation runs, and the number of replications to be made of each run. (Chapters 11 and 12 discuss issues associated with the experimental design, and Sanchez [2007] discusses this topic extensively.)

9. Production runs and analysis Production runs and their subsequent analysis, are used to estimate measures of performance for the system designs that are being simulated. (Chapters 11 and 12 discuss the analysis of simulation experiments, and Chapter 4 discusses software to aid in this step, including AutoStat (in AutoMod), OptQuest (in several pieces of simulation software), and SimRunner (in ProModel).

10. More runs? Given the analysis of runs that have been completed, the analyst determines whether additional runs are needed and what design those additional experiments should follow.

11. Documentation and reporting There are two types of documentation: program and progress. Program documentation is necessary for numerous reasons. If the program is going to be used again by the same or different analysts, it could be necessary to understand how the program operates. This will create confidence in the program, so that model users and policymakers can make decisions based on the analysis. Also, if the program is to be modified by the same or a different analyst, this step can be greatly facilitated by adequate documentation. One experience with an inadequately documented program is usually enough to convince an analyst of the necessity of this important step. Another reason for documenting a program is so that model users can change parameters at will in an effort to learn the relationships between input parameters and output measures of performance or to discover the input parameters that "optimize" some output measure of performance.

Musselman [1998] discusses progress reports that provide the important, written history of a simulation project. Project reports give a chronology of work done and decisions made. This can prove to be of great value in keeping the project on course. Musselman suggests frequent reports (monthly, at least) so that even those not involved in the day-to-day operation can be kept abreast. The awareness of these others can often enhance the successful completion of the project by surfacing misunderstandings early, when the problem can be solved easily. Musselman also suggests maintaining a project log to provide a comprehensive record of accomplishments, change requests, key decisions, and other items of importance.

On the reporting side, Musselman suggests frequent deliverables. These may or may not be the results of major accomplishments. His maxim is that "it is better to work with many intermediate milestones than with one absolute deadline." Possibilities prior to the final report include a model specification, prototype demonstrations, animations, training results, intermediate analyses, program documentation, progress reports, and presentations. He suggests that these deliverables should be timed judiciously over the life of the project.

The results of all the analysis should be reported clearly and concisely in a final report. This will allow the model users (now the decision makers) to review the final formulation, the alternative systems that were addressed, the criteria by which the alternatives were compared, the results of the experiments, and the recommended solution(s) to the problem. Furthermore, if decisions have to be justified at a higher level, the final report should provide a vehicle of certification for the model user/decision maker and add to the credibility of the model and of the model-building process.

12. Implementation   The success of the implementation phase depends on how well the previous eleven steps have been performed. It is also contingent upon how thoroughly the analyst has involved the ultimate model user during the entire simulation process. If the model user has been involved during the entire model-building process and if the model user understands the nature of the model and its outputs, the likelihood of a vigorous implementation is enhanced [Pritsker, 1995]. Conversely, if the model and its underlying assumptions have not been properly communicated, implementation will probably suffer, regardless of the simulation model's validity.

The simulation-model building process shown in Figure 1.3 can be broken down into four phases. The first phase, consisting of steps 1 (Problem formulation) and 2 (Setting of objective and overall design), is a period of discovery or orientation. The initial statement of the problem is usually quite "fuzzy," the initial objectives will usually have to be reset, and the original project plan will usually have to be fine-tuned. These recalibrations and clarifications could occur in this phase or perhaps will occur after or during another phase (i.e., the analyst might have to restart the process).

The second phase is related to model building and data collection and includes steps 3 (Model conceptualization), 4 (Data collection), 5 (Model translation), 6 (Verification), and 7 (Validation). A continuing interplay is required among the steps. Exclusion of the model user during this phase can have dire implications at the time of implementation.

The third phase concerns the running of the model. It involves steps 8 (Experimental design), 9 (Production runs and analysis), and 10 (More runs). This phase must have a comprehensively conceived plan for experimenting with the simulation model. A discrete-event stochastic simulation is, in fact, a statistical experiment. The output variables are estimates that contain random error, and therefore a proper statistical analysis is required. Such a philosophy is in contrast to that of the analyst who makes a single run and draws an inference from that single data point.

The fourth phase, implementation, involves steps 11 (Documentation and reporting) and 12 (Implementation). Successful implementation depends on continual involvement of the model user and on the successful completion of every step in the process. Perhaps the most crucial point in the entire process is Step 7 (Validation), because an invalid model is going to lead to erroneous results, which, if implemented, could be dangerous, costly, or both.

REFERENCES

BANKS, J. [2008], "Some Burning Questions about Simulation." *ICS Newsletter*, INFORMS Computing Society, Spring.

BANKS, J., and R. R. GIBSON [1997], "Don't Simulate When: 10 Rules for Determining when Simulation Is Not Appropriate," *IIE Solutions*, September.

GORDON, G. [1978], *System Simulation*, 2d ed., Prentice-Hall, Englewood Cliffs, NJ.

HENDERSON, S. G. [2003], "Input Model Uncertainty: Why Do We Care and What Should We Do About It?" in *Proceedings of the Winter Simulation Conference*, eds. S. Chick, P. J. Sánchez, D. Ferrin, and D. J. Morrice, New Orleans, LA, Dec. 7–10, pp. 90–100.

KLEIJNEN, J. P. C. [1998], "Experimental Design for Sensitivity Analysis, Optimization, and Validation of Simulation Models," in *Handbook of Simulation*, ed. J. Banks, John Wiley, New York.

LAW, A. M. [2007], *Simulation Modeling and Analysis*, 4th ed., McGraw–Hill, New York.

MORRIS, W. T. [1967], "On the Art of Modeling," *Management Science*, Vol. 13, No. 12.

MUSSELMAN, K. J. [1998], "Guidelines for Success," in *Handbook of Simulation*, ed. J. Banks, John Wiley, New York.

NAYLOR, T. H., J. L. BALINTFY, D. S. BURDICK, and K. CHU [1966], *Computer Simulation Techniques*, Wiley, New York.

PEGDEN, C. D., R. E. SHANNON, and R. P. SADOWSKI [1995], *Introduction to Simulation Using SIMAN*, 2d ed., McGraw–Hill, New York.

PRITSKER, A. A. B. [1995], *Introduction to Simulation and SLAM II*, 4th ed., Wiley & Sons, New York.

PRITSKER, A. A. B. [1998], "Principles of Simulation Modeling," in *Handbook of Simulation*, ed. J. Banks, John Wiley, New York.

SANCHEZ, S.R. [2007], "Work Smarter, Not Harder: Guidelines for Designing Simulation Experiments," in *Proceedings of the 2007 Winter Simulation Conference*, eds. S. G. Henderson, B. Biller, M.-H. Hsieh, J. Shortle, J. D. Tew, and R. R. Barton, Washington, DC, Dec. 9–12, pp. 84–94.

SARGENT, R.G. [2007], "Verification and Validation of Simulation Models," in *Proceedings of the 2007 Winter Simulation Conference*, eds. S. G. Henderson, B. Biller, M.-H. Hsieh, J. Shortle, J. D. Tew, and R. R. Barton, Washington, DC, Dec. 9–12, pp. 124–137.

SHANNON, R. E. [1975], *Systems Simulation: The Art and Science*, Prentice-Hall, Englewood Cliffs, NJ.

SHANNON, R. E. [1998], "Introduction to the Art and Science of Simulation," in *Proceedings of the Winter Simulation Conference*, eds. D. J. Medeiros, E. F. Watson, J. S. Carson, and M. S. Manivannan, Washington, DC, Dec. 13–16, pp. 7–14.

EXERCISES

1. Name several entities, attributes, activities, events, and state variables for the following systems:

   (a) A cafeteria
   (b) A grocery store
   (c) A laundromat
   (d) A fast-food restaurant
   (e) A hospital emergency room
   (f) A taxicab company with 10 taxis
   (g) An automobile assembly line

2. Consider the simulation process shown in Figure 1.3.

   (a) Reduce the steps by at least two by combining similar activities. Give your rationale.
   (b) Increase the steps by at least two by separating current steps or enlarging on existing steps. Give your rationale.

3. A simulation of a major traffic intersection is to be conducted, with the objective of improving the current traffic flow. Provide three iterations, in increasing order of complexity, of steps 1 and 2 in the simulation process of Figure 1.3.

4. A simulation is to be conducted of cooking a spaghetti dinner, to discover at what time a person should start in order to have the meal on the table by 7:00 P.M. Download a recipe from the web for preparing a spaghetti dinner (or ask a friend or relative for the recipe). As best you can, trace what you understand to be needed, in the data-collection phase of the simulation process of Figure 1.3, in order to perform a simulation in which the model includes each step in the recipe. What are the events, activities, and state variables in this system?

5. What are the events and activities associated with the use of your checkbook?

6. Numerous reasons for simulation were given in Section 1.1. But there may be other reasons. Look at applications in the current *WSC Proceedings* and see if you can discover additional reasons. (*WSC Proceedings* are found at www.informs-cs.org/wscpapers.html)

7. In the current *WSC Proceedings* available at www.informs-cs.org/wscpapers.html, read an article on the application of simulation related to your major area of study or interest, and prepare a report on how the author accomplishes the steps given in Figure 1.3.

8. Get a copy of a recent *WSC Proceedings* from the URL in Exercise 7 and report on the different applications discussed in an area of interest to you.

9. Get a copy of a recent *WSC Proceedings* from the URL in Exercise 7 and report on the most unusual application that you can find.

10. Locate one of the cases in Section 1.12 and describe the methodology used.

11. Go to the Winter Simulation Conference website at www.wintersim.org and address the following:

    (a) What advanced tutorials were offered at the previous WSC or are planned at the next WSC?
    (b) Where and when will the next WSC be held?

12. Go to the Winter Simulation Conference website at www.wintersim.org and address the following:

    (a) When was the largest (in attendance) WSC, and how many attended?
    (b) In what calendar year, from the beginning of WSC, was there no conference?
    (c) What was the largest expanse of time, from the beginning of WSC, between occurrences of the Conference?
    (d) Beginning with the 25th WSC, can you discern a pattern for the location of the conference?

13. What is the procedure for contributing a paper to the WSC?

14. Classify the list of vendors at a recent WSC as software vendors and others.

15. What is the purpose and history of the WSC Foundation?

16. Using your favorite search engine, search the web for 'discrete event simulation output analysis' and prepare a report discussing what you find.

17. Using your favorite search engine, search the web for 'supply chain simulation' and prepare a report discussing what you find.

18. Using your favorite search engine, search the web for 'web based simulation' and prepare a report discussing what you find.

2

Simulation Examples in a Spreadsheet

This chapter uses spreadsheet examples to provide an introduction to simulation. The examples are relatively simple; many can be carried out manually as well as in a spreadsheet. Our main purpose is to introduce and illustrate some of the key concepts in simulation; it is not to teach how to formulate or develop a spreadsheet model, but rather to use a few simple spreadsheet models to illustrate, by example, some of the key elements of any simulation. These key elements range from the components of a model to experimentation with the model.

Although some types of real-world simulation models, such as those for risk analysis, financial analysis, some reliability problems and others, are amenable to a spreadsheet formulation and solution, the spreadsheet has severe limitations for most complex real-world dynamic, event-based simulations. More often, these complex models are developed in a simulation package or general purpose programming language, not in a spreadsheet, as will be discussed in Chapter 3.

The spreadsheet examples include both Monte Carlo simulations and dynamic simulations that track events over time. Spreadsheet simulations are carried out by devising a simulation table that provides a method for tracking system state over time. Although the simulation tables contain a number of common elements, in general they are ad hoc, meaning they are custom designed for the problem at hand. Due to its ad hoc nature, a general framework for carrying out dynamic discrete-event simulations is needed; the provision of such a framework is left for Chapter 3.

The key elements introduced in this chapter include the use of random numbers to represent uncertainty, the descriptive statistics used for predicting system performance in queueing and inventory models, and the advantage of thinking of a model in terms of inputs, activities, events, states, outputs and responses. These topics are covered in a more systematic manner in later chapters; here they are introduced via example.

Section 2.1 covers the basics, such as generating random samples and the basic components of a simulation model; subsequent examples depend on the concepts and techniques learned here. The examples in the remaining sections cover application areas such as coin tossing, queueing or waiting lines, inventory policies, reliability, hitting a target (bombing or a dart board), and project network analysis. In the examples, we put limited focus on model implementation; instead, we focus on identifying model components (inputs, events, states, and so on) and conducting a simple experiment, the key concept being trials or replications.

All of the examples in this chapter have a spreadsheet solution in Excel on the book's web site, www.bcnn.net. The spreadsheets include notes and comments, which add technical detail to each example, including implementation details, which may prove of value to a student modifying an existing model or developing a new spreadsheet simulation model, as requested in some of the exercises.

2.1   The Basics of Spreadsheet Simulation

In this section we cover the basic material that allows us to carry out the example simulations in the remainder of the chapter. We will learn how to generate random numbers in a spreadsheet simulation, and where to obtain them for a manual simulation. We will learn how to generate random samples for a few simple situations, such as tossing a coin, service and arrival times for queueing simulations, and random demand for inventory models. The methods learned here are sufficient for any of the simple discrete statistical distributions used in the examples in this chapter.

2.1.1   How to Simulate Randomness

Almost all of the simulation models considered in this text contain one or more random variables. Clearly with coin tossing, some mechanism is needed to simulate a random toss that will result in a head or a tail. In queueing models, both service times and interarrival times may be defined by a statistical distribution when the arrival times and service times are not known in advance and cannot be predicted but follow a statistical pattern.

How do we get random quantities? We leave for Chapters 7 and 8 the theory and general explanations behind the generation of random numbers and random variables. In this chapter, we take a pragmatic, practical point of view: How do we get random quantities for a spreadsheet or manual simulation?

Generating any random quantity begins with generating a random number in the interval from 0 to 1. The term "random number" always refers to a randomly generated quantity in the unit interval between 0 and 1. In contrast, the term "random variable" refers to any randomly generated quantity with a specified statistical distribution. Any method for generating a sequence of random numbers is called a Random Number Generator (RNG); these methods are studied in Chapter 7.

The generation method for random numbers should produce a sequence of numbers that have two important statistical properties:

1. The numbers should be uniformly distributed between 0 and 1.
2. Subsequent numbers generated should be statistically independent of all previous numbers generated.

Uniformity implies, among other things, that the probability of falling in an interval (a, b) (with a and b between 0 and 1) equals the length of the interval, $b - a$. Statistical independence implies, among other things, that knowing the value of previously generated numbers does not help in any way in predicting the value of numbers generated subsequently. In addition, it implies that the correlation between successive numbers is zero.

The resulting numbers are often called pseudo-random numbers. One reason is the paradox that, while we want the generated numbers to be statistically uniform and independent, we also want control. We want to be able to reproduce the sequence at any time so that simulation experiments can be controlled and reproduced, and for another practical reason: to assist in model development and debugging.

In general, how do we get these random numbers? Where do they come from? (In what follows, VBA stands for Visual Basic for Applications, the implementation of the VB programming language in Excel and other Microsoft Office applications.) Here are a number of standard sources for random numbers:

1. In Excel, we can use the built-in worksheet function RAND in a formula in a cell. RAND generates values between 0 and 1. Typical examples of formulas using RAND include generating a random value between 0 and 1, as follows:

   =RAND()

   or in a more complex expression such as the following for generating one of two values, 0 or 1:

   =IF(RAND()<=0.5,0,1)

   Note that the IF worksheet function checks a condition in its first argument, and if the condition is true returns the first value; otherwise it returns the second value.
2. Excel also provides the worksheet function, RANDBETWEEN(), that returns uniformly distributed integer values between a specified low and high value. Excel, or some of its add-ins, provide other functions to generate random values. We use none of these.
3. In a worksheet cell in Excel, we can call a user-written VBA function that, in turn, calls an internal VBA library function such as Rnd() or the VBA function (MRG32k3a) described in Chapter 7 and available on the book's web site. The spreadsheet examples in this chapter all have the VBA function Rnd01() that can be called from a cell. (Rnd01() simply calls the internal VBA library function Rnd().)
4. Simulation packages provide methods to generate both random numbers and random variables from a long list of standard statistical distributions as well as from data.

5. Most general-purpose programming languages provide a built-in library function to generate random numbers, and many provide facilities to generate samples from a variety of statistical distributions. Lacking this, the model developer can program one or more of the algorithms from Chapters 7 and 8.

6. For manual (paper-and-pencil) simulations, you could use a physical RNG such as tossing one or more coins, or throwing dice. These being impractical for all but the simplest cases, another possibility is a table of random numbers or random variables from a specific distribution, such as Tables A.1 (for uniform) and A.2 (for the standard normal) in the Appendix.

Neither the worksheet function RAND() nor the VB function Rnd() should be used in professional work; they both have known deficiencies. For example, the VB function Rnd() returns a single precision value (about 5 decimal place of accuracy), while all good RNG return double precision (about 15 decimal places); secondly, Rnd() has a short period—it repeats after generating a relatively small number of values. Its period is 2^{24}, about 16.77 million. The RAND() function has an even shorter period and more serious deficiencies regarding the two key properties of uniformity and independence. (See Chapter 7 for a detailed look at RNG period and other important issues regarding uniformity and statistical independence of RNG. Note also that Microsoft changed the RAND() function in Office 2003; reportedly, it has both bugs and deficiencies and has been widely criticized.) We use RAND() only in Example 2.1, merely for instructional purposes. Since much better generators are readily available, including the VBA function MRG32k3a, available on the book's web site in the Excel file, "RandomNumberTools.xls" and described in Chapters 7 and 8, we feel strongly that there is no reason (or excuse) to use a deficient generator for professional work.

Most RNG can be initialized by a user-specified number called a *seed*. The VBA code behind the 'Reset Seed & Run' button in the spreadsheet examples shows how to set the seed for VBA's Rnd() generator. Apparently, there is no way to set the seed for Excel's RAND generator (another reason not to use it). The MRG32k3a generator (in "RandomNumberTools.xls") comes with the VBA function, InitializeRNSeed(), for setting that generator's seed. By specifying the seed, the user can make a random number generator repeat a sequence of numbers. This is valuable for debugging simulation models and repeating simulation experiments. Remember that we are simulating randomness; the control given by repeatability is a feature, not a bug.

2.1.2   The Random Generators Used in the Examples

Each example spreadsheet has a number of VBA functions for generating random variables in a few simple cases. We advise starting with an existing spreadsheet model when developing a new model; in this way, all the VBA functions written for the examples will be available for the new model.

The VBA functions and Excel worksheet functions for generating random values in a cell in a worksheet include the following:

1. Rnd01(), to generate a random number between 0 and 1;
2. DiscreteUniform(min, max), to generate a random integer from the minimum value, min, to the maximum value, max, where min and max can be numbers or a single-cell range containing the number;

3. DiscreteEmp(rCumProb, rValues), to generate samples from a discrete distribution specified by two Excel ranges (ranges of cells), rCumProb for the cumulative probabilities, and rValues for the values;
4. Uniform(low, high), to generate a real value (not just integers) from the continuous uniform distribution between low and high, where low and high are numbers or a cell containing a number;
5. NORMSINV(Rnd01()), to generate a standard normal with mean 0 and standard deviation 1.

All of these functions, except NORMSINV(), are VBA functions written for the examples in this chapter; NORMSINV() is an Excel worksheet function that can be used (in the form given above) to generate a standard normal. We recommend not using the worksheet functions RAND() and RANDBETWEEN(). All spreadsheet solutions have access to any of the functions, whether used or not in the supplied solution; therefore, the example spreadsheets are good starting points for some of the exercises.

At this early point in the text, we do not expect the reader to completely understand how all of these work; that is left to Chapters 7 and 8. Our goal, however, is that you understand the nature of the random values generated by each function, and be able to use them in the exercises if modifying an example or creating a new spreadsheet model. Each of these functions is explained in more detail as they are used in an example in later sections.

2.1.3   How to Use the Spreadsheets

The spreadsheet simulation models are in the 'One Trial' worksheet of each solution workbook. This sheet contains the definition and specification of model inputs, the simulation table containing the formulas to generate each row to carry the simulation forward, and some summary statistics (and usually a histogram) representing the results of one simulation run. The two buttons on the worksheet control the simulation:

1. On each click, the 'Generate New Trial' button makes a new run of the simulation, using a new sequence of random numbers each time.
2. On each click, the 'Reset Seed & Run' button sets the RNG seed to the value specified by the user in an adjacent cell, and then runs the simulation using this seed. The default seed is the arbitrary value 12,345.

You can always return to the initial results by clicking the 'Reset Seed & Run' button.

The 'Experiment' worksheet allows a user to conduct a simulation experiment by repeatedly executing the 'One Trial' worksheet, each time with different random numbers, and recording the designated model response (from the Link cell) to the Response Table. The user specifies the number of trials, for example, 100, meaning that the 'One Trial' worksheet will be executed 100 times and the 100 response values recorded in the Response Table. This is called replication; the number of replications is the number of trials.

The experimental results over all trials are summarized in the "Multi-Trial Summary" table on the 'Experiment' sheet. These summary statistics include frequency distributions and an accompanying histogram, and standard statistics such as sample average, median, minimum, and maximum. An example is shown in Table 2.6 in Section 2.2.

To run the experiment on the 'Experiment' worksheet, use the two buttons as follows:

1. On each click, the 'Run Experiment' button makes one run of the experiment for the specified number of trials. Each subsequent click runs the experiment using different random numbers.
2. On each click, the 'Reset Seed & Run' button sets the RNG seed to the specified value in an adjacent cell (default 12,345) and makes one run of the experiment. Each click reproduces the exact same original results.

The results given in all the examples in this chapter come from clicking the 'Reset Seed & Run' button; this guarantees that a reader can reproduce these same results. Each spreadsheet solution contains comments and explanations on different aspects of the solution, including the implementation of model logic; some solutions have a worksheet named 'Explain' with lengthier explanations of certain aspects of the model.

2.1.4   How to Simulate a Coin Toss

How can we simulate a coin toss in a spreadsheet model? The solution shows how to use a uniform random number generator in a spreadsheet simulation.

Example 2.1:   Coin Tossing

We want to simulate a sequence of 10 coin tosses. The coin is a "fair" coin, meaning that the chance of getting a head is the same as getting a tail. Each has probability 0.5 of occurring. You should run the simulation multiple times and compare the results with your expectations from real-life coin tossing. This is an example of a Monte Carlo simulation; there are no events or clock times being tracked.

The solution, shown in Table 2.1, is in the spreadsheet "Example2.1CoinToss.xls". In fact, there are four solutions in the spreadsheet, the first two using the RAND() random number generator, and

Table 2.1   Simulation Table for Coin Tossing

	B	C	D	E	F	G	H	I
11		Solution #1A		Solution #1B		Solution #2A		Solution #2B
12			Solution using RN from ColC	Solution using RAND()			Solution using RN from ColG	Solution using Rnd01()
13								
14	Toss	RAND()			Toss	Rnd01()		
15	1	0.7317089	T	T	1	0.9871114	T	T
16	2	0.8285837	T	T	2	0.0225598	H	H
17	3	0.0701168	H	T	3	0.0008356	H	T
18	4	0.2147295	H	H	4	0.2127686	H	T
19	5	0.8613179	T	H	5	0.8586159	T	H
20	6	0.4159098	H	H	6	0.5503362	T	T
21	7	0.7492317	T	H	7	0.5243730	T	H
22	8	0.9671661	T	T	8	0.8884351	T	H
23	9	0.4823089	H	T	9	0.2111118	H	T
24	10	0.6738897	T	H	10	0.7680427	T	H

the last two using the supplied VBA function Rnd01(). It is left as an exercise (at the end of the chapter) to devise a justification for the statistical equivalence of the 4 solutions, even though on any click of the button 'Generate New Trial', the 4 sets of results are different.

The spreadsheet logic for coin tossing is simple. The second solution (Column E) is:

```
=IF(RAND()<=0.5,"H","T")
```

while the fourth solution (Column I) is:

```
=IF(Rnd01()<=0.5,"H","T")
```

Each call to RAND() or Rnd01() produces a new random number. The 'One Trial' worksheet records the frequency for heads versus tails in the 10 tosses and graphs the results in a histogram.

The first and third solutions are two steps, an approach that is useful when a generated random number is used more than once. Columns C and G contain the generated random number, and columns D and H the result ("H" or "T"). The fourth solution is the basis for the Coin Tossing Game in Example 2.4.

For all future spreadsheet simulations, we use the supplied VBA function, Rnd01(), or one of the VBA functions based on Rnd01(), instead of RAND(). However, for serious professional work, we recommend using the generator implemented in the VBA function, MRG32k3a, from the file "RandomNumberTools.xls," described in Chapter 7. As previously mentioned, both the Excel worksheet function, RAND(), and the VBA function, Rnd(), have serious limitations and deficiencies, and should only be used for casual or educational purposes.

2.1.5   How to Simulate a Random Service Time

The following example shows how to generate random samples from an arbitrary discrete distribution. The example is simple, having just 3 possible values of service time. The key principle, namely, the transformation from a random number to a service time of a prescribed probability, easily generalizes to any discrete distribution with a finite number of possible values.

The methods described here and in Section 2.1.6 for arrival times are used in almost all of the subsequent examples in the chapter.

Example 2.2:   Random Service Times ──

An automated telephone information service spends either 3, 6, or 10 minutes with each caller. The proportion of calls for each service length is 30%, 45%, and 25%, respectively. We want to simulate these service times in a spreadsheet. Our actual purpose is to learn how to generate random samples from a discrete distribution, in preparation for the queueing, inventory, and other examples to follow. This also is an example of a Monte Carlo simulation.

Table 2.2 shows the input specification and a portion of the simulation table (the first 4 callers), taken from the spreadsheet model, "Example2.2ServiceTimes.xls". The input specification defines the probabilities for the service times, and also includes the cumulative probabilities. Cumulative probabilities always increase to 1.0; as we shall see, the generation method uses the cumulative probabilities. The method always starts with a random number and ends with the desired value, in

Table 2.2   Input Specification and Simulation Table for Service Times

	A	B	C	D
4		Service		Cumulative
5		Time	Probability	Probability
6				
7		3	0.30	0.30
8		6	0.45	0.75
9		10	0.25	1.00
10				
11		**Number of Callers=**		25
12		**Simulation Table**		
13		**Step**	**Activity**	
14				
15		**Caller**	**Service Time**	
16		1	6	
17		2	6	
18		3	10	
19		4	6	

this case, a random service time with the specified distribution. The simulation table in the spreadsheet shows the resulting frequency distribution (or histogram) of the generated service times for 25 callers.

Figure 2.1 illustrates how to transform a random number to a service time. Think of it as throwing a special dart at a special dart board; the dart will always land somewhere on the unit interval, with equal probability of landing at any of the points on the line between 0 and 1. The subinterval where the dart lands determines the value generated.

In Figure 2.1, the probabilities (at the top) are represented by non-overlapping segments on a unit interval, and must add up to 1. The cumulative probabilities (just below the line) are represented by points on the line. The arrows represent the transformation. The algorithm works as follows: First, choose a random number, say R. If R is in the first interval, that is, if R is less than or equal to 0.30, the procedure generates a sample of 3 minutes for service time. Similarly, if R is between 0.30 and 0.75, the service time is 6 minutes, and finally, if R is greater than 0.75, then the service time is 10 minutes. This algorithm easily generalizes to any discrete distribution.

To illustrate the procedure, we first generate 5 samples of random numbers, either in Excel (using RAND or Rnd01 or any other random number generator), or take 5 samples from a table of random numbers such as Table A.1 in the Appendix. Suppose the 5 samples are:

$$0.9871 \quad 0.0226 \quad 0.0008 \quad 0.2128 \quad 0.8586$$

After the transformation illustrated in Figure 2.1, the resulting service times are:

$$10 \qquad 3 \qquad 3 \qquad 3 \qquad 10$$

In a small sample (here, size 5), we cannot expect the observed frequencies of occurrence of each value to be close to the probabilities; however, with a large enough sample, the sample

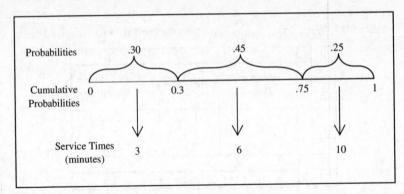

Figure 2.1 "Random Dart" Transformation from Random Number to Service Time.

frequencies should be fairly close to the probabilities. You can experiment with the spreadsheet example, "Example2.2ServiceTimes.xls", to see how close or far small samples can be from the assumed probabilities.

In Excel, we might think we could implement this transformation, at least for distributions with a small number of values (here, 3), with a nested IF worksheet function, as in:

```
=IF(Rnd01()<=0.30,3,IF(Rnd01()<=0.75,6),10)      (WRONG !!)
```

This attempt at a solution introduces a grave error. Each reference to the Rnd01() function generates a new random number; the transformation procedure depends on using just one random number. (You may want to modify the spreadsheet solution in "Example2.2ServiceTimes.xls" with this incorrect solution to see the resulting probabilities.) This approach can be modified to use two cells, the first to hold the result from Rnd01, the second with a modified IF statement.

A better approach is to use the VBA function called DiscreteEmp() (for discrete empirical, meaning defined by data) supplied with all the examples. To see how this function is used, see the spreadsheet solution, "Example2.2ServiceTimes.xls"; a typical cell for generating a random service time appears as follows:

```
=DiscreteEmp($D$7:$D$9, $B$7:$B$9)
```

where $D$7:$D$9 is the range of cells containing the cumulative probabilities and $B$7:$B$9 is the range of cells containing the desired values for service times. Discrete means that only the values listed will be generated. Each of the two worksheet ranges must be a column vector (meaning all data in one column), with each having the same length, that is, the same number of cells. The "$" in each address in front of each row and column indicates an absolute address rather than a relative address; this makes it easier to copy and paste the formula into all the cells in the "Service Time" column after typing it into just one cell.)

The method used here with the supplied VBA function, DiscreteEmp(), easily generalizes to most of the examples (and exercises) in the chapter that use simple discrete distributions for model inputs. To use DiscreteEmp(), supply two ranges in the worksheet, one for cumulative probabilities and the second for the desired values. It works with any number of values in the distribution.

In future examples, keep Figure 2.1 in mind; it illustrates exactly how `DiscreteEmp()` works. Although its VBA implementation involves a loop and may appear complex to a non-programmer, its basic logic is the simple logic of Figure 2.1 — the simple logic of throwing a dart at a special dart board, the unit interval, and a transformation.

2.1.6   How to Simulate a Random Arrival Time

In this section our goal is to learn how to generate random interarrival times (times between successive arrivals) and compute arrival times of customers or other entities to a system. The interarrival times, as well as the service times in Example 2.2, are examples of simulation-generated time durations, called *activity times* or *activities*. These are distinguished from *arrival times*, which are times on a simulation clock representing the time-of-arrival to a facility.

A simulation clock is a key component of every dynamic discrete-event simulation. In all of the Excel spreadsheet solutions, the clock times are labeled just above the appropriate column titles of the simulation table, so that you can easily distinguish between the activity times (service or interarrival times, for example) and the computed clock times. In all cases, a clock time represents the time of occurrence of an event, such as an arrival, a service beginning, or a service completion in a queueing model.

Example 2.3:   Random Arrival Times

Telephone calls to the telephone information service, where service times are defined in Example 2.2, occur at random times defined by a discrete distribution for which the interarrival times have values 1, 2, 3, or 4 minutes, all with equal probability. Our purpose here is to show how to generate both interarrival times and arrival times.

Since this example has one event, namely, the arrival event, it is our first example of a dynamic, event-based model (despite its simplicity). *Dynamic* simply means time-based, with system state changing over time; dynamic, event-based means that the model tracks the progression of event occurrences over time.

The interarrival times can be generated randomly in exactly the same manner as done in Example 2.2 for service times. The spreadsheets, however, supply a second VBA function to simplify random generation from a discrete uniform distribution. *Uniform* means that all the values have equal probabilities; in this example, all the interarrival times may have one of the values 1, 2, 3, or 4 minutes, each occurring with probability 0.25. For such discrete uniform distributions, the spreadsheets supply the VBA function, `DiscreteUniform()`, which takes two arguments: either the cell containing the minimum value followed by the cell containing the maximum value of the desired range of integers, or the actual low and high values, as in:

```
= DiscreteUniform($G$5,$G$6)
```

or

```
= DiscreteUniform(1,4)
```

The `DiscreteUniform(low, high)` function is implemented by a single line of VBA code, as follows:

```
DiscreteUniform = low + Int((high - low + 1)*Rnd())
```

which becomes, for this example:

```
DiscreteUniform = 1 + Int(4 * Rnd())
```

where `Rnd()` generates a random number between 0 and 1 (in VBA, it is always less than one but may equal zero), and Int() is a VBA function that truncates (rounds down) its argument. (Note that 4 `* Rnd()` generates real values between 0.0 and 3.99999, so that truncating and then adding 1 yields the desired range 1 to 4. Why do we expect equal probabilities?)

Table 2.3 shows the input specification for interarrival times and a portion of the simulation table, taken from the spreadsheet model in "Example2.3ArrivalTimes.xls". As shown in Step 1 (row 14), we assume that the first arrival occurs at simulation time 0. (The assumed arrival time for the first customer is model-dependent; it could be any other specified time, constant or random.) The interarrival times are randomly and independently generated using the `DiscreteUniform(1, 4)` function. The arrival times are computed by a simple formula, by adding the current customer's interarrival time to the previous customer's arrival time. (You should verify the spreadsheet formula for arrival times.)

Table 2.3   Input Specification and Simulation Table for Arrival Times.

	A	B	C	D
4		**Interarrival Times**		
5		**(minutes)**		
6		**Minimum**	1	
7		**Maximum**	4	
8				
9		**Number of Callers=**		25
10		**Simulation Table**		
11		**Step**	**Activity**	**Clock**
12			**Interarrival**	
13		**Caller**	**Time**	**Arrival Time**
14		1		0
15		2	2	2
16		3	1	3
17		4	3	6

Example 2.3 is the first that contains an event, namely, the arrival event, and the event time associated with that event, namely, the CLOCK time for the time of arrival. Note that Table 2.3 distinguishes the activity times from the CLOCK times.

2.1.7   A Framework for Spreadsheet Simulation

The simulation table provides the structure, and the steps outlined below provide a set of general guidelines for carrying out a spreadsheet simulation. We have seen several examples of a simulation table, with Table 2.3 in Example 2.3 being the best so far—since it contains a clock time for the arrival event, making it one of the simplest possible dynamic, event-based simulations.

First, the model developer must define the model inputs, the system states, and the model outputs. Inputs are the exogenous variables that are (usually) defined independently of other system characteristics; examples include the probability of a head on a coin toss, the service time and the interarrival time distributions in a queueing system, or the demand distribution in an inventory model. These input specifications are used to generate random activity times and other random variables. Other inputs, such as policy or decision parameters, may be constants.

Outputs are used to compute measures of system performance (also known as *model responses*, or simply *responses*). For example, an output could be an individual customer's waiting time in a queue, or the cost of an individual transaction in an inventory system. The associated responses might be average waiting time in a queue, or the average cost per unit time in an inventory system, respectively.

Each simulation table is custom designed for the problem at hand. Each column in the table is one of the following types:

1. An activity time associated with a model input;
2. Any other random variable defined by a model input;
3. A system state;
4. An event, or the clock time of an event;
5. A model output;
6. Sometimes, a model response.

The components of a model—activities, states, events and so on—are expanded and explained in more detail in Chapter 3. For the spreadsheet simulations here, the above-listed items are sufficient.

In general, model responses, also known as *measures of system performance*, are computed outside the simulation table after the simulation has been completed. For example, in a waiting line model, the delay in queue of each individual customer is a *model output*, while the average delay over all customers is a *model response*. The model output is tracked in the simulation table; the response is computed afterwards.

An activity time is a time defined by a constant or a statistical distribution, such as a service time, or an interarrival time. An activity is a duration of time, in contrast to event times, which are clock times.

System states include, for example, server status (busy or idle) in a queueing model and inventory levels in an inventory model. In general, the set of system states includes any information needed for the simulation to proceed through time (or from one step to the next), as well as any quantity or information that makes the simulation logic simpler or more transparent and thus easier to understand. The greatest simplicity possible, as long as the model logic is correct, assists in an ease of understanding that is of tremendous value, especially when trying to verify that a model's logic is indeed correct (a topic covered in Chapter 10).

Table 2.4 A Generic Simulation Table

	Activities and System States						*Outputs*
Step	x_{i1}	x_{i2}	$\cdots$	x_{ij}	$\cdots$	x_{ip}	y_i
1							
2							
3							
.							
.							
.							
n							

Each row in the simulation table represents a step in the simulation and is usually associated with the occurrence of one or more events, or the progress of one entity (such as a customer) through the system. The simulation table is designed so that a given step depends solely on model inputs and/or one or more previous steps or previously computed values in the current step.

In general, a modeler develops and runs a spreadsheet or manual simulation by following these guidelines:

1. Determine the characteristics of each of the inputs to the simulation. Often encountered types of inputs include constants as well as probability distributions.
2. Determine the activities, events, and system states relevant to the problem.
3. Determine model responses or summary performance measures, depending on objectives and specific questions to meet those objectives.
4. Determine the model outputs needed for computing model responses.
5. Construct a simulation table. A generic representation of a simulation table is shown in Table 2.4. In this table, there are p activities and system states, $x_{ij}, j = 1, 2, \ldots, p$, and one output, y_i, for each simulation step $i = 1, 2, \ldots, n$. Initialize the table by filling in the data for Step 1. When doing a spreadsheet model, the initial step or steps (usually just Step 1) are typically different from subsequent steps, which are defined by formulas.
6. For each Step i, generate a value for each of the activities, then compute the system states and outputs y_i. Random input values may be computed by sampling values from the distributions chosen in Step 1. An output typically depends on the inputs, system states, and possibly other outputs.
7. When the simulation is finished, use the outputs to compute the model responses or measures of performance.

In the following sections, we will see several examples of spreadsheet simulations carried out in a custom-designed simulation table. These include examples of a coin tossing game, plus examples in application areas such as queueing, inventory, reliability, and network analysis. The coin tossing game asks an easy question yet has an answer that we believe will be surprising to most readers.

The two queueing examples provide illustrations of single-server and two-server systems. The first of the inventory examples involves a problem that has a closed-form solution; thus, the simulation solution can be compared to the mathematical solution. The second inventory example pertains to the classic order-level model.

The remaining examples include a model of hitting a target, a reliability model, a Monte Carlo model to estimate the distribution of a complex random variable, namely, lead-time demand, and finally, a model of a project activity network. The target hitting model introduces the use of normally distributed random numbers.

2.2   A Coin Tossing Game

In a coin tossing game, is it more likely that you and your opponent each share the lead about 1/2 of the time (that is, on about 1/2 of the tosses), compared to, say, one of you being ahead most of the time?

With the following example, we explain in more detail the layout of the spreadsheet solutions, the two main worksheets ('One Trial' and 'Experiment'), the buttons to run the simulations, and the terminology used: *trials* (or replications) and *experiments*. The details of model logic (formulas used, functions called, and so on) are explained in the spreadsheet itself, all of which can be found at the book's web site, www.bcnn.net.

Example 2.4:   A Coin Tossing Game

Charlize tosses a coin for her friends, Harry and Tom, exactly 100 times. Harry wins $1.00 when a head lands up, and Tom loses $1.00. Tom wins when the coin lands with tails up, with Harry paying $1. Charlize tracks their respective wins and losses as they play the game. On any toss, Harry could be ahead, or Tom could be ahead, or they could be even.

On any game (a game being exactly 100 tosses), is it more likely that both Harry and Tom are each ahead on about 1/2 of the tosses, or is it more likely that one of the two is ahead most of the time?

Note that we are not asking about winnings at the end of the game; we are not asking about the proportion of heads versus tails (which, as you know, should be close to 50:50 in the long run). The question concerns how often Harry or Tom is ahead or behind during the course of the game. Note also that the two possibilities in the question are not the only two possible results of a game; therefore, their probabilities will not add up to 1.

To be specific, which of these is more likely, and by about how much?

(A) Harry and Tom are each ahead on about 1/2 of the tosses; specifically, Harry is ahead on 45 to 55 of the tosses, and thus Tom is also ahead on 45 to 55 of the tosses; or

(B) Harry is ahead on 95 or more tosses; or

(C) Harry is ahead on 5 or fewer tosses.

Guess before looking at the results. Table 2.5 shows the input specifications and the first 14 tosses in the simulation table, taken from the spreadsheet model in "Example2.4CoinTossGame.xls" when using the default seed (12,345) and clicking the button 'Reset Seed & Run'. The 'One Trial' worksheet has the logic and results for one game of 100 tosses.

Table 2.5   Input Specification and Simulation Table for the Coin Tossing Game

A	B	C	D
4			
5	**Coin**	**Probability**	**Cumulative Probability**
6			
7	Head	0.50	0.50
8	Tail	0.50	1.00
9			
10	**Numer of Tosses = 100**		
11	**Simulation Table**		
12	**Step**	**State**	**Output**
13	**Toss**	**Result of Toss**	**Harry's Winnings**
14	1	T	-$1
15	2	H	0
16	3	T	-1
17	4	H	0
18	5	H	1
19	6	T	0
20	7	H	1
21	8	T	0
22	9	T	-1
23	10	H	0
24	11	T	-1
25	12	H	0
26	13	T	-1

If you inspect the 'One Trial' worksheet in the spreadsheet solution (for the trial using the default seed for the RNG), you will see that on toss 13, Harry's winnings are negative (he's losing money) and they stay negative for the remainder of the game; in fact, in this game, Tom is ahead on 92 of the 100 tosses, yet his winnings at the end are just $6.00. Harry is ahead on just 2 of the 100 tosses, and they are even on 6 of the 100 tosses (the last time on toss 12).

Now, we cannot answer the questions asked by playing just one game (equivalent to clicking the 'Generate New Trial' button once, since each trial is one game of 100 tosses). To play many games, we use the 'Experiment' worksheet and set the Number of Trials (cell C3) to 400. (Any fairly large number of trials will do for our present purposes.) Then click the button labeled 'Reset Seed & Run', which sets the seed to its default value (12,345) and runs the experiment with this seed. Running the experiment means that the 'One Trial' worksheet is executed 400 times, and after each trial, the result of the trial is recorded in the 'Experiment' worksheet. The spreadsheet contains comments and explanations for the formulas, functions, and other technical details for implementing model logic and collecting the desired outputs.

Table 2.6   Experiment Worksheet for the Coin Tossing Game

	A	B	C	D	E	F
3		**Number of Trials:**	**400**			
4		**Link to Measure of Performance:**				
5		**Name of Measure:**		**Link**		
6		**Number of Tosses out of 100 on which Harry is Ahead.**		**31**		
7						
8						
9		**Response Table**		**Multi-Trial Summary**		
10	**Trial**	**Number Tosses w/ Harry Ahead**		**Bins**	**Frequency**	**Relative Frequency**
11	1	2		0 to 5	73	18.3%
12	2	91		45 to 55	28	7.0%
13	3	1		95 to 100	46	11.5%
14	4	76				
15	5	17				
16	6	11		**Average**	48.2	
17	7	94		**Median**	50	
18	8	95				
19	9	78		**Minimum**	0	
20	10	0		**Maximum**	100	
21	11	40				

Table 2.6 shows the experimental specification (400 trials and the response) plus the Response Table for the first 11 trials (first 11 games), plus the Multi-Trial Summary. Figure 2.2 shows the histogram based on the frequency results in the Multi-Trial Summary.

Based on the experiment with 400 trials (or replications), the results may appear surprising to many people:

Harry behind most of the time   In 73 of the 400 games (18.3%), Harry is ahead on 5 or fewer of the 100 tosses.

Harry and Tom about the same   In 28 of the 400 games (7.0%), Harry and Tom share the lead about equally.

Harry ahead most of the time   In 46 of the 400 games (11.5%), Harry is ahead on 95 or more of the 100 tosses.

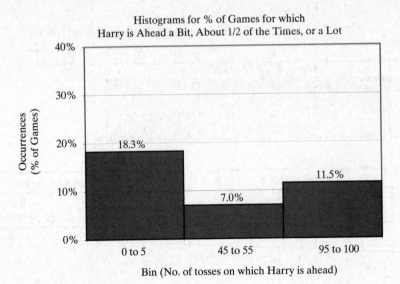

Figure 2.2   Histogram for Harry's Chances of Being Ahead.

Of course, these three possibilities are not the only possibilities; that is why the frequencies do not add up to 400 games or 100%. You may wonder: What explains that the result "0 to 5" appears to be more likely than "95 to 100", and that both are more likely than Tom and Harry sharing the lead about half of the time? What about the magnitude of this difference? Do these results contradict the assumption that the coins are "fair" and heads and tails have equal probability? Do these results say anything about Harry or Tom's final winnings or losses?

You may experiment by clicking the 'Run Experiment' button several times, to see how the results vary from one experiment to the next. With 400 games constituting each experiment, it seems intuitively that the results would not change too much. One run of the experiment is identical to any other except that a new set of random numbers is used. (You can always return to the original random numbers and original results presented here by clicking the 'Reset Seed & Run' button.) To verify the results quoted here based on 400 games (i.e., 400 trials), you could run an experiment with a very large number, say, 4000 trials. In Chapters 11 and 12, we will learn statistical methods for estimating the accuracy of the results from a given set of trials, and see that accuracy increases as the number of trials increases.

2.3   Queueing Simulation in a Spreadsheet

We begin by describing the components of queueing or waiting line models. We follow with two examples: a single-server queue and a two-server queue; both are dynamic, event-based models.

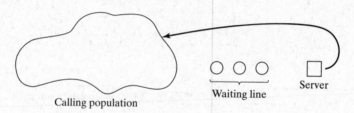

Figure 2.3   Single-channel queueing system.

2.3.1   Waiting Line Models

A queueing system is described by its calling population, the nature of the arrivals, the service mechanism, the system capacity, and the queueing discipline. These attributes of a queueing system are described in detail in Chapter 5. A simple single-channel queueing system is portrayed in Figure 2.3. The members of the calling population may be, for example, customers, telephone calls, or jobs in a repair shop; here, we refer to them as *units*.

In the simple models in this chapter, we assume an unlimited potential calling population and a constant arrival rate. That is, no matter how many units have already arrived or are still present in the system, the interarrival time until the next arrival follows the same statistical distribution. Arrivals for service occur one at a time in a random fashion; once they join the waiting line, they are eventually served. In addition, service times are of some random length according to a probability distribution which does not change over time. The system capacity has no limit, meaning that any number of units can wait in line. Finally, units are served in the order of their arrival (often called FIFO: first in, first out, or FCFS: first come, first served) by a single server or multiple parallel servers.

Services and arrivals are defined by the distribution of service times and the distribution of time between arrivals, respectively, as illustrated by Examples 2.2 and 2.3. For any simple single- or multi-channel queue, the overall effective arrival rate must be less than the total service rate, or the waiting line will grow without bound. When queues grow without bound, they are termed *explosive* or unstable. (In some re-entrant queueing networks in which units return a number of times to the same server before finally exiting from the system, the condition that arrival rate be less than service rate might not guarantee stability. See Harrison and Nguyen [1995] for more explanation. Interestingly, this type of instability was noticed first, not in theory, but in actual manufacturing in semiconductor manufacturing plants.) More complex situations can occur—for example, arrival rates that are greater than service rates for short periods of time, or networks of queues with routing. However, this chapter sticks to the most basic queues.

How do the concepts of system state, events, and simulation clock apply to queueing models? System state for a queueing model typically consists of the number of units in the system and server status, either busy or idle. An event is a set of circumstances that causes an instantaneous change in the state of the system. In a single-channel (or any multi-channel) queueing system, the events are the arrival event, the service beginning event, and the departure event, also called the service completion event. The simulation clock is used to track simulated time.

When a departure event occurs, the simulation proceeds in the manner shown in the flow diagram of Figure 2.4. The departure event causes the system state to change. If there are no units waiting in

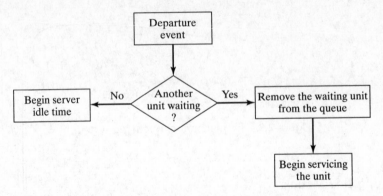

Figure 2.4   Departure event flow diagram.

the queue, server status changes to idle; if there are one or more units waiting, the next unit begins service. In both cases, the number of units in the system decreases by one.

The flow diagram for the arrival event is shown in Figure 2.5. The unit finds the server either idle or busy; therefore, either the unit begins service immediately, or it enters the queue. The unit follows the course of action shown in Figure 2.6. If the server is busy, the unit enters the queue. If the server is idle and the queue is empty, the unit begins service. It is not possible for the server to be idle while the queue is nonempty.

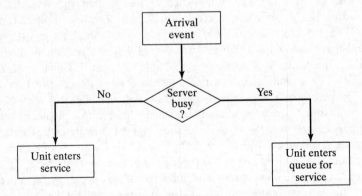

Figure 2.5   Arrival event flow diagram.

		Queue status	
		Not empty	Empty
Server status	Busy	Enter queue	Enter queue
	Idle	Impossible	Enter service

Figure 2.6   Potential unit actions upon arrival.

		Queue status	
		Not empty	Empty
Server outcomes	Busy	////////	Impossible
	Idle	Impossible	////////

Figure 2.7   Server outcomes after the completion of service.

After a service completion, the server either becomes idle or remains busy with the next unit. The relationship of these two outcomes to queue status is shown in Figure 2.7. If the queue is not empty, another unit engages the server, which then remains in the busy state. If the queue is empty, the server becomes idle. These two possibilities are shown as the shaded portions of Figure 2.7.

In most simulations, many events occur at random times, the randomness imitating uncertainty in real life. For example, it is not known with certainty when the next customer will arrive at a grocery checkout counter, or how long the bank teller will take to complete a transaction. In these cases, a statistical model of the data is developed either from data collected and analyzed or from subjective estimates and assumptions. For our purposes, we will assume simple discrete probability distributions for interarrival and service times.

Now, how can the events described here occur in simulated time? Simulations of queueing systems generally require the maintenance of an event list for determining what happens next. The event list tracks the future times at which the different types of events will occur. Simulations based on an event list are described in Chapter 3. This chapter simplifies the simulation by tracking each unit explicitly. Note that this ad hoc approach is satisfactory for simple simulations in a spreadsheet, but the more systematic approach introduced in Chapter 3 is essential for the more complex real-world simulations (and even the more interesting student simulations).

Table 2.7 was designed as a simulation table specifically for a single-channel queue that serves customers on a first-in–first-out (FIFO) basis. It keeps track of the clock time at which each event occurs. (Note that it contains no columns for outputs; these will be added in the next example.) This simulation table computes arrival times from an assumed sequence of interarrival times:

$$2 \qquad 4 \qquad 1 \qquad 2 \qquad 6$$

Columns B, C, and E are simulation clock times. Column B, the arrival times, are computed solely from the generated interarrival times. Column E, the time for the departure (or service completion) event, is easily computed by the generated input in Column D and the time service begins in Column C. Thus we see that computing Column C, the time service begins, is the main challenge. For an arriving unit, service begins at the later of two times: its arrival time or the departure time of the previous unit. (For 2-server and other multi-server queues, it is not quite so easy to compute the service completion time.)

As shown in Table 2.7, the first customer arrives at clock time 0 and immediately begins service, which requires two minutes. Service is completed at clock time 2. The second customer arrives at clock time 2 and is finished at clock time 3. Note that the fourth customer arrived at clock time 7, but service could not begin until clock time 9. This occurred because customer 3 did not finish service until clock time 9.

Table 2.7   Simulation Table Emphasizing Clock Times

A	B	C	D	E
Customer Number	Arrival Time (Clock)	Time Service Begins (Clock)	Service Time (Activity)	Time Service Ends (Clock)
1	0	0	2	2
2	2	2	1	3
3	6	6	3	9
4	7	9	2	11
5	9	11	1	12
6	15	15	4	19

Table 2.8   Chronological Ordering of Events

Event Type	Customer Number	Clock Time
Arrival	1	0
Departure	1	2
Arrival	2	2
Departure	2	3
Arrival	3	6
Arrival	4	7
Departure	3	9
Arrival	5	9
Departure	4	11
Departure	5	12
Arrival	6	15
Departure	6	19

The occurrence of the arrival and departure events in chronological order is shown in Table 2.8 and Figure 2.8. While Table 2.8 is ordered by clock time, the events may or may not be ordered by customer number. The chronological ordering of events is the basis of the approach to discrete-event simulation described in Chapter 3.

Figure 2.8 depicts the number of customers in the system over time. It is a visual image of the event listing of Table 2.8. Customer 1 is in the system from clock time 0 to clock time 2. Customer 2 arrives at clock time 2 and departs at clock time 3. No customers are in the system from clock time 3 to clock time 6. During some time periods, two customers are in the system, such as at clock time 8, when customers 3 and 4 are both in the system. Also, there are times when events occur simultaneously, such as at clock time 9, when customer 5 arrives and customer 3 departs.

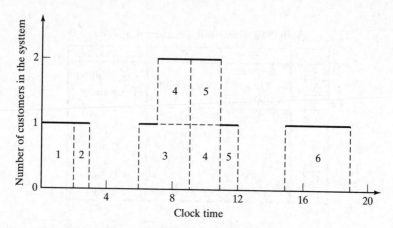

Figure 2.8   Number of customers in the system.

2.3.2   Simulating a Single-Server Queue

The first queueing example is a single-server queue. We illustrate how to compute typical outputs and performance measures, such as the customer's average waiting time and the proportion of time the server is idle.

Example 2.5:   The Grocery Checkout, a Single-Server Queue

A small grocery store has one checkout counter. Customers arrive at the checkout counter at random times that range from 1 to 8 minutes apart. We assume that interarrival times are integer-valued with each of the 8 values having equal probability; this is a discrete uniform distribution, as shown in Table 2.9. The service times vary from 1 to 6 minutes (also integer-valued), with the probabilities shown in Table 2.10. Our objective is to analyze the system by simulating the arrival and service of 100 customers and to compute a variety of typical measures of performance for queueing models.

In actuality, 100 customers may be too small a sample size to draw reliable conclusions. Depending on our objectives, the accuracy of the results may be enhanced by increasing the sample size (number of customers), or by running multiple trials (or replications), as illustrated on the 'Experiment' worksheet, topics covered in Chapter 11. A second issue, also discussed in Chapter 11, is that of initial conditions. A simulation of a grocery store that starts with an empty system may or may not be realistic unless the intention is to model the system from startup or to model until

Table 2.9   Distribution of Time Between Arrivals

	G	H
4	**Interarrival Times**	
5	**(minutes)**	
6	**Minimum**	1
7	**Maximum**	8

Table 2.10   Distribution of Service Times

	A	B	C	D
4		**Service**		**Cumulative**
5		**Times**	**Probability**	**Probability**
6		**(Minutes)**		
7		1	0.10	0.10
8		2	0.20	0.30
9		3	0.30	0.60
10		4	0.25	0.85
11		5	0.10	0.95
12		6	0.05	1.00

steady-state operation is reached. Here, to keep calculations simple, the starting conditions are an empty grocery, and any concerns are overlooked.

How to generate service times and arrival times from the service time and interarrival time distributions has been covered in Sections 2.1.5 and 2.1.6, respectively; the same methods apply here and in subsequent examples. For manual simulations, use the random numbers in Table A.1 in the Appendix. It is good practice to start at a random position in Table A.1 and proceed in a systematic direction, never reusing the same stream of numbers in a given problem. If the same pattern is used repeatedly, statistical bias or other odd effects could affect the results.

Table 2.11 shows a portion of the simulation table for the single-channel queue, taken from "Example2.5SingleServer.xls"; it also shows totals and averages for selected columns (above the simulation table). The first step is to initialize the table by filling in cells for the first customer, who is assumed to arrive at time 0 with service beginning immediately. The customer finishes at time 2 minutes (the first random service time). After the first customer, subsequent rows in the table are based on the randomly generated values for interarrival time and service time of the current customer, the completion time of the previous customer, and, for the spreadsheet solution, the simulation logic incorporated in the formulas in the cells of the simulation table.

Continuing, the second customer arrives at time 5 minutes and service begins immediately. Therefore, the second customer has no wait in the queue and the server has 3 minutes of idle time.

Skipping down, we see that the fifth customer arrives at time 16 minutes and finds the server busy. At time 18, the previous customer leaves and the fifth customer begins service, having waited 2 minutes in the queue. This process continues for all 100 customers. The spreadsheet solution in "Example2.5SingleServer.xls" contains the formulas that implement the steps described here. To best understand these formulas, you should first attempt to do a manual simulation of a few steps, figure out the logic, and then study how the formulas implement the logic.

Columns G, I, and J have been added to collect three model outputs, namely, each customer's time in queue and in system, and the server's idle time (if any) between the previous customer's departure and this customer's arrival time. These model outputs are, in turn, used to compute several measures of performance, namely, the customer's average time in queue and in system, and the proportion of time the server is idle.

In order to compute the system performance measures, we can use the calculated responses in Table 2.11, which shows totals and averages for interarrival times, service times, the time customers

Table 2.11 Model Responses and Simulation Table (first 11 customers) for the Grocery Store Simulation

	A	B	C	D	E	F	G	H	I	J
15		TOTALS	420		320		163		483	106
16		AVERAGES	4.24		3.20		1.63		4.83	1.07
17			Number of Customers=	100						
18						Simulation Table				
19		Step	Activity	Clock	Activity	Clock	Output	Clock	Output	Output
20			Interarrival		Service	Time	Waiting Time	Time	Time Customer	Idle Time
21			Time		Time	Service	in Queue	Service	Spends in System	of Server
22		Customer	(Minutes)	Arrival Time	(Minutes)	Begins	(Minutes)	Ends	(Minutes)	(Minutes)
23		1	0	0	2	0	0	2	2	
24		2	5	5	2	5	0	7	2	3
25		3	5	10	4	10	0	14	4	3
26		4	4	14	4	14	0	18	4	0
27		5	2	16	3	18	2	21	5	0
28		6	8	24	2	24	0	26	2	3
29		7	7	31	3	31	0	34	3	5
30		8	8	39	5	39	0	44	5	5
31		9	5	44	1	44	0	45	1	0
32		10	2	46	6	46	0	52	6	1
33		11	1	47	4	52	5	56	9	0

wait in the queue, the time customers spend in the system, and idle time of the server. These results are taken from the first trial run on the 'One Trial' worksheet of the spreadsheet model when using the default RNG seed (12,345). With these results, we can illustrate a number of typical performance measures for waiting line models:

1. The average waiting time for a customer is 1.63 minutes, computed as follows:

$$\frac{\text{Average waiting time}}{\text{(minutes)}} = \frac{\text{total time customers wait in queue (minutes)}}{\text{total numbers of customers}}$$

$$= \frac{163}{100} = 1.63 \text{ minutes}$$

2. The probability that a customer has to wait in the queue is 0.46, computed as follows:

$$\text{Probability (wait)} = \frac{\text{numbers of customers who wait}}{\text{total number of customers}}$$

$$= \frac{46}{100} = 0.46$$

3. The server is idle about 25% of the time, computed as follows:

$$\frac{\text{Probability of idle}}{\text{server}} = \frac{\text{total idle time of server (minutes)}}{\text{total run time of simulation (minutes)}}$$

$$= \frac{106}{426} = 0.25$$

Therefore, the server is busy about 75% of the time.

4. The average service time is 3.20 minutes, computed as follows:

$$\frac{\text{Average service time}}{\text{(minutes)}} = \frac{\text{total service time (minutes)}}{\text{total number of customers}}$$

$$= \frac{320}{100} = 3.20 \text{ minutes}$$

This result can be compared with the expected service time by finding the mean of the service-time distribution, using the equation

$$E(S) = \sum_{s=0}^{\infty} sp(s)$$

Applying the expected-value equation to the distribution in Table 2.10 gives

Expected service time =
$$1(0.10) + 2(0.20) + 3(0.30) + 4(0.25) + 5(0.10) + 6(0.05) = 3.2 \text{ minutes}$$

The expected service time is exactly the same as the average service time from this trial of the simulation. In general, this will not be true; here it is a pure coincidence.

5. The average time between arrivals is 4.24 minutes, computed as follows:

$$\text{Average time between arrivals (minutes)} = \frac{\text{sum of all times between arrivals (minutes)}}{\text{number of arrivals} - 1}$$

$$= \frac{420}{99} = 4.24 \text{ minutes}$$

Since the first arrival occurs at time 0, we divide by 99 rather than 100. This result can be compared to the expected time between arrivals by finding the mean of the discrete uniform distribution whose endpoints are $a = 1$ and $b = 8$. The mean is given by

$$E(A) = \frac{a+b}{2} = \frac{1+8}{2} = 4.5 \text{ minutes}$$

The expected time between arrivals is slightly higher than the average. However, as the simulation becomes longer, the average value of the time between arrivals tends to become closer to the theoretical mean, $E(A)$.

6. The average waiting time of those who wait is 3.54 minutes, computed as follows:

$$\text{Average waiting time of those who wait (minutes)} = \frac{\text{total time customers wait in queue (minutes)}}{\text{total number of customers who wait}}$$

$$= \frac{163}{46} = 3.54 \text{ minutes}$$

Note that this is different from the average waiting time of all customers, many of whom (54 in this simulation) have no wait at all before their service begins. Over all customers, the average waiting time is $\frac{163}{100} = 1.63$ minutes. The number who wait, 46, comes from the simulation table, column G under "Waiting Time in Queue," by counting the number of nonzero waiting times, or alternately from Figure 2.9, which displays a histogram of the 100 waiting times.

7. The average time a customer spends in the system is 4.83 minutes, computed in two ways. First:

$$\text{Average time customer spends in the system (minutes)} = \frac{\text{total time customers spend in the system (minutes)}}{\text{total number of customers}}$$

$$= \frac{483}{100} = 4.83 \text{ minutes}$$

The second way is to realize that the following relationship must hold:

$$\begin{matrix} \text{Average time} \\ \text{customer spends} \\ \text{in the system} \\ \text{(minutes)} \end{matrix} = \begin{matrix} \text{average time} \\ \text{customer spends} \\ \text{waiting in the} \\ \text{queue (minutes)} \end{matrix} + \begin{matrix} \text{average time} \\ \text{customer spends} \\ \text{in service} \\ \text{(minutes)} \end{matrix}$$

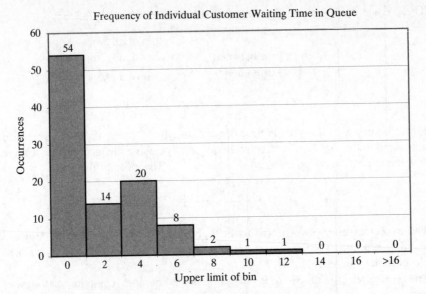

Figure 2.9   Frequency of waiting time in queue.

From findings 1 and 4, this results in

$$\text{Average time customer spends in the system} = 1.63 + 3.20 = 4.83 \text{ minutes}$$

We are now interested in running an experiment to address the question of how the average time in queue for the first 100 customers varies from day to day. Running the 'One Trial' worksheet once with 100 customers corresponds to one day. Running it 50 times corresponds to using the 'Experiment' worksheet with "Number of Trials" set to 50, where each trial represents one day.

For Example 2.5, the frequency of waiting time in queue for the first trial of 100 customers is shown in Figure 2.9. (Note: In all histograms in the remainder of this chapter, the upper limit of the bin is indicated on the legend on the x-axis, even if the legend is shown centered within the bin.) As shown in the histogram, 54% of the customers did not have to wait, and of those who did wait, 34% waited less than four minutes (but more than zero minutes). These results and the histogram in Figure 2.9 represent the results of one trial, and by themselves do not constitute an answer to our question.

To answer our question, we use the 'Experiment' sheet, setting "Number of Trials" to 50 and click the button labeled 'Reset Seed & Run'. This results in 50 estimates of average time in queue (each averaged over 100 customers). The overall average waiting time over 50 trials was 1.32 minutes. Figure 2.10 shows a histogram of the 50 average waiting times for the 50 trials. This shows how average waiting time varies day-by-day over a sample of 50 days.

By experimenting with the spreadsheet solution and doing some of the related exercises, you can discover the effects of randomness and of the assumed input data. For example, what if you run 400 trials, or 10 trials, instead of 50? How much does the shape of the distribution in Figure 2.10 change the number of trials? Why is the range (0 to 12 minutes) in Figure 2.9 so much wider than the range (0.5 to 3 minutes) in Figure 2.10?

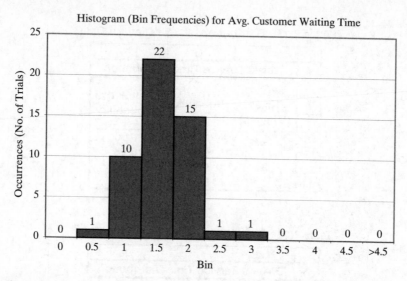

Figure 2.10   Frequency distribution of average waiting times for 50 trials.

2.3.3   Simulating a Queue with Two Servers

The second queueing example is a queue with two servers that have different service times.

Example 2.6:   The Able–Baker Call Center Problem _____

A computer technical support center is staffed by two people, Able and Baker, who take calls and try to answer questions and solve computer problems. The time between calls ranges from 1 to 4 minutes, with distribution as shown in Table 2.12. Able is more experienced and can provide service faster than Baker. The distributions of their service times are shown in Tables 2.13 and 2.14.

When both are idle, Able takes the call. If both are busy, the call goes on hold; this is basically the same as a customer having to wait in line in any queueing model. In general, arrival times, service times, and hence clock times in a simulation should be continuous (real-valued) variables; the fact

Table 2.12   Interarrival Distribution of Calls for Technical Support

	B	C	D
4	**Interarrival Distribution of Calls**		
5	**Interarrival**		
6	**Time**	**Probability**	**Cumulative**
7	**(Minutes)**		**Probability**
8	1	0.25	0.25
9	2	0.40	0.65
10	3	0.20	0.85
11	4	0.15	1.00

Table 2.13   Distribution of Able's Service Time

	F	G	H
4	**Able's Service Time Distribution**		
5	**Service**		**Cumulative**
6	**Times**	**Probability**	**Probability**
7	**(Minutes)**		
8	2	0.30	0.30
9	3	0.28	0.58
10	4	0.25	0.83
11	5	0.17	1.00

Table 2.14   Distribution of Baker's Service Time

	I	J	K
4	**Baker's Service Time Distribution**		
5	**Service**		**Cumulative**
6	**Times**	**Probability**	**Probability**
7	**(Minutes)**		
8	3	0.35	0.35
9	4	0.25	0.60
10	5	0.20	0.80
11	6	0.20	1.00

that they are discrete, integer-valued here (and in other examples) is simply for ease of explanation as well as understanding the random variable generation techniques explained earlier.

The simulation proceeds in a manner similar to the single-server case in Example 2.5, except that the underlying logic (the spreadsheet formulas in the simulation table) is more complex for two reasons: There are two servers, and the two servers have different service times. The complete solution is in the workbook "Example2.6AbleBaker.xls", a portion of which is shown in Table 2.15, along with some totals needed to estimate caller delay (hold time). The workbook provides a detailed explanation of the simulation steps, including the underlying logic (formulas) for computing each step. (To understand the steps, it would first be helpful for you to do a manual simulation, perhaps by starting with a simulation table containing Step 1 (row 20) and the columns for interarrival and service times, and attempting to match the results in Table 2.15.)

Our objective is to estimate caller delay (hold time) waiting for Able or Baker to answer. To estimate this and other desired measures of system performance, a simulation of the first 100 callers is made.

Looking at Table 2.15, we see that Caller 1, by assumption, arrives at clock time 0. Able is idle, so Caller 1 begins service immediately. The service time, 4 minutes, is generated from the distribution given in Table 2.13 by following the procedure (throwing the "random dart") in Example 2.2. Thus, Caller 1 completes service at clock time 4 minutes and is not delayed.

Table 2.15 Simulation Table for Call-Center Example (first 10 of 100 callers)

	B	C	D	E	F	G	H	I	J	K	L	M
13	TOTALS											
14	Number of Callers= 100										73	432
15							Seed for Random Numbers		12,345			
16							**Simulation Table**					
17	Step	Activity	Clock	Clock	Clock	State	Activity	Clock	Clock	Clock	Output	Output
18	Caller Number	Interarrival Time (Minutes)	Arrival Time	When Able Available	When Baker Available	Server Chosen	Service Time (Minutes)	Time Service Begins	Service Completion Time		Output	Output
19									Able	Baker	Caller Delay (Minutes)	Time in System (Minutes)
20	1		0	0	0	Able	4	0	4		0	4
21	2	1	1	4	0	Baker	4	1		5	0	4
22	3	1	2	4	5	Able	3	4	7		2	5
23	4	3	5	7	5	Baker	3	5		8	0	3
24	5	2	7	7	8	Able	5	7	12		0	5
25	6	2	9	12	8	Baker	5	9		14	0	5
26	7	1	10	12	14	Able	5	12	17		2	7
27	8	2	12	17	14	Baker	4	14		18	2	6
28	9	4	16	17	18	Able	2	17	19		1	3
29	10	1	17	19	18	Baker	3	18		21	1	4

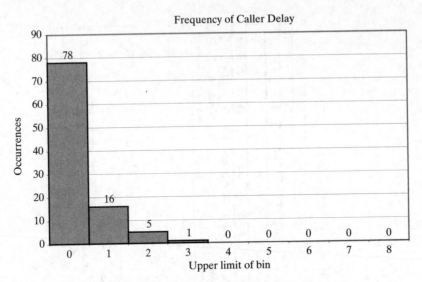

Figure 2.11   Frequency of caller delay for first trial.

An interarrival time of 1 minute, generated from Table 2.12 by following the procedure in Example 2.3, makes Caller 2 arrive at clock time 1 minute. Able is busy, so Caller 2 is served by Baker.

Looking at Caller 3, serviced by Able from clock time 4 minutes to clock time 7 minutes, we see that Caller 3 arrived at time 2 minutes to find both servers busy, hence had to wait.

For the single trial of 100 callers, the frequency chart in Figure 2.11 shows that 78 of 100 (78%) of the callers had no delay, 16% had a delay of one minute, while a few had to wait up to 3 minutes. Other trials, using different random numbers, might get similar or vastly different results; only experimentation can determine the inherent variability when simulating 100 customers.

The distribution of caller delay in Figure 2.11 is skewed to the right. On the 'One Trial' sheet, click the button 'Generate New Trial' repeatedly and note that the caller delay distribution often has a long, thin tail to the right. Along with the tall bars (high frequency) close to zero, what does this say about the nature of caller delay? One trial does not provide sufficient data on which to form reliable conclusions, but this is one of the great advantages to having the spreadsheet—the effect of variability is quite evident—and the simulation can be repeated any number of times using the 'Experiment' worksheet.

In Table 2.15, total customer delay is 73 minutes, or about 0.73 minutes per caller on the average. The histogram in Figure 2.11 shows, however, that the average of 0.73 minutes by itself is not a good indicator of how long a caller has to wait for Able or Baker. While 78% of the customers in this trial experienced no delay, a few experienced delays up to 3 minutes. This illustrates the importance of choosing the proper measures of system performance, depending on the model's goals and objectives. After all, the callers who do not have to wait will not complain about waiting, but those who wait a long time will not care that the average was 0.73 minutes. (In actuality, in some later trials, the average delay is considerably higher, and the distribution is skewed even more to the right. You

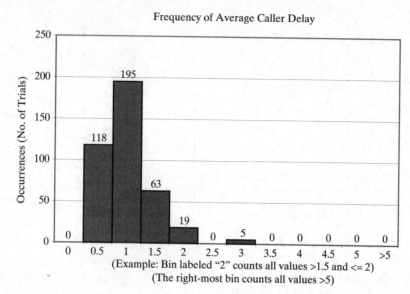

Figure 2.12   Frequency of average caller delay for experiment of 400 trials.

should click 'Generate New Trial' several times to see how much random variability this system may exhibit. Each trial represents a different day.)

Next, we ask the question: How does average caller delay (for the first 100 customers) vary from day to day? Using the 'Experiment' worksheet, and associating each day as one trial, we set the "Number of Trials" to 400 (representing 400 days). The resulting distribution of average caller delay is shown in the histogram in Figure 2.12. Note that this histogram is not a histogram of individual caller delay, as in Figure 2.11.

From Figure 2.12, we see that about 22% of the average delays (87 of 400) are longer than one minute, and about 78% (313 of 400) are 1 minute or less. Only about 1.25% (5 of 400) are longer than 2 minutes. We see that the result of the first trial (an average caller delay of 0.73 minutes) falls in the most frequent bin, but there are days that have much higher average delays. This example shows that a one-day (one trial) simulation may well be unrepresentative of long-term system performance.

In summary, whether, and to what extent, Able and Baker can handle the demand for service depends on your criteria, and the cost or value of the delays customers experience. We can guess that adding an additional server may well reduce caller delay (hold time) to nearly zero; however, the benefit of reduced customer delay would have to be traded off against the cost of an additional server.

2.4   Inventory Simulation in a Spreadsheet

An important class of simulation problems involves inventory systems. The simple inventory system shown in Figure 2.13 is called an (M, N) system, which has an inventory review every N time periods

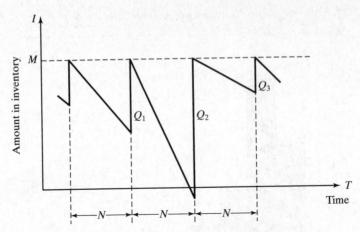

Figure 2.13   Probabilistic order-level inventory system.

and maximum inventory level M. At each N time units, the inventory level is checked and an order is made to bring the inventory up to the level M.

For simplicity, the lead time (i.e., the length of time between the placement and receipt of an order) is shown as zero in Figure 2.13; orders arrive immediately after being placed. In reality, lead time may vary greatly and would need to be modeled as a known constant or as a random variable.

Demands are not usually known with certainty, so the order quantities are modeled by a probability distribution. In the graph of Figure 2.13, demand is shown as being uniform over the time period. In actuality, demands are not usually uniform and do fluctuate over time.

At the end of the first review period, an order quantity, Q_1, is placed to bring the inventory level up to M. This is repeated at every inventory review.

Notice that, in the second cycle, the amount in inventory drops below zero, indicating a shortage. In Figure 2.13, these units are backordered; when the order arrives, the demand for the backordered items is satisfied first. To avoid shortages, a buffer, or safety, stock would need to be carried. In some inventory systems, a portion of sales may be lost rather than backordered when inventory runs out.

Inventory systems have a number of potential sources of revenue and costs:

1. Revenue from sales;
2. Cost of items sold;
3. Carrying cost, the cost of carrying stock in inventory;
4. Ordering cost, the cost associated with placing orders;
5. Cost of lost sales or backorders;
6. Cost due to scrap;
7. Salvage revenue from scrap or other damaged goods.

The carrying cost can be attributed to the interest paid on funds borrowed to buy the items (or equivalently, the loss from not having the funds available for other investment purposes). Carrying costs also include the cost for handling and storage space, hiring of guards, and other similar costs. Ordering cost may include shipping and transportation, and discounts depending on order quantity.

When trying to improve an inventory system, you face a number of trade-offs. An alternative to carrying high inventory (with high carrying costs) is to make more frequent reviews and, consequently, more frequent purchases or replenishments. This has an associated cost, the ordering cost. There may also be costs associated with running out of goods, with such shortages causing customers to become angry and a subsequent loss of good will. Larger inventories decrease the probability of a shortage. By varying the policy or decision parameters, the costs may be traded off in order to minimize the total cost of an inventory system.

The total cost (or total profit) of an inventory system is the measure of performance. This can be affected by the policy alternatives. For example, in Figure 2.13, the decision maker can control the maximum inventory level, M, and the length of the cycle, N.

Inventory systems and policies have a number of parameters, some controllable and some not. The controllable variables are the decision or policy variables. For each item in inventory, these include:

1. The maximum inventory level, M
2. The review period, N
3. The order quantity (for replenishment), Q
4. Lead time for replenishment, L

Lead time may be random when it cannot be controlled; in other situations, lead time may be tightly controlled, in effect a known constant (as in a just-in-time manufacturing system when ordering parts or components), in which case it becomes a policy variable.

In a model of an (M, N) inventory system, the events that may occur are the demand for items in the inventory, the review of the inventory position and resulting decision to place a replenishment order, and the arrival of replenishment stock. When the lead time is assumed to be zero, as in Figure 2.13, the last two events occur at the same time.

In the examples and exercises, we simulate a variety of inventory systems and policies.

2.4.1   Simulating the News Dealer's Problem

In the following example, a news dealer must decide how many newspapers to buy. In this and similar problems, only a single time period of specified length is relevant, and only a single procurement is made. Inventory remaining at the end of the single time period is sold for scrap, discarded or recycled. A wide variety of real-world problems have similar characteristics, including the stocking of certain spare parts, perishable items, style goods, and special seasonal items [Hadley and Whitin, 1963].

Example 2.7:   The News Dealer's Problem _____

A news dealer buys papers for 33 cents each and sells them for 50 cents each. Newspapers not sold at the end of the day are sold as scrap for 5 cents each. Newspapers can be purchased in bundles of 10. Thus, the newsstand can buy 50 or 60 or 70 papers, and so on. The order quantity, Q, is the only policy decision. Unlike some inventory problems, the order quantity Q is fixed since ending inventory is always zero due to scrapping leftover papers.

Table 2.16   Distribution of Daily Newspaper Demand, by Type of Newsday

	B	C	D	E	F	G	H
4	**Distribution of Newspapers Demanded**						
5	**Demand**	**Demand Probabilities**			**Cumulative Probabilities**		
6		**Good**	**Fair**	**Poor**	**Good**	**Fair**	**Poor**
7	40	0.03	0.10	0.44	0.03	0.10	0.44
8	50	0.05	0.18	0.22	0.08	0.28	0.66
9	60	0.15	0.40	0.16	0.23	0.68	0.82
10	70	0.20	0.20	0.12	0.43	0.88	0.94
11	80	0.35	0.08	0.06	0.78	0.96	1.00
12	90	0.15	0.04	0.00	0.93	1.00	1.00
13	100	0.07	0.00	0.00	1.00	1.00	1.00

There are three types of newsdays: "good", "fair", and "poor", but the news dealer cannot predict which type will occur on any given day. Table 2.16 gives the distribution of newspapers demanded by type of day. Table 2.17 provides the distribution for type of newsday.

Our objective is to compute the optimal number of papers the newsstand should purchase (the order quantity, Q). This will be accomplished by simulating demands for 20 days (in the 'One Trial' worksheet) and recording profits from sales each day for a given order quantity. After a trial is completed, total profit (the model response or performance measure) is calculated. To complete the experiment, we use the 'Experiment' sheet to replicate some number of times for a given order quantity (for example, 70 papers). Finally, we vary the order quantity over some reasonable range (40, 50, 60, 70, and so on) and compare different order policies based on total profit over the 20-day period.

Daily profit is given by the following relationship:

$$\text{Profit} = \binom{\text{revenue}}{\text{from sales}} - \binom{\text{cost of}}{\text{newspapers}} - \binom{\text{lost profit from}}{\text{excess demand}} + \binom{\text{salvage from sale}}{\text{of scrap papers}}$$

Since the revenue from sales is 50 cents per paper, and the cost to the dealer is 33 cents per paper, the lost profit from excess demand is 17 cents for each paper demanded that could not be provided.

Table 2.17   Distribution of Type of Newsday

	J	K	L
4	**Type of Newsday**		
5	**Type**	**Probability**	**Cumulative Probability**
6			
7	Good	0.35	0.35
8	Fair	0.45	0.80
9	Poor	0.20	1.00

Table 2.18 Simulation Table for Purchase of 70 Newspapers

	B	C	D	E	F	G	HI	I
16				Simulation Table				
17								
18 19	Day	Type of Newsday	Demand	Revenue from Sales	Lost Profit from Excess Demand	Salvage from Sale of Scrap	Daily Cost	Daily Profit
20	1	Fair	50	$25.00	$0.00	$1.00	$23.10	$2.90
21	2	Fair	50	$25.00	$0.00	$1.00	$23.10	$2.90
22	3	Fair	70	$35.00	$0.00	$0.00	$23.10	$11.90
23	4	Good	80	$35.00	$1.70	$0.00	$23.10	$10.20
24	5	Poor	40	$20.00	$0.00	$1.50	$23.10	-$1.60
25	6	Fair	50	$25.00	$0.00	$1.00	$23.10	$2.90
26	7	Poor	50	$25.00	$0.00	$1.00	$23.10	$2.90
27	8	Fair	70	$35.00	$0.00	$0.00	$23.10	$11.90
28	9	Good	40	$20.00	$0.00	$1.50	$23.10	-$1.60
29	10	Good	100	$35.00	$5.10	$0.00	$23.10	$6.80
30	11	Fair	70	$35.00	$0.00	$0.00	$23.10	$11.90
31	12	Poor	50	$25.00	$0.00	$1.00	$23.10	$2.90
32	13	Fair	50	$25.00	$0.00	$1.00	$23.10	$2.90
33	14	Poor	40	$20.00	$0.00	$1.50	$23.10	-$1.60
34	15	Good	80	$35.00	$1.70	$0.00	$23.10	$10.20
35	16	Good	70	$35.00	$0.00	$0.00	$23.10	$11.90
36	17	Good	80	$35.00	$1.70	$0.00	$23.10	$10.20
37	18	Poor	40	$20.00	$0.00	$1.50	$23.10	-$1.60
38	19	Fair	70	$35.00	$0.00	$0.00	$23.10	$11.90
39	20	Good	100	$35.00	$5.10	$0.00	$23.10	$6.80
40							TOTAL PROFIT =	$114.70

Such a shortage cost is somewhat controversial, but makes the problem much more interesting. The salvage value of scrap papers is 5 cents each.

The simulation table for the decision to purchase 70 newspapers, taken from the solution workbook, "Example2.7NewsDealer.xls", is shown in Table 2.18. On day 1, the demand is for 50 newspapers, so with 70 available, there are 20 to scrap. The revenue from the sale of the 50 newspapers is $25.00, and the scrap value is $1.00, so the first-day profit is computed as follows:

$$\text{First-day Profit} = \$25.00 - \$23.10 - 0 + \$1.00 = \$2.90$$

The profit from other days is easily computed in the same manner. The profit for the 20-day period is the sum of the daily profits, $114.70, as shown in Table 2.18. In general, because the results of one day are independent of previous days, inventory problems of this type are much easier than queueing problems to solve in a spreadsheet.

With the 'Experiment' worksheet from "Example2.7NewsDealer.xls", we set up an experiment with 400 trials or replications for the order policy of ordering 70 newspapers. Figure 2.14 shows

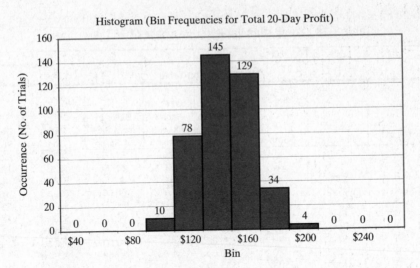

Figure 2.14   Frequency of total (20-day) profit when news dealer orders 70 papers per day.

the result of these 400 trials, each trial being a 20-day simulation run. The average total (20-day) profit was \$135.49 (averaged over the 400 trials). The minimum 20-day profit was \$86.60 and the maximum was \$198.00. Figure 2.14 shows that only 38 of the 400 trials resulted in a total 20-day profit of more than \$160. (Recall that the number beneath a bin is the upper limit of the bin.)

The single trial shown in Table 2.18 had a profit of \$114.70, which is on the low side of the distribution in Figure 2.14. This illustrates once again that the result for one 20-day trial may not be representative of long-term performance, and shows the necessity of simulating a number of trials as well as conducting a careful statistical analysis (as we will study in Chapters 11 and 12) before drawing conclusions.

On the 'One Trial' sheet, look how Daily Profit varies when clicking the button 'Generate New Trial' multiple times. The results vary quite a bit both in the histogram entitled 'Frequency of Daily Profit' (showing what happened on each of the 20 days) and in the total profit over those 20 days. Figure 2.15 shows four typical histograms for daily profit from the first four trials, whose shape varies quite a bit more than the shape for the histogram for total 20-day profit.

Exercise 28 asks you to conduct an experiment to determine the optimum number of newspapers for the news dealer to order. To solve this problem by simulation requires setting a policy of buying a certain number of papers each day and running the experiment for a large number of trials. The policy (number of newspapers purchased) is changed to other values in a reasonable range (say, 40, 50, 60, and so on) and the simulation repeated until the best value is found. To estimate with confidence whether a simulated difference is likely to be real and not a random fluke, requires the statistical methods discussed in Chapters 11 and 12.

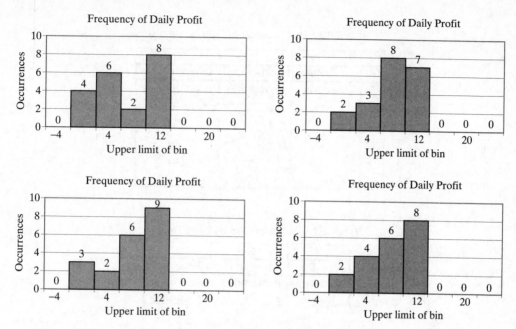

Figure 2.15   Four typical histograms of daily profit for the News Dealer's Problem.

2.4.2   Simulating an (M, N) Inventory Policy

The next example is the classic order-up-to-level (M, N) inventory policy with a fixed review period (N), a maximum inventory level (M), random demand, and random lead time.

Example 2.8:   A Refrigerator Inventory Problem

A company selling refrigerators maintains inventory by conducting a review after a fixed number of days (say N) and making a decision about how many to order to replenish the store's inventory. The current policy is to order up to a level (the *order up to* level—say, M), using the following relationship:

$$\text{Order quantity} = (\text{Order-up-to level}) - (\text{Ending inventory}) + (\text{Shortage quantity})$$

For example, if the order-up-to level (M) is 11 refrigerators, the review period (N) is five days, and the ending inventory on day 5 is 3 refrigerators, then on the fifth day of the cycle, 8 refrigerators will be ordered from the supplier. On the other hand, if there had been a shortage of two refrigerators on the fifth day, then 13 refrigerators would have been ordered; the first two received would be provided to the customers who placed an order and were willing to wait (a so-called backorder). (In contrast, the *lost sales* case occurs when customer demand is lost if inventory is not available.)

The daily *demand*, or number of refrigerators purchased each day by customers, is randomly distributed as shown in Table 2.19. After the company places an order to replenish their supply, the *lead time* is 1 to 3 days, randomly distributed, as shown in Table 2.20. Assume that the orders are placed at the end of the day. With a lead time of zero, the order from the supplier will arrive the next

Table 2.19 Distribution of Daily Demand

	B	C	D
4	**Distribution of Daily Demand**		
5	**Demand**	**Probability**	**Cumulative**
6			**Probability**
7	0	0.10	0.10
8	1	0.25	0.35
9	2	0.35	0.70
10	3	0.21	0.91
11	4	0.09	1.00

Table 2.20 Distribution of Lead Time

	F	G	H
4	**Distribution of Lead Time**		
5	**Lead Time**	**Probability**	**Cumulative**
6	**(days)**		**Probability**
7	1	0.60	0.60
8	2	0.30	0.90
9	3	0.10	1.00

morning, and the refrigerators will be available for distribution that next day. With a lead time of one day (the smallest possibility in Table 2.20), the order from the supplier arrives the second morning after, and will be available for distribution that day.

The simulation table for the policy to order-up-to 11 refrigerators, shown in Table 2.21, is taken from the spreadsheet solution in "Example2.8RefrigInventory.xls". The simulation was started with assumed initial conditions as follows: a beginning inventory of 3 refrigerators and a previous replenishment order for 8 refrigerators to arrive in 2 days' time. (These values can be changed in the spreadsheet solution, if desired.)

Following the simulation table for several selected days indicates the simulation logic. The order for 8 refrigerators is available on the morning of the third day of the first cycle, raising the inventory level from zero refrigerators to 8 refrigerators. Demands during the remainder of the first cycle reduced the ending inventory level to 3 refrigerators at the end of the fifth day. Thus, an order for 8 refrigerators was placed; the lead time turns out to be 2 days. This order is added to inventory on the morning of day 7 in cycle 2.

Note that on day 10, there is a shortage, resulting in a backorder and an order for 13 refrigerators. Since the lead time is 2 days, additional shortages occur before the order arrives, so that when the order of 13 refrigerators does arrive, the backorders plus current demand take inventory down to 6.

From the simulation of 25 days, representing 5 inventory review periods, the average ending inventory is approximately 2.76 (69/25) units. On 5 of 25 days, a shortage condition existed.

Table 2.21 Simulation Table for Inventory Policy with Order-Up-To 11 Refrigerators

	Clock	State	State	State	Input	State	State	State	Activity	State
Step	Day	Cycle	Day within Cycle	Beginning Inventory	Demand	Ending Inventory	Shortage Quantity	Pending Order (Quantity)	Lead Time (days)	Days until Order Arrives
Simulation Table										
	0	0	5	-	-	3	0	8	2	2
	1	1	1	3	2	1	0	8		1
	2	1	2	1	1	0	0			
	3	1	3	8	2	6	0			
	4	1	4	6	1	5	0			
	5	1	5	5	2	3	0	8	1	1
	6	2	1	3	3	0	0			
	7	2	2	8	2	6	0			
	8	2	3	6	3	3	0			
	9	2	4	3	2	1	0			
	10	2	5	1	3	0	2	13	2	2
	11	3	1	0	1	0	3	13		1
	12	3	2	0	2	0	5			
	13	3	3	13	2	6	0			
	14	3	4	6	3	3	0			
	15	3	5	3	1	2	0	9	1	1
	16	4	1	2	0	2	0			
	17	4	2	11	4	7	0			
	18	4	3	7	2	5	0			
	19	4	4	5	3	2	0			
	20	4	5	2	3	0	1	12	1	1
	21	5	1	0	2	0	3			
	22	5	2	12	1	8	0			
	23	5	3	8	4	4	0			
	24	5	4	4	1	3	0			
	25	5	5	3	1	2	0	9	1	1
TOTAL						69	14			
AVERAGE					2.04	2.76	0.56			

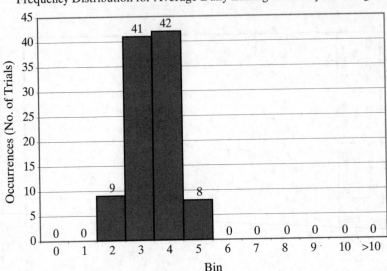

Frequency Distribution for Average Daily Ending Inventory of Refrigerators

Figure 2.16   Histogram of 25-day average ending inventory.

In Example 2.8, there cannot be more than one order outstanding from the supplier at any one time, but there are situations where lead times are so long that the relationship shown so far needs to be modified as follows:

Order quantity = (Order-up-to level) − (Ending inventory) − (On order) + (Shortage quantity)

The inclusion of the *on-order* quantity guarantees that double ordering does not occur.

On the 'One Trial' worksheet, repeatedly clicking on 'Generate New Trial' will run the next trial or replication with different random numbers, showing that, for a 25-day simulation, results are highly variable for daily ending inventory. (Watch the histogram change as you run new trials.) In contrast, setting the number of trials to 100 in the 'Experiment' worksheet and running the experiment repeatedly, by clicking the 'Run Experiment' button, produces much less variability in the average daily ending inventory (the average being taken over one trial of 25 days). The 25-day average is usually in the range from 1 to 5 refrigerators, as shown in the resulting histogram in Figure 2.16. This difference in variability illustrates a well-known fact: averages vary less than individual observations. Which to use in any specific analysis depends on the objectives and questions being asked.

The spreadsheet solution in "Example2.8RefrigInventory.xls" allows you to change the ordering policy (i.e., the values of M and N). It also allows you to make some changes to the statistical distributions for daily demand and lead time, as long as certain restrictions are met: the lead time must be one or more days, the review period must be larger than the largest lead time, and possibly other limitations; otherwise, the formulas implementing the logic may be incorrect.

2.5   Other Examples of Simulation

This section includes four miscellaneous examples: a reliability problem, a bombing mission, the estimation of the distribution of lead-time demand, and a project activity network.

2.5.1   Simulation of a Reliability Problem

The next example compares two replacement policies for replacing bearings in a milling machine. The example is a Monte Carlo simulation, not a dynamic event-based model, since events and clock times are not included. Each bearing life is randomly generated 15 times and the resulting costs computed.

Example 2.9:   Replacing Bearings in a Milling Machine ————————————————

A milling machine has three different bearings that fail in service. The distribution of the life of each bearing is identical, as shown in Table 2.22. When a bearing fails, the mill stops, a mechanic is called, and he or she installs a new bearing (costing $32 per bearing). The delay time for the mechanic to arrive varies randomly, having the distribution given in Table 2.23. Downtime for the mill is estimated to cost $10 per minute. The direct on-site cost of the mechanic is $30 per hour. The mechanic takes 20 minutes to change one bearing, 30 minutes to change two bearings, and 40 minutes to change three bearings. The engineering staff has proposed a new policy to replace all three bearings whenever one bearing fails. Management needs an evaluation of the proposal, using total cost per 10,000 bearing-hours as the measure of performance.

Table 2.22   Distribution for Bearing Life

	B	C	D
3	**Distribution of Bearing-Life**		
4	**Bearing**	**Probability**	**Cumulative**
5	**Life**		**Probability**
6	1000	0.10	0.1
7	1100	0.13	0.23
8	1200	0.25	0.48
9	1300	0.13	0.61
10	1400	0.09	0.70
11	1500	0.12	0.82
12	1600	0.02	0.84
13	1700	0.06	0.90
14	1800	0.05	0.95
15	1900	0.05	1.00

Table 2.23   Distribution of Delay until Mechanic Arrives

	F	G	H
4	**Distribution of Delay Time**		
5	**Delay Time**	**Probability**	**Cumulative Probability**
6			
7	5	0.60	0.60
8	10	0.30	0.90
9	15	0.10	1.00

Table 2.24   Bearing Replacement under Current Method

	B	C	D	E	F	G	H
17	**Simulation Table**						
18		**Bearing 1**		**Bearing 2**		**Bearing3**	
19		**Life (Hours)**	**Delay (minutes)**	**Life (Hours)**	**Delay (minutes)**	**Life (Hours)**	**Delay (minutes)**
20	**Step**						
21	1	1000	5	1700	10	1300	10
22	2	1200	5	1100	5	1100	5
23	3	1200	10	1000	10	1300	10
24	4	1500	5	1000	10	1100	15
25	5	1700	5	1900	15	1200	5
26	6	1200	5	1200	10	1500	10
27	7	1300	5	1500	5	1100	10
28	8	1700	5	1700	5	1400	15
29	9	1000	5	1300	5	1800	15
30	10	1800	10	1300	5	1200	5
31	11	1200	5	1100	5	1500	5
32	12	1100	5	1800	5	1100	10
33	13	1300	10	1200	5	1700	10
34	14	1300	10	1100	5	1300	10
35	15	1100	5	1300	5	1300	10
36	**TOTALS**	19,600	95	20,200	105	19,900	145

The simulation table for the current policy, shown in Table 2.24 and taken from the spreadsheet "Example2.9CurrentBearings.xls", represents a simulation of 15 bearing changes under the current policy of replacing only the single bearing that fails. Note that there are instances where more than one bearing fails at the same time. This is unlikely to occur in practice and is due to the crude simplifying assumption that bearing life is a multiple of 100 hours (so that we could have a simple discrete distribution rather than a more realistic one). Here, we assume that the times are never exactly the same and thus no more than one bearing is changed at any breakdown.

From this single trial (or replication) of the simulation, the cost of the current system is estimated as follows:

$$\text{Cost of bearing} = 45 \text{ bearings} \times \$32/\text{bearing}$$
$$= \$1,440$$

$$\text{Cost of delay time} = (95 + 105 + 145) \text{ minutes} \times \$10/\text{minute}$$
$$= \$3,450$$

$$\text{Cost of downtime during repair} = 45 \text{ bearings} \times 20 \text{ minutes/bearing} \times \$10/\text{minute}$$
$$= \$9,000$$

$$\text{Cost of mechanics} = 45 \text{ bearings} \times 20 \text{ minutes/bearing} \times \$30/60 \text{ minutes}$$
$$= \$450$$

$$\text{Total cost} = \$1,440 + \$3,450 + \$9,000 + \$450$$
$$= \$14,340$$

The total life of all 45 bearings is $(19,600 + 20,200 + 19,900) = 59,700$ hours. Therefore, the total cost per 10,000 bearing-hours is ($\$14,340/5.97) = \$2,402$.

The simulation table for the proposed method, shown in Table 2.25, is taken from the spreadsheet "Example2.9ProposedBearings.xls". For the first set of bearings, the earliest failure is at 1,100 hours.

Table 2.25   Bearing Replacement under Proposed Method

	B	C	D	E	F	G
17		Simulation Table				
18		**Bearing 1**	**Bearing 2**	**Bearing 3**	**First Failure**	
19		Life	Life	Life		**Delay**
20	**Step**	**(Hours)**	**(Hours)**	**(Hours)**	**(Hours)**	**(minutes)**
21	1	1300	1100	1300	1,100	10
22	2	1100	1200	1500	1,100	5
23	3	1100	1400	1800	1,100	15
24	4	1200	1500	1100	1,100	10
25	5	1700	1300	1300	1,300	5
26	6	1700	1100	1000	1,000	10
27	7	1000	1900	1200	1,000	10
28	8	1500	1700	1300	1,300	15
29	9	1300	1100	1800	1,100	5
30	10	1200	1100	1300	1,100	5
31	11	1000	1200	1200	1,000	15
32	12	1500	1700	1200	1,200	10
33	13	1300	1700	1000	1,000	10
34	14	1800	1200	1100	1,100	10
35	15	1300	1300	1100	1,100	10
36	**TOTALS**				16600	145

All three bearings are replaced at that time, even though the remaining bearings had more life in them. For example, Bearing 1 would have lasted 200 additional hours.

From this single trial (or replication) of the model, the cost of the proposed policy is estimated as follows:

$$
\begin{aligned}
\text{Cost of bearings} &= 45 \text{ bearings} \times \$32/\text{bearing} \\
&= \$1,440 \\
\text{Cost of delay time} &= 145 \text{ minutes} \times \$10/\text{minute} \\
&= \$1,450 \\
\text{Cost of downtime during repairs} &= 15 \text{ sets} \times 40 \text{ minutes/set} \times \$10/\text{minute} \\
&= \$6,000 \\
\text{Cost of mechanics} &= 15 \text{ sets} \times 40 \text{ minutes/set} \times \$30/60 \text{ minutes} \\
&= \$300 \\
\text{Total cost} &= \$1,440 + \$1,450 + \$6,000 + \$300 \\
&= \$9,190
\end{aligned}
$$

The total life of the bearings is $(16,600 \times 3) = 49,800$ hours. Therefore, the total cost per 10,000 bearing-hours is $(\$9,190/4.98) = \$1,845$. The new policy generates a savings of $557 per 10,000 hours of bearing-life.

The spreadsheet solutions for Example 2.9 are in: "Example2.9CurrentBearings.xls" for the current policy and "Example2.9ProposedBearings.xls" for the proposed policy. In each spreadsheet, a user can change a number of inputs: the distribution of bearing life, the distribution of delay time until mechanic arrival, the cost parameters, and the repair time per bearing or set of bearings.

2.5.2 Simulation of Hitting a Target

The next example simulates a bombing mission where the actual location a bomb hits varies from its target by a random amount defined by a normal distribution. The normal distribution is reviewed in Chapter 5; methods for generating random samples from a normal distribution are covered in Chapter 8. The spreadsheet solution has a method for generating these normal random variables; the manual solution uses standard normal numbers from Table A.2 in the Appendix.

For our purposes, recall that a normal distribution (the bell-shaped curve) is symmetric around its mean with a likely range determined by its standard deviation; its use in the next example is a classic use of the normal distribution to represent an error, in this case the deviation from a target. This example is a pure Monte Carlo sampling experiment, since it contains neither events nor clock times; therefore, it is not a dynamic, discrete-event model.

Example 2.10: A Bombing Mission ──────────────────────────

Consider a bomber attempting to destroy an ammunition depot, as shown in Figure 2.17. (This bomber has conventional rather than laser-guided weapons.) If a bomb falls anywhere inside the target, a hit is scored; otherwise, the bomb is a miss. (Note that when a bomb appears visually to have touched a

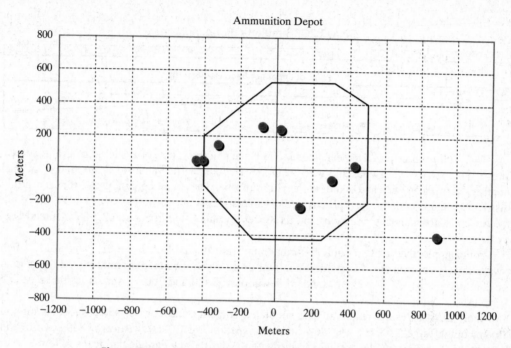

Figure 2.17   Ammunition depot with 10 bomb locations.

boundary line, it may or may not have hit the target; the model determines mathematically whether a hit has occurred, using the (X, Y) coordinates and the equations of the piecewise-linear boundary of the depot.)

The bomber flies in the horizontal direction and carries 10 bombs. The aiming point is (0,0). The actual point of impact is assumed to be normally distributed around the aiming point with a standard deviation of 400 meters in the direction of flight and 200 meters in the perpendicular direction. The problem is to simulate the operation and estimate the number of bombs on target.

Recall that the standardized normal variate Z, having mean 0 and standard deviation 1, is distributed as

$$Z = \frac{X - \mu}{\sigma}$$

where X is a normal random variable, μ is the mean of the distribution of X, and σ is the standard deviation of X. Then, with mean zero and standard deviations given by $\sigma_X = 400$ and $\sigma_Y = 200$, we have

$$X = 400Z_i$$

$$Y = 200Z_j$$

where (X, Y) are the simulated coordinates where the bomb hits.

Table 2.26 Simulated Bombing Run

	F	G	H	I	J	K	L	M	N	O	P	Q
3						**Point of Impact of Each Bomb(1-10)**						
4	**Bomb**	1	2	3	4	5	6	7	8	9	10	**Number Hits**
5	**X**	891.8	-1257.3	429.6	24.5	-321.0	-77.3	131.4	304.5	-442.7	-403.9	
6	**Y**	-400.7	-159.4	25.3	243.6	146.5	260.7	-228.3	-62.0	49.7	51.0	
7	**Hit ?**	Miss	Miss	Hit	Hit	Hit	Hit	Hit	Hit	Miss	Miss	6

The i and j subscripts have been added to indicate that the values of Z should be different for x and y. What are these Z values and where can they be found? The values of Z are random normal numbers with mean zero and standard deviation of one. These can be generated from uniformly distributed random numbers, a topic covered in Chapter 7. For our purposes, we can use Table A.2 in the Appendix for manual simulations and the Excel formula listed in Section 2.1.2 for spreadsheet models.

For example, the first two random normal numbers from Table A.2, are 0.23 and -0.17, which results in (x, y) coordinates for the first bomb of $(92.0, 34.0)$, obtained by multiplying the normal random number by the respective standard deviations of 400 and 200, as shown in the preceding formula.

Table 2.26 shows the results of one simulated bombing run. For our spreadsheet solution in "Example2.10Target.xls", we use the worksheet function, NORMSINV(Rnd01()), described in Section 2.1.2, to generate the normal variates with the desired standard deviation, as follows:

```
=$E5*NORMSINV(Rnd01())
```

for the x coordinate (where cell E5 contains the standard deviation in the x direction). A similar formula works for the y coordinate.

Using the spreadsheet solution in "Example2.10Target.xls" with the assumed shape and standard deviations, we ran an experiment with 400 trials (each trial being 10 bombs). The histogram in Figure 2.18 gives the frequency of number of hits, that is, the number of bombing runs on which there were a given number of hits (with number of possible hits ranging from 0 to 10).

From the "Multi-Trial Summary" on the 'Experiment' sheet of the spreadsheet model, note that the actual results range from 2 hits to 10 hits, and the average is 6.77 hits. (Clearly, no hits or one hit are possible. Since we made 400 replications, the experiment suggests, however, that one hit or no hits are extremely unlikely to occur on a bombing mission having the assumed characteristics.) If we simulate only one mission (that is, one trial), a very misleading result could occur, but Figure 2.18 provides useful descriptive information. For instance, in 17% (68 of 400) of the bombing runs there are 5 or fewer hits. In about 60% of the runs (239 of 400), there were 7 or more hits.

The spreadsheet model in "Example2.10Target.xls" allows you to change the target's shape by editing the (X, Y) coordinates of each endpoint of each side of the target, so long as a convex shape is maintained. You can also change the standard deviations, with a smaller standard deviation increasing the likelihood of being closer to the target and a larger one increasing the chances of being further from the target. In the spreadsheet solution, whether or not a bomb with a simulated (X, Y) landing spot hits the target is determined by linear inequalities using elementary algebra; the details are explained in the spreadsheet.

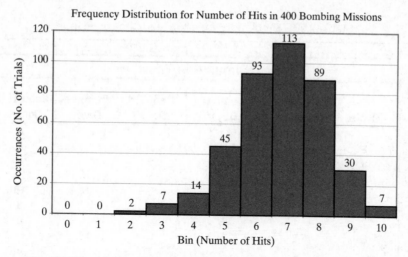

Figure 2.18   Histogram for number of hits on a bombing mission.

2.5.3   Estimating the Distribution of Lead-Time Demand

The purpose of the next example is to determine the distribution of lead-time demand, that is, the total demand received over the lead time. Lead time is the time from placement of a replenishment order until the order is received. During this period, demands continue to arrive from customers. The total customer demand over this period is the lead-time demand.

This model is an example of a Monte Carlo simulation; it is not a dynamic, event-based simulation. Essentially, there are no events and no clock times; it is a pure sampling experiment whose purpose is to estimate a complex statistical distribution. The distribution of lead-time demand is often of use in devising better inventory policies to reduce backorders and lost sales for the least cost.

We assume that both lead time and demand are random variables. Lead-time demand is thus a random variable defined as the sum of the demands over the lead time, or $\sum_{i=0}^{T} D_i$, where T is the lead time and D_i is the demand during the ith time period. The distribution of lead-time demand is found by simulating many cycles of lead time and for each period in the lead time, generating the random demand and totalling it, and building a histogram based on the results.

Example 2.11:   Lead-Time Demand

A firm sells bulk rolls of newsprint. The daily demand is given by the following probability distribution:

Daily Demand (Rolls)	3	4	5	6
Probability	0.20	0.35	0.30	0.15

Table 2.27   Simulation Table for Lead-Time Demand

	B	C	D	E	F	G
13	**Simulation Table**					
14,15,16	**Cycle**	**Lead Time**	**Demand Day1**	**Demand Day2**	**Demand Day3**	**Lead Time Demand**
17	1	2	3	5		8
18	2	1	4			4
19	3	2	3	5		8
20	4	3	4	5	4	13
21	5	3	4	5	6	15
22	6	1	3			3
23	7	1	6			6
24	8	2	5	6		11
25	9	2	5	3		8
26	10	3	4	3	5	12
27	11	1	4			4
28	12	1	5			5
29	13	3	4	4		14
30	14	1	6			6
31	15	1	3			3
32	16	2	5	3		8
33	17	2	4	6		10
34	18	3	4	5	5	14
35	19	3	4	3	6	13
36	20	2	5	6		11

The lead time (the number of days from placing an order until the firm receives the order from the supplier), is a random variable given by the following distribution:

Lead Time (Days)	1	2	3
Probability	0.36	0.42	0.22

Table 2.27 shows the simulation table, taken from "Example2.11LeadTimeDemand.xls". One feature of this simulation, different from previous ones, is that the number of random demands generated in one cycle is itself random. If a lead time of 2 days is generated, then two random demands are generated and summed; if the lead time is 3 days, then three random demands must be generated and summed.

Using the spreadsheet solution, you can generate many cycles, sum the demands over each cycle to get the lead-time demand, and with the histogram, see the resulting distribution of lead-time demand, as illustrated in Figure 2.19. This result was obtained by using the 'One Trial' worksheet to

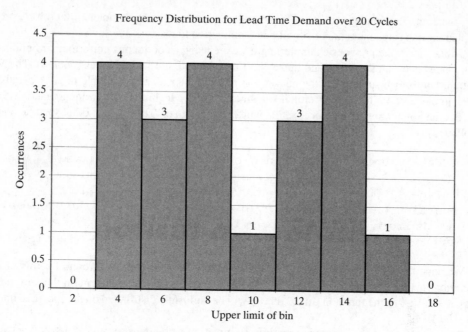

Figure 2.19   Frequency of lead-time demand.

simulate 20 cycles of lead-time demand by clicking 'Reset Seed & Run' in the spreadsheet solution "Example2.11LeadTimeDemand.xls". By clicking the 'Generate New Trial' button several times, you can see how variable the lead-time demand can be over just 20 cycles.

Compare these results over 20 cycles to the results in the spreadsheet model on the 'Experiment' sheet, which shows the results for 200 trials, each trial being one lead-time demand. This comparison shows that, due to the small sample, the histogram in Figure 2.19 may not be representative of the shape of the distribution of lead-time demand. This illustrates the importance of having a sample size (number of trials) large enough to obtain accurate estimates of the desired response variable.

2.5.4   Simulating an Activity Network

Suppose you have a project that requires the completion of a number of activities. Each activity takes a certain amount of time, either constant or random. Some activities must be carried out sequentially, that is, in a specified order; other activities can be done in parallel. The project can be represented by a network of activities, as illustrated by Figure 2.20. The arcs represent activities and the nodes represent the start or end of an activity.

In the network in Figure 2.20, there are three paths, each going from the node labeled "Start" to the node labeled "Finish". Each path represents a sequence of activities that must be completed in order. The activities on different paths can be carried out in parallel. The time for each activity is noted on the arc, where, for example, U(2, 4) represents a random variable that is uniformly distributed between 2 and 4 minutes. The time to complete all activities on a path is the sum of the activity times along the path. To complete the entire project, all activities must be completed; therefore, project completion time is the maximum over all path completion times.

Unlike any of the previous examples, here we are using a continuous uniform distribution; that is, all values between the lower and upper limit are allowed, not just the integer values. The VBA function, Uniform(), supplied in the add-in file "BCNNvba.xla" and available to any spreadsheet model, implements uniform generation in one line of VBA code, in a simple formula that transforms the VBA random number generator Rnd(), which is uniform on 0 to 1, to any other specified range, as follows:

```
'  In the Uniform function, low is the lower limit,
'                           high is the upper limit.
'  Type ''Double'' is a real value (double precision).
Function Uniform(low As Double, high As Double) As Double
    Uniform = low + (high - low) * Rnd()
End Function
```

We hope that this simple formula has some intuitive appeal. The background for this random generation method is covered in Chapter 7. Basically, since Rnd() is uniform on the unit interval (0, 1), it follows that the function Uniform() will be uniformly distributed on the specified interval (*low, high*).

Pritsker [1995] was a pioneer in applying simulation techniques to project activity networks. While Example 2.12 is a Monte Carlo simulation, some activity network models, due to their greater complexity, may be implemented in a discrete-event simulation package or a specialized package designed for such models.

Example 2.12:   Project Simulation

Suppose that three friends want to cook bacon, eggs, and toast for breakfast for some weekend visitors. They divide the work, with each friend preparing one item. Figure 2.20 represents the sequence of activities, with each path representing the tasks for one person, as follows:

Top path:	Start	→	A	Crack eggs
	A	→	B	Scramble eggs
	B	→	Finish	Cook eggs
Middle path:	Start	→	C	Make toast
	C	→	Finish	Butter toast
Bottom path:	Start	→	Finish	Fry bacon

The time to accomplish each activity, being somewhat uncertain, is represented by the uniform distribution shown on the associated arc in Figure 2.20. The friends want to estimate the distribution of total preparation time so that they can tell the visitor when to be at the breakfast table. With the distribution, and not just the mean or average total preparation time, they can estimate the probability of preparing breakfast within a specified amount of time. The model is in the spreadsheet "Example2.12Project.xls".

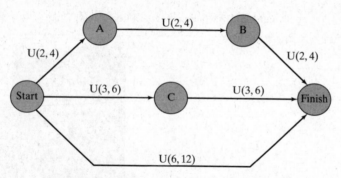

Figure 2.20   Activity network for Example 2.12.

Since the mean time for a continuous uniform is the midpoint between the lower and upper limit, you can see that each of the 3 paths in the project network of Figure 2.20 has a mean of 9 minutes. Based on the lower limit of each uniform, you can see that the minimum time for each path is 6 minutes; similarly, the maximum time is 12 minutes. Do these times provide any help in predicting the total project completion time?

For breakfast preparation, the project completion time is the maximum time over all of the paths. With the spreadsheet solution, "Example2.12Project.xls", we can sample the task times for a large number of trials and get an idea of the range and distribution of possible project completion times. The 'One Trial' worksheet provides the solution for one trial or replication of the network simulation. The logic and formulas implementing it are much simpler than many of the previous examples and we leave those for the student.

With the 'Experiment' worksheet, we can run many trials and estimate the likely range for project completion time, as well as other statistical measures. We set "Number of Trials" to 400 and run the experiment by clicking the 'Reset Seed & Run' button once. With 400 trials and using the default seed, the results are as follows:

Sample Mean:	10.21 minutes
Minimum:	6.65 minutes
Maximum:	11.99 minutes

The so-called critical path is the path that turns out to be the longest on a particular trial; that is, that path's time is the project completion time. Over 400 trials with each using different random numbers, each path will have some probability of being the critical path. For each of the 400 trials, the experiment determines which path was critical, with these results:

Top path:	25.50% of the trials
Middle path:	32.25%
Bottom path:	42.25%

We conclude that the chance of the bacon being the last item ready is approximately 42.25%. Why aren't the paths each represented about 1/3 of the time? In addition, how accurate is 42.25%? You can click the 'Run Experiment' button several times to get an idea of how variable this estimate

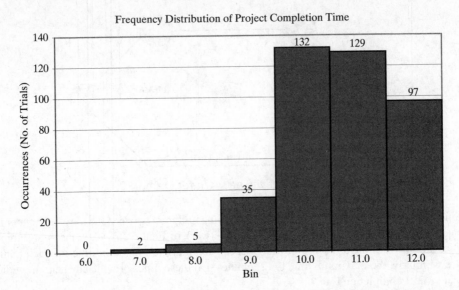

Frequency Distribution of Project Completion Time

Figure 2.21   Histogram of project completion times based on 400 trials.

tends to be; it varies by a few percent on each click. The magnitude of this random variation depends on the number of trials (400 in this example) in the experiment. In Chapter 11, we will cover statistical analysis techniques for estimating the likely statistical error in our estimates. (This technique is called a confidence interval.)

The project completion times were placed in a frequency chart or histogram, shown in Figure 2.21. From the chart, we draw the following inferences:

The chances are about 10.5% (42 of 400) that breakfast will be ready in 9 minutes or less.
The chances are about 24.25% (97 of 400) that it will take from 11 to 12 minutes.

Note that the bin labeled "12.0" is actually the interval from 11.0 to 12.0 minutes, since the bin label is always the upper limit of the bin.

Recalling that the mean time along each of the 3 paths is 9 minutes, the histogram tells us that the chance of taking longer than 9 minutes is approximately 89.5%. In other words, the mean of each path tells us little about the distribution of project completion time, which, after all, is determined by the slowest sequence of tasks.

2.6   Summary

The purpose of this chapter was to introduce a few key simulation concepts by means of examples, to illustrate a few general areas of application, and to motivate the remaining chapters.

We used the simulation table, custom designed for each problem, as the organizing principle. The columns in the simulation table represented activity times or other random input (such as demand), clock times for the occurrence of an event, system states, and model outputs. The rows represented a simulation step, for example, a customer or one day's demand. The formulas in each cell implement the simulation logic, or compute model outputs. The examples illustrate the need for working out the characteristics of the input data, generating random variables from the input assumptions, defining and computing model responses (measures of performance), and analyzing the resulting responses.

Model inputs were defined by constant values for policy and decision parameters, or simple statistical distributions for such variables as service times, interarrival times, lead time, and demand. Separately from the simulation table, but using the model outputs and other data, model responses, also known as system performance measures, were computed based on the results of one trial or one run of the simulation. These model responses included totals, averages, and frequency distributions (used for the histograms).

In Chapter 9, we will cover the topic of using statistical distributions to represent model inputs that vary randomly, and how to choose an appropriate distribution. In the examples, we covered how to generate random samples from four distribution types: the discrete empirical, the discrete uniform, the continuous uniform, and the normal distributions. (Empirical means that the distribution is based on data and specified probabilities.) The spreadsheet examples, available on the book's web site at www.bcnn.net, provide VBA functions to generate each type of distribution. You should use these as a starting point for any of the exercises at the end of the chapter. The general topic of random numbers and random variable generation is covered in Chapters 7 and 8.

Events in a simulation table were generated by using the random variables and the logic (formulas) of the particular model. The queueing examples, especially the two-channel queue, illustrate some of the complex dependencies that can occur—in the examples discussed, between subsequent customers visiting the queue. Because of these complexities, the ad-hoc simulation table approach fails or becomes unbearably complex, even with relatively simple networks of queues and other moderately complex models. For this and other reasons, a more systematic methodology, such as the event-scheduling approach described in Chapter 3, is needed. These subjects are treated in more detail in the remaining chapters of the text.

We introduced the idea, and necessity, of running an experiment consisting of multiple trials or replications. We saw the resulting statistical variability, and the need to conduct some kind of further analysis—the topics of Chapters 11 and 12.

We hope that the examples presented in this chapter provide some insight into the need for the topics covered by the remaining chapters.

REFERENCES

HADLEY G., AND T.M. WHITIN [1963], *Analysis of Inventory Systems*, Prentice-Hall, Englewood Cliffs, NJ.

HARRISON, J.M., AND V. NGUYEN [1995], "Some Badly Behaved Closed Queueing Networks," *Proceedings of IMA Workshop on Stochastic Networks*, eds. F. Kelly and R.J. Williams.

PRITSKER, A. A. B. [1995], *Introduction to Simulation and SLAM II*, 4th ed., John Wiley, New York.

SEILA, A., V. CERIC, AND P. TADIKAMALLA [2003], *Applied Simulation Modeling*, Duxbury, Belmont, CA.

EXERCISES

Spreadsheet and Manual Exercises

For hints on implementation in Excel, study the spreadsheet solutions at www.bcnn.net and read the 'Explain' worksheet for any example of interest. You may also need one or more of the VBA functions in the spreadsheet solutions. Therefore, to have these VBA functions available, it is easiest to start with an existing spreadsheet model, say, one of the examples closest to the exercise, and to have the Excel add-in file, "BCNNvba.xla", in the same folder as your spreadsheet. The add-in file can be found on the book's web site. As long as a VBA function is in a VB module in Excel's VBA environment (in the spreadsheet itself, or an opened or referenced add-in), you can call it in a formula from a spreadsheet cell.

    For any exercise where you are constructing a spreadsheet model, start with one of the examples, delete, modify or add to the existing inputs and simulation table in the 'One Trial' worksheet, and use what remains. On the 'Experiment' worksheet, change the response in the cell just below the word "Link", and change any of the text in the titles and column headers as desired. The VBA implementations of all the functions described in Section 2.1.5 are in the add-in file "BCNNvba.xla". Each function is available to every example spreadsheet, whether or not the example currently uses the function.

1. Consider the following continuously operating job shop. Interarrival times of jobs are distributed as follows:

Time Between Arrivals (Hours)	Probability
0	.23
1	.37
2	.28
3	.12

   Processing times for jobs are normally distributed, with mean 50 minutes and standard deviation 8 minutes. Construct a simulation table and perform a simulation for 10 new customers. Assume that, when the simulation begins, there is one job being processed (scheduled to be completed in 25 minutes) and there is one job with a 50-minute processing time in the queue.

   (a) What was the average time in the queue for the 10 new jobs?
   (b) What was the average processing time of the 10 new jobs?
   (c) What was the maximum time in the system for the 10 new jobs?

2. A baker is trying to figure out how many dozens of bagels to bake each day. The probability distribution of the number of bagel customers is as follows:

Number of Customers/Day	8	10	12	14
Probability	0.35	0.30	0.25	0.10

Customers order 1, 2, 3, or 4 dozen bagels according to the following probability distribution.

Number of Dozen Ordered/Customer	1	2	3	4
Probability	0.4	0.3	0.2	0.1

Bagels sell for $8.40 per dozen. They cost $5.80 per dozen to make. All bagels not sold at the end of the day are sold at half-price to a local grocery store. Based on 5 days of simulation, how many dozen (to the nearest 5 dozen) bagels should be baked each day?

3. Develop and interpret flow diagrams analogous to Figures 2.4 and 2.5 for a queueing system with i channels.

4. Smalltown Taxi operates one vehicle during the 9:00 A.M. to 5:00 P.M. period. Currently, consideration is being given to the addition of a second vehicle to the fleet. The demand for taxis follows the distribution shown:

Time Between Calls (Minutes)	15	20	25	30	35
Probability	0.14	0.22	0.43	0.17	0.04

The distribution of time to complete a service is as follows:

Service Time (Minutes)	5	15	25	35	45
Probability	0.12	0.35	0.43	0.06	0.04

Simulate 5 individual days of operation of the current system and of the system with an additional taxicab. Compare the two systems with respect to the waiting times of the customers and any other measures that might shed light on the situation.

5. The random variables X, Y, and Z are distributed as follows:

$$X \sim N(\mu = 100, \sigma^2 = 100)$$
$$Y \sim N(\mu = 300, \sigma^2 = 225)$$
$$Z \sim N(\mu = 40, \sigma^2 = 64)$$

Simulate 50 values of the random variable

$$W = \frac{X + Y}{Z}$$

Prepare a histogram of the resulting values, using class intervals of width equal to 3.

6. Given A, B, and C, which are uncorrelated random variables. Variable A is normally distributed with $\mu = 100$ and $\sigma^2 = 400$. Variable B is discrete uniformly distributed with a probability distribution given by $p(b) = 1/5$ with $b = 0, 1, 2, 3$ and 4. Variable C is distributed in accordance with the following table:

Value of C	Probability
10	.10
20	.25
30	.50
40	.15

Use simulation to estimate the mean of a new variable D, that is defined as

$$D = (A - 25B)/(2C)$$

Use a sample of size 100.

7. Estimate, by simulation, the average number of lost sales per week for an inventory system that functions as follows:

   (a) Whenever the inventory level falls to or below 10 units, an order is placed. Only one order can be outstanding at a time.
   (b) The size of each order is equal to $20 - I$, where I is the inventory level when the order is placed.
   (c) If a demand occurs during a period when the inventory level is zero, the sale is lost.
   (d) Daily demand is normally distributed, with a mean of 5 units and a standard deviation of 1.5 units. (Round off demands to the closest integer during the simulation and, if a negative value results, give it a demand of zero.)
   (e) Lead time is distributed uniformly between zero and 5 days—integers only.
   (f) The simulation will start with 18 units in inventory.
   (g) For simplicity, assume that orders are placed at the close of the business day and received after the lead time has occurred. Thus, if lead time is one day, the order is available for distribution on the morning of the second day of business following the placement of the order.
   (h) Let the simulation run for 5 weeks.

8. An elevator in a manufacturing plant carries exactly 400 kilograms of material. There are three kinds of material packaged in boxes that arrive for a ride on the elevator. These materials and their distributions of time between arrivals are as follows:

Material	Weight (kilograms)	Interarrival Time (Minutes)
A	200	5 ± 2 (uniform)
B	100	6 (constant)
C	50	$P(2) = 0.33$
		$P(3) = 0.67$

It takes the elevator 1 minute to go up to the second floor, 2 minutes to unload, and 1 minute to return to the first floor. The elevator does not leave the first floor unless it has a full load. Simulate 1 hour of operation of the system. What is the average transit time for a box of material A (time from its arrival until it is unloaded)? What is the average waiting time for a box of material B? How many boxes of material C made the trip in 1 hour?

9. The random variables X and Y are real-valued (not integers) and uniformly distributed over intervals as follows:

$$X \sim 10 \pm 10 \text{ (uniform)}$$
$$Y \sim 10 \pm 8 \text{ (uniform)}$$

(a) Simulate 200 values of the random variable

$$Z = XY$$

Prepare a histogram of the resulting values. What is the range of Z and what is its average value?

(b) Same as (a), except that

$$Z = X/Y$$

10. Consider the assembly of two steel plates, each plate having a hole drilled in its center. The plates are to be joined by a pin. The plates are aligned for assembly relative to the bottom left corner (0,0). The hole placement is centered at (3,2) on each plate. The standard deviation in each direction is 0.0045. The hole diameter is normally distributed, with a mean of 0.3 and a standard deviation of 0.005. The pin diameter is also distributed normally, with a mean of 0.29 and a standard deviation of 0.004. What fraction of pins will go through the assembled plates? Base your answer on a simulation of 50 observations.

[*Hint*]: Clearance $= \text{Min}(h_1, h_2) - [(x_1 - x_2)^2 + (y_1 - y_2)^2]^{.5} - p$, where

$h_i = $ hole diameter, $i = $ plate 1, 2
$p = $ pin diameter
$x_i = $ distance to center of plate hole, horizontal direction, $i = $ plate 1, 2
$y_i = $ distance to center of plate hole, vertical direction, $i = $ plate 1, 2

11. In the exercise above, the pin will wobble if it is too loose. Wobbling occurs if $\text{Min}(h_1, h_2) - p \geq 0.006$. What fraction of the assemblies wobble? (Conditional probability—i.e., given that the pins go through.)

12. Three points are chosen at random on the circumference of a circle. Estimate the probability that they all lie on the same semicircle, by Monte Carlo sampling methods. Perform 5 replications.

13. Two theorems from statistics are as follows:

Theorem 1   Let $Z_1, Z_2, \ldots, Z_k$ be normally and independently distributed random variables with mean $\mu = 0$ and variance $\sigma^2 = 1$. Then the random variable

$$\chi^2 = Z_1^2 + Z_2^2 + \cdots + Z_k^2$$

is said to follow the chi-squared distribution with k degrees of freedom, abbreviated χ_k^2.

Theorem 2   Let $Z \sim N(0, 1)$ and V be a chi-squared random variable with k degrees of freedom. If Z and V are independent, then the random variable

$$T = \frac{Z}{\sqrt{V/k}}$$

is said to follow the t distribution with k degrees of freedom, abbreviated t_k.

Generate one sample of a t-distributed random variate with 3 degrees of freedom. Use as many of the following random values as needed, in the order shown:

random numbers	random normal numbers
0.6729	1.06
0.1837	−0.72
0.2572	0.28
0.8134	−0.18
0.5251	−0.63

How many random numbers and standard random normals are needed to generate one sample from a t-distribution?

14. A bank has one drive-in teller and room for one additional customer to wait. Customers arriving when the queue is full park and go inside the bank to transact business. The times between arrivals and the service-time distributions follow:

Time Between Arrivals (Minutes)	Probability	Service Time (Minutes)	Probability
0	0.09	1	0.20
1	0.17	2	0.40
2	0.27	3	0.28
3	0.20	4	0.12
4	0.15		
5	0.12		

Simulate the operation of the drive-in teller for 10 new customers. The first of the 10 new customers arrives at a time determined at random. Start the simulation with one customer being served, leaving at time 3, and one in the queue. How many customers went into the bank to transact business?

15. The number of acres covered by Cedar Bog Lake is being estimated. The lake is shown next, with the scale shown in feet:

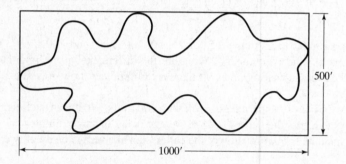

Use simulation to estimate the size of the lake. (One acre = 43,560 square feet.)

16. In Example 2.6, assume that the average delay with another server, Charlie, is virtually zero. But another server costs $20/hour. If the average caller delay costs $200 per hour, at what point is it advisable to add another server?

17. In Example 2.12, why isn't each path represented about 1/3 of the time?

18. In Example 2.11, the only way to have a lead-time demand of 3 or 4 is to have lead time of one day and a demand of 3 or 4. The probability of that happening is 0.36(0.20 + 0.35) = 0.198. Using this logic, what is the probability of having lead-time demand equal 11 or 12? What is the probability of having it equal 18? (Use computation, not simulation.)

19. Run the experiment in Example 2.4 (the Coin Tossing Game) for 4000 trials. Explain why the probability for Harry being ahead on 5 or fewer of the 100 tosses should be larger than the probability that Harry is ahead for 95 or more of the tosses. And why you can say the same thing about Tom—with no contradiction.

Exercises based on the Spreadsheet Examples

20. Modify Example 2.2 in Section 2.1.5 to use the "WRONG !!" formula to generate the service times. Generate a large number of service times, say, 400, and develop a histogram of the resulting frequencies. How close are the sample frequencies to the specified probabilities for each value of service time? Explain why the results differ. That is, explain why the formula is wrong.

21. In Example 2.5, let the arrival distribution be uniformly distributed between 1 and 10 minutes (integers only). Run the experiment for 50 trials. Is there any difference between the bin frequencies shown and those of Figure 2.10?

22. In Example 2.5, let the service-time distribution be changed to the following:

Service Time (Minutes)	1	2	3	4	5	6
Probability		0.05	0.10	0.20	0.30	0.25	0.10

Run the experiment for 50 trials. Is there any difference between the average waiting time here and that in Example 2.5 before the change was made? If there is a difference, to what do you attribute it?

23. Run the experiment in Example 2.5 for 25, 50, 100, 200, and 400 trials. (All trials are with different sets of random numbers.) What are the differences in the minimum and maximum values of the average waiting times? If there are differences, how do you explain them?

24. In Example 2.6, run 10 experiments, each with 100 trials. What was the fraction of trials in each experiment such that the callers had an average delay of three minutes or less? For the 1000 trials just simulated, what fraction of the callers had an average delay of two minutes or less?

25. In Example 2.6, conduct an experiment of 400 trials. Explain the large spread between the minimum average delay and the maximum average delay.

26. In Example 2.6, run 10 experiments with 50 trials each and 10 experiments with 400 trials each. Explain the differences in output that you observe, if any, between the two experiments.

27. In Example 2.6, modify the spreadsheet so that the number of calls taken by Able and Baker can be displayed. What fraction did each take in an experiment of 400 trials?

28. In Example 2.7, determine the best policy for ordering newspapers based on experiments of 50 trials.

29. Run 10 experiments, each with 400 trials, in Example 2.7. What is the largest difference that you got between the maximum and minimum daily profit?

30. Use the spreadsheet for Example 2.8. For 100 trials, what is the average ending inventory if $M = 11$ and $N = 4, 5, 6$?

31. Use the spreadsheet for Example 2.8. For 100 trials, what is the average ending inventory if $N = 5$ and $M = 10, 11, 12$?

32. Starting with the solution to Example 2.8, modify the solution to model the case where sales are lost; that is, instead of a shortage, those sales are lost. Keep track of the lost sales.

33. At what bearing cost in Example 2.9 is the total cost per 10,000 hours virtually the same for the current and proposed systems? Base your estimate on 10 experiments of the current system and of the proposed system, each experiment consisting of 400 trials.

34. Change the bin width for the histogram on the Experiment sheet for Example 2.9 (current or proposed) to a width of $50 beginning below the minimum value (for example, if the minimum value is $1528.46, let the first cell begin at $1500). What is the advantage of doing this?

35. Using the spreadsheet for Example 2.9 (proposed), run 10 experiments of 40 trials each, and record the range (maximum value − minimum value) of the results. Next, compute the average range. Then do the same as before, but use 400 trials in each experiment. If there is a difference, how do you explain it?

36. Set $\sigma_x = 600$ meters and $\sigma_y = 300$ meters in the spreadsheet for Example 2.10. Leave the target intact. Conduct a simulation with 200 trials. What was the average number of hits?

37. Set $\sigma_x = 300$ meters and $\sigma_y = 600$ meters in the spreadsheet for Example 2.10. Leave the target intact. Conduct a simulation with 200 trials. What was the average number of hits?

38. Set $\sigma_x = 300$ meters and $\sigma_y = 150$ meters in the spreadsheet for Example 2.10. Leave the target intact. Conduct a simulation with 200 trials. What was the average number of hits? Explain the difference between this exercise and the previous one.

39. Set $\sigma_x = 2\sigma_y$ meters in the spreadsheet for Example 2.10. Leave the target intact. What is the value of σ_x if the average number of hits is to be about 6.0, based on one experiment of 400 trials.

40. In Example 2.11, suppose you wanted a better estimate of the distribution of lead-time demand: Would you (1) increase the number of cycles on the 'One Trial' worksheet or (2) increase the number of trials on the 'Experiment' worksheet? Explain your answer.

41. In Example 2.11, suppose you wanted a better estimate of the mean lead-time demand? What would you do?

42. In Example 2.11, let the demand probabilities be equal for the possible values 3, 4, 5, and 6. Run an experiment with 400 trials. Compare the average lead-time demand from using the original input data and from using the new input data.

43. In Example 2.11, let the lead-time probabilities be equal for the possible values 1, 2, and 3. Run an experiment with 400 trials. Compare the average lead-time demand from using the original input data and from using the new input data.

44. In Example 2.12, run the experiment 20 times, each with 400 trials. Record the number of times that the middle path was the longest. What is your best estimate of the probability that the middle path is taken?

45. In the above exercise, what is the smallest value and the largest value encountered for the number of times that the middle path was selected? What if you had conducted one simulation, gotten the smallest (or largest) value, and reported that as the result? How could you argue yourself out of that jam?

46. In Example 2.12, suppose the third pathway (the bottom one in Figure 2.20) is changed so that it consists of six U(1,2) activities. Modify the spreadsheet to accommodate this. Which is the most frequently occurring path now? What insight does this exercise provide?

47. Using Excel, generate 1000 values in a column, using the formula =-10*LN(RAND()).

    (a) Compute descriptive statistics about the data in that column, including the minimum value, the maximum value, the mean, the median, and the standard deviation.
    (b) Tabulate the values into 10 bins, each of width equal to 5, the first bin beginning at 0, and the eleventh bin for overflow (if any).
    (c) Does the histogram resemble any distribution with which you are familiar? If so, what is its name?

    Hint: Use FREQUENCY or COUNT in an Excel formula to count bin frequencies.

48. Using Excel, generate 12 columns, each with 250 values, using the formula =RAND().

    In cell M1, place the formula =SUM(A1:L1)-6 and copy it to the 249 cells below M1 in column M.

    (a) Compute descriptive statistics about the data in column M, including minimum value, maximum value, mean, median, and standard deviation.
    (b) Tabulate the values with 9 bins. The first bin will include all values less than or equal to −3.5; the next 6 bins are of width one; the last bin will include all values greater than 3.5.
    (c) Does the histogram resemble any distribution with which you are familiar? If so, what is its name?

    Hint 1: Use FREQUENCY or COUNT in an Excel formula to count bin frequencies.

    Hint 2: The values in Column M can be used as a rough approximation to those in Table A.2.

49. Using Excel, generate a table that can be used instead of Table A.1, using the formula =INT(RAND()*100000).

50. Use Excel to devise a Monte Carlo experiment to estimate the value of π, as in the equation for the area of a circle,

$$A = \pi r^2$$

    Hint: Consider a quarter circle with unit radius inside a unit square. Throw a random dart onto the unit square.

51. A college student is making financial plans for the next year. Expenses include the following known costs:

Tuition	$8,400
Dormitory	$5,400

There are also a lot of uncertain expenses, with ranges as follows:

Meals	$900–$1,350
Entertainment	$600–$1,200
Transportation	$200–$600
Books	$400–$800

The student foresees income to pay for next year. The only certain amounts are as follows:

State scholarship	$3,000
Parental contribution	$4,000

Some of the income is variable, with ranges as follows:

Waiting tables	$3,000–$5,000
Library aide	$2,000–$3,000

The student predicts that a loan will be needed. Develop a spreadsheet model and use it with 1,000 trials to forecast the probability distribution for the size of the loan. Treat any amounts that are variable as uniformly distributed over the range indicated.

52. The 22-stop Bally 1090 slot machine had the following number of symbols on each reel:

Symbols	Reel 1	Reel 2	Reel 3
Cherries	2	5	2
Oranges	2	3	7
Plums	5	1	10
Bells	10	2	1
Bars	2	10	1
7's	1	1	1

Payoffs were as follows:

Combination	Payoff	Combination	Payoff
Left cherry	2	Three plums	20
Right cherry	2	Plum plum bar	14
Any two cherries	5	Bar plum plum	14
Three cherries	20	Three bells	20
Three oranges	20	Bell bell bar	18
Orange orange bar	10	Bar bell bell	18
Bar orange orange	10	Three bars	50
		Three 7's	100

Using Excel, simulate 1,000 plays of this slot machine, each for $1.

How much was the gain or loss?

3

General Principles

This chapter develops a common framework for the modeling of complex systems by using discrete-event simulation. It covers the basic building blocks of all discrete-event simulation models: entities and attributes, activities, and events. In discrete-event simulation, a system is modeled in terms of its state at each point in time; of the entities that pass through the system and the entities that represent system resources; and of the activities and events that cause the system state to change. Discrete-event models are appropriate for those systems for which changes in system state occur only at discrete points in time.

The simulation languages and software (collectively called simulation packages) described in Chapter 4 are fundamentally packages for discrete-event simulation. A few of the packages also include the capability to model continuous variables in a purely continuous simulation or a mixed discrete–continuous model. The discussion in this chapter focuses on the discrete-event concepts and methodologies. The discussion in Chapter 4 focuses more on the capabilities of the individual packages and on some of their higher-level constructs.

This chapter introduces and explains the fundamental concepts and methodologies underlying all discrete-event simulation packages. These concepts and methodologies are not tied to any particular package. Many of the packages use different terminology from that used here, and most have a number of higher-level constructs designed to make modeling simpler and more straightforward for their application domain. For example, this chapter discusses the fundamental abstract concept of an entity, but Chapter 4 discusses more concrete realizations of entities, such as machines, conveyors, and vehicles that are built into some of the packages to facilitate modeling in the manufacturing, material handling, or other domains.

Applications of simulation in specific contexts are discussed in Part Five of this text. Topics covered include the simulation of manufacturing and material handling systems in Chapter 13 and the simulation of networked computer systems in Chapter 14.

Section 3.1 covers the general principles and concepts of discrete-event simulation, the event scheduling/time advance algorithm, and the three prevalent world views: event scheduling, process interaction, and activity scanning. Section 3.2 introduces some of the notions of list processing, one of the more important methodologies used in discrete-event simulation software. Chapter 4 covers the implementation of the concepts in a number of the more widely used simulation packages.

3.1 Concepts in Discrete-Event Simulation

The concept of a system and a model of a system were discussed briefly in Chapter 1. This chapter deals exclusively with dynamic, stochastic systems (i.e., involving time and containing random elements) that change in a discrete manner. This section expands on these concepts and proposes a framework for the development of a discrete-event model of a system. The major concepts are briefly defined and then illustrated by examples:

System   A collection of entities (e.g., people and machines) that interact together over time to accomplish one or more goals.

Model   An abstract representation of a system, usually containing structural, logical, or mathematical relationships that describe a system in terms of state, entities and their attributes, sets, processes, events, activities, and delays.

System state   A collection of variables that contain all the information necessary to describe the system at any time.

Entity   Any object or component in the system that requires explicit representation in the model (e.g., a server, a customer, a machine).

Attributes   The properties of a given entity (e.g., the priority of a waiting customer, the routing of a job through a job shop).

List   A collection of (permanently or temporarily) associated entities, ordered in some logical fashion (such as all customers currently in a waiting line, ordered by "first come, first served," or by priority).

Event   An instantaneous occurrence that changes the state of a system (such as an arrival of a new customer).

Event notice   A record of an event to occur at the current or some future time, along with any associated data necessary to execute the event; at a minimum, the record includes the event type and the event time.

Event list   A list of event notices for future events, ordered by time of occurrence; also known as the future event list (FEL).

Activity   A duration of time of specified length (e.g., a service time or interarrival time), which is known when it begins (although it may be defined in terms of a statistical distribution).

Delay A duration of time of unspecified indefinite length, which is not known until it ends (e.g., a customer's delay in a last-in–first-out waiting line which, when it begins, depends on future arrivals).

Clock A variable representing simulated time, called CLOCK in the examples to follow.

Different simulation packages use different terminology for the same or similar concepts—for example, lists are sometimes called sets, queues, or chains. Sets or lists are used to hold both entities and event notices. The entities on a list are always ordered by some rule, such as first-in–first-out or last-in–first-out, or are ranked by some entity attribute, such as priority or due date. The future event list is always ranked by the event time recorded in the event notice. Section 3.2 discusses a number of methods for handling lists and introduces some of the methodologies for efficient processing of ordered sets or lists.

An activity typically represents a service time, an interarrival time, or any other processing time whose duration has been characterized and defined by the modeler. An activity's duration may be specified in a number of ways:

a. Deterministic—for example, always exactly 5 minutes;
b. Statistical—for example, as a random draw from the set {2, 5, 7} with equal probabilities;
c. A function depending on system variables and/or entity attributes—for example, loading time for an iron ore ship as a function of the ship's allowed cargo weight and the loading rate in tons per hour.

However it is characterized, the duration of an activity is computable from its specification at the instant it begins. Its duration is not affected by the occurrence of other events (unless, as is allowed by some simulation packages, the model contains logic to cancel or postpone an activity in progress). To keep track of activities and their expected completion time, at the simulated instant that an activity duration begins, an event notice is created having an event time equal to the activity's completion time. For example, if the current simulated time is CLOCK = 100 minutes and an inspection time of exactly 5 minutes is just beginning, then an event notice is created that specifies the type of event (an end-of-inspection event), and the event time ($100 + 5 = 105$ minutes).

In contrast to an activity, a delay's duration is not specified by the modeler ahead of time, but rather is determined by system conditions. Quite often, a delay's duration is measured and is one of the desired outputs of a model run. Typically, a delay ends when some set of logical conditions becomes true or one or more other events occur. For example, a customer's delay in a waiting line may be dependent on the number and duration of service of other customers ahead in line as well as the availability of servers and equipment.

A delay is sometimes called a *conditional wait*, an activity an *unconditional wait*. The completion of an activity is an event, often called a *primary event*, that is managed by placing an event notice on the FEL. In contrast, delays are managed by placing the associated entity on another list, perhaps representing a waiting line, until such time as system conditions permit the processing of the entity. The completion of a delay is sometimes called a conditional or secondary event, but such events are not represented by event notices, nor do they appear on the FEL.

The systems considered here are dynamic, that is, changing over time. Therefore, system state, entity attributes and the number of active entities, the contents of sets, and the activities and delays currently in progress are all functions of time and are constantly changing over time. Time itself is represented by a variable called CLOCK.

Example 3.1:   Call Center, Revisited

Consider the Able–Baker call center system of Example 2.6. A discrete-event model has the following components:

System state

$L_Q(t)$, the number of callers waiting to be served at time t;

$L_A(t)$, 0 or 1 to indicate Able as being idle or busy at time t;

$L_B(t)$, 0 or 1 to indicate Baker as being idle or busy at time t.

Entities Neither the callers nor the servers need to be explicitly represented, except in terms of the state variables, unless certain caller averages are desired (compare Examples 3.4 and 3.5).

Events

Arrival event;

Service completion by Able;

Service completion by Baker.

Activities

Interarrival time, defined in Table 2.12;

Service time by Able, defined in Table 2.13;

Service time by Baker, defined in Table 2.14.

Delay

A caller's wait in queue until Able or Baker becomes free.

___

The definition of the model components provides a static description of the model. In addition, a description of the dynamic relationships and interactions between the components is also needed. Some questions that need answers include:

1. How does each event affect system state, entity attributes, and set contents?
2. How are activities defined (i.e., deterministic, probabilistic, or some other mathematical equation)? What event marks the beginning or end of each activity? Can the activity begin regardless of system state, or is its beginning conditioned on the system being in a certain state? (For example, a machining "activity" cannot begin unless the machine is idle, not broken, and not in maintenance.)
3. Which events trigger the beginning (and end) of each type of delay? Under what conditions does a delay begin or end?
4. What is the system state at time 0? What events should be generated at time 0 to "prime" the model—that is, to get the simulation started?

A discrete-event simulation is the modeling over time of a system all of whose state changes occur at discrete points in time—those points when an event occurs. A discrete-event simulation (hereafter called a simulation) proceeds by producing a sequence of system snapshots (or system images) that represent the evolution of the system through time. A given snapshot at a given time

(CLOCK $= t$) includes not only the system state at time t, but also a list (the FEL) of all activities currently in progress and when each such activity will end, the status of all entities and current membership of all sets, plus the current values of cumulative statistics and counters that will be used to calculate summary statistics at the end of the simulation. A prototype system snapshot is shown in Figure 3.1. (Not all models will contain every element exhibited in Figure 3.1. Further illustrations are provided in the examples in this chapter.)

3.1.1   The Event Scheduling/Time Advance Algorithm

The mechanism for advancing simulation time and guaranteeing that all events occur in correct chronological order is based on the future event list (FEL). This list contains all event notices for events that have been scheduled to occur at a future time. Scheduling a future event means that, at the instant an activity begins, its duration is computed or drawn as a sample from a statistical distribution; and that the end-activity event, together with its event time, is placed on the future event list. In the real world, most future events are not scheduled but merely happen—such as random breakdowns or random arrivals. In the model, such random events are represented by the end of some activity, which in turn is represented by a statistical distribution.

At any given time t, the FEL contains all previously scheduled future events and their associated event times (called $t_1, t_2, \ldots$ in Figure 3.1). The FEL is ordered by event time, meaning that the events are arranged chronologically—that is, the event times satisfy

$$t < t_1 \leq t_2 \leq t_3 \leq \cdots \leq t_n.$$

Time t is the value of CLOCK, the current value of simulated time. The event associated with time t_1 is called the imminent event; that is, it is the next event that will occur. After the system snapshot at simulation time CLOCK $= t$ has been updated, the CLOCK is advanced to simulation time CLOCK $= t_1$, the imminent event notice is removed from the FEL, and the event is executed. Execution of the imminent event means that a new system snapshot for time t_1 is created, one based on the old snapshot at time t and the nature of the imminent event. At time t_1, new future

CLOCK	System state	Entities and attributes	Set 1	Set 2	...	Future event list (FEL)	Cumulative statistics and counters
t	$(x, y, z, \ldots)$					$(3, t_1)$ — Type 3 event to occur at time t_1 $(1, t_2)$ — Type 1 event to occur at time t_2 . . .	

Figure 3.1 Prototype system snapshot at simulation time t.

events may or might not be generated, but if any are, they are scheduled by creating event notices and putting them into their proper position on the FEL. After the new system snapshot for time t_1 has been updated, the clock is advanced to the time of the new imminent event and that event is executed. This process repeats until the simulation is over. The sequence of actions that a simulator (or simulation language) must perform to advance the clock and build a new system snapshot is called the *event-scheduling/time-advance algorithm*, whose steps are listed in Figure 3.2 (and explained thereafter).

The length and contents of the FEL are constantly changing as the simulation progresses, and thus its efficient management in a simulation software package will have a major impact on the model's computer runtime. The management of a list is called *list processing*. The major list processing operations performed on a FEL are removal of the imminent event, addition of a new event to the list, and occasionally removal of some event (called cancellation of an event). As the imminent event is usually at the top of the list, its removal is as efficient as possible. Addition of a new event (and cancellation of an old event) requires a search of the list. The efficiency of this search depends on the logical organization of the list and on how the search is conducted. In addition to the FEL, all the sets in a model are maintained in some logical order, and the operations of addition and removal of entities from the set also require efficient list-processing techniques. A brief introduction to list processing in simulation is given in Section 3.2.

The removal and addition of events from the FEL is illustrated in Figure 3.2. Event 3 with event time t_1 represents, say, a service completion event at server 3. Since it is the imminent event at time t, it is removed from the FEL in Step 1 (Figure 3.2) of the event-scheduling/time-advance algorithm. When event 4 (say, an arrival event) with event time t^* is generated at Step 4, one possible way to determine its correct position on the FEL is to conduct a top-down search:

If $t^* < t_2$,	place event 4 at the top of the FEL.
If $t_2 \leq t^* < t_3$,	place event 4 second on the list.
If $t_3 \leq t^* < t_4$,	place event 4 third on the list.
$\vdots$	
If $t_n \leq t^*$,	place event 4 last on the list.

(In Figure 3.2, it was assumed that t^* was between t_2 and t_3.) Another way is to conduct a bottom-up search. The least efficient way to maintain the FEL is to leave it as an unordered list (additions placed arbitrarily at the top or bottom), which would require at Step 1 of Figure 3.2 a complete search of the list for the imminent event before each clock advance. (The imminent event is the event on the FEL with the lowest event time.)

The system snapshot at time 0 is defined by the initial conditions and the generation of the so-called exogenous events. The specified initial conditions define the system state at time 0. For example, in Figure 3.2, if $t = 0$, then the state $(5, 1, 6)$ might represent the initial number of customers at three different points in the system. An exogenous event is a happening "outside the system" that impinges on the system. An important example is an arrival to a queueing system. At time 0, the first arrival event is generated and is scheduled on the FEL (meaning that its event notice is placed on the FEL). The interarrival time is an example of an activity. When the clock eventually is advanced to the time of this first arrival, a second arrival event is generated. First, an interarrival time, call it a^*, is

Old system snapshot at time t

CLOCK	System state	$\cdots$	Future event list	$\cdots$
t	(5, 1, 6)		$(3, t_1)$ – Type 3 event to occur at time t_1 $(1, t_2)$ – Type 1 event to occur at time t_2 $(1, t_3)$ – Type 1 event to occur at time t_3 $\quad\quad\quad \cdot \quad\quad\quad \cdot \quad\quad\quad \cdot$ $\quad\quad\quad \cdot \quad\quad\quad \cdot \quad\quad\quad \cdot$ $\quad\quad\quad \cdot \quad\quad\quad \cdot \quad\quad\quad \cdot$ $(2, t_n)$ – Type 2 event to occur at time t_n	

Event-scheduling/time-advance algorithm

Step 1. Remove the event notice for the imminent event (event 3, time t_1) from FEL.

Step 2. Advance CLOCK to imminent event time (i.e., advance CLOCK from t to t_1).

Step 3. Execute imminent event: update system state, change entity attributes, and set membership as needed.

Step 4. Generate future events (if necessary) and place their event notices on FEL, ranked by event time. (Example: Event 4 to occur at time t^*, where $t_2 < t^* < t_3$.)

Step 5. Update cumulative statistics and counters.

New system snapshot at time t_1

CLOCK	System state	$\cdots$	Future event list	$\cdots$
t_1	(5, 1, 5)		$(1, t_2)$ – Type 1 event to occur at time t_2 $(4, t^*)$ – Type 4 event to occur at time t^* $(1, t_3)$ – Type 1 event to occur at time t_3 $\quad\quad\quad \cdot \quad\quad\quad \cdot \quad\quad\quad \cdot$ $\quad\quad\quad \cdot \quad\quad\quad \cdot \quad\quad\quad \cdot$ $\quad\quad\quad \cdot \quad\quad\quad \cdot \quad\quad\quad \cdot$ $(2, t_n)$ – Type 2 event to occur at time t_n	

Figure 3.2   Advancing simulation time and updating system image.

generated; it is added to the current time, CLOCK $= t$; the resulting (future) event time, $t + a^* = t^*$, is used to position the new arrival event notice on the FEL. This method of generating an external arrival stream is called *bootstrapping*; it provides one example of how future events are generated in Step 4 of the event-scheduling/time-advance algorithm. Bootstrapping is illustrated in Figure 3.3.

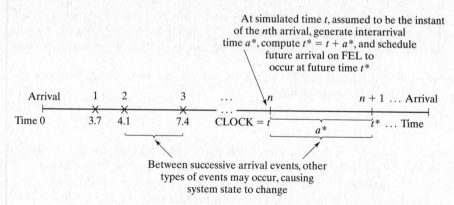

Figure 3.3 Generation of an external arrival stream by bootstrapping.

The first three interarrival times generated are 3.7, 0.4, and 3.3 time units. The end of an interarrival interval is an example of a primary event.

A second example of how future events are generated (Step 4 of Figure 3.2) is provided by a service completion event in a queueing simulation. When one customer completes service, at current time $\text{CLOCK} = t$, if the next customer is present, then a new service time, s^*, will be generated for the next customer. The next service completion event will be scheduled to occur at future time $t^* = t + s^*$, by placing onto the FEL a new event notice, of type *service completion*, with event time t^*. In addition, a service-completion event will be generated and scheduled at the time of an arrival event provided that, upon arrival, there is at least one idle server in the server group. A service time is an example of an activity. Beginning service is a conditional event, because its occurrence is triggered only on the condition that a customer be present and a server be free. Service completion is an example of a primary event. Note that a conditional event, such as beginning service, is triggered by a primary event's occurring and by certain conditions prevailing in the system. Only primary events appear on the FEL.

A third important example is the alternate generation of runtimes and downtimes for a machine subject to breakdowns. At time 0, the first runtime will be generated, and an end-of-runtime event will be scheduled. Whenever an end-of-runtime event occurs, a downtime will be generated, and an end-of-downtime event will be scheduled on the FEL. When the CLOCK is eventually advanced to the time of this end-of-downtime event, a runtime is generated, and an end-of-runtime event is scheduled on the FEL. In this way, runtimes and downtimes continually alternate throughout the simulation. A runtime and a downtime are examples of activities, and *end of runtime* and *end of downtime* are primary events.

Every simulation must have a stopping event, here called E, which defines how long the simulation will run. There are generally two ways to stop a simulation:

1. At time 0, schedule a stop simulation event at a specified future time T_E. Thus, before simulating, it is known that the simulation will run over the time interval $[0, T_E]$. Example: Simulate a job shop for $T_E = 40$ hours.
2. Run length T_E is determined by the simulation itself. Generally, T_E is the time of occurrence of some specified event E. Examples: T_E could be the time of the 100th service completion

at a certain service center. T_E could be the time of breakdown of a complex system. T_E could be the time of disengagement or total kill (whichever occurs first) in a combat simulation. T_E could be the time at which a distribution center ships the last carton in a day's orders.

In case 2, T_E is not known ahead of time. Indeed, it could be one of the statistics of primary interest to be produced by the simulation.

3.1.2   World Views

When using a simulation package or even when doing a manual simulation, a modeler adopts a world view or orientation for developing a model. The most prevalent world views are the event-scheduling world view, as discussed in the previous section, the process-interaction world view, and the activity-scanning world view. Even if a particular package does not directly support one or more of the world views, understanding the different approaches could suggest alternative ways to model a given system.

To summarize the previous discussion, when using the event-scheduling approach, a simulation analyst concentrates on events and their effect on system state. This world view will be illustrated by the manual simulations of Section 3.1.3 and the Java simulation in Chapter 4.

When using a package that supports the process-interaction approach, a simulation analyst thinks in terms of processes. The analyst defines the simulation model in terms of entities or objects and their life cycle as they flow through the system, demanding resources and queueing to wait for resources. More precisely, a process is the life cycle of one entity. This life cycle consists of various events and activities. Some activities might require the use of one or more resources whose capacities are limited. These and other constraints cause processes to interact, the simplest example being an entity forced to wait in a queue (on a list) because the resource it needs is busy with another entity. The process-interaction approach is popular because it has intuitive appeal and because the simulation packages that implement it allow an analyst to describe the process flow in terms of high-level block or network constructs, while the interaction among processes is handled automatically.

In more precise terms, a process is a time-sequenced list of events, activities, and delays, including demands for resources, that define the life cycle of one entity as it moves through a system. An example of a "customer process" is shown in Figure 3.4. In this figure, we see the interaction between two customer processes as customer $n + 1$ is delayed until the previous customer's "end service event" occurs. Usually, many processes are active simultaneously in a model, and the interaction among processes could be quite complex.

Underlying the implementation of the process interaction approach in a simulation package, but usually hidden from a modeler's view, events are being scheduled on a future event list and entities are being placed onto lists whenever they face delays, causing one process to temporarily suspend its execution while other processes proceed. It is important that the modeler have a basic understanding of the concepts and, for the simulation package being used, a detailed understanding of the built-in but hidden rules of operation. Schriber and Brunner [2003] provide understanding in this area.

Both the event-scheduling and the process-interaction approach use a variable time advance—that is, when all events and system state changes have occurred at one instant of simulated time, the simulation clock is advanced to the time of the next imminent event on the FEL. The activity-scanning approach, in contrast, uses a fixed time increment and a rule-based approach to decide whether any activities can begin at each point in simulated time.

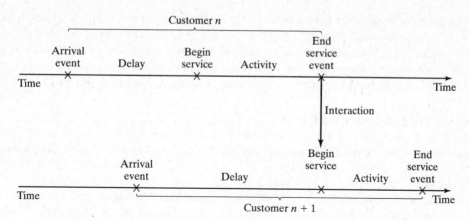

Figure 3.4 Two interacting customer processes in a single-server queue.

With the activity-scanning approach, a modeler concentrates on the activities of a model and those conditions, simple or complex, that allow an activity to begin. At each clock advance, the conditions for each activity are checked, and, if the conditions are true, then the corresponding activity begins. Proponents claim that the activity-scanning approach is simple in concept and leads to modular models that are more easily maintained, understood, and modified by other analysts at later times. They admit, however, that the repeated scanning to discover whether an activity can begin results in slow runtime on computers. Thus, the pure activity-scanning approach has been modified (and made conceptually somewhat more complex) by what is called the three-phase approach, which combines some of the features of event scheduling with activity scanning to allow for variable time advance and the avoidance of scanning when it is not necessary, but keeps the main advantages of the activity-scanning approach.

In the three-phase approach, events are considered to be activities of duration zero time units. With this definition, activities are divided into two categories, which are called B and C.

B activities   activities bound to occur; all primary events and unconditional activities.
C activities   activities or events that are conditional upon certain conditions being true.

The B-type activities and events can be scheduled ahead of time, just as in the event-scheduling approach. This allows variable time advance. The FEL contains only B-type events. Scanning to learn whether any C-type activities can begin or C-type events occur happens only at the end of each time advance, after all B-type events have been completed. In summary, with the three-phase approach, the simulation proceeds with repeated execution of the 3 phases until it is completed:

Phase A   Remove the imminent event from the FEL and advance the clock to its event time. Remove from the FEL any other events that have the same event time.
Phase B   Execute all B-type events that were removed from the FEL. (This could free a number of resources or otherwise change system state.)
Phase C   Scan the conditions that trigger each C-type activity and activate any whose conditions are met. Rescan until no additional C-type activities can begin and no events occur.

The three-phase approach improves the execution efficiency of the activity-scanning method. In addition, proponents claim that the activity-scanning and three-phase approaches are particularly good at handling complex resource problems in which various combinations of resources are needed to accomplish different tasks. These approaches guarantee that all resources being freed at a given simulated time will be freed before any available resources are reallocated to new tasks.

Example 3.2:   Call Center, Back Again _____

For the Call Center in Example 3.1, if we were to adopt the three-phase approach to the activity-scanning world view, the conditions for beginning each activity in Phase C would be as follows:

Activity	Condition
Service time by Able	A caller is in queue and Able is idle
Service time by Baker	A caller is in queue, Baker is idle and Able is busy

If it were to take the process-interaction world view, we would view the model from the viewpoint of a caller and its "life cycle." Considering a life cycle as beginning upon arrival, a customer process is pictured in Figure 3.4.

_____

In summary, as will be illustrated in Chapter 4, the process-interaction approach has been adopted by the simulation packages most popular in the USA. On the other hand, a number of activity-scanning packages are popular in the UK and Europe. Some of the packages allow portions of a model to be event-scheduling-based, if that orientation is convenient, mixed with the process-interaction approach. Finally, some of the packages are based on a flow chart, block diagram, or network structure, which upon closer examination turns out to be a specific implementation of the process-interaction concept.

3.1.3 Manual Simulation Using Event Scheduling

In the conducting of an event-scheduling simulation, a simulation table is used to record the successive system snapshots as time advances.

Example 3.3:   Single-Channel Queue _____

Reconsider the grocery store with one checkout counter that was simulated in Example 2.5 by an ad hoc method. The system consists of those customers in the waiting line plus the one (if any) checking out. A stopping time of 60 minutes is set for this example. The model has the following components:

System state $(LQ(t), LS(t))$, where $LQ(t)$ is the number of customers in the waiting line, and $LS(t)$ is the number being served (0 or 1) at time t.

Entities   The server and customers are not explicitly modeled, except in terms of the state variables.

Events
Arrival (A);
Departure (D);
Stopping event (E), scheduled to occur at time 60.

Event notices

(A, t), representing an arrival event to occur at future time t;

(D, t), representing a customer departure at future time t;

(E, 60), representing the simulation stop event at future time 60.

Activities

Interarrival time, defined in Table 2.9;

Service time, defined in Table 2.10.

Delay   Customer time spent in waiting line.

The event notices are written as (event type, event time). In this model, the FEL will always contain either two or three event notices. The effect of the arrival and departure events was first shown in Figures 2.4 and 2.5 and is shown in more detail in Figures 3.5 and 3.6.

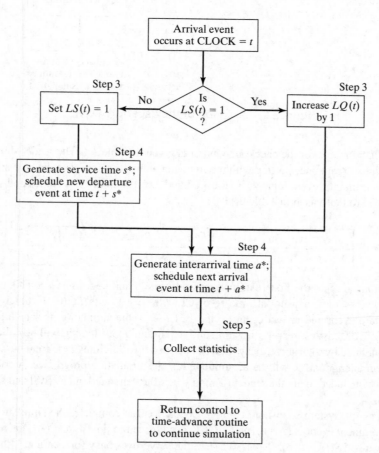

Figure 3.5 Execution of the arrival event.

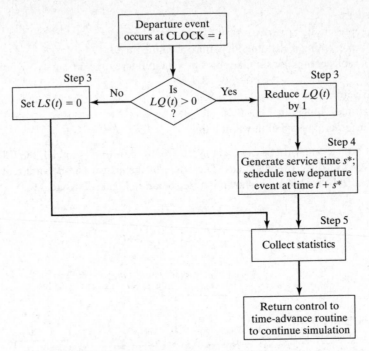

Figure 3.6   Execution of the departure event.

The simulation table for the checkout counter is given in Table 3.1. The reader should cover all system snapshots except one, starting with the first, and attempt to construct the next snapshot from the previous one and the event logic in Figures 3.5 and 3.6. The interarrival times and service times will be identical to those used in Table 2.11:

Interarrival Times	1	1	6	3	7	5	2	4	1···
Service Times		4	2	5	4	1	5	4	1 4···

Initial conditions are that the first customer arrive at time 0 and begin service. This is reflected in Table 3.1 by the system snapshot at time zero (CLOCK = 0), with $LQ(0) = 0, LS(0) = 1$, and both a departure event and arrival event on the FEL. Also, the simulation is scheduled to stop at time 60. Only two statistics, server utilization and maximum queue length, will be collected. Server utilization is defined by total server busy time (B) divided by total time (T_E). Total busy time, B, and maximum queue length, MQ, will be accumulated as the simulation progresses. A column headed "Comments" is included to aid the reader (a^* and s^* are the generated interarrival and service times, respectively).

As soon as the system snapshot at time CLOCK = 0 is complete, the simulation begins. At time 0, the imminent event is (A,1). The CLOCK is advanced to time 1, and (A,1) is removed from the FEL. Because $LS(t) = 1$ for $0 \le t \le 1$ (i.e., the server was busy for 1 minute), the cumulative busy time is increased from $B = 0$ to $B = 1$. By the event logic in Figure 3.6, set $LS(1) = 1$ (the

Table 3.1 Simulation Table for Checkout Counter

CLOCK	System state $LQ(t)$	$LS(t)$	Future event list	Comment	Cumulative statistics B	MQ
0	0	1	(A, 1) (D, 4) (E, 60)	First A occurs ($a^* = 1$) Schedule next A ($s^* = 4$) Schedule first D	0	0
1	1	1	(A, 2) (D, 4) (E, 60)	Second A occurs: (A, 1) ($a^* = 1$) Schedule next A (Customer delayed)	1	1
2	2	1	(D, 4) (A, 8) (E, 60)	Third A occurs: (A, 2) ($a^* = 6$) Schedule next A (Two customers delayed)	2	2
4	1	1	(D, 6) (A, 8) (E, 60)	First D occurs: (D, 4) ($s^* = 2$) Schedule next D (Customer delayed)	4	2
6	0	1	(A, 8) (D, 11) (E, 60)	Second D occurs: (D, 6) ($s^* = 5$) Schedule next D	6	2
8	1	1	(D, 11) (A, 11) (E, 60)	Fourth A occurs: (A, 8) ($a^* = 3$) Schedule next A (Customer delayed)	8	2
11	1	1	(D, 15) (A, 18) (E, 60)	Fifth A occurs: (A, 11) ($a^* = 7$) Schedule next A Third D occurs: (D, 11) ($s^* = 4$) Schedule next D (Customer delayed)	11	2
15	0	1	(D, 16) (A, 18) (E, 60)	Fourth D occurs: (D, 15) ($s^* = 1$) Schedule next D	15	2
16	0	0	(A, 18) (E, 60)	Fifth D occurs: (D, 16)	16	2
18	0	1	(D, 23) (A, 23) (E, 60)	Sixth A occurs ($a^* = 5$) Schedule next A ($s^* = 5$) Schedule next D	16	2
23	0	1	(A, 25) (D, 27) (E, 60)	Seventh A occurs: (A, 23) ($a^* = 2$) Schedule next Arrival Sixth D occurs: (D, 23)	21	2

server becomes busy). The FEL is left with three future events, (A, 2), (D, 4), and (E, 60). The simulation CLOCK is next advanced to time 2, and an arrival event is executed. The interpretation of the remainder of Table 3.1 is left to the reader.

The simulation in Table 3.1 covers the time interval [0,23]. At simulated time 23, the system empties, but the next arrival also occurs at time 23. The server was busy for 21 of the 23 time units simulated, and the maximum queue length was two. This simulation is, of course, too short to draw any reliable conclusions. Exercise 1 asks the reader to continue the simulation and to compare the results with those in Example 2.5. Note that the simulation table gives the system state at all times, not just the listed times. For example, from time 11 to time 15, there is one customer in service and one in the waiting line.

When an event-scheduling algorithm is implemented in simulation software, the software maintains just one set of system states and other attribute values, that is, just one snapshot (the current one or partially updated one). With the idea of implementing event scheduling in Java or some other general-purpose language, the following rule should be followed. A new snapshot can be derived only from the previous snapshot, newly generated random variables, and the event logic (Figures 3.5 and 3.6). Past snapshots should be ignored for advancing of the clock. The current snapshot must contain all information necessary to continue the simulation.

Example 3.4:   The Checkout-Counter Simulation, Continued

Suppose that, in the simulation of the checkout counter in Example 3.3, the simulation analyst desires to estimate mean response time and mean proportion of customers who spend 5 or more minutes in the system. A response time is the length of time a customer spends in the system. In order to estimate these customer averages, it is necessary to expand the model in Example 3.3 to represent the individual customers explicitly. In addition, to be able to compute an individual customer's response time when that customer departs, it will be necessary to know that customer's arrival time. Therefore, a customer entity with arrival time as an attribute will be added to the list of model components in Example 3.3. These customer entities will be stored in a list to be called "CHECKOUT LINE"; they will be called $C1, C2, C3, \ldots$ Finally, the event notices on the FEL will be expanded to indicate which customer is affected. For example, (D,4,$C1$) means that customer $C1$ will depart at time 4. The additional model components are the following:

Entities
   (Ci, t), representing customer Ci who arrived at time t.
Event notices
   (A, t, Ci), the arrival of customer Ci at future time t;
   (D, t, Cj), the departure of customer Cj at future time t,
Set
   "CHECKOUT LINE," the set of all customers currently at the checkout counter (being served or waiting to be served), ordered by time of arrival.

Three new cumulative statistics will be collected: S, the sum of customer response times for all customers who have departed by the current time; F, the total number of customers who spend 5 or more minutes at the checkout counter; and N_D, the total number of departures up to the current simulation time. These three cumulative statistics will be updated whenever the departure event

Table 3.2 Simulation Table for Example 3.4

CLOCK	System state		CHECKOUT LINE	Future event list	Cumulative statistics		
	$LQ(t)$	$LS(t)$			S	N_D	F
0	0	1	$(C1,0)$	$(A,1,C2)$ $(D,4,C1)$ $(E,60)$	0	0	0
1	1	1	$(C1,0)$ $(C2,1)$	$(A,2,C3)$ $(D,4,C1)$ $(E,60)$	0	0	0
2	2	1	$(C1,0)$ $(C2,1)$ $(C3,2)$	$(D,4,C1)$ $(A,8,C4)$ $(E,60)$	0	0	0
4	1	1	$(C2,1)$ $(C3,2)$	$(D,6,C2)$ $(A,8,C4)$ $(E,60)$	4	1	0
6	0	1	$(C3,2)$	$(A,8,C4)$ $(D,11,C3)$ $(E,60)$	9	2	1
8	1	1	$(C3,2)$ $(C4,8)$	$(D,11,C3)$ $(A,11,C5)$ $(E,60)$	9	2	1
11	1	1	$(C4,8)$ $(C5,11)$	$(D,15,C4)$ $(A,18,C6)$ $(E,60)$	18	3	2
15	0	1	$(C5,11)$	$(D,16,C5)$ $(A,18,C6)$ $(E,60)$	25	4	3
16	0	0		$(A,18,C6)$ $(E,60)$	30	5	4
18	0	1	$(C6,18)$	$(D,23,C6)$ $(A,23,C7)$ $(E,60)$	30	5	4
23	0	1	$(C7,23)$	$(A,25,C8)$ $(D,27,C7)$ $(E,60)$	35	6	5

occurs; the logic for collecting these statistics would be incorporated into Step 5 of the departure event in Figure 3.6.

The simulation table for Example 3.4 is shown in Table 3.2. The same data for interarrival and service times will be used again; so Table 3.2 essentially repeats Table 3.1, except that the new components are included (and the comment column has been deleted). These new components are needed for the computation of the cumulative statistics S, F, and N_D. For example, at time 4, a departure event occurs for customer $C1$. The customer entity $C1$ is removed from the list called "CHECKOUT LINE"; the attribute "time of arrival" is noted to be 0, so the response time for this customer was 4 minutes. Hence, S is incremented by 4 minutes. N_D is incremented by one customer, but F is not incremented, for the time in system was less than five minutes. Similarly, at time 23, when the departure event $(D, 23, C6)$ is being executed, the response time for customer $C6$ is computed by

$$\text{Response time} = \text{CLOCK TIME} - \text{attribute "time of arrival"}$$
$$= 23 - 18$$
$$= 5 \text{ minutes}$$

Then S is incremented by 5 minutes, and F and N_D by one customer.

For a simulation run length of 23 minutes, the average response time was $S/N_D = 35/6 = 5.83$ minutes, and the observed proportion of customers who spent 5 or more minutes in the system was $F/N_D = 0.83$. Again, this simulation was far too short to regard these estimates as having any degree of accuracy. The purpose of Example 3.4, however, was to illustrate the notion that, in many simulation models, the information desired from the simulation (such as the statistics S/N_D and F/N_D) to some extent dictates the structure of the model.

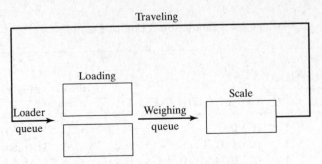

Figure 3.7   Dump truck problem.

Example 3.5:   The Dump-Truck Problem

Six dump trucks are used to haul coal from the entrance of a small mine to the railroad. Figure 3.7 provides a schematic of the dump-truck operation. Each truck is loaded by one of two loaders. After a loading, the truck immediately moves to the scale, to be weighed as soon as possible. Both the loaders and the scale have a first-come–first-served waiting line (or queue) for trucks. Travel time from a loader to the scale is considered negligible. After being weighed, a truck begins a travel time (during which time the truck unloads) and then afterward returns to the loader queue.

The distributions of loading time, weighing time, and travel time are given in Tables 3.3, 3.4, and 3.5, respectively. These activity times are generated in exactly the same manner as service times in Section 2.1.5, Example 2.2, using the cumulative probabilities to divide the unit interval into subintervals whose lengths correspond to the probabilities of each individual value. As before, random numbers come from Table A.1 or one of Excel's random number generators. A random number is drawn and the interval it falls into determines the next random activity time.

The purpose of the simulation is to estimate the loader and scale utilizations (percentage of time busy). The model has the following components:

System state

$[LQ(t), L(t), WQ(t), W(t)]$, where

$LQ(t)$ = number of trucks in loader queue;

$L(t)$ = number of trucks (0, 1, or 2) being loaded

$WQ(t)$ = number of trucks in weigh queue;

$W(t)$ = number of trucks (0 or 1) being weighed, all at simulation time t.

Entities

The six dump trucks ($DT1, \dots, DT6$).

Event notices

(ALQ, t, DTi), DTi arrives at loader queue (ALQ) at time t;

(EL, t, DTi), DTi ends loading (EL) at time t;

(EW, t, DTi), DTi ends weighing (EW) at time t.

Table 3.3   Distribution of Loading Time for the Dump Trucks

Loading Time	Probability	Cumulative Probability	Random Number Interval
5	0.30	0.30	$0.0 \leq R \leq 0.3$
10	0.50	0.80	$0.3 < R \leq 0.8$
15	0.20	1.00	$0.8 < R \leq 1.0$

Table 3.4   Distribution of Weighing Time for the Dump Trucks

Weighing Time	Probability	Cumulative Probability	Random Number Interval
12	0.70	0.70	$0.0 \leq R \leq 0.7$
16	0.30	1.00	$0.7 < R \leq 1.0$

Table 3.5   Distribution of Travel Time for the Dump Trucks

Travel Time	Probability	Cumulative Probability	Random Number Interval
40	0.40	0.40	$0.0 \leq R \leq 0.4$
60	0.30	0.70	$0.4 < R \leq 0.7$
80	0.20	0.90	$0.7 < R \leq 0.9$
100	0.10	1.00	$0.9 < R \leq 1.0$

Lists

Loader queue, all trucks waiting to begin loading, ordered on a first-come–first-served basis; Weigh queue, all trucks waiting to be weighed, ordered on a first-come–first-served basis.

Activities

Loading time, weighing time, and travel time.

Delays

Delay at loader queue, and delay at scale.

The simulation table is given in Table 3.6. To initialize the table's first row, we assume that, at time 0, five trucks are at the loaders and one is at the scale. For simplicity, we take the (randomly generated) activity times from the following list as needed:

Loading Time	10	5	5	10	15	10	10
Weighing Time	12	12	12	16	12	16	
Travel Time	60	100	40	40	80		

When an end-loading (*EL*) event occurs, say, for dump truck j at time t, other events might be triggered. If the scale is idle [$W(t) = 0$], dump truck j begins weighing and an end-weighing event (*EW*) is scheduled on the FEL; otherwise, dump truck j joins the weigh queue. If, at this time, there is another truck waiting for a loader, it will be removed from the loader queue and will begin loading by the scheduling of an *end-loading* event (*EL*) on the FEL. Both this logic for the occurrence of the end-loading event and the appropriate logic for the other two events, should be incorporated into an event diagram, as in Figures 3.5 and 3.6 of Example 3.3. The construction of these event-logic diagrams is left as an exercise for the reader (Exercise 2).

As an aid to the reader, in Table 3.6, whenever a new event is scheduled, its event time is written as "$t +$ (activity time)." For example, at time 0, the imminent event is an *EL* event with event time 5. The clock is advanced to time $t = 5$, dump truck 3 joins the weigh queue (because the scale is occupied), and dump truck 4 begins to load. Thus, an *EL* event is scheduled for truck 4 at future time 10, computed by (present time) + (loading time) = $5 + 5 = 10$.

In order to estimate the loader and scale utilizations, two cumulative statistics are maintained:

$$B_L = \text{total busy time of both loaders from time 0 to time } t$$
$$B_S = \text{total busy time of the scale from time 0 to time } t$$

Both loaders are busy from time 0 to time 20, so $B_L = 40$ at time $t = 20$—but, from time 20 to time 24, only one loader is busy; thus, B_L increases by only 4 minutes over the time interval [20, 24]. Similarly, from time 25 to time 36, both loaders are idle ($L(25) = 0$), so B_L does not change. For the relatively short simulation in Table 3.6, the utilizations are estimated as follows:

$$\text{Average loader utilization} = \frac{49/2}{76} = 0.32$$

$$\text{Average scale utilization} = \frac{76}{76} = 1.00$$

These estimates cannot be regarded as accurate estimates of the long-run "steady-state" utilizations of the loader and scale; a considerably longer simulation would be needed to reduce the effect of the assumed conditions at time 0 (five of the six trucks at the loaders) and to realize accurate estimates. On the other hand, if the analyst were interested in the so-called transient behavior of the system over a short period of time (say 1 or 2 hours), given the specified initial conditions, then the results in Table 3.6 can be considered representative (or constituting one sample) of that transient behavior. Additional samples can be obtained by conducting additional simulations, each one having the same initial conditions but using a different stream of random numbers to generate the activity times.

Table 3.6, the simulation table for the dump-truck operation, could have been simplified somewhat by not explicitly modeling the dump trucks as entities—that is, the event notices could be written as (*EL*, t), and so on, and the state variables used to keep track merely of the number of trucks

Table 3.6 Simulation Table for Dump-Truck Operation

CLOCK t	System state				Lists		Future event list	Cumulative statistics	
	$LQ(t)$	$L(t)$	$WQ(t)$	$W(t)$	Loader queue	Weigh queue		B_L	B_S
0	3	2	0	1	DT4 DT5 DT6		(EL, 5, DT3) (EL, 10, DT2) (EW, 12, DT1)	0	0
5	2	2	1	1	DT5 DT6	DT3	(EL, 10, DT2) (EL, 5 + 5, DT4) (EW, 12, DT1)	10	5
10	1	2	2	1	DT6	DT3 DT2	(EL, 10, DT4) (EW, 12, DT1) (EL, 10 + 10, DT5)	20	10
10	0	2	3	1		DT3 DT2 DT4	(EW, 12, DT1) (EL, 20, DT5) (EL, 10 + 15, DT6)	20	10
12	0	2	2	1		DT2 DT4	(EL, 20, DT5) (EW, 12 + 12, DT3) (EL, 25, DT6) (ALQ, 12 + 60, DT1)	24	12
20	0	1	3	1		DT2 DT4 DT5	(EW, 24, DT3) (EL, 25, DT6) (ALQ, 72, DT1)	40	20
24	0	1	2	1		DT4 DT5	(EL, 25, DT6) (EW, 24 + 12, DT2) (ALQ, 72, DT1) (ALQ, 24 + 100, DT3)	44	24
25	0	0	3	1		DT4 DT5 DT6	(EW, 36, DT2) (ALQ, 72, DT1) (ALQ, 124, DT3)	45	25
36	0	0	2	1		DT5 DT6	(EW, 36 + 16, DT4) (ALQ, 72, DT1) (ALQ, 36 + 40, DT2) (ALQ, 124, DT3)	45	36
52	0	0	1	1		DT6	(EW, 52 + 12, DT5) (ALQ, 72, DT1) (ALQ, 76, DT2) (ALQ, 52 + 40, DT4) (ALQ, 124, DT3)	45	52
64	0	0	0	1			(ALQ, 72, DT1) (ALQ, 76, DT2) (EW, 64 + 16, DT6) (ALQ, 92, DT4) (ALQ, 124, DT3) (ALQ, 64 + 80, DT5)	45	64
72	0	1	0	1			(ALQ, 76, DT2) (EW, 80, DT6) (EL, 72 + 10, DT1) (ALQ, 92, DT4) (ALQ, 124, DT3) (ALQ, 144, DT5)	45	72
76	0	2	0	1			(EW, 80, DT6) (EL, 82, DT1) (EL, 76 + 10, DT2) (ALQ, 92, DT4) (ALQ, 124, DT3) (ALQ, 144, DT5)	49	76

in each part of the system, not which trucks were involved. With this representation, the same utilization statistics could be collected. On the other hand, if mean "system" response time, or proportion of trucks spending more than 30 minutes in the "system," were being estimated, where "system" refers to the loader queue and loaders and the weigh queue and scale, then dump truck entities (DTi), together with an attribute equal to arrival time at the loader queue, would be indispensable. Whenever a truck left the scale, that truck's response time could be computed as current simulation time (t) minus the arrival-time attribute. This new response time would be used to update the cumulative statistics: S = total response time of all trucks that have been through the "system" and F = number of truck response times that have been greater than 30 minutes. This example again illustrates the notion that, to some extent, the complexity of the model depends on the performance measures being estimated.

Example 3.6:   The Dump-Truck Problem (revisited) ───────────────────────────────────

For the Dump Truck problem in Example 3.5, if we were to adopt the activity-scanning approach, the conditions for beginning each activity would be as follows:

Activity	Condition
Loading time	Truck is at front of loader queue, and at least one loader is idle.
Weighing time	Truck is at front of weigh queue, and weigh scale is idle.
Travel time	Truck has just completed a weighing.

If we were to take the process-interaction world view, we would view the model from the viewpoint of one dump truck and its "life cycle." Considering a life cycle as beginning at the loader queue, we can picture a dump-truck process as in Figure 3.8.

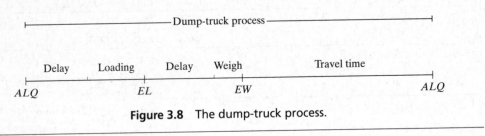

Figure 3.8   The dump-truck process.

3.2   List Processing

List processing deals with methods for handling lists of entities and the future event list. Simulation packages provide, both explicitly for an analyst's use and covertly in the simulation mechanism behind the language, facilities for an analyst or the model itself to use lists and to perform the basic operations on lists.

Section 3.2.1 describes the basic properties and operations performed on lists. Section 3.2.2 discusses the use of arrays for processing lists and the use of array indices to create linked lists, arrays being a simpler mechanism for describing the basic operations than the more general dynamically allocated linked lists discussed in section 3.2.3. Finally, section 3.2.4 briefly introduces some of the more advanced techniques for managing lists.

The purpose of this discussion of list processing is not to prepare the reader to implement lists and their processing in a general-purpose language such as Visual Basic, C, C++, or Java, but rather to increase the reader's understanding of lists and of their underlying concepts and operations.

3.2.1 Basic Properties and Operations Performed on Lists

As has previously been discussed, lists are a set of ordered or ranked records. In simulation, each record represents one entity or one event notice.

Lists are ranked, so they have a *top* or *head* (the first item on the list), some way to traverse the list (to find the second, third, etc. items on the list), and a *bottom* or *tail* (the last item on the list). A head pointer is a variable that points to or indicates the record at the top of the list. Some implementations of lists also have a tail pointer that points to the bottom item on the list.

For purposes of discussion, an entity, along with its attributes or an event notice, will be referred to as a *record*. An entity identifier and its attributes are *fields* in the entity record; the event type, event time, and any other event-related data are fields in the event-notice record. Each record on a list will also have a field that holds a "next pointer" that points to the next record on the list, providing a way to traverse the list. Some lists also require a "previous pointer," to allow for traversing the list from bottom to top.

For either type of list, the main activities in list processing are adding a record to a list and removing a record from a list. More specifically, the main operations on a list are the following:

a. removing a record from the top of the list;
b. removing a record from any location on the list;
c. adding a record to the top or bottom of the list;
d. adding a record at an arbitrary position in the list, one specified by the ranking rule.

The first and third operations, removing or adding a record to the top or bottom of the list, can be carried out in minimal time by adjusting two record pointers and the head or tail pointer; the other two operations require at least a partial search through the list. Making these two operations efficient is the goal of list-processing techniques.

In the event-scheduling approach, when time is advanced and the imminent event is due to be executed, the removal operation takes place first—namely, the event at the top of the FEL is removed from the FEL. If an arbitrary event is being canceled, or an entity is removed from a list based on some of its attributes (say, for example, its priority and due date) to begin an activity, then the second removal operation is performed. When an entity joins the back of a first-in–first-out queue implemented as a list, then the third operation, adding an entity to the bottom of a list, is performed. Finally, if a queue has the ranking rule *earliest due date first*, then, upon arrival at the queue, an entity must be added to the list not at the top or bottom but at the position determined by the due-date ranking rule.

For simulation on a computer, whether via a general-purpose language (such as Visual Basic, C, C++, or Java) or via a simulation package, each entity record and event notice is stored in a physical location in computer memory. There are two basic possibilities: (a) All records are stored in arrays. Arrays hold successive records in contiguous locations in computer memory. They therefore can be referenced by an array index that can be thought of as a row number in a matrix. (b) All entities and event notices are represented by structures (as in C) or classes (as in Java), allocated from RAM memory as needed, and tracked by pointers to a record or structure.

Most simulation packages use dynamically allocated records and pointers to keep track of lists of items; as arrays are conceptually simpler, the concept of linked lists is first explained through arrays and array indices in Section 3.2.2 and then applied to dynamically allocated records and pointers in Section 3.2.3.

3.2.2 Using Arrays for List Processing

The array method of list storage may be implemented as an array of records; that is, each array element is a record that contains the fields needed to track the properties of the event notices or entries in the list. For convenience, we use the notation $R(i)$ to refer to the i-th record in the array, however it may be stored in the language being used. Most modern simulation packages do not use arrays for list storage, but rather, use dynamically allocated records—that is, records that are created upon first being needed and subsequently destroyed when they are no longer needed.

Arrays are advantageous in that any specified record, say the i-th, can be retrieved quickly without searching, merely by referencing $R(i)$. Arrays are disadvantaged when items are added to the middle of a list or the list must be rearranged. In addition, arrays typically have a fixed size, determined at compile time or upon initial allocation when a program first begins to execute. In simulation, the maximum number of records for any list could be difficult (or impossible) to predict, and the current number of them in a list may vary widely over the course of the simulation run. Worse yet, most simulations require more than one list; if they are kept in separate arrays, each would have to be dimensioned to the largest the list would ever be, potentially using excessive amounts of computer memory.

In the use of arrays for storing lists, there are two basic methods for keeping track of the ranking of records in a list. One method is to store the first record in $R(1)$, the second in $R(2)$, and so on, and the last in $R(tailptr)$, where *tailptr* is used to refer to the last item in the list. Although simple in concept and easy to understand, this method will be extremely inefficient for all except the shortest lists, those of less than five or so records, because adding an item, for example, in position 41 in a list of 100 items, will require that the last 60 records be physically moved down one array position to make space for the new record. Even if the list were a first-in–first-out list, removing the top item from the list would be inefficient, as all remaining items would have to be physically moved up one position in the array. The physical rearrangement method of managing lists will not be discussed further. What is needed is a method to track and rearrange the logical ordering of items in a list without having to move the records physically in computer memory.

In the second method, a variable called a head pointer, with name *headptr*, points to the record at the top of the list. For example, if the record in position $R(11)$ were the record at the top of the list, then *headptr* would have the value 11. In addition, each record has a field that stores the index or pointer of the next record in the list. For convenience, let $R(i, \text{next})$ represent the next index field.

Example 3.7: A List for the Dump Trucks at the Weigh Queue

In Example 3.5, the dump-truck problem, suppose that a waiting line of three dump trucks occurred at the weigh queue, specifically, $DT3$, $DT2$, and $DT4$, in that order, at exactly CLOCK time 10 in Table 3.6. Suppose further that the model is tracking one attribute of each dump truck: its arrival time at the weigh queue, updated each time it arrives. Finally, suppose that the entities are stored in records in an array dimensioned from 1 to 6, one record for each dump truck. Each entity is represented by a record with 3 fields: the first is an entity identifier; the second is the arrival time at the weigh queue; the last is a pointer field to "point to" the next record, if any, in the list representing the weigh queue, as follows:

$$[DTi, \text{ arrival time at weigh queue, next index}]$$

Before its first arrival at the weigh queue, and before being added to the weigh queue list, a dump truck's second and third fields are meaningless. At time 0, the records would be initialized as follows:

$$R(1) = [DT1, 0.0, 0]$$
$$R(2) = [DT2, 0.0, 0]$$
$$\vdots$$
$$R(6) = [DT6, 0.0, 0]$$

Then, at CLOCK time 10 in the simulation in Table 3.6, the list of entities in the weigh queue would be defined by

$$\textit{headptr} = 3$$
$$R(1) = [DT1, 0.0, 0]$$
$$R(2) = [DT2, 10.0, 4]$$
$$R(3) = [DT3, 5.0, 2]$$
$$R(4) = [DT4, 10.0, 0]$$
$$R(5) = [DT5, 0.0, 0]$$
$$R(6) = [DT6, 0.0, 0]$$
$$\textit{tailptr} = 4$$

To traverse the list, start with the head pointer, go to that record, retrieve that record's next pointer, and proceed, to create the list in its logical order—for example,

$$\textit{headptr} = 3$$
$$R(3) = [DT3, 5.0, 2]$$
$$R(2) = [DT2, 10.0, 4]$$
$$R(4) = [DT4, 10.0, 0]$$

The zero entry for next pointer in R(4), as well as $\textit{tailptr} = 4$, indicates that DT4 is at the end of the list.

Using next pointers for a first-in–first-out list, such as the weigh queue in this example, makes the operations of adding and removing entity records, as dump trucks join and leave the weigh queue, particularly simple. At CLOCK time 12, dump truck DT3 begins weighing and thus leaves the weigh queue. To remove the DT3 entity record from the top of the list, update the head pointer by setting it equal to the next pointer value of the record at the top of the list:

$$\textit{headptr} = R(\textit{headptr}, \text{next})$$

In this example, we get

$$headptr = R(3, \text{next}) = 2$$

meaning that dump truck DT2 in R(2) is now at the top of the list.

Similarly, at CLOCK time 20, dump truck DT5 arrives at the weigh queue and joins the rear of the queue. To add the DT5 entity record to the bottom of the list, the following steps are taken:

$R(tailptr, \text{next}) = 5$ (update the next pointer field of the previously last item)
$tailptr = 5$ (update the value of the tail pointer)
$R(tailptr, \text{next}) = 0$ (for the new tail item, set its next pointer field to zero)

This approach becomes slightly more complex when a list is a ranked list, such as the future event list, or an entity list ranked by an entity attribute. For ranked lists, to add or remove an item anywhere except to the head or tail of the list, searching is usually required. See Example 3.8.

Note that, in the dump-truck problem, the loader queue could also be implemented as a list using the same six records and the same array, because each dump-truck entity will be on at most one of the two lists and, while loading, weighing or traveling, a dump truck will be on neither list.

3.2.3   Using Dynamic Allocation and Linked Lists

In procedural languages, such as C++ and Java, and in most simulation languages, entity records are dynamically created when an entity is created and event-notice records are dynamically created whenever an event is scheduled on the future-event list. The languages themselves, or the operating systems on which they are running, maintain a linked list of free chunks of computer memory and allocate a chunk of desired size upon request to running programs. (Another use of linked lists!) When an entity "dies," that is, exits from the simulated system, and also after an event occurs and the event notice is no longer needed, the corresponding records are freed, making that chunk of computer memory available for later reuse; the language or operating system adds the chunk to the list of free memory.

In this text, we are not concerned with the details of allocating and freeing computer memory, so we will assume that the necessary operations occur as needed. With dynamic allocation, a record is referenced by a pointer instead of by an array index. When a record is allocated in C++ or Java, the allocation routine returns a pointer to the allocated record, which must be stored in a variable or a field of another record for later use. A pointer to a record can be thought of as the physical or logical address in computer memory of the record.

In our example, we will use a notation for records identical to that in the previous section (3.2.2):

Entities: [ID, attributes, next pointer]
Event notices: [event type, event time, other data, next pointer]

but we will not reference them by the array notation $R(i)$ as before, because it would be misleading. If for some reason we wanted the third item on the list, we would have to traverse the list, counting items until we reached the third record. Unlike in arrays, there is no way to retrieve the i-th record in a linked list directly, because the actual records could be stored at any arbitrary location in computer memory and are not stored contiguously, as arrays are.

Example 3.8:   The Future Event List and the Dump-Truck Problem _____

Beginning from Table 3.6, event notices in the dump-truck problem of Example 3.5 are expanded to include a pointer to the next event notice on the future event list and can be represented by

$$[\text{event type, event time, DT}i, \textit{nextptr}],$$

—for example,

$$[\text{EL, 10, DT3, }\textit{nextptr}]$$

where EL is the end-loading event to occur at future time 10 for dump truck DT3, and the field *nextptr* points to the next record on the FEL. Keep in mind that the records may be stored anywhere in computer memory, and in particular are not necessarily stored contiguously. Figure 3.9 represents the future event list at CLOCK time 10, taken from Table 3.6. The fourth field in each record is a pointer value to the next record in the future event list.

The C++ and Java languages, and other general-purpose languages, use different notation for referencing data from pointer variables. For discussion purposes, if R is a pointer to a record, then

$$R \rightarrow \text{eventtype}, R \rightarrow \text{eventtime}, R \rightarrow \text{next}$$

are the event type, the event time and the next record for the event notice that R points to. For example, if R is set equal to the head pointer for the FEL at CLOCK time 10, then

$$R \rightarrow \text{eventtype} = \text{EW}$$
$$R \rightarrow \text{eventtime} = 12$$
$$R \rightarrow \text{next is the pointer to the second event notice on the FEL,}$$

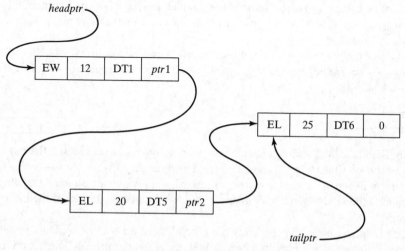

Figure 3.9 The dump-truck future event list as a linked list.

so that

$$R \rightarrow next \rightarrow eventtype = EL$$
$$R \rightarrow next \rightarrow eventtime = 20$$
$$R \rightarrow next \rightarrow next \text{ is the pointer to the third event notice on the FEL}$$

If one of the pointer fields is zero (or null), then that record is the last item in the list, and the pointer variable *tailptr* points to that last record, as depicted in Figure 3.9.

What we have described are called singly-linked lists, because there is a one-way linkage from the head of the list to its tail. The tail pointer is kept mostly for convenience and efficiency, especially for lists for which items are added at the bottom of the list. Of course, a tail pointer is not strictly necessary, because the last item can always be found by traversing the list, but it does make some operations more efficient.

For some purposes, it is desirable to traverse or search a list by starting at the tail in addition to the head. For such purposes, a doubly-linked list can be used. Records on a doubly-linked list have two pointer fields, one for the next record and one for the previous record. Good references that discuss arrays, singly- and doubly-linked lists, and searching and traversing of lists are Cormen, *et al.* [2001] and Sedgewick [1998].

3.2.4   Advanced Techniques

Many of the modern simulation packages use techniques and representations of lists that are more efficient than searching through a doubly-linked list. Most of these topics are too advanced for this text. The purpose of this section is to introduce some of the more advanced ideas briefly.

One idea to speed up processing of doubly-linked lists is to use a middle pointer in addition to a head and tail pointer. With special techniques, the *mid* pointer will always point to the approximate middle of the list. Then, when a new record is being added to the list, the algorithm first examines the middle record to decide whether to begin searching at the head of the list or the middle of the list. Theoretically, except for some overhead due to maintenance of the mid pointer, this technique should cut search times in half. A few advanced techniques use one or more mid pointers, depending on the length of the list.

Some advanced algorithms use list representations other than a doubly-linked list, such as heaps or trees. These topics are beyond the scope of this text. Good references are Cormen, *et al.* [2001] and Sedgewick [1998].

3.3   Summary

This chapter introduced the major concepts and building blocks in simulation, the most important being entities and attributes, events, and activities. The three major world views—event scheduling, process interaction, activity scanning—were discussed. Finally, to gain an understanding of one of the most important underlying methodologies, Section 3.2 introduced some of the basic notions of list processing.

The next chapter will provide a survey of some of the most widely used and popular simulation packages, most of which either use exclusively or allow the process-interaction approach to simulation.

REFERENCES

CORMEN, T. H., C. E. LEISEROON, AND R. L. RIVEST [2001], *Introduction to Algorithms*, 2nd ed., McGraw-Hill, New York.

SCHRIBER, T. J., AND D. T. BRUNNER [2003], "Inside Simulation Software: How it Works and Why it Matters," *Proceedings of the 2003 Winter Simulation Conference*, S. Chick, P. J. Sánchez, D. Ferrin, and D. J. Morrice, eds., New Orleans, LA, Dec. 7–10, pp. 113–123.

SEDGEWICK, R. [1998], *Algorithms in C++*, 3d ed., Addison-Wesley, Reading, MA.

EXERCISES

Instructions to the reader: For most exercises, the reader should first construct a model by explicitly defining the following:

1. system state;

2. system entities and their attributes;

3. sets, and the entities that may be put into the sets;

4. events and activities; event notices;

5. variables needed to collect cumulative statistics.

Second, the reader should either (1) develop the event logic (as in Figures 3.5 and 3.6 for Example 3.3) in preparation for using the event-scheduling approach, or (2) develop the system processes (as in Figure 3.4) in preparation for using the process-interaction approach.

Most problems contain activities that are uniformly distributed over an interval $[a, b]$. When conducting a manual simulation, assume that $a, a + 1, a + 2, \ldots, b$ are the only possible values; that is, the activity time is a *discrete* random variable. The discreteness assumption will simplify the manual simulation.

1. (a) Using the event-scheduling approach, continue the (manual) checkout-counter simulation in Example 3.3, Table 3.1. Use the same interarrival and service times that were previously generated and used in Table 2.11, the simulation table for Example 2.5. (This refers to columns C and E of Table 2.11.) Continue until all arriving customers receive service. After using the last interarrival time, assume there are no more arrivals.

   (b) Do exercise 1(a) again, adding the model components necessary to estimate mean response time and proportion of customers who spend 5 or more minutes in the system. [*Hint:* See Example 3.4, Table 3.2.]

   (c) Comment on the relative merits of manual versus computerized simulations.

2. Construct the event-logic diagrams for the dump-truck problem, Example 3.5.

3. In the dump-truck problem of Example 3.5, it is desired to estimate mean response time and the proportion of response times which are greater than 30 minutes. A response time for a truck begins when that truck arrives at the loader queue and ends when the truck finishes weighing. Add the model components and cumulative statistics needed to estimate these two measures of system performance. Simulate for 8 hours.

4. Prepare a table in the manner of Table 3.2, until the CLOCK reaches time 15, using the interarrival and service times given below in the order shown. The stopping event will be at time 30.

   Interarrival times: 1   5   6   3   8

   Service times: 3   5   4   1   5

5. Continue Table 3.2 until caller C7 departs.

6. The data for Table 3.2 are changed to the following:

   Interarrival times: 4   5   2   8   3   7

   Service times: 5   3   4   6   2   7

   Prepare a table in the manner of Table 3.2 with a stopping event at time 25.

7. Redo Example 2.6 (the Able–Baker call center problem) by a manual simulation, using the event-scheduling approach.

8. Redo Example 2.8 (the (M, N) inventory system) by a manual simulation, using the event-scheduling approach.

9. Redo Example 2.9 (the bearing-replacement problem) by a manual simulation, using the event-scheduling approach.

10. Rework Example 3.5, but using the following data:

   Loading times: 10   5   10   10   5   10   5

   Weigh times: 12   16   12   12   16   12   12

   Travel times: 40   60   40   80   100   40

4

Simulation Software

In this chapter, we first discuss the history of simulation software and speculate on its future. Simulation software has a history that is just reaching middle age. We base this history on our collective experience, articles written by Professor Richard Nance, panel discussions at the annual Winter Simulation Conference, and a recent article by one of the co-authors.

Next, we discuss features and attributes of simulation software. If you were about to purchase simulation software, what would concern you? Would it be the cost, the ease of learning, the ease of use, or would it be the power to model the kind of system with which you are concerned? Or would it be the animation capabilities? Following the discussion of features, we discuss other issues and concerns related to the selection of simulation software.

Software used to develop simulation models can be divided into three categories. First, there are the general-purpose programming languages, such as C, C++, and Java. Second, there are simulation programming languages, examples being GPSS/H and SIMAN V®. Third, there are the simulation environments. This category includes many products that are distinguished one way or another (by, for example, cost, application area, or type of animation), but have common characteristics, such as a graphical user interface and an environment that supports all (or most) aspects of a simulation study. Many simulation environments contain a simulation programming language, but some take a graphical approach similar to process-flow diagramming.

In the first category, we discuss simulation in Java. Java is a general-purpose programming language that was not specifically designed for use in simulation. Java was chosen since it is widely used and widely available. Today very few people writing discrete-event simulation models are using programming languages alone; however, in certain application areas, some people are using packages based on Java or on some other general-purpose language. Understanding how to develop a model in a general-purpose language helps to understand how the basic concepts and algorithms discussed in Chapter 3 are implemented.

In the second category, we discuss GPSS/H, a highly structured process-interaction simulation language. GPSS was designed for relatively easy simulation of queueing systems, as in job shops, but it has been used to simulate systems of great complexity. It was first introduced by IBM; today, there are various implementations of GPSS, GPSS/H being one of the most widely used.

In the third category, we have selected a number of simulation software packages for discussion. There are many simulation packages currently available; we have selected a few that have survived and thriven for a number of years, to represent different approaches for model building.

One of the important components of a simulation environment is the output analyzer, which is used to conduct experimentation and assist with analyses. To illustrate the range of desirable characteristics, we look at four tools incorporated into some of the simulation environments. Typically these statistical analysis tools compute summary statistics, confidence intervals, and other statistical measures. Some support warmup determination, design of experiments, and sensitivity analyses. Many packages now offer optimization techniques based on genetic algorithms, evolutionary strategies, tabu search, scatter search, and other recently developed heuristic methods. In addition to the support for statistical analysis and optimization, the simulation environments offer data management, scenario definition, and run management. Data management offers support for managing all the input and output data associated with the analyses.

4.1 History of Simulation Software

Our discussion of the history of simulation software is based on Nance [1995], who breaks the years from 1955 through 1986 into five periods. Additional historical information is taken from a panel discussion at the 1992 Winter Simulation Conference entitled "Perspectives of the Founding Fathers" [Wilson et al., 1992], during which eight early users of simulation presented their historical perspectives. We add a sixth and most recent period, and then speculate on the future:

1. 1955–1960   The Period of Search
2. 1961–1965   The Advent
3. 1966–1970   The Formative Period
4. 1971–1978   The Expansion Period
5. 1979–1986   The Period of Consolidation and Regeneration
6. 1987–2008   The Period of Integrated Environments
7. 2009–2011   The Future

The following subsections provide a brief presentation of this history. As indicated in Nance [1995], there were at least 137 simulation programming languages reported as of 1981, and many more since then. This brief history is far from comprehensive. The languages and packages we mention have stood the test of time by surviving to the present day or were the historical forerunner of a package in present use.

4.1.1   The Period of Search (1955–1960)

In the early years, simulation was conducted in FORTRAN or other general-purpose programming language, without the support of simulation-specific routines. In their first period, much effort was expended on the search for unifying concepts and the development of reusable routines to facilitate simulation. The General Simulation Program of K.D. Tocher and D.G. Owen [1960] is considered the first "language effort." Tocher identified and developed routines that could be reused in subsequent simulation projects.

4.1.2   The Advent (1961–1965)

The forerunners of the simulation programming languages (SPLs) in use today appeared in the period 1961–1965. As Harold Hixson said in Wilson et al., [1992], "in the beginning there were FORTRAN, ALGOL, and GPSS"—that is, there were the FORTRAN-based packages (such as SIMSCRIPT and GASP), the ALGOL descendent SIMULA, and GPSS.

The first process-interaction SPL, GPSS was developed by Geoffrey Gordon at IBM and appeared in about 1961. Gordon developed GPSS (General Purpose Simulation System) for quick simulations of communications and computer systems, but its ease of use quickly spread its popularity to other application areas. GPSS is based on a block-diagram representation (similar to a process-flow diagram) and is suited for queueing models of all kinds. As reported by Reitman in Wilson et al., [1992], as early as 1965 GPSS was connected to an interactive display terminal that could interrupt and display intermediate results, a foreshadowing of the interactive simulations of today, but far too expensive at the time to gain widespread use.

Harry Markowitz (later to receive a Nobel Prize for his work in portfolio theory) provided the major conceptual guidance for SIMSCRIPT, first appearing in 1963. The RAND Corporation developed the language under sponsorship of the U.S. Air Force. SIMSCRIPT originally was heavily influenced by FORTRAN, but in later versions, its developers broke from its FORTRAN base and created its own SPL. The initial versions were based on event scheduling.

Phillip J. Kiviat of the Applied Research Laboratory of the United States Steel Corporation began the development of GASP (General Activity Simulation Program) in 1961. Originally, it was based on the general-purpose programming language ALGOL, but later a decision was made to base it on FORTRAN. GASP, like GPSS, used flowchart symbols familiar to engineers. It was not a language proper, but rather a collection of FORTRAN routines to facilitate simulation in FORTRAN.

Numerous other SPLs were developed during this time. Notably, they included SIMULA, an extension of ALGOL, developed in Norway and widely used in Europe, and the Control and Simulation Language (CSL), which took an activity-scanning approach.

4.1.3   The Formative Period (1966–1970)

During this period, concepts were reviewed and refined to promote a more consistent representation of each language's worldview. The major SPLs matured and gained wider usage.

Rapid hardware advancements and user demands forced some languages, notably GPSS, to undergo major revisions. GPSS/360, with its extensions to earlier versions of GPSS, emerged for the IBM 360 computer. Its popularity motivated at least six other hardware vendors and other groups to produce their own implementations of GPSS or a look-alike.

SIMSCRIPT II represented a major advancement in SPLs. In its free-form English-like language and "forgiving" compiler, an attempt was made to give the user major consideration in the language design.

ECSL, a descendent of CSL, was developed and became popular in the UK. In Europe, SIMULA added the concept of classes and inheritance, thus becoming a precursor of the modern object-oriented programming languages.

4.1.4   The Expansion Period (1971–1978)

Major advances in GPSS during this period came from outside IBM. Julian Reitman of Norden Systems headed the development of GPSS/NORDEN, a pioneering effort that offered an interactive, visual online environment. James O. Henriksen of Wolverine Software developed GPSS/H, released in 1977 for IBM mainframes, later for mini-computers and the PC. It was notable for being compiled and reportedly 5 to 30 times faster than standard GPSS. With the addition of new features, including an interactive debugger, it has become the principal version of GPSS in use today.

Alan Pritsker at Purdue made major changes to GASP, with GASP IV appearing in 1974. It incorporated state events in addition to time events, thus adding support for the activity-scanning worldview in addition to the event-scheduling worldview.

Efforts were made during this period to try and simplify the modeling process. Using SIMULA, an attempt was made to develop a system definition from a high-level user perspective that could be translated automatically into an executable model. Similar efforts included interactive program generators, the "Programming by Questionnaire," and natural-language interfaces, together with automatic mappings to the language of choice. As did earlier over-optimistic beliefs in automatic programming, these efforts ran into severe limitations in the generality of what could be modeled—that is, they ran into the unavoidable complexity of real-world systems. Nevertheless, efforts to simplify simulation modeling continue, with the most success seen in simulation systems designed for application to narrow domains.

4.1.5   The Period of Consolidation and Regeneration (1979–1986)

The fifth period saw the beginnings of SPLs written for, or adapted to, desktop computers and the microcomputer. During this period, the predominant SPLs extended their implementations to many computers and microprocessors while maintaining their basic structure.

Two major descendants of GASP appeared: SLAM II and SIMAN. The Simulation Language for Alternative Modeling (SLAM), produced by Pritsker and Associates, Inc., sought to provide multiple modeling perspectives and combined modeling capabilities [Pritsker and Pegden, 1979]—that is, it had an event-scheduling perspective based on GASP, a network worldview (a variant of the process-interaction perspective), and a continuous component. With SLAM, you could select one worldview or use a mix of all three.

SIMAN (SIMulation ANalysis) possessed a general modeling capability found in SPLs such as GASP IV, but also had a block-diagram component similar in some respects to that in SLAM and GPSS. C. Dennis Pegden developed SIMAN as a one-person faculty project over a period of about two years; he later founded Systems Modeling Corporation to market SIMAN. SIMAN was the first major simulation language executable on the IBM PC and designed to run under MS-DOS constraints. Similar to GASP, SIMAN allowed an event-scheduling approach by programming in FORTRAN with a supplied collection of FORTRAN routines, a block-diagram approach (another variant of the process-interaction worldview) analogous in some ways to that of GPSS and SLAM, and a continuous component.

4.1.6   The Period of Integrated Environments (1987–2008)

This period is notable for the growth of SPLs on the personal computer and the emergence of simulation environments with graphical user interfaces, animation, and other visualization tools. AutoMod and Extend were two of the earliest software packages to offer integrated simulation environments. Today, virtually all simulation packages come with environments that integrate model building, model debugging, animation, and interactive running of models. Many of these environments also contain input-data analyzers and output analyzers. Some packages attempt to simplify the modeling process by the use of process-flow or block diagramming and of "fill-in-the-blank" windows that avoid the need to learn programming syntax. Animation ranges from schematic-like representations to 2-D and 3-D scale drawings.

Recent advancements have been made in web-based simulation. Simulation has been deemed to have a role in supply-chain management. The combination of simulation and emulation shows promise. (Emulation is the imitation of a device such as those material handling devices illustrated in www.demo3d.com.)

4.1.7   The Future (2009–2011)

In an article in the *ICS Newsletter* [Banks, 2008], one of the co-authors asked a panel of simulationists, including well-known experts in their field, their opinion or best guess regarding future developments in discrete-event simulation. The first question was: "What remarkable results will we observe in simulation software in the near term, say, within three years?"

The most-often mentioned future developments, including more use of virtual reality, included an improved user interface, better animation, and a paradigm shift such as a major adoption of agent-based modeling in discrete-event modeling. Although, to date, agent-based modeling has been widely used in modeling collective behavior (civil violence, war, even determining what elements cause a standing ovation) and social phenomenon (seasonal migration, pollution, transmission of disease, and sexual reproduction, to name a few) its use in industrial and commercial simulations is just beginning.

Agent-based modeling is a method for simulating the actions and interactions of autonomous individuals (the agents) in a network, with a view to assessing their effects on the system as a whole. Agents may be people or animals or other entities that have agency, meaning that they are not passive, they actively make decisions, retain memory of past situations and decisions, and exhibit learning. In summary, agents are active and may exhibit cognition, memory, and learning. They may or may not be goal-directed.

Each agent's behavior is governed by a set of local rules; that is, the agent responds in a prescribed way to local, near-by conditions. The interest is in patterns of behavior that emerge for the group of agents, called *emergent behavior*.

An interesting and well known example is the "boids" (that is, simulated birds flying in a given 2-D or 3-D area) in which flocking behavior emerges. (You can find many examples of the boids model on the Internet, each based on similar but not identical assumptions or rules. For one, see www.navgen.com/3d_boids.) The flocking behavior is not scripted or explicitly programmed, but rather it emerges naturally (and somewhat surprisingly) from a set of fairly simple rules. The basic rule is that if a boid comes close to one or more other boids, then it adjusts its direction and speed to the other boids. In most of the boids simulations found on the Internet, the boids are not goal-directed; that is, they have no destination but simply fly. In at least one boids example, the boids do have a goal; there are two or more flocks of boids, and members of one flock attack and kill members of other flocks.

Software packages and toolkits for agent-based modeling include

- AnyLogic (www.xjtek.com) discussed in Section 4.7.1,
- Ascape (www.nutechsolutions.com),
- MASON (cs.gmu.edu/%7Eeclab/projects/mason),
- NetLogo (ccl.northwestern.edu/netlogo),
- StarLogo (education.mit.edu/starlogo),
- Swarm (www.swarm.org), and
- RePast (www.repast.sourceforge.net),

to name a few.

Information about various software packages is given in Section 4.7, including the websites of the vendors. A view of current developments in simulation software is available from these websites.

4.2   Selection of Simulation Software

This chapter includes a brief introduction to a number of simulation-software packages. Every two years, the journal *OR/MS Today* publishes a simulation-software survey [Swain, 2007]. The 2007 issue had 48 products, including simulation support packages such as input-data analyzers.

There are many features that are relevant when selecting simulation software [Banks, 1996]. Some of these features are shown, along with a brief description, in Tables 4.1 to 4.5. We offer the following advice when evaluating and selecting simulation software:

1. Do not focus on a single issue, such as ease of use. Consider the accuracy and level of detail obtainable, ease of learning, vendor support, and applicability to your applications.
2. Execution speed is important. Do not think exclusively in terms of experimental runs that take place at night and over the weekend. Speed affects development time. During debugging, an analyst might have to wait for the model to run up to the point in simulated time where an error occurs many times before the error is identified.
3. Beware of advertising claims and demonstrations. Many advertisements exploit positive features of the software only. Similarly, the demonstrations solve the test problem very well, but perhaps not your problem.
4. Ask the vendor to solve a small version of your problem.

Table 4.1 Model-Building Features

Feature	Description
Modeling worldview	Process interaction, event perspectives, and continuous modeling, depending on needs
Input-data analysis capability	Estimate empirical or statistical distributions from raw data
Graphical model building	Process-flow, block-diagram, or network approach
Conditional routing	Route entities based on prescribed conditions or attributes
Simulation programming	Capability to add procedural logic through a high-level powerful simulation language
Syntax	Easily understood, consistent, unambiguous, English-like
Input flexibility	Accepts data from external files, databases, spreadsheets, or interactively
Modeling conciseness	Powerful actions, blocks, or nodes
Randomness	Random-variate generators for all common distributions, e.g.,
	exponential
	triangular
	uniform
	normal
Specialized components and templates	Material handling: vehicles, conveyors, bridge cranes, AS/RS, etc.
	Handling of liquids and bulk materials
	Communication systems
	Computer systems
	Call centers
	Others
User-built custom objects	Reusable objects, templates, and submodels
Continuous flow	Tanks and pipes or bulk conveyors
Interface with general-programming language	Link code in C, C++, Java, or other general-programming language

Table 4.2   Runtime Environment

Feature	Description
Execution speed	Many runs needed for scenarios and replications. Impacts development as well as experimentation
Model size; number of variables and attributes	Should have no built-in limits
Interactive debugger	Monitor the simulation in detail as it progresses. Ability to break, trap, run until, step; to display status, attributes and variables; etc.
Model status and statistics	Display at any time during simulation
Runtime license	Ability to change parameters and run a model (but not to change logic or build a new model)

5. Beware of "checklists" with "yes" and "no" as the entries. For example, many packages claim to have a conveyor entity. However, implementations have considerable variation and level of fidelity. Implementation and capability are what is important. As a second example, most packages offer a runtime license, but these vary considerably in price and features.

Table 4.3   Animation and Layout Features

Feature	Description
Type of animation	True to scale or iconic (such as process-flow diagram)
Import drawing and object files	From CAD (vector formats) drawings or icons (bit-mapped or raster graphics)
Dimension	2-D, 2-D with perspective, 3-D
Movement	Motion of entities or status indicators
Quality of motion	Smooth or jerky
Libraries of common objects	Extensive predrawn graphics
Navigation	Panning, zooming, rotation
Views	User defined, named
Display step	Control of animation speed
Selectable objects	Dynamic status and statistics displayed upon selection
Hardware requirements	Standard or special video card

Table 4.4   Output Features

Feature	Description
Scenario manager	Create user-defined scenarios to simulate
Run manager	Make all runs (scenarios and replications) and save results for future analyses
Warmup capability	For steady-state analysis
Independent replications	Using a different set of random numbers
Optimization	Genetic algorithms, tabu search, etc.
Standardized reports	Summary reports including averages, counts, minimum and maximum, etc.
Customized reports	Tailored presentations for managers
Statistical analysis	Confidence intervals, designed experiments, etc.
Business graphics	Bar charts, pie charts, time lines, etc.
Costing module	Activity-based costing included
File export	Input to spreadsheet or database for custom processing and analysis
Database maintenance	Store output in an organized manner

Table 4.5   Vendor Support and Product Documentation

Feature	Description
Training	Regularly scheduled classes of high quality
Documentation	Quality, completeness, online
Help system	General or context-sensitive
Tutorials	For learning the package or specific features
Support	Telephone, e-mail, web
Upgrades, maintenance	Regularity of new versions and maintenance releases that address customer needs
Track record	Stability, history, customer relations

6. Simulation users ask whether the simulation model can link to and use code or routines written in external languages such as C, C++, or Java. This is a good feature, especially when the external routines already exist and are suitable for the purpose at hand. However, the more important question is whether the simulation package and language are sufficiently powerful to avoid having to write logic in any external language.

7. There may be a significant trade-off between the graphical model-building environments and ones based on a simulation language. Graphical model building removes the learning curve

due to language syntax, but it does not remove the need for procedural logic in most real-world models and the debugging to get it right. Beware of "no programming required" unless either the package is a near-perfect fit to your problem domain or programming (customized procedural logic) is possible with the supplied blocks, nodes, or process-flow diagram—in which case "no programming required" refers to syntax only and not the development of procedural logic.

4.3 An Example Simulation

Example 4.1:   The Checkout Counter: A Typical Single-Server Queue _____

The system, a grocery checkout counter, is modeled as a single-server queue. The simulation will run until 1000 customers have been served. In addition, assume that the interarrival times of customers are exponentially distributed with a mean of 4.5 minutes, and that the service times are (approximately) normally distributed with a mean of 3.2 minutes and a standard deviation of 0.6 minute. (The approximation is that service times are always positive.) When the cashier is busy, a queue forms with no customers turned away. This example was simulated manually in Examples 3.3 and 3.4 by using the event-scheduling point of view. The model contains two events: the arrival and departure events. Figures 3.5 and 3.6 provide the event logic.

The following three sections illustrate the simulation of this single-server queue in Java, GPSS/H, and SSF (Scalable Simulation Framework). Although this example is much simpler than models that arise in the study of complex systems, its simulation contains the essential components of all discrete-event simulations.

4.4 Simulation in Java

Java is a widely available programming language that has been used extensively in simulation. It does not, however, provide any facilities directly aimed at aiding the simulation analyst, who therefore must program all details of the event-scheduling/time-advance algorithm, the statistics-gathering capability, the generation of samples from specified probability distributions, and the report generator. Nevertheless, the runtime library does provide a random-number generator. Unlike FORTRAN or C, the object-orientedness of Java does support modular construction of large models. For the most part, the special-purpose simulation languages hide the details of event scheduling, whereas in Java all the details must be explicitly programmed. To a certain extent though, simulation libraries such as SSF [Cowie, 1999] alleviate the development burden by providing access to standardized simulation functionality and by hiding low-level scheduling minutiae.

There are many online resources for learning Java; we assume a prior working knowledge of the language. Any discrete-event simulation model written in Java contains the components discussed in Section 4.3: system state, entities and attributes, sets, events, activities and delays, and the components listed below. To facilitate development and debugging, it is best to organize the Java model in a modular fashion by using methods. The following components are common to almost all models written in Java:

Clock A variable defining simulated time

Initialization method A method to define the system state at time 0

Min-time event method A method that identifies the imminent event, that is, the element of the future event list (`FutureEventList`) that has the smallest time-stamp

Event methods For each event type, a method to update system state (and cumulative statistics) when that event occurs

Random-variate generators Methods to generate samples from desired probability distributions

Main program To maintain overall control of the event-scheduling algorithm

Report generator A method that computes summary statistics from cumulative statistics and prints a report at the end of the simulation

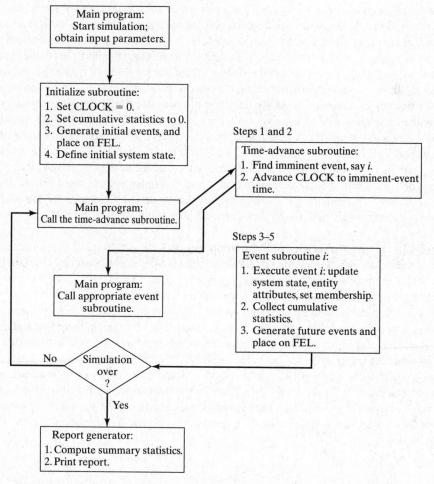

Figure 4.1 Overall structure of an event-scheduling simulation program.

The overall structure of a Java simulation program is shown in Figure 4.1. This flowchart is an expansion of the event-scheduling/time-advance algorithm outlined in Figure 3.2. (The steps mentioned in Figure 4.1 refer to the five steps in Figure 3.2.)

The simulation begins by setting the simulation `Clock` to zero, initializing cumulative statistics to zero, generating any initial events (there will always be at least one) and placing them on the `FutureEventList`, and defining the system state at time 0. The simulation program then cycles, repeatedly passing the current least-time event to the appropriate event methods until the simulation is over. At each step, after finding the imminent event but before calling the event method, the simulation `Clock` is advanced to the time of the imminent event. (Recall that, during the simulated time between the occurrence of two successive events, the system state and entity attributes do not change in value. Indeed, this is the definition of discrete-event simulation: The system state changes only when an event occurs.) Next, the appropriate event method is called to execute the imminent event, update cumulative statistics, and generate future events (to be placed on the `FutureEventList`). Executing the imminent event means that the system state, entity attributes, and set membership are changed to reflect the fact that the event has occurred. Notice that all actions in an event method take place at one instant of simulated time. The value of the variable `Clock` does not change in an event method. If the simulation is not over, control passes again to the time-advance method, then to the appropriate event method, and so on. When the simulation is over, control passes to the report generator, which computes the desired summary statistics from the collected cumulative statistics and prints a report.

The efficiency of a simulation model in terms of computer runtime is determined to a large extent by the techniques used to manipulate the `FutureEventList` and other sets. As discussed earlier in Section 4.3, removal of the imminent event and addition of a new event are the two main operations performed on the `FutureEventList`. Java includes general, efficient data structures for searching and priority lists; it is usual to build a customized interface to these to suit the application. In the example to follow, we use customized interfaces to implement the event list and the list of waiting customers. The underlying priority-queue organization is efficient, in the sense of having access costs that grow only in the logarithm of the number of elements in the list.

Example 4.2:   Single-Server Queue Simulation in Java

The grocery checkout counter, defined in detail in Example 4.1, is now simulated by using Java. A version of this example was simulated manually in Examples 3.3 and 3.4, where the system state, entities and attributes, sets, events, activities, and delays were analyzed and defined.

Class `Event` represents an event. It stores a code for the event type (`arrival` or `departure`), and the event time-stamp. It has associated methods (functions) for creating an event and accessing its data. It also has an associated method **compareTo**, which compares the event with another (passed as an argument) and reports whether the first event should be considered to be smaller, equal, or larger than the argument event. The methods for this model and the flow of control are shown in Figure 4.2, which is an adaptation of Figure 4.1 for this particular problem. Table 4.6 lists the variables used for system state, entity attributes and sets, activity durations, and cumulative and summary statistics; the functions used to generate samples from the exponential and normal distributions; and all the other methods needed.

Table 4.6 Definitions of Variables, Functions, and Methods in the Java Model of the Single-Server Queue

Variables	Description
System state	
QueueLength	Number of customers queued (but not in service) at current simulated time
NumberInService	Number being served at current simulated time
Entity attributes and sets	
Customers	FCFS queue of customers in system
Future event list	
FutureEventList	Priority-ordered list of pending events
Activity durations	
MeanInterArrivalTime	The interarrival time between the previous customer's arrival and the next arrival
MeanServiceTime	The service time of the most recent customer to begin service
Input parameters	
MeanInterarrivalTime	Mean interarrival time (4.5 minutes)
MeanServiceTime	Mean service time (3.2 minutes)
SIGMA	Standard deviation of service time (0.6 minute)
TotalCustomers	The stopping criterion—number of customers to be served (1000)
Simulation variables	
Clock	The current value of simulated time
Statistical Accumulators	
LastEventTime	Time of occurrence of the last event
TotalBusy	Total busy time of server (so far)
MaxQueueLength	Maximum length of waiting line (so far)
SumResponseTime	Sum of customer response times for all customers who have departed (so far)
NumberOfDepartures	Number of departures (so far)
LongService	Number of customers who spent 4 or more minutes at the checkout counter (so far)
Summary statistics	
RHO =BusyTime/Clock	Proportion of time server is busy (here the value of Clock is the final value of simulated time)
AVGR	Average response time (equal to SumResponseTime/TotalCustomers)
PC4	Proportion of customers who spent 4 or more minutes at the checkout counter

continues...

Table 4.6   Continued

Functions	Description
`exponential(mu)`	Function to generate samples from an exponential distribution with mean `mu`
`normal(xmu,SIGMA)`	Function to generate samples from a normal distribution with mean `xmu` and standard deviation `SIGMA`

Methods	Description
`Initialization`	Initialization method
`ProcessArrival`	Event method that executes the arrival event
`ProcessDeparture`	Event method that executes the departure event
`ReportGeneration`	Report generator

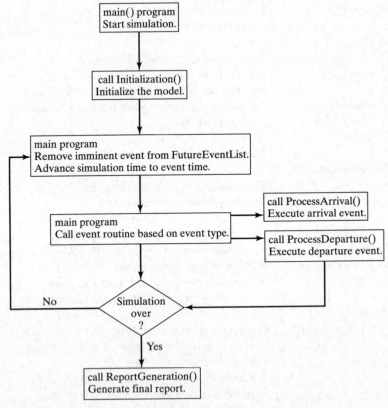

Figure 4.2 Overall structure of Java simulation of a single-server queue.

```
class Sim {

// Class Sim variables
public static double Clock, MeanInterArrivalTime, MeanServiceTime,
        SIGMA, LastEventTime, TotalBusy, MaxQueueLength, SumResponseTime;
public static long  NumberOfCustomers, QueueLength, NumberInService,
        TotalCustomers, NumberOfDepartures, LongService;

public final static int arrival = 1;
public final static int departure = 2;

public static EventList FutureEventList;
public static Queue Customers;
public static Random stream;

public static void main(String argv[]) {

  MeanInterArrivalTime = 4.5; MeanServiceTime = 3.2;
  SIGMA                = 0.6; TotalCustomers  = 1000;
  long seed            = Long.parseLong(argv[0]);
  stream = new Random(seed);             // initialize rng stream
  FutureEventList = new EventList();
  Customers = new Queue();

  Initialization();

  // Loop until first "TotalCustomers" have departed
  while(NumberOfDepartures < TotalCustomers ) {
    Event evt = (Event)FutureEventList.getMin();  // get imminent event
    FutureEventList.dequeue();                     // be rid of it
    Clock = evt.get_time();                        // advance in time
    if( evt.get_type() == arrival ) ProcessArrival(evt);
    else  ProcessDeparture(evt);
    }
  ReportGeneration();
 }
```

Figure 4.3   Java main program for the single-server queue simulation.

The entry point of the program and the location of the control logic is through class Sim, shown in Figure 4.3. Variables of classes EventList and Queue are declared. As these classes are all useful for programs other than Sim, their declarations are given in other files, per Java rules. A variable of the Java built-in class Random is also declared; instances of this class provided random-number streams. The main method controls the overall flow of the event-scheduling/time-advance algorithm.

```
public static void Initialization()    {
  Clock = 0.0;
  QueueLength = 0;
  NumberInService = 0;
  LastEventTime = 0.0;
  TotalBusy = 0 ;
  MaxQueueLength = 0;
  SumResponseTime = 0;
  NumberOfDepartures = 0;
  LongService = 0;

  // create first arrival event
  Event evt =
   new Event(arrival, exponential( stream, MeanInterArrivalTime));
  FutureEventList.enqueue( evt );
}
```

Figure 4.4   Java initialization method for the single-server queue simulation.

The main program method first gives values to variables describing model parameters; it creates instances of the random-number generator, event list, and customer queue; and then it calls method `Initialization` to initialize other variables, such as the statistics-gathering variables. Control then enters a loop which is exited only after `TotalCustomers` customers have received service. Inside the loop, a copy of the imminent event is obtained by calling the `getMin` method of the priority queue, and then that event is removed from the event list by a call to `dequeue`. The global simulation time `Clock` is set to the time-stamp contained in the imminent event, and then either `ProcessArrival` or `ProcessDeparture` is called, depending on the type of event. When the simulation is finally over, a call is made to method `ReportGeneration` to create and print out the final report.

A listing for the `Sim` class method `Initialization` is given in Figure 4.4. The simulation clock, system state, and other variables are initialized. Note that the first arrival event is created by generating a local `Event` variable whose constructor accepts the event's type and time. The event time-stamp is generated randomly by a call to `Sim` class method `exponential` and is passed to the random-number stream to use with the mean of the exponential distribution from which to sample. The event is inserted into the future event list by calling method `enqueue`. This logic assumes that the system is empty at simulated time `Clock = 0`, so that no departure can be scheduled. It is straightforward to modify the code to accommodate alternative starting conditions by adding events to `FutureEventList` and `Customers` as needed.

Figure 4.5 gives a listing of `Sim` class method `ProcessArrival`, which is called to process each arrival event. The basic logic of the arrival event for a single-server queue was given in Figure 3.5 (where LQ corresponds to `QueueLength` and LS corresponds to `NumberInService`). First, the new arrival is added to the queue `Customers` of customers in the system. Next, if the server is idle (`NumberInService == 0`) then the new customer is to go immediately into service, so `Sim` class method `ScheduleDeparture` is called to do that scheduling. An arrival to an idle queue does not update the cumulative statistics, except possibly the maximum queue length. An arrival to a

```
public static void ProcessArrival(Event evt) {
  Customers.enqueue(evt);
  QueueLength++;
  // if the server is idle, fetch the event, do statistics
  // and put into service
  if( NumberInService == 0) ScheduleDeparture();
  else TotalBusy += (Clock - LastEventTime);  // server is busy

  // adjust max queue length statistics
  if (MaxQueueLength < QueueLength) MaxQueueLength = QueueLength;

  // schedule the next arrival
  Event next_arrival =
   new Event(arrival, Clock+exponential(stream,MeanInterArrivalTime));
  FutureEventList.enqueue( next_arrival );
  LastEventTime = Clock;
}
```

Figure 4.5   Java arrival event method for the single-server queue simulation.

busy queue does *not* cause the scheduling of a departure, but does increase the total busy time by the amount of simulation time between the current event and the one immediately preceding it (because, if the server is busy now, it had to have had at least one customer in service by the end of processing the previous event). In either case, a new arrival is responsible for scheduling the next arrival, one random interarrival time into the future. An arrival event is created with simulation time equal to the current Clock value plus an exponential increment, that event is inserted into the future event list, the variable LastEventTime recording the time of the last event processed is set to the current time, and control is returned to the main method of class Sim.

Sim class method ProcessDeparture, which executes the departure event, is listed in Figure 4.6, as is method ScheduleDeparture. A flowchart for the logic of the departure event was given in Figure 3.6. After removing the event from the queue of all customers, the number in service is examined. If there are customers waiting, then the departure of the next one to enter service is scheduled. Then, cumulative statistics recording the sum of all response times, sum of busy time, number of customers who used more than 4 minutes of service time, and number of departures are updated. (Note that the maximum queue length cannot change in value when a departure occurs.) Notice that customers are removed from Customers in FIFO order; hence, the response time response of the departing customer can be computed by subtracting the arrival time of the job leaving service (obtained from the copy of the arrival event removed from the Customers queue) from the current simulation time. After the incrementing of the total number of departures and the saving of the time of this event, control is returned to the main program.

Figure 4.6 also gives the logic of method ScheduleDeparture, called by both Process-Arrival and ProcessDeparture to put the next customer into service. The Sim class method normal, which generates normally distributed service times, is called until it produces a nonnegative sample. A new event with type departure is created, with event time equal to the current simulation time plus the service time just sampled. That event is pushed onto FutureEventList, the number

```
public static void ScheduleDeparture() {
  double ServiceTime;
  // get the job at the head of the queue
  while (( ServiceTime = normal(stream,MeanServiceTime, SIGMA)) < 0 );
  Event depart = new Event(departure,Clock+ServiceTime);
  FutureEventList.enqueue( depart );
  NumberInService = 1;
  QueueLength--;
}

public static void ProcessDeparture(Event e) {
  // get the customer description
  Event finished = (Event) Customers.dequeue();
  // if there are customers in the queue then schedule
  // the departure of the next one
  if( QueueLength > 0 ) ScheduleDeparture();
  else NumberInService = 0;
  // measure the response time and add to the sum
  double response = (Clock - finished.get_time());
  SumResponseTime += response;
  if( response > 4.0 ) LongService++; // record long service
  TotalBusy += (Clock - LastEventTime );
  NumberOfDepartures++;
  LastEventTime = Clock;
}
```

Figure 4.6   Java departure event method for the single-server queue simulation.

in service is set to one, and the number waiting (QueueLength) is decremented to reflect the fact that the customer entering service is waiting no longer.

The report generator, Sim class method ReportGeneration, is listed in Figure 4.7. The summary statistics, RHO, AVGR, and PC4, are computed by the formulas in Table 4.6; then the input parameters are printed, followed by the summary statistics. It is a good idea to print the input parameters at the end of the simulation, in order to verify that their values are correct and that these values have not been inadvertently changed.

Figure 4.8 provides a listing of Sim class methods exponential and normal, used to generate random variates. Both of these functions call method nextDouble, which is defined for the built-in Java Random class generates a random number uniformly distributed on the (0,1) interval. We use Random here for simplicity of explanation; superior random-number generators can be built by hand, as described in Chapter 7. The techniques for generating exponentially and normally distributed random variates, discussed in Chapter 8, are based on first generating a $U(0,1)$ random number. For further explanation, the reader is referred to Chapters 7 and 8.

The output from the grocery-checkout-counter simulation is shown in Figure 4.9. It should be emphasized that the output statistics are estimates that contain random error. The values shown are influenced by the particular random numbers that happened to have been used, by the initial

```
public static void ReportGeneration() {
double RHO    = TotalBusy/Clock;
double AVGR   = SumResponseTime/TotalCustomers;
double PC4    = ((double)LongService)/TotalCustomers;


System.out.print( "SINGLE SERVER QUEUE SIMULATION ");
System.out.println( "- GROCERY STORE CHECKOUT COUNTER ");
System.out.println( "\tMEAN INTERARRIVAL TIME                                   "
   + MeanInterArrivalTime );
System.out.println( "\tMEAN SERVICE TIME                                        "
   + MeanServiceTime );
System.out.println( "\tSTANDARD DEVIATION OF SERVICE TIMES                      "
   + SIGMA );
System.out.println( "\tNUMBER OF CUSTOMERS SERVED                               "
   + TotalCustomers );
System.out.println();
System.out.println( "\tSERVER UTILIZATION                                       "
   + RHO );
System.out.println( "\tMAXIMUM LINE LENGTH                                      "
   + MaxQueueLength );
System.out.println( "\tAVERAGE RESPONSE TIME                                    "
   + AVGR + "  MINUTES" );
System.out.println( "\tPROPORTION WHO SPEND FOUR ");
System.out.println( "\t MINUTES OR MORE IN SYSTEM                               "
   + PC4 );
System.out.println( "\tSIMULATION RUNLENGTH                                     "
   + Clock + " MINUTES" );
System.out.println( "\tNUMBER OF DEPARTURES                                     "
   + TotalCustomers );
 }
```

Figure 4.7   Java report generator for the single-server queue simulation.

conditions at time 0, and by the run length (in this case, 1000 departures). Methods for estimating the standard error of such estimates are discussed in Chapter 11.

In some simulations, it is desired to stop the simulation after a fixed length of time, say TE = 12 hours = 720 minutes. In this case, an additional event type, stop event, is defined and is scheduled to occur by scheduling a stop event as part of simulation initialization. When the stopping event does occur, the cumulative statistics will be updated and the report generator called. The main program and method Initialization will require minor changes. Exercise 1 asks the reader to make these changes. Exercise 2 considers the additional change that any customer at the checkout counter at simulated time Clock = TE should be allowed to exit the store, but no new arrivals are allowed after time TE.

```
public static double exponential(Random rng, double mean) {
 return -mean*Math.log( rng.nextDouble() );
}

public static double SaveNormal;
public static int  NumNormals = 0;
public static final double  PI = 3.1415927 ;

public static double normal(Random rng, double mean, double sigma) {
        double ReturnNormal;
        // should we generate two normals?
        if(NumNormals == 0 ) {
          double r1 = rng.nextDouble();
          double r2 = rng.nextDouble();
          ReturnNormal = Math.sqrt(-2*Math.log(r1))*Math.cos(2*PI*r2);
          SaveNormal   = Math.sqrt(-2*Math.log(r1))*Math.sin(2*PI*r2);
          NumNormals = 1;
        } else {
          NumNormals = 0;
          ReturnNormal = SaveNormal;
        }
        return ReturnNormal*sigma + mean ;
 }
```

Figure 4.8   Random-variate generators for the single-server queue simulation.

```
SINGLE SERVER QUEUE SIMULATION - GROCERY STORE CHECKOUT COUNTER
        MEAN INTERARRIVAL TIME                  4.5
        MEAN SERVICE TIME                       3.2
        STANDARD DEVIATION OF SERVICE TIMES     0.6
        NUMBER OF CUSTOMERS SERVED              1000

        SERVER UTILIZATION                      0.671
        MAXIMUM LINE LENGTH                     9.0
        AVERAGE RESPONSE TIME                   6.375 MINUTES
        PROPORTION WHO SPEND FOUR
          MINUTES OR MORE IN SYSTEM             0.604
        SIMULATION RUNLENGTH                    4728.936 MINUTES
        NUMBER OF DEPARTURES                    1000
```

Figure 4.9   Output from the Java single-server queue simulation.

4.5 Simulation in GPSS

GPSS is a highly structured, special-purpose simulation programming language based on the process-interaction approach and oriented toward queueing systems. A block diagram provides a convenient way to describe the system being simulated. There are over 40 standard blocks in GPSS. Entities called transactions may be viewed as flowing through the block diagram. Blocks represent events, delays, and other actions that affect transaction flow. Thus, GPSS can be used to model any situation where transactions (entities, customers, units of traffic) are flowing through a system (e.g., a network of queues, with the queues preceding scarce resources). The block diagram is converted to block statements, control statements are added, and the result is a GPSS model.

The first version of GPSS was released by IBM in 1961. It was the first process-interaction simulation language and became popular; it has been implemented anew and improved by many parties since 1961, with GPSS/H being the most widely used version today. Example 4.3 is based on GPSS/H.

GPSS/H is a product of Wolverine Software Corporation, www.wolverinesoftware.com [Banks, Carson, and Sy, 1995; Henriksen, 1999]. It is a flexible, yet powerful tool for simulation. Unlike the original IBM implementation, GPSS/H includes built-in file and screen I/O, use of an arithmetic expression as a block operand, an interactive debugger, faster execution, expanded control statements, ordinary variables and arrays, a floating-point clock, built-in math functions, and built-in random-variate generators.

The animator for GPSS/H is Proof Animation™, another product of Wolverine Software Corporation [Henriksen, 1999]. Proof Animation provides a 2-D or 3-D animation, usually based on a scale drawing. It can run in postprocessed mode (after the simulation has finished running) or concurrently. In postprocessed mode, the animation is driven by two files: the layout file for the static background, and a trace file that contains commands to make objects move and produce other dynamic events. It can work with any simulation package that can write the ASCII trace file. Alternately, it can run concurrently with the simulation by sending the trace file commands as messages, or it can be controlled directly by using its DLL (dynamic link library) version.

Example 4.3:   Single-Server Queue Simulation in GPSS/H _____

Figure 4.10 exhibits the block diagram and Figure 4.11 the GPSS program for the grocery-store checkout-counter model described in Example 4.2. Note that the program (Figure 4.11) is a translation of the block diagram together with additional definition and control statements. (In the discussion that follows, all nonreserved words are shown in italics.)

In Figure 4.10, the GENERATE block represents the arrival event, with the interarrival times specified by RVEXPO(1,&*IAT*). RVEXPO stands for "random variable, exponentially distributed," the 1 indicates the random-number stream to use, and &*IAT* indicates that the mean time for the exponential distribution comes from a so-called ampervariable &*IAT*. Ampervariable names begin with the "&" character; Wolverine added ampervariables to GPSS because the original IBM implementation had limited support for ordinary global variables, with no user freedom for naming them.

The next block is a QUEUE with a queue named *SYSTIME*. It should be noted that the QUEUE block is not needed for queues or waiting lines to form in GPSS. The true purpose of the QUEUE block

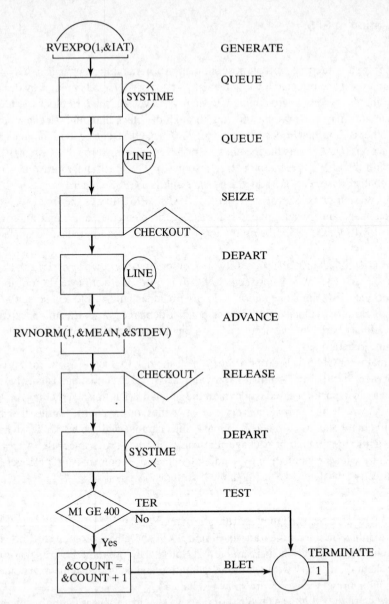

RVEXPO(1,&IAT)   GENERATE

QUEUE

SYSTIME

QUEUE

LINE

SEIZE

CHECKOUT

DEPART

LINE

ADVANCE

RVNORM(1,&MEAN,&STDEV)

CHECKOUT   RELEASE

DEPART

SYSTIME

TEST

M1 GE 400   TER

No

Yes

&COUNT = &COUNT + 1   BLET

TERMINATE

1

Figure 4.10   GPSS block diagram for the single-server queue simulation.

```
          SIMULATE
*
*         Define Ampervariables
*
          INTEGER        &LIMIT
          REAL           &IAT,&MEAN,&STDEV,&COUNT
          LET            &IAT=4.5
          LET            &MEAN=3.2
          LET            &STDEV=.6
          LET            &LIMIT=1000
*
*         Write Input Data to File
*
          PUTPIC         FILE=OUT,LINES=5,(&IAT,&MEAN,&STDEV,&LIMIT)
Mean interarrival time                      **.**   minutes
Mean service time                           **.**   minutes
Standard deviation of service time          **.**   minutes
Number of customers to be served            *****

*
*         GPSS/H Block Section
*
          GENERATE       RVEXPO(1,&IAT)   Exponential arrivals
          QUEUE          SYSTIME          Begin response time data collection
          QUEUE          LINE             Customer joins waiting line
          SEIZE          CHECKOUT         Begin checkout at cash register
          DEPART         LINE             Customer starting service leaves queue
          ADVANCE        RVNORM(1,&MEAN,&STDEV)  Customer's service time
          RELEASE        CHECKOUT         Customer leaves checkout area
          DEPART         SYSTIME          End response time data collection
          TEST GE        M1,4,TER         Is response time GE 4 minutes?
          BLET           &COUNT=&COUNT+1  If so, add 1 to counter
TER       TERMINATE      1
*
          START          &LIMIT           Simulate for required number
*
*         Write Customized Output Data to File
*
          PUTPIC         FILE=OUT,LINES=7,(FR(CHECKOUT)/1000,QM(LINE),_
QT(SYSTIME),&COUNT/N(TER),AC1,N(TER))
Server utilization                   .***
Maximum line length                  **
Average response time                **.**     minutes
Proportion who spend four minutes    .***
      or more in the system
Simulation runlength                 ****.** minutes
Number of departures                 ****
*
          END
```

Figure 4.11 GPSS/H program for the single-server queue simulation.

is to work in conjunction with the DEPART block to collect data on queues or any other subsystem. In Example 4.3, we want to measure the system response time—that is, the time a transaction spends in the system. Placing a QUEUE block at the point that transactions enter the system and placing its counterpart the DEPART block, at the point that the transactions complete their processing causes the response times to be collected automatically. The purpose of the DEPART block is to signal the end of data collection for an individual transaction. The QUEUE and DEPART block combination is not necessary for queues to be modeled, but rather is used for statistical data collection.

The next QUEUE block (with queue named *LINE*) begins data collection for the waiting line before the cashier. The customers may or may not have to wait for the cashier. Upon arrival at an idle checkout counter, or after advancing to the head of the waiting line, a customer captures the cashier, as represented by the SEIZE block with the resource named *CHECKOUT*. Once the transaction representing a customer captures the cashier represented by the resource *CHECKOUT*, the data collection for the waiting-line statistics ends, as represented by the DEPART block for the queue named *LINE*. The transaction's service time at the cashier is represented by an ADVANCE block. RVNORM indicates "random variable, normally distributed." Again, random-number stream 1 is being used, the mean time for the normal distribution is given by ampervariable *&MEAN*, and its standard deviation is given by ampervariable *&STDEV*. Next, the customer gives up the use of the facility *CHECKOUT* with a RELEASE block. The end of the data collection for response times is indicated by the DEPART block for the queue *SYSTIME*.

Next, there is a TEST block that checks to see whether the time in the system, M1, is greater than or equal to 4 minutes. (Note that M1 is a reserved word in GPSS/H; it automatically tracks transaction total time in system.) In GPSS/H, the maxim is "if true, pass through." Thus, if the customer has been in the system four minutes or longer, the next BLET block (for block LET) adds one to the counter *&COUNT*. If not true, the escape route is to the block labeled *TER*. That label appears before the TERMINATE block, whose purpose is the removal of the transaction from the system. The TERMINATE block has a value "1", indicating that one more transaction is added toward the limiting value (or "transactions to go").

The control statements in this example are all of those lines in Figure 4.11 that precede or follow the block section. (There are eleven blocks in the model from the GENERATE block to the TERMINATE block.) The control statements that begin with an "*" are comments, some of which are used for spacing purposes. The control statement SIMULATE tells GPSS/H to conduct a simulation; if it is omitted, GPSS/H compiles the model and checks for errors only. The ampervariables are defined as integer or real by control statements INTEGER and REAL. It seems that the ampervariable *&COUNT* should be defined as an integer; however, it will be divided later by a real value. If it is integer, the result of an integer divided by a real value is truncation, and that is not desired in this case. The four assignment statements (LET) provide data for the simulation. These four values could have been placed directly into the program; however, the preferred practice is to place them in ampervariables at the top of the program so that changes can be made more easily or the model can be modified to read them from a data file.

To ensure that the model data is correct, and for the purpose of managing different scenarios simulated, it is good practice to echo the input data. This is accomplished with a PUTPIC (for "put picture") control statement. The five lines following PUTPIC provide formatting information, with the asterisks being markers (called picture formatting) in which the values of the four ampervariables

```
Mean interarrival time                    4.50    minutes
Mean service time                         3.20    minutes
Standard deviation of service time        0.60    minutes
Number of customers to be served          1000

Server utilization                        0.676
Maximum line length                       7
Average response time                     6.33      minutes
Proportion who spend four minutes         0.646
     or more in the system
Simulation runlength                      4767.27 minutes
Number of departures                      1000
```

Figure 4.12   Customized GPSS/H output report for the single-server queue simulation.

replace the asterisks when PUTPIC is executed. Thus, "**.**" indicates a value that may have two digits following the decimal point and up to two before it.

The START control statement controls simulation execution. It starts the simulation, sets up a "termination-to-go" counter with initial value its operand (*&LIMIT*), and controls the length of the simulation.

After the simulation completes, a second PUTPIC control statement is used to write the desired output data to the same file *OUT*. The printed statistics are all gathered automatically by GPSS. The first output in the parenthesized list is the server utilization. FR*(CHECKOUT)*/1000 indicates that the fractional utilization of the resource *CHECKOUT* is printed. Because FR*(CHECKOUT)* is in parts per thousand, the denominator is provided to compute fractional utilization. QM*(LINE)* is the maximum value in the queue *LINE* during the simulation. QT*(SYSTIME)* is the average time in the queue *SYSTIME*. *&COUNT*/N*(TER)* is the number of customers who had a response time of four or more minutes divided by the number of customers that went through the block with label *TER*, or N*(TER)*. AC1 is the clock time, whose last value gives the length of the simulation.

The contents of the customized output file *OUT* are shown in Figure 4.12. The standard GPSS/H output file is displayed in Figure 4.13. Although much of the same data shown in the file *OUT* can be found in the standard GPSS/H output, the customized file is more compact and uses the language of the problem rather than GPSS jargon. There are many other reasons that customized output files are useful. For example, if 50 replications of the model are to be made and the lowest, highest, and average value of a response are desired, this can be accomplished by using control statements, with the results in a very compact form, rather than extracting the desired values from 50 standard output files.

4.6   Simulation in SSF

The Scalable Simulation Framework (SSF) is an Application Program Interface (API) that describes a set of capabilities for object-oriented, process-view simulation. The API is sparse and was designed to allow implementations to achieve high performance (e.g., on parallel computers). SSF APIs exist

RELATIVE CLOCK: 4767.2740   ABSOLUTE CLOCK: 4767.2740

BLOCK	CURRENT	TOTAL	BLOCK	CURRENT	TOTAL
1		1003	TER		1000
2		1003			
3	3	1003			
4		1000			
5		1000			
6		1000			
7		1000			
8		1000			
9		1000			
10		646			

FACILITY	TOTAL TIME	AVAIL TIME	UNAVL TIME	ENTRIES	AVERAGE TIME/XACT	CURRENT STATUS	PERCENT AVAIL	SEIZING XACT	PREEMPTING XACT
		--AVG-UTIL-DURING--							
CHECKOUT	0.676			1000	3.224	AVAIL			

QUEUE	MAXIMUM CONTENTS	AVERAGE CONTENTS	TOTAL ENTRIES	ZERO ENTRIES	PERCENT ZEROS	AVERAGE TIME/UNIT	$AVERAGE TIME/UNIT	QTABLE NUMBER	CURRENT CONTENTS
SYSTIME	8	1.331	1003	0		6.325	6.235		3
LINE	7	0.655	1003	334	33.3	3.111	4.665		3

RANDOM STREAM	ANTITHETIC VARIATES	INITIAL POSITION	CURRENT POSITION	SAMPLE COUNT	CHI-SQUARE UNIFORMITY
1	OFF	100000	103004	3004	0.83

Figure 4.13 Standard GPSS/H output report for the single-server queue simulation.

for both C++ and Java, and implementations exist in both languages. SSF has a wide user base—particularly in network simulation by using the add-on framework SSFNet (www.ssfnet.org). Our chapter on network simulation uses SSFNet.

The SSF API defines five base classes. `process` is a class that implements threads of control; the `action` method of a derived class contains the execution body of the thread. The `Entity` class is used to describe simulation objects. It contains state variables, processes, and communication endpoints. The `inChannel` and `outChannel` classes are communication endpoints. The `Event` class defines messages sent between entities. One model entity communicates with another by "writing" an `Event` into an `outChannel`; at some later time, it is available at one or more `inChannels`. A `process` that expects input on an `inChannel` can suspend, waiting for an event on it. These points, and others, will be elaborated upon as we work through an SSF implementation of the single-server queue.

Source code given in Figure 4.14 expresses the logic of arrival generation in SSF for the single-server queue example. The example is built on two SSF processes. One of these generates jobs and adds them to the system, the other services the enqueued jobs. Class `SSQueue` is a class that contains the whole simulation experiment. It uses the auxiliary classes `Random` (for random-number generation) and `Queue` (to implement FIFO queueing of general objects). `SSQueue` defines experimental constants ("public static final" types) and contains SSF communication endpoints `out` and `in`, through which the two processes communicate. `SSQueue` also defines an inner class `arrival`, which stores the identity and arrival time of each job.

Class `Arrivals` is an SSF process. Its constructor stores the identity of the entity that owns it, and creates a random-number generator that is initialized with the seed passed to it. For all but the initial call, method `action` generates and enqueues a new arrival, then blocks (via SSF method `waitFor`) for an inter-arrival time; on the first call, it by-passes the job-generation step and blocks for an initial interarrival time. The call to `waitFor` highlights details needing explanation. An `SSQueue` object calls the `Arrival` constructor and is saved as the "owner." This class contains an auxiliary method `exponential`, which samples an exponential random variable with specified mean by using a specified random-number stream. It also contains methods `d2t` and `t2d` that translate between a discrete "tick"-based integer clock and a double-precision floating-point representation. In the `waitFor` call, we use the same code seen earlier to sample the exponential in double-precision format, but then use `d2t` to convert it into the simulator's integer clock format. The specific conversion factor is listed as an `SSQueue` constant, 10^9 ticks per unit time.

SSF interprocess communication is used sparingly in this example. Because service is nonpreemptive, when a job's service completes, the process providing service can examine the list of waiting customers (in variable `owner.Waiting`) to see whether it needs to give service to another customer. Thus, the only time the server process needs to be told that there is a job waiting is when a job arrives to an empty system. This is reflected in `Arrivals.action` by use of its owner's `out` channel.

A last point of interest is that `Arrivals` is, in SSF terminology, a "simple" process. This means that every statement in `action` that might suspend the process would be the last statement executed under normal execution semantics. The `Arrivals` class tells SSF that it is simple by overriding a default method `isSimple` to return the value true, rather than the default value (false). The key reason for using simple processes is performance—they require that no state be saved, only the condition under which the process ought to be reanimated. And, when it is reanimated, it starts executing at the first line of `action`.

```
// SSF MODEL OF JOB ARRIVAL PROCESS
class SSQueue extends Entity {

  private static Random rng;
  public static final double MeanServiceTime = 3.2;
  public static final double SIGMA = 0.6;
  public static final double MeanInterarrivalTime = 4.5;
  public static final long ticksPerUnitTime = 1000000000;
  public long generated=0;
  public Queue Waiting;
  outChannel out;
  inChannel in;

  public static long TotalCustomers=0, MaxQueueLength=0, TotalServiceTime=0;
  public static long LongResponse=0, SumResponseTime=0, jobStart;

  class arrival {
   long id,  arrival_time;
   public arrival(long num, long a) { id=num; arrival_time = a; }
  }

class Arrivals extends process {
    private Random rng;
    private SSQueue owner;
    public Arrivals (SSQueue _owner, long seed) {
        super(_owner); owner = _owner;
        rng = new Random(seed);
    }
    public boolean isSimple() { return true; }
    public void action() {
      if ( generated++ > 0 ) {
        // put a new Customer on the queue with the present arrival time
        int Size = owner.Waiting.numElements();
        owner.Waiting.enqueue( new arrival(generated, now()));
        if( Size == 0) owner.out.write( new Event() );    // signal start of burst
      }
      waitFor(owner.d2t( owner.exponential(rng, owner.MeanInterarrivalTime)) );
   }
  }
}
```

Figure 4.14   SSF Model of Job-Arrival Process.

Figure 4.15 illustrates the code for the Server process. Like process Arrival, its constructor is called by an instance of SSQueue and is given the identity of that instance and a random-number seed. Like Arrival, it is a simple process. It maintains state variable in_service to remember the specifics of a job in service and state variable service_time to remember the value of the service time sampled for the job in service. When the SSF kernel calls action, either a job has completed service, or the Arrival process has just signaled Server though the inChannel. We distinguish the cases by looking at variable in_service, which will be nonnull if a job is in service, just now completed. In this case, some statistics are updated. After this task is done, a test is made for customers waiting for service. The first waiting customer is dequeued from the waiting list and is

```
// SSF MODEL OF SINGLE SERVER QUEUE : ACCEPTING JOBS
class Server extends process {
 private Random rng;
 private SSQueue owner ;
 private arrival in_service;
 private long service_time;

 public Server(SSQueue _owner, long seed) {
       super(_owner);
       owner = _owner;
       rng = new Random(seed);
 }
 public boolean isSimple() { return true; }

 public void action() {
    //  executes due to being idle and getting a job, or by service time expiration.
    //  if there is a job awaiting service, take it out of the queue
    //  sample a service time, do statistics, and wait for the service epoch


    // if in_service is not null, we entered because of a job completion
    if( in_service != null ) {
        owner.TotalServiceTime += service_time;
        long in_system = (now() -in_service.arrival_time);
        owner.SumResponseTime  +=  in_system;
        if( owner.t2d(in_system) > 4.0 ) owner.LongResponse++;
        in_service = null;
        if( owner.MaxQueueLength < owner.Waiting.numElements() + 1 )
             owner.MaxQueueLength = owner.Waiting.numElements() + 1;
        owner.TotalCustomers++;
    }
    if( owner.Waiting.numElements() > 0 ) {
        in_service = (arrival)owner.Waiting.dequeue();
        service_time = -1;
        while ( service_time < 0.0 )
          service_time = owner.d2t(owner.normal( rng, owner.MeanServiceTime, owner.SIGMA));
                                                    // model service time
        waitFor( service_time );
    } else {
        waitOn( owner.in );  // we await a wake-up call
    }
  }
 }
}
```

Figure 4.15   SSF Model of Single-Server Queue : Server.

copied into the in_service variable; the process then samples a service time and suspends through a waitFor call. If no customer was waiting, the process suspends on a waitOn statement until an event from the Arrival process awakens it.

SSF bridges the gap between models developed in pure Java and models developed in languages specifically designed for simulation. It provides the flexibility offered by a general-programming language, yet has essential support for simulation.

4.7   Simulation Environments

All the simulation packages described in later subsections run on a PC under Microsoft Windows 2000 or XP. Most run under Vista. Although in terms of specifics the packages all differ, generally they have many things in common.

Common characteristics include a graphical user interface, animation, and automatically collected outputs to measure system performance. In virtually all packages, simulation results may be displayed in tabular or graphical form in standard reports and interactively while running a simulation. Outputs from different scenarios can be compared graphically or in tabular form. Most provide statistical analyses that include confidence intervals for performance measures and comparisons, plus a variety of other analysis methods. Some of the statistical-analysis modules are described in Section 4.8.

All the packages described here take the process-interaction worldview. A few also allow event-scheduling models and mixed discrete-continuous models. For animation, some emphasize scale drawings in 2-D or 3-D; others emphasize iconic-type animations based on schematic drawings or process-flow diagrams. A few offer both scale drawing and schematic-type animations. Almost all offer dynamic business graphing in the form of time lines, bar charts, and pie charts.

In addition to the information contained in this chapter, the websites given below can be investigated:

AnyLogic	www.anylogic.com
Arena	www.arenasimulation.com
AutoMod	www.automod.com
Enterprise Dynamics	www.incontrol.ne
ExtendSim	www.extendsim.com
Flexsim	www.flexsim.com
ProModel	www.promodel.com
SIMUL8	www.simul8.com

4.7.1   AnyLogic

AnyLogic is a multi-method simulation software that supports discrete-event (process-centric) modeling, agent-based modeling, and system dynamics. The main idea behind AnyLogic is that the modeler can choose the method (or, perhaps, a combination of methods) and abstraction level best fitting the problem and thus capture the complexity and heterogeneity of manufacturing, logistics, business processes, human resources, or customer behavior, as well as macro level dynamics of economic and social systems, and their interactions.

AnyLogic includes a library of objects for process modeling (Source, Delay, Queue, Resource-Pool, etc.), supports space, mobility, networks, interaction, and various behavior definition forms for agents, and stock-and-flow diagrams for system dynamics models. The AnyLogic simulation engine supports hybrid (discrete plus continuous) time.

AnyLogic models (as well as the model development environment) are based on Java and thus are cross-platform. Models can be published as Java applets on the web so that remote users can execute them within their web browsers. The models have many "extension points" where the user can enter expressions and actions in Java.

The animation editor is a part of the AnyLogic model development environment. The editor supports a large variety of graphical shapes, controls (sliders, buttons, text inputs, etc.), images, GIS (Geographical Information System) map, and imported CAD drawings. Scalable and hierarchical animations can be developed. For example, the user can create a global view of a manufacturing process with some aggregated indicators as well as detailed animations of particular operations, and switch between those views.'

The input data for the model can be processed with the distribution fitting software Stat::Fit (`www.geerms.com`), which supports AnyLogic syntax for probability distributions. The simulation output can be statistically processed and visualized within the model (a variety of charts and histograms is offered), or exported to a database, text file, or a spreadsheet.

The AnyLogic experiment framework includes simple simulation, parameter variation, compare runs, sensitivity analysis, Monte Carlo, calibration, optimization, and free-form custom experiments. The OptQuest optimization engine (see Section 4.8.2) is fully integrated into AnyLogic.

AnyLogic is available in two editions, Advanced and Professional. The Professional edition allows the user to export models as standalone Java applications, embed GIS maps and CAD drawings: It integrates with version control software for better teamwork, includes a full-featured debugger, and allows the user to save the state of the model during a simulation run and restore it at a later time. It also includes Pedestrian Library for high-precision physical level modeling of areas with dense pedestrian traffic.

4.7.2   Arena

Arena Basic and Professional Editions are offered by Rockwell Automation. Arena is a general-purpose simulation software that can be used for simulating discrete and continuous systems.

The Arena Basic Edition is targeted at modeling business processes and other systems in support of high-level analysis needs. It represents process dynamics in a hierarchical flowchart and stores system information in data spreadsheets. It has built-in activity-based costing and is closely integrated with the flowcharting software Visio.

The Arena Professional Edition (PE) is designed for more detailed models of discrete and continuous systems. First released in 1993, Arena PE employs an object-based design for entirely graphical model development. Simulation models are built from graphical objects called *modules* to define system logic and such physical components as machines, operators, clerks, or doctors. Modules are represented by icons plus associated data entered in a dialog window. These icons are connected to represent entity flow. Modules are organized into collections called *templates*. The template is the core collection of modules providing general-purpose features for modeling all types of applications. In addition to standard features, such as resources, queues, process logic, and system data, the Arena templates include modules focused on specific aspects of manufacturing, material-handling, and flow process systems. Arena PE can also be used to model combined discrete/continuous systems, such as pharmaceutical and chemical production, through its built-in continuous-modeling capabilities and specialized flow template. This version also gives the user the capability to craft custom simulation objects that mirror components of the real system, including terminology, process logic, data, performance metrics, and animation.

The Arena family also includes products designed specifically to model call centers and high-speed production lines, namely Arena Contact Center and Arena Packaging Edition. These products can be used in combination with the Professional Edition.

At the heart of Arena is the SIMAN simulation language. Arena's open architecture, including embedded Visual Basic for Applications (VBA), enables data transfer with other applications as well as custom interface development. For animating simulation models, Arena's core modeling constructs are accompanied by standard graphics that show queues, resource status, and entity flow. Arena's 2-D animations are created by using Arena's built-in drawing tools and by incorporating clip art, AutoCAD, Visio, and other graphics. The Arena 3DPlayer can be used to create 3D imagery for viewing 3-D animations of Arena models.

Arena's Input Analyzer automates the process of selecting the proper distribution and its parameters for representing existing data, such as process and interarrival times. The OptQuest optimization engine (Section 4.8.2) is fully integrated into Arena. The Output Analyzer and Process Analyzer (Section 4.8.2) automate comparison of different design alternatives.

4.7.3   AutoMod

The AutoMod Product Suite is offered by Applied Materials. It includes the Automod simulation package, AutoStat for experimentation and analysis, and AutoView for making AVI movies of the 3-D animation. The main focus of the AutoMod simulation product is manufacturing and material-handling systems. AutoMod's strength is in detailed, large models used for planning, operational decision support, and control-systems testing.

AutoMod has built-in templates for most common material-handling systems, including vehicle systems, conveyors, automated storage and retrieval systems, bridge cranes, power and free conveyors, and kinematic (robotic) systems. It also includes a powerful model communication module, which allows models to communicate with control systems, and other simulation models.

AutoMod's pathmover vehicle system can be used to model lift trucks, humans walking or pushing carts, automated guided vehicles, tuggers, trains, overhead transport vehicles, monorails, trucks, and cars. All the movement templates are based on a 3-D scale drawing (drawn or imported from CAD as 2-D or 3-D). All the components of a template are highly parameterized. For example, the conveyor template contains conveyor sections, motors that power the conveyor sections, stations for load induction or removal, and photo-eyes. Sections are defined by length, width, speed, acceleration, and type (accumulating or non-accumulating), plus other specialized parameters. As product is moved by the conveyor system, the size of the product (loads), along with the attributes of the conveyor sections, are all accurately accounted for, providing tools for extremely accurate simulation results.

In addition to the material-handling templates, AutoMod contains a full simulation programming language. An AutoMod model consists of one or more systems. A system can be either a process system, in which flow and control logic are defined, or a movement system, based on one of the material-handling templates. A model may contain any number of systems (there are no size limitations), which can be saved and reused as objects in other models. Processes can contain complex logic to control the flow of either manufacturing materials (loads) or control messages. The process logic, executed by loads, can be as simple or as complex as needed to accomplish the desired simulation behavior. Its 3-D animation can be viewed from any angle or perspective in real time or at an accelerated time scale. The user can freely zoom, pan, or rotate the 3-D world of the simulation model.

In the AutoMod worldview, loads (products, parts, etc.) move from process to process and compete for resources (equipment, operators, vehicles, conveyor space, and queues). The load is the active entity, executing action statements in each process. Loads can be sent from process to process either directly, or by using a conveyor or vehicle movement system.

AutoStat, described in Section 4.8.2, works with AutoMod models to provide a complete environment for the user to define scenarios, determine warmup, conduct experimentation including design of experiments, and perform analyses. It also offers optimization based on evolution strategies.

4.7.4   Enterprise Dynamics

Enterprise Dynamics is a simulation platform, offered by Incontrol Simulation Software, Inc. Enterprise Dynamics is an object-oriented event-scheduling discrete-event simulator with its own built-in functional simulation programming language, 4DScript. Enterprise Dynamics comes with a library of Atoms, allowing fast model building in core application areas, manufacturing and material-handling systems, logistics, and transportation. A separate suite of Atoms and applications allows an airport simulator, including baggage and passenger flows.

Enterprise Dynamics has an OpenGL-based 3D visualization engine that provides 3D visualization in "virtual reality." The 3D visualization can be viewed during both simulation and modeling phases.

When factory-provided Atom libraries are not sufficient for specific modeling needs, additional aspects of the models appearance and behavior can be customized in its simulation language 4DScript, using more than 1500 available functions. The model code can be changed "on-the fly" without the need to recompile or stop the simulation. Other platform features aimed at supporting model development include a CAD drawing import wizard, a database interface (both ODBC and ADO) to enable storage of model structure and data for larger models in RDBMS (SQL Server, Oracle, MySQL, Access), XML support to enable information exchange with other systems, and ActiveX interfaces that enable the development of customized simulation solutions using Enterprise Dynamics' simulation engine invisible behind the scenes.

A relatively new addition to Enterprise Dynamics is the Emulation add-on, which allows an interface of the simulated model of the production system with the logistics software managing it. The user can then test the logistics software before the actual hardware becomes available. This emulation interface is available on various levels, starting from direct PLC communication (using OPC) for testing of PLC circuits, up to high level interfaces with logistics software.

Results can be exported to MS Office applications for further analysis if necessary. The OptQuest optimization engine (see Section 4.8.2) is fully integrated into Enterprise Dynamics.

4.7.5   ExtendSim

The ExtendSim family of products is offered by Imagine That Inc. [Krahl, 2008]. ExtendSim OR is for simulating discrete-event and mixed discrete-continuous systems. ExtendSim AT adds a discrete rate module for rate-based batch systems, and ExtendSim Suite adds 3D animation.

ExtendSim CP is for continuous modeling only. ExtendSim combines a drag-and-drop user interface with a development environment for creating custom components. Graphical objects, called blocks, represent logical as well as physical elements. Models can be built graphically by placing and connecting blocks, then entering parameters in the block's dialog. Models can also be created programmatically using the ExtendSim scripting capability.

Each ExtendSim block has an icon, user interface, animation, precompiled code, and online help. Elemental blocks include Create, Queue, Activity, Resource Pool, and Exit. The active entities, called *items* in ExtendSim, originate at Create blocks and move from block to block by way of item connectors. Separate value connectors allow the attachment of a calculation to a block parameter or the retrieval of statistical information for reporting purposes. Input parameters can be changed interactively during a model run and can come from internal or external sources. Outputs are displayed dynamically and in graphical and tabular format. Blocks are collectively organized into libraries of general-purpose elements as well as libraries for specific applications such as high-speed packaging lines and chemical production. Third-party developers have created ExtendSim libraries for vertical markets, including supply-chain dynamics, communication systems, and pulp and paper processing.

ExtendSim provides iconic process-flow animation of the block diagram. The Suite product includes 3D animation that is separate from, but integrated with, the logical simulation model; 3D animation can run concurrently or be buffered. Collections of blocks representing a submodel, such as a subassembly line or a functional process, can be grouped into a hierarchical block on the model worksheet. ExtendSim supports multiple layers of hierarchy, and hierarchical blocks can be stored in a library for reuse. Clones of critical parameters can be grouped and displayed at the worksheet level for user access to a model dashboard. Agent-based modeling, activity-based costing, and statistical analysis of output data with confidence intervals are also supported. ExtendSim includes an Evolutionary Optimizer as well as an integrated, relational database. Block parameters and data tables dynamically link with ExtendSim database tables for centralized data management.

ExtendSim supports the Microsoft component object model (COM/ActiveX), open database connectivity (ODBC), and file transfer protocol (FTP) for internet data exchange. New blocks can be created either hierarchically or by using the compiled C-based programming environment. The message-based language includes simulation-specific functions and custom interface development tools. ExtendSim has an open architecture – source code is provided for block modification and custom development. The architecture also supports linking to and using code and routines written in external languages.

4.7.6   Flexsim

Flexsim simulation software was developed and is owned by Flexsim Software Products, Inc. of Orem, Utah [Nordgren, 2003]. Flexsim is a discrete-event, object-oriented simulator developed in C++, using Open GL technology. Animation is shown in 3-D and as virtual reality. All views can be shown concurrently during the model development or run phase. Flexsim integrates Microsoft's Visual C++ IDE and compiler within a graphical 3-D click-and-drag simulation environment.

In 2007, Flexsim introduced a precompiled language to Flexsim called Flexscript that allows programming without the need to compile before the model run. Flexsim currently offers both Flexscript and C++ for modeling complex algorithms. A simulation model of any flow system or process can be created in Flexsim by using drag-and-drop model-building objects. Flexsim provides the ability to customize objects for specific needs. Users can add modified objects to their own user library which can be saved and shared within their organization. Robust defaults allow a modeler to have a model up and running quickly. Flexsim provides object libraries for discrete event, continuous, and agent-based modeling.

In 2008 Flexsim introduced discrete-event simulation (DS) software for running large models over a computer network or internet with the processing of the model distributed across any number of processors. Flexsim DS can be used to model large and complex systems using hundreds of computers linked together. Flexsim DS allows multi-user model collaboration. Teams of modelers can work together via the internet from a remote location to build a model, analyze results, and run experiments. Flexsim DS enables evolving models and collaborative design, all in a virtual reality environment.

4.7.7   ProModel

ProModel is an offering of ProModel Corporation, founded in 1988. It is a simulation and animation tool designed primarily for modeling manufacturing systems. Though ProModel can be and often is used for modeling non-manufacturing systems, the company offers similar products for healthcare simulation (MedModel) and service system simulation (ServiceModel).

ProModel is a Microsoft Gold Certified Partner and offers a plug-in to Microsoft Visio called Process Simulator for simulating process charts, value stream maps, and facility layouts for any industry. It also offers a plug-in to Microsoft Project for modeling projects of all types that may have variable task times and shared resources.

In addition to providing solutions aimed at modeling a broad range of processes and projects, ProModel offers several focused simulation products aimed at specific types of planning problems. These include Emergency Department Simulator (EDS) for determining resource requirements and analyzing patient flows in emergency departments; and Portfolio Simulator for visualizing, analyzing, and optimizing product and project portfolios. Additional products have been developed and are being successfully used in the U.S. Department of Defense for resource capacity planning.

ProModel provides both 2-D and 3-D animation as well as dynamic plots and counters that update during the simulation. The animation layout is created automatically as the model is developed, using graphics from predefined libraries or imported by the user.

Models are built using a graphical user interface and modeling constructs. ProModel's capabilities including constructs for modeling operators, fork trucks, conveyors, cranes, automatic-guided vehicle systems, and pumps and tanks. ProModel automatically tracks cost data based on material, labor, and equipment costs entered by the user.

ProModel interfaces with Microsoft Excel and other commercial database programs for developing integrated solutions. It also has a COM interface for customization and creating connectivity with other applications. ProModel supports collaborative modeling in which models can be built and maintained separately and then run in concert when desired.

In addition to generating both standard and customized output reports, ProModel exports data to Excel for more extensive analysis as well as to Minitab for automatically creating six-sigma six-pack charts and process capability diagrams. ProModel's runtime interface allows a user to define multiple scenarios for experimentation and side-by-side comparison. SimRunner, discussed in Section 4.8.2, adds the capability of performing simulation optimization. It is based on an evolutionary strategy algorithm, a variant of the genetic algorithm approach. The OptQuest optimization engine (discussed in Section 4.8.2) is fully integrated into ProModel.

4.7.8   SIMUL8

SIMUL8 is provided by SIMUL8 Corporation and was first introduced in 1995. In SIMUL8, models are created by drawing the flow of work with the computer mouse, using icons and arrows to represent the resources and queues in the system. Default values are provided for all properties of the icons so that the animation can be viewed very early in the modeling process. Drilling down in property boxes opens up progressively more detailed properties. The main focus of SIMUL8 is service industries where people are processing transactions or are themselves the transactions (for example, in a bank, call center, or hospital).

Like some other packages SIMUL8 has the concepts of "Templates" and "Components." Templates, or prebuilt simulations, focus on particular recurring decision types that can be quickly parameterized to fit a specific company issue. Components are user-defined icons that can be reused and shared across a company's simulations. This reduces time-to-build simulations, standardizes how some situation are handled across a corporation, and often removes much of the data collection phase of a simulation study.

SIMUL8 Corporation's approach is to spread simulation very widely across businesses, rather than concentrate it in the hands of dedicated and highly trained simulation professionals. This means they have very different pricing and support policies, but it also means the software has to contain features that watch how the product is being used, and provides assistance if some potentially invalid analysis is conducted.

SIMUL8 saves its simulation model and data in XML format so that it is easy to transfer it to and from other applications. It has an interface for building "front ends" for customized data entry and supports COM/ActiveX so that external applications can build and control SIMUL8 simulations.

The product has a web-based version that allows anyone with an internet browser to run models and is available in two levels, Standard and Professional. The two levels provide the same simulation features, but Professional adds 3-D, "virtual reality" views of the simulation, database links to corporate databases, an optimizer, version tracking, and other features likely only to be useful to full-time simulation modelers.

4.8   Experimentation and Statistical-Analysis Tools

4.8.1   Common Features

Virtually all simulation packages offer various degrees of support for statistical analysis of simulation outputs. In recent years, many packages have added optimization as one of the analysis tools. To support analysis, most packages provide scenario definition, run-management capabilities, and data export to spreadsheets and other external applications.

Optimization is used to find a "near-optimal" solution. The user must define an objective or fitness function, usually a cost or cost-like function that incorporates the trade-off between additional throughput and additional resources. Until recently, the methods available for optimizing a system had difficulty coping with the random and nonlinear nature of most simulation outputs. Advances in the field of metaheuristics have offered new approaches to simulation optimization, ones based on artificial intelligence, neural networks, genetic algorithms, evolutionary strategies, tabu search, and scatter search.

4.8.2   Products

This section briefly discusses Arena's Output and Process Analyzer, AutoStat for AutoMod, OptQuest (which is used in a number of simulation products), and SimRunner for ProModel.

Arena's Output and Process Analyzer

Arena comes with the Output Analyzer and Process Analyzer. In addition, Arena uses OptQuest for optimization.

The Output Analyzer provides confidence intervals, comparison of multiple systems, and warm-up determination to reduce initial condition biases. It creates various plots, charts, and histograms, smoothes responses, and does correlation analysis.

The Process Analyzer adds sophisticated scenario-management capabilities to Arena for comprehensive design of experiments. It allows a user to define scenarios, make the desired runs, and analyze the results. It allows an arbitrary number of controls and responses. Responses can be added after runs have been completed. It will rank scenarios by any response and provide summaries and statistical measures of the responses. A user can view 2-D and 3-D charts of response values across either replications or scenarios.

AutoStat

AutoStat is the run manager and statistical-analysis product in the AutoMod product family. AutoStat provides a number of analyses, including warm-up determination for steady-state analysis, absolute and comparison confidence intervals, design of experiments, sensitivity analysis, and optimization via an evolutionary strategy.

With AutoStat, an end-user can define any number of scenarios by defining factors and their range of values. Factors include model input parameters, such as resource capacity, the number of vehicles, or vehicle speed. Data files (either a specific cell or the entire file) can also be defined as input factors. By allowing a data file to be a factor, a user can experiment with, for example, alternate production schedules, customer orders for different days, different labor schedules, or any other numerical inputs typically specified in a data file. Any standard or custom output can be designated as a response. For each defined response, AutoStat computes descriptive statistics (average, standard deviation, minimum, and maximum) and confidence intervals. New responses can be defined after runs are made, because AutoStat saves and compresses the standard and custom outputs from all runs. Various charts and plots are available to provide graphical comparisons.

AutoStat helps to determine which results and alternatives are statistically significant, which means that with high probability, the observed results are not caused by random fluctuations. AutoStat supports correlated sampling (see Chapter 12) using common random numbers. This sampling technique minimizes variation between paired samples, giving a better indication of the true effects of model changes.

AutoStat is capable of distributing simulation runs across a local area network and pulling back all results to the user's computer. Support for multiple computers and CPUs gives users the ability to make many more runs of the simulation than would otherwise be possible, by using idle computers during off-hours. This is especially useful in multifactor analysis and optimization studies, both of which could require large numbers of runs. AutoStat also has a diagnostics capability that automatically detects "unusual" runs, where the definition of "unusual" is determined by the user.

AutoStat also works with AutoSched AP, a rule-based simulation package for finite-capacity scheduling in the semiconductor industry.

OptQuest

OptQuest was developed by Fred Glover, James Kelly, and Manuel Laguna of the University of Colorado, cofounders of OptTek Systems, Inc. [April et al., 2003]. OptQuest is based on a combination of methods: Scatter search, tabu search, linear/integer programming, and data mining, including neural networks. Scatter search is a population-based approach where existing solutions are combined to create new solutions. Tabu search is an adaptive memory search method that is then superimposed to drive the method away from past solutions toward better ones, and data mining additionally screens out solutions likely to be poor. The combination of methods allows the search process to escape local optimality in the quest for the best solution. Some of the features of OptTek's methods include:

- The ability to handle nonlinear and discontinuous relationships that are not specifiable by the kinds of equations and formulas used in standard mathematical programming formulations;
- The ability to handle multiple kinds of constraints (linear and nonlinear);
- The ability to handle multiple goals;
- The ability to create an efficient frontier of solutions to support decision-making analysis;
- The ability to solve problems that involve uncertain supplies, demands, prices, costs, flow rates, and queueing rates.

The core OptQuest technology has also been embedded in many simulation-based applications ranging from Project Portfolio Management to Workforce Optimization.

SimRunner

SimRunner was developed by PROMODEL Corporation out of the simulation-optimization research of Royce Bowden, Mississippi State University [Harrell et al., 2003]. It is available for ProModel, MedModel, and ServiceModel.

SimRunner uses genetic algorithms and evolution strategies, which are variants of evolutionary algorithms. Evolutionary algorithms manipulate a population of solutions to a problem in such a way that poor solutions fade away and good solutions continually evolve toward an optimum solution. These population-based direct-search techniques have the ability to avoid being trapped in locally optimal solutions. Evolutionary algorithms have proved robust in that they have been successfully used to solve a wide variety of difficult real-world problems that are nonlinear, multimodal, multidimensional, discontinuous, and stochastic.

Before invoking the evolutionary algorithms in SimRunner, a user first specifies input factors (integer or real-valued decision variables) composed of ProModel macros and then specifies an objective function composed of simulation-output responses. SimRunner manipulates the input factors within constraints specified by the user as it seeks to minimize, maximize, or achieve a user-specified target value for the objective function. The optimization-output report includes a confidence interval on the mean value of the objective function for each solution evaluated over the course of the optimization and displays 3-D plots of the simulation's output-response surface for the solutions evaluated. The solutions can be sorted by the value of the objective function (default), by the individual simulation-output responses used to define the objective function, or by any decision variable.

In addition to the multivariable optimization module, SimRunner has modules for helping users estimate the end of the warm-up phase (initialization bias) of a steady-state simulation and the number of replications needed to obtain an estimate of the objective function's mean value to within a specified percentage error and confidence level. Both modules are valuable for achieving accurate estimates of an objective function's mean value, which is necessary for successful decision making and simulation optimization.

REFERENCES

APRIL, J., F. GLOVER, J. P. KELLY, AND M. LAGUNA [2003], "Practical Introduction to Simulation Optimization," *Proceedings of the 2003 Winter Simulation Conference*, eds. S. Chick, P. J. Sánchez, D. Ferrin, and D. J. Morrice, New Orleans, LA, Dec. 7–10, pp. 71–78.

BANKS, J., J. S. CARSON, AND J. N. SY [1995], *Getting Started with GPSS/H*, 2d ed., Wolverine Software Corporation, Annandale, VA.

BANKS, J. [1996], "Interpreting Software Checklists," *OR/MS Today*, Vol. 22, No. 3, pp. 74–78.

BANKS, J. [2008], "Some burning questions about simulation," *ICS Newsletter*, Spring.

COWIE, J. [1999], "Scalable Simulation Framework API Reference Manual," www.ssfnet.org/SSFdocs /ssfapiManual.pdf.

HARRELL, C. R., B. K. GHOSH, AND R. BOWDEN [2003], *Simulation Using ProModel*, 2d ed., New York: McGraw-Hill.

HENRIKSEN, J. O. [1999], "General-Purpose Concurrent and Post-Processed Animation with Proof," *Proceedings of the 1999 Winter Simulation Conference*, eds. P. A. Farrington, H. B. Nembhard, D. T. Sturrock, G. W. Evans, Phoenix, AZ, Dec. 5–8, pp. 176–181.

KRAHL, D. [2008], "ExtendSim 7" *Proceedings of the 2008 Winter Simulation Conference*, eds. S.J. Mason, R.R. Hill, L. Mönch, O. Rose, T. Jefferson, and J.W. Fowler, Miami, FL, Dec. 7–10 pp. 215–221.

NANCE, R. E. [1995], "Simulation Programming Languages: An Abridged History," *Proceedings of the 1995 Winter Simulation Conference*, eds. C. Alexopoulos, K. Kang, W. R. Lilegdon, and D. Goldsman, Arlington, VA, Dec. 13–16, pp. 1307–1313.

NORDGREN, W. B. [2003], "Flexsim Simulation Environment," *Proceedings of the 2003 Winter Simulation Conference*, eds. S. Chick, P. J. Sánchez, D. Ferrin, and D. J. Morrice, New Orleans, LA, Dec. 7–10, pp. 197–200.

PRITSKER, A. A. B., AND C. D. PEGDEN [1979], *Introduction to Simulation and SLAM*, John Wiley, New York.

SWAIN, J. J. [2007], "New Frontiers in Simulation: Biennial Survey of Discrete-Event Simulation Software Tools," *OR/MS Today*, October, Vol.34, No. 5, pp. 23–43.

TOCHER, D. D., AND D. G. OWEN [1960], "The Automatic Programming of Simulations," *Proceedings of the Second International Conference on Operational Research*, eds. J. Banbury and J. Maitland, pp. 50–68.

WILSON, J. R., et al. [1992], "The Winter Simulation Conference: Perspectives of the Founding Fathers," *Proceedings of the 1992 Winter Simulation Conference*, eds. J. Swain, D. Goldsman, R. C. Crain, and J. R. Wilson, Arlington, VA, Dec. 13–16, pp.37–62.

EXERCISES

For the exercises below, the reader should code the model in a general-purpose language (such as C, C++, or Java), a special-purpose simulation language (such as GPSS/H), or any desired simulation environment.

Most problems contain activities that are uniformly distributed over an interval $[a, b]$. Assume that all values between a and b are possible; that is, the activity time is a *continuous* random variable. The uniform distribution is denoted by $U(a, b)$, where a and b are the endpoints of the interval, or by $m \pm h$, where m is the mean and h is the "spread" of the distribution. These four parameters are related by the equations

$$m = \frac{a+b}{2} \quad h = \frac{b-a}{2}$$

$$a = m - h \quad b = m + h$$

Some of the uniform-random-variate generators available require specification of a and b; others require m and h.

Some problems have activities that are assumed to be normally distributed, as denoted by $N(\mu, \sigma^2)$, where μ is the mean and σ^2 the variance. (Since activity times are nonnegative, the normal distribution is appropriate only if $\mu \geq k\sigma$, where k is at least 4 and preferably 5 or larger. If a negative value is generated, it is discarded.) Other problems use the exponential distribution with some rate λ or mean $1/\lambda$. Chapter 5 reviews these distributions; Chapter 8 covers the generation of random variates having these distributions. All of the languages have a facility to easily generate samples from these distributions. For C, C++, or Java simulations, the student may use the functions given in Section 4.4 for generating samples from the normal and exponential distributions.

1. Make the necessary modifications to the Java model of the checkout counter (Example 4.2) so that the simulation will run for exactly 60 hours.

2. In addition to the change in Exercise 1, also assume that any customers still at the counter at time 60 hours will be served, but no arrivals after time 60 hours are allowed. Make the necessary changes to the Java code and run the model.

3. Implement the changes in Exercises 1 and 2 in any of the simulation packages.

4. Ambulances are dispatched at a rate of one every 15 ± 10 minutes in a large metropolitan area. Fifteen percent of the calls are false alarms, which require 12 ± 2 minutes to complete. All other calls can be one of two kinds. The first kind are classified as serious. They constitute 15% of the non-false alarm calls and take 25 ± 5 minutes to complete. The remaining calls take 20 ± 10 minutes to complete. Assume that there are a very large number of available ambulances, and that any number can be on call at any time. Simulate the system until 500 calls are completed.

5. In Exercise 4, estimate the mean time that an ambulance takes to complete a call.

6. (a) In Exercise 4, suppose that there is only one ambulance available. Any calls that arrive while the ambulance is out must wait. Can one ambulance handle the work load? (b) Simulate with x ambulances, where $x = 1, 2, 3,$ or 4, and compare the alternatives on the basis of length of time a call must wait, percentage of calls that must wait, and percentage of time the ambulance is out on call.

7. There are four agents at the check-in counter for Drafty Airlines at the Dusty Plains Airport. Passengers arrive for check-in every 30 ± 30 seconds. It takes 100 ± 30 seconds to check in a passenger. Drafty Airlines is going to replace the agents with automatic passenger check-in devices (APCID). An APCID takes 120 ± 45 seconds to check in a passenger. How many APCIDs are needed to achieve an average waiting time not greater than the current waiting time with the four agents? Run for 10 hours of simulated time.

8. A superhighway connects one large metropolitan area to another. A vehicle leaves the first city every 20 ± 15 seconds. Twenty percent of the vehicles have 1 passenger, 30% of the vehicles have 2 passengers, 10% have 3 passengers, and 10% have 4 passengers. The remaining 30% of the vehicles are buses, which carry 40 people. It takes 60 ± 10 minutes for a vehicle to travel between the two metropolitan areas. How long does it take for 5000 people to arrive in the second city?

9. People arrive at a meat counter at a rate of one every 25 ± 10 seconds. There are two sections: one for beef and one for chicken. People want goods from them in the following proportion: beef only, 50%; chicken only, 30%; beef and chicken, 20%. It takes 45 ± 20 seconds for a butcher to serve one customer for one order. All customers place one order, except "beef and chicken" customers place two orders. Assume that there are enough butchers available to handle all customers present at any time. How long does it take for 200 customers to be served?

10. In Exercise 9, what is the maximum number of butchers needed during the course of simulation? Would this number always be sufficient to guarantee that no customer ever had to wait?

11. In Exercise 9, simulate with x butchers, where $x = 1, 2, 3, 4$. When all butchers are busy, a line forms. For each value of x, estimate the mean number of busy butchers.

12. A one-chair unisex hair shop has arrivals at the rate of one every 20 ± 15 minutes. One-half of the arriving customers want a dry cut, 30% want a style, and 20% want a trim only. A dry cut takes 15 ± 5 minutes, a style cut takes 25 ± 10 minutes, and a trim takes 10 ± 3 minutes. Simulate 400 customers coming through the hair shop. Compare the given proportion of service requests of each type with the simulated outcome. Are the results reasonable?

13. An airport has two concourses. Concourse 1 passengers arrive at a rate of one every 15 ± 2 seconds. Concourse 2 passengers arrive at a rate of one every 10 ± 5 seconds. It takes 30 ± 5 seconds to walk down concourse 1 and 35 ± 10 seconds to walk down concourse 2. Both concourses empty into the main lobby, adjacent to the baggage claim. It takes 10 ± 3 seconds to reach the baggage claim area from the main lobby. Only 60% of the passengers go to the

baggage claim area. Simulate the passage of 500 passengers through the airport system. How many of these passengers went through the baggage claim area? In this problem, the expected number through the baggage claim area can be computed by $0.60(500) = 300$. How close is the simulation estimate to the expected number? Why the difference if a difference exists?

14. In a multiphasic screening clinic, patients arrive at a rate of one every 5 ± 2 minutes to enter the audiology section. The examination takes 3 ± 1 minutes. Eighty percent of the patients were passed on to the next test with no problems. Of the remaining 20%, one-half require simple procedures that take 2 ± 1 minutes and are then sent for reexamination with the same probability of failure. The other half are sent home with medication. Simulate the system to estimate how long it takes to screen and pass 200 patients. (*Note:* Patients sent home with medication are not considered "passed.")

15. Consider a bank with four tellers. tellers 3 and 4 deal only with business accounts; Tellers 1 and 2 deal only with general accounts. Clients arrive at the bank at a rate of one every 3 ± 1 minutes. Of the clients, 33% are business accounts. Clients randomly choose between the two tellers available for each type of account. (Assume that a customer chooses a line without regard to its length and does not change lines.) Business accounts take 15 ± 10 minutes to complete, and general accounts take 6 ± 5 minutes to complete. Simulate the system for 500 transactions to be completed. What percentage of time is each type of teller busy? What is the average time that each type of customer spends in the bank?

16. Repeat Exercise 15, but assume that customers join the shortest line for the teller handling their type of account.

17. In Exercises 15 and 16, estimate the mean delay of business customers and of general customers. (Delay is time spent in the waiting line, and is exclusive of service time.) Also estimate the mean length of the waiting line, and the mean proportion of customers who are delayed longer than 1 minute.

18. Three different machines are available for machining a special type of part for 1 hour of each day. The processing-time data is as follows:

Machine	Time to Machine One Part
1	20 ± 4 seconds
2	10 ± 3 seconds
3	15 ± 5 seconds

Assume that parts arrive by conveyor at a rate of one every 15 ± 5 seconds for the first 3 hours of the day. Machine 1 is available for the first hour, machine 2 for the second hour, and machine 3 for the third hour of each day. How large a storage area is needed for parts waiting for a machine? Do parts "pile up" at any particular time? Why?

19. People arrive at a self-service cafeteria at the rate of one every 30 ± 20 seconds. Forty percent go to the sandwich counter, where one worker makes a sandwich in 60 ± 30 seconds. The rest go to the main counter, where one server spoons the prepared meal onto a plate in 45 ± 30 seconds. All customers must pay a single cashier, which takes 25 ± 10 seconds. For all customers, eating takes 20 ± 10 minutes. After eating, 10% of the people go back for dessert, spending an additional 10 ± 2 minutes altogether in the cafeteria. Simulate until 100 people have left the cafeteria. How many people are left in the cafeteria, and what are they doing, at the time the simulation stops?

20. Thirty trucks carrying bits and pieces of a C-5N cargo plane leave Atlanta at the same time for the Port of Savannah. From past experience, it is known that it takes 6 ± 2 hours for a truck to make the trip. Forty percent of the drivers stop for coffee, which takes an additional 15 ± 5 minutes. (a) Model the situation as follows: For each driver, there is a 40% chance of stopping for coffee. When will the last truck reach Savannah? (b) Model it so that exactly 40% of the drivers stop for coffee. When will the last truck reach Savannah?

21. Customers arrive at the Last National Bank every 40 ± 35 seconds. Currently, the customers pick one of two tellers at random. A teller services a customer in 75 ± 25 seconds. Once a customer joins a line, the customer stays in that line until the transaction is complete. Some customers want the bank to change to the single-line method used by the Lotta Trust Bank. Which is the faster method for the customer? Simulate for one hour before collecting any summary statistics, then simulate for an 8-hour period. Compare the two queueing disciplines on the basis of teller utilization (percentage of busy time), mean delay of customers, and proportion of customers who must wait (before service begins) more than 1 minute, and more than 3 minutes.

22. Loana Tool Company rents chain saws. Customers arrive to rent chain saws at the rate of one every 30 ± 30 minutes. Dave and Betty handle these customers. Dave can rent a chain saw in 14 ± 4 minutes. Betty takes 10 ± 5 minutes. Customers returning chain saws arrive according to the same distribution as those renting chain saws. Dave and Betty spend 2 minutes with a customer to check in the returned chain saw. Service is first-come-first-served. When no customers are present, or Betty alone is busy, Dave gets these returned saws ready for re-renting. For each saw, this maintenance and cleanup takes him 6 ± 4 minutes and 10 ± 6 minutes, respectively. Whenever Dave is idle, he begins the next maintenance or cleanup. Upon finishing a maintenance or cleanup, Dave begins serving customers if one or more is waiting. Betty is always available for serving customers. Simulate the operation of the system starting with an empty shop at 8:00 A.M., closing the doors at 6:00 P.M., and getting chain saws ready for re-renting until 7:00 P.M. From 6:00 until 7:00 P.M., both Dave and Betty do maintenance and cleanup. Estimate the mean delay of customers who are renting chain saws.

23. In Exercise 22, change the shop rule regarding maintenance and cleanup to get a chain saw ready for re-rental. Now Betty does all this work. Upon finishing a cleanup on a saw, she helps Dave if a line is present. (That is, Dave and Betty both serve new customers and check in returned saws until Dave alone is busy, or the shop is empty.) Then she returns to her maintenance and cleanup duties. (a) Estimate the mean delay of customers who are renting chain saws. Compare the two shop rules on this basis. (b) Estimate the proportion of customers who must wait more

than 5 minutes. Compare the two shop rules on this basis. (c) Discuss the pros and cons of the two criteria in parts (a) and (b) for comparing the two shop rules. Suggest other criteria.

24. U of LA (University of Lower Altoona) has only one color printer for student use. Students arrive at the color printer every 15 ± 10 minutes to use it for 12 ± 6 minutes. If the color printer is busy, 60% will come back in 10 minutes to try again. If the color printer is still busy, 50% (of the 60%) will return in 15 minutes. How many students fail to use the color printer compared to 500 that actually finish? Demand and service occur 24 hours a day.

25. A warehouse holds 1000 cubic meters of cartons. These cartons come in three sizes: little (1 cubic meter), medium (2 cubic meters), and large (3 cubic meters). The cartons arrive at the following rates: little, every 10 ± 10 minutes; medium, every 15 minutes; and large, every 8 ± 8 minutes. If no cartons are removed, use simulation to determine how long will it take to fill an empty warehouse?

26. Go Ape! buys a Banana II computer to handle all of its web-browsing needs. Web-browsing employees arrive every 10 ± 10 minutes to use the computer. Web-browsing takes 7 ± 7 minutes. The monkeys that run the computer cause a system failure every 60 ± 60 minutes. The failure lasts for 8 ± 4 minutes. When a failure occurs, the web-browsing that was being done resumes processing from where it left off. Simulate the operation of this system for 24 hours. Estimate the mean system response time. (A system response time is the length of time from arrival until web-browsing is completed.) Also estimate the mean delay for those web-browsing employees who are in service when a computer system failure occurs.

27. Able, Baker, and Charlie are three carhops at the Sonic Drive-In (service at the speed of sound!). Cars arrive every 5 ± 5 minutes. The carhops service customers at the rate of one every 10 ± 6 minutes. However, the customers prefer Able over Baker, and Baker over Charlie. If the carhop of choice is busy, the customers choose the first available carhop. Simulate the system for 1000 service completions. Estimate Able's, Baker's, and Charlie's utilization (percentage of time busy).

28. Jiffy Car Wash is a five-stage operation that takes 2 ± 1 minutes for each stage. There is room for 6 cars to wait to begin the car wash. The car wash facility holds 5 cars, which move through the system in order, one car not being able to move until the car ahead of it does. Cars arrive every 2.5 ± 2 minutes for a wash. If the car cannot get into the system, it drives across the street to Speedy Car Wash. Estimate the balking rate per hour. That is, how many cars drive off per hour? Simulate for one 12-hour day.

29. Workers come to an internal supply store at the rate of one every 10 ± 4 minutes. Their requisitions are handled by one of three clerks; a clerk takes 22 ± 10 minutes to handle a requisition. All requisitions are then passed to a single cashier, who spends 7 ± 6 minutes per requisition. Simulate the system for 120 hours. (a) Estimate the utilization of each clerk, based on the 120-hour simulation. (b) How many workers are completely served? How many do the three clerks serve? How many workers arrive? Are all three clerks ever busy at the same time? What is the average number of busy clerks?

30. People arrive at a barbershop at the rate of one every 4.5 minutes. If the shop is full (it can hold five people altogether), 30% of the potential customers leave and come back in 60 ± 20 minutes. The others leave and do not return. One barber gives a haircut in 8 ± 2 minutes; the second talks a lot and takes 12 ± 4 minutes. If both barbers are idle, a customer prefers the first barber. (Treat customers trying to reenter the shop as if they are new customers.) Simulate this system until 300 customers have received a haircut. (a) Estimate the balking rate, that is, the number turned away per minute. (b) Estimate the number turned away per minute who do not try again. (c) What is the average time spent in the shop? (d) What is the average time spent getting a haircut (not including delay)? (e) What is the average number of customers in the shop?

31. People arrive at a microscope exhibit at a rate of one every 8 ± 2 minutes. Only one person can see the exhibit at a time. It takes 5 ± 2 minutes to see the exhibit. A person can buy a "privilege" ticket for $1 which gives him or her priority in line over those who are too cheap to spend the buck. Some 50% of the viewers are willing to do this, but they make their decision to do so only if one or more people are in line when they arrive. The exhibit is open continuously from 10:00 A.M. to 4:00 P.M. Simulate the operation of the system for one complete day. How much money is generated from the sale of privilege tickets?

32. Two machines are available for drilling parts (A-type and B-type). A-type parts arrive at a rate of one every 10 ± 3 minutes, B-type parts at a rate of one every 3 ± 2 minutes. For B-type parts, workers choose an idle machine, or if both drills, the Dewey and the Truman, are busy, they choose a machine at random and stay with their choice. A-type parts must be drilled as soon as possible; therefore, if a machine is available, preferably the Dewey, it is used; otherwise the part goes to the head of the line for the Dewey drill. All jobs take 4 ± 3 minutes to complete. Simulate the completion of 100 A-type parts. Estimate the mean number of A-type parts waiting to be drilled.

33. A computer center has two color printers. Students arrive at a rate of one every 8 ± 2 minutes to use the color printer. They can be interrupted by professors, who arrive at a rate of one every 12 ± 2 minutes. There is one systems analyst who can interrupt anyone, but students are interrupted before professors. The systems analyst spends 6 ± 4 minutes on the color printer and then returns in 20 ± 5 minutes. Professors and students spend 4 ± 2 minutes on the color printer. If a person is interrupted, that person joins the head of the queue and resumes service as soon as possible. Simulate for 50 professor-or-analyst jobs. Estimate the interruption rate per hour, and the mean length of the waiting line of students.

34. Parts are machined on a drill press. They arrive at a rate of one every 5 ± 3 minutes, and it takes 3 ± 2 minutes to machine them. Every 60 ± 60 minutes, a rush job arrives, which takes 12 ± 3 minutes to complete. The rush job interrupts any nonrush job. When the regular job returns to the machine, it stays only for its remaining process time. Simulate the machining of 10 rush jobs. Estimate the mean system response time for each type of part. (A response time is the total time that a part spends in the system.)

35. E-mail messages are generated at a rate of one every 35 ± 10 seconds for transmission one at a time. Transmission takes 20 ± 5 seconds. At intervals of 6 ± 3 minutes, urgent e-mail messages with transmission times of 10 ± 3 seconds take over the transmission line. Any e-mail message in progress must be reprocessed for 2 minutes before it can be resubmitted for transmission. When resubmitted, it goes to the head of the line. Simulate for 90 minutes. Estimate the percentage of time the line is busy with ordinary e-mail messages.

36. A worker packs boxes that arrive at a rate of one every 15 ± 3 minutes. It takes 10 ± 3 minutes to pack a box. Once every hour, the worker is interrupted to wrap specialty orders that take 16 ± 3 minutes to pack. Upon completing the specialty order, the worker then completes the interrupted order. Simulate for 40 hours. Estimate the mean proportion of time that the number of boxes waiting to be packed is more than five and the average number of boxes waiting to be packed.

37. A patient arrives at the Emergency Room at Hello Hospital about 40 ± 19 minutes. Each patient will be treated by either Doctor Slipup or Doctor Gutcut. Twenty percent of the patients are classified as NIA (need immediate attention) and the rest as CW (can wait). NIA patients are given the highest priority (3), see a doctor as soon as possible for 40 ± 37 minutes, but then their priority is reduced to 2 and they wait until a doctor is free again, when they receive further treatment for 30 ± 25 minutes and are then discharged. CW patients initially receive the priority 1 and are treated (when their turn comes) for 15 ± 14 minutes; their priority is then increased to 2, they wait again until a doctor is free and receive 10 ± 8 minutes of final treatment, and are then discharged. Simulate for 20 days of continuous operation, 24 hours per day. Precede this by a 2-day initialization period to load the system with patients. Report conditions at times 0 days, 2 days, and 22 days. Does a 2-day initialization appear long enough to load the system to a level reasonably close to steady-state conditions? (a) Measure the average and maximum queue length of NIA patients from arrival to first seeing a doctor. What percent do not have to wait at all? Also tabulate and plot the distribution of this initial waiting time for NIA patients. What percent wait less than 5 minutes before seeing a doctor? (b) Tabulate and plot the distribution of total time in system for all patients. Estimate the 90% quantile—that is, 90% of the patients spend less than x amount of time in the system. Estimate x. (c) Tabulate and plot the distribution of remaining time in system from after the first treatment to discharge, for all patients. Estimate the 90% quantile. (*Note:* Most simulation packages provide the facility to automatically tabulate the distribution of any specified variable.)

38. People arrive at a newspaper stand with an interarrival time that is exponentially distributed with a mean of 0.5 minute. Fifty-five percent of the people buy just the morning paper, 25% buy the morning paper and a *Wall Street Journal*. The remainder buy only the *Wall Street Journal*. One clerk handles the *Wall Street Journal* sales, another clerk morning-paper sales. A person buying both goes to the *Wall Street Journal* clerk. The time it takes to serve a customer is normally distributed with a mean of 40 seconds and a standard deviation of 4 seconds for all transactions. Collect statistics on queues for each type of transaction. Suggest ways for making the system more efficient. Simulate for 4 hours.

39. Bernie remodels houses and makes room additions. The time it takes to finish a job is normally distributed with a mean of 17 elapsed days and a standard deviation of 3 days. Homeowners sign contracts for jobs at exponentially distributed intervals having a mean of 20 days. Bernie has only one crew. Estimate the mean waiting time (from signing the contract until work begins) for those jobs where a wait occurs. Also estimate the percentage of time the crew is idle. Simulate until 100 jobs have been completed.

40. Parts arrive at a machine in random fashion with exponential interarrival times having a mean of 60 seconds. All parts require 5 seconds to prepare and align for machining. There are three different types of parts, in the proportions shown below. The times to machine each type of part are normally distributed with mean and standard deviation as follows:

Part Type	Percent	Mean (seconds)	σ (seconds)
1	50	48	8
2	30	55	9
3	20	85	12

Find the distribution of total time to complete processing for all types of parts. What proportion of parts take more than 60 seconds for complete processing? How long do parts have to wait, on average? Simulate for one 8-hour day.

41. Shopping times at a department store have been found to have the following distribution:

Shopping Time (minutes)	Number of Shoppers
0–10	90
10–20	120
20–30	270
30–40	145
40–50	88
50–60	28

After shopping, the customers choose one of six checkout counters. Checkout times are normally distributed with a mean of 5.1 minutes and a standard deviation of 0.7 minutes. Interarrival times are exponentially distributed with a mean of 1 minute. Gather statistics for each checkout counter (including the time waiting for checkout). Tabulate the distribution of time to complete shopping and the distribution of time to complete shopping and checkout procedures. What proportion of customers spend more than 45 minutes in the store? Simulate for one 16-hour day.

42. The interarrival time for parts needing processing is given as follows:

Interarrival Time (seconds)	Proportion
10–20	0.20
20–30	0.30
30–40	0.50

There are three types of parts: A, B, and C. The proportion of each part, and the mean and standard deviation of the normally distributed processing times are as follows:

Part Type	Proportion	Mean	Standard Deviation
A	0.5	30 seconds	3 seconds
B	0.3	40 seconds	4 seconds
C	0.2	50 seconds	7 seconds

Each machine processes any type of part, one part at a time. Use simulation to compare one machine with two machines working in parallel, and two machines with three machines working in parallel. What criteria would be appropriate for such a comparison?

43. Orders are received for one of four types of parts. The interarrival time between orders is exponentially distributed with a mean of 10 minutes. The table that follows shows the proportion of the parts by type and the time needed to fill each type of order by the single clerk.

Part Type	Percentage	Service Time (minutes)
A	40	N(6.1, 1.3)
B	30	N(9.1, 2.9)
C	20	N(11.8, 4.1)
D	10	N(15.1, 4.5)

Orders of types A and B are picked up immediately after they are filled, but orders of types C and D must wait 10 ± 5 minutes to be picked up. Tabulate the distribution of time to complete delivery for all orders combined. What proportion take less than 15 minutes? What proportion take less than 25 minutes? Simulate for an 8-hour initialization period, followed by a 40-hour run. Do not use any data collected in the 8-hour initialization period.

44. Three independent widget-producing machines all require the same type of vital part, which needs frequent maintenance. To increase production it is decided to keep two spare parts on

hand (for a total of $2 + 3 = 5$ parts). After 2 hours of use, the part is removed from the machine and taken to a single technician, who can do the required maintenance in 30 ± 20 minutes. After maintenance, the part is placed in the pool of spare parts, to be put into the first machine that requires it. The technician has other duties, namely, repairing other items which have a higher priority and which arrive every 60 ± 20 minutes requiring 15 ± 15 minutes to repair. Also, the technician takes a 15-minute break in each 2-hour time period. That is, the technician works 1 hour 45 minutes, takes off 15 minutes, works 1 hour 45 minutes, takes off 15 minutes, and so on. (a) What are the model's initial conditions—that is, where are the parts at time 0 and what is their condition? Are these conditions typical of "steady state"? (b) Make each replication of this experiment consist of an 8-hour initialization phase followed by a 40-hour data-collection phase. Make four statistically independent replications of the experiment all in one computer run (i.e., make four runs with each using a different set of random numbers). (c) Estimate the mean number of busy machines and the proportion of time the technician is busy. (d) Parts are estimated to cost the company $50 per part per 8-hour day (regardless of how much they are in use). The cost of the technician is $20 per hour. A working machine produces widgets worth $100 for each hour of production. Develop an expression to represent total cost per hour which can be attributed to widget production (i.e., not all of the technician's time is due to widget production). Evaluate this expression, given the results of the simulation.

45. The Wee Willy Widget Shop overhauls and repairs all types of widgets. The shop consists of five work stations, and the flow of jobs through the shop is as depicted here:

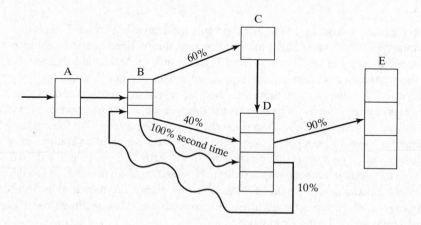

Regular jobs arrive at station A at the rate of one every 15 ± 13 minutes. Rush jobs arrive every 4 ± 3 hours and are given a higher priority, except at station C, where they are put on a conveyor and sent through a cleaning and degreasing operation along with all other jobs. For jobs the first time through a station, processing and repair times are as follows:

Station	Number of Machines or Workers	Processing and/or Repair Times (minutes)	Description
A	1	12 ± 2	Receiving clerk
B	3	40 ± 20	Disassembly and parts replacement
C	1	20	Degreaser
D	4	50 ± 40	Reassembly and adjustments
E	3	40 ± 5	Packing and shipping

The times listed above hold for all jobs that follow one of the two sequences A $\longrightarrow$ B $\longrightarrow$ C $\longrightarrow$ D $\longrightarrow$ E or A $\longrightarrow$ B $\longrightarrow$ D $\longrightarrow$ E. However, about 10% of the jobs coming out of station D are sent back to B for further work (which takes 30 ± 10 minutes) and then are sent to D and finally to E. The path of these jobs is as follows:

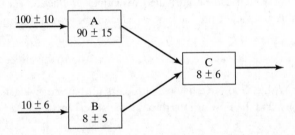

Every 2 hours, beginning 1 hour after opening, the degreasing station C shuts down for routine maintenance, which takes 10 ± 1 minute. However, this routine maintenance does not begin until the current widget, if any, has completed its processing. (a) Make three independent replications of the simulation model, where one replication equals an 8-hour simulation run, preceded by a 2-hour initialization run. The three sets of output represent three typical days. The main performance measure of interest is mean response time per job, where a response time is the total time a job spends in the shop. The shop is never empty in the morning, but the model will be empty without the initialization phase. So run the model for a 2-hour initialization period and collect statistics from time 2 hours to time 10 hours. This "warm-up" period will reduce the downward bias in the estimate of mean response time. Note that the 2-hour warm-up is a device to load a simulation model to some more realistic level than empty. From each of the three independent replications, obtain an estimate of mean response time. Also obtain an overall estimate, the sample average of the three estimates.

(b) Management is considering putting one additional worker at the busiest station (A, B, D, or E). Would this significantly improve mean response time? (c) As an alternative to part (b), management is considering replacing machine C with a faster one that processes a widget in only 14 minutes. Would this significantly improve mean response time?

46. A building-materials firm loads trucks with two payloader tractors. The distribution of truck-loading times has been found to be exponential, with a mean loading time of 6 minutes. The truck interarrival time is exponentially distributed with an arrival rate of 16 per hour. The waiting time of a truck and driver is estimated to cost $50 per hour. How much (if any) could the firm save (per 10-hour day) if an overhead hopper system that would fill any truck in a constant time of 2 minutes is installed? (Assume that the present tractors could and would adequately service the conveyors loading the hoppers.)

47. A milling-machine department has 10 machines. The runtime until failure occurs on a machine is exponentially distributed with a mean of 20 hours. Repair times are uniformly distributed between 3 and 7 hours. Select an appropriate run length and appropriate initial conditions. (a) How many repair people are needed to ensure that the mean number of machines running is greater than eight? (b) If there are two repair people, estimate the number of machines that are either running or being serviced.

48. Forty people are waiting to pass through a turnstile that takes 2.5 ± 1.0 seconds to traverse. Simulate this system 10 times, each one independent of the others, and determine the range and the average time for all 40 people to traverse.

49. People borrow *Gone with the Wind* from the local library and keep it for 21 ± 10 days. There is only one copy of the book in the library. You are the sixth person on the reservation list (five are ahead of you). Simulate 50 independent cycles of book borrowing to determine the probability that you will receive the book within 100 days.

50. Jobs arrive every 300 ± 30 seconds to be handled by a process that consists of four operations: OP10 requires 50 ± 20 seconds, OP20 requires 70 ± 25 seconds, OP30 requires 60 ± 15 seconds, OP40 requires 90 ± 30 seconds. Simulate this process until 250 jobs are completed; then combine the four operations of the job into one with the distribution 240 ± 100 seconds and simulate the process with this distribution. Does the average time in the system change for the two alternatives?

51. Two types of jobs arrive to be processed on the same machine. Type 1 jobs arrive every 50 ± 30 seconds and require 35 ± 20 seconds for processing. Type 2 jobs arrive every 100 ± 40 seconds and require 20 ± 15 seconds for processing. For an 8-hour simulation, what is the average number of jobs waiting to be processed?

52. Two types of jobs arrive to be processed on the same machine. Type 1 jobs arrive every 80 ± 30 seconds and require 35 ± 20 seconds for processing. Type 2 jobs arrive every 100 ± 40 seconds and require 20 ± 15 seconds for processing. Engineering has judged that there is excess capacity on the machine. For a simulation of 8 hours of operation of the system, find X for Type 3 jobs that arrive every $X \pm 0.4X$ seconds and require a time of 30 seconds on the machine so that the average number of jobs waiting to be processed is two or less.

53. Using spreadsheet software, generate 1000 uniformly distributed random values with mean 10 and spread 2. Plot these values with intervals of width 0.5 between 8 and 12. How close did the simulated set of values come to the expected number in each interval?

54. Using a spreadsheet, generate 1000 exponentially distributed random values with a mean of 10. What is the maximum of the simulated values? What fraction of the generated values is less than the mean of 10? Plot a histogram of the generated values. [*Hint*: If you cannot find an exponential generator in the spreadsheet you use, use the formula $-10*LOG(1 - R)$, where R is a uniformly distributed random number from 0 to 1 and LOG is the natural logarithm. The rationale for this formula is explained in Chapter 8 on random-variate generators.]

55. There are many "boids" simulations on the internet. Search the internet for two examples of a boids simulation. For each one, find a description of the rules that each boid follows. Compare and contrast the two sets of rules. At a minimum, answer the following questions regarding each model. (a) Are the boids flying in a 2-D or 3-D space? Are they flying in a world with boundary, or without boundary? [*Hint*: Do they turn at edges or walls? Or do they re-appear on the other side? If so, this implies that two apparent boundaries are glued together and are not really boundaries at all.] Describe the world or universe in which the boids are flying. (b) Does the model have user-adjustable parameters? Such parameters may include the number of boids, the closeness factor (how close one boid must be to other boids before it's affected by the other's behavior), speed, and so forth. Make a list of all such parameters. (c) How long does it take for flocking behavior to emerge?

Part II

Mathematical and Statistical Models

5

Statistical Models in Simulation

In modeling real-world phenomena, there are few situations where the actions of the entities within the system under study can be predicted completely. The world the model builder sees is probabilistic rather than deterministic. There are many causes of variation. The time it takes a repairperson to fix a broken machine is a function of the complexity of the breakdown, whether the repairperson brought the proper replacement parts and tools to the site, whether another repairperson asks for assistance during the course of the repair, whether the machine operator receives a lesson in preventive maintenance, and so on. To the model builder, these variations appear to occur by chance and cannot be predicted. However, some statistical model might well describe the time to make a repair.

An appropriate model can be developed by sampling the phenomenon of interest. Then, through educated guesses (or using software for the purpose), the model builder would select a known distribution form, make an estimate of the parameter(s) of this distribution, and then test to see how good a fit has been obtained. Through continued efforts in the selection of an appropriate distribution form, a postulated model could be accepted. This multistep process is described in Chapter 9.

Section 5.1 contains a review of probability terminology and concepts. Some typical applications of statistical models, or distribution forms, are given in Section 5.2. Then, a number of selected discrete and continuous distributions are discussed in Sections 5.3 and 5.4. The selected distributions are those that describe a wide variety of probabilistic events and, further, appear in different contexts in other chapters of this text. Additional discussion about the distribution forms appearing in this chapter, and about distribution forms mentioned but not described, is available from a number of sources [Devore, 1999; Hines and Montgomery, 1990; Law and Kelton, 2000; Papoulis, 1990; Ross, 2002; Walpole and Myers, 2002]. Section 5.5 describes the Poisson process and its relationship to the exponential distribution. Section 5.6 discusses empirical distributions.

5.1   Review of Terminology and Concepts

5.1.1   Discrete random variables

Let X be a random variable. If the number of possible values of X is finite, or countably infinite, X is called a discrete random variable. The possible values of X may be listed as $x_1, x_2, \ldots$. In the finite case, the list terminates; in the countably infinite case, the list continues indefinitely.

Example 5.1 _____

The number of jobs arriving each week at a job shop is observed. The random variable of interest is X, where

$$X = \text{number of jobs arriving each week}$$

The possible values of X are given by the range space of X, which is denoted by R_X. Here $R_X = \{0, 1, 2, \ldots\}$.

Let X be a discrete random variable. With each possible outcome x_i in R_X, a number $p(x_i) = P(X = x_i)$ gives the probability that the random variable equals the value of x_i. The numbers $p(x_i), i = 1, 2, \ldots$, must satisfy the following two conditions:

    a. $p(x_i) \geq 0$, for all i
    b. $\sum_{i=1}^{\infty} p(x_i) = 1$

The collection of pairs $(x_i, p(x_i))$, $i = 1, 2, \ldots$ is called the probability distribution of X, and $p(x_i)$ is called the probability mass function (pmf) of X.

Example 5.2 _____

Consider the experiment of tossing a single die. Define X as the number of spots on the up face of the die after a toss. Then $R_X = \{1, 2, 3, 4, 5, 6\}$. Assume the die is loaded so that the probability that a given face lands up is proportional to the number of spots showing. The discrete probability distribution for this random experiment is given by

x_i	1	2	3	4	5	6
$p(x_i)$	1/21	2/21	3/21	4/21	5/21	6/21

The conditions stated earlier are satisfied—that is, $p(x_i) \geq 0$ for $i = 1, 2, \ldots, 6$ and $\sum_{i=1}^{\infty} p(x_i) = 1/21 + \cdots + 6/21 = 1$. The distribution is shown graphically in Figure 5.1.

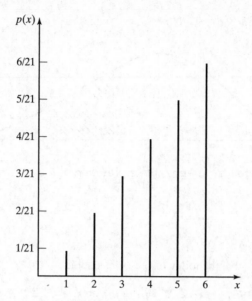

Figure 5.1   Probability mass function for loaded-die example.

5.1.2   Continuous random variables

If the range space R_X of the random variable X is an interval or a collection of intervals, X is called a continuous random variable. For a continuous random variable X, the probability that X lies in the interval $[a, b]$ is given by

$$P(a \leq X \leq b) = \int_a^b f(x)\, dx \tag{5.1}$$

The function $f(x)$ is called the probability density function (pdf) of the random variable X. The pdf satisfies the following conditions:

    a. $f(x) \geq 0$ for all x in R_X
    b. $\int_{R_X} f(x)dx = 1$
    c. $f(x) = 0$ if x is not in R_X

As a result of Equation (5.1), for any specified value x_0, $P(X = x_0) = 0$, because

$$\int_{x_0}^{x_0} f(x)\, dx = 0$$

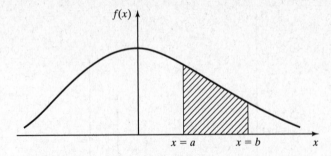

Figure 5.2   Graphical interpretation of $P(a < X < b)$.

$P(X = x_0) = 0$ also means that the following equations hold:

$$P(a \le X \le b) = P(a < X \le b) = P(a \le X < b) = P(a < X < b) \qquad (5.2)$$

The graphical interpretation of Equation (5.1) is shown in Figure 5.2. The shaded area represents the probability that X lies in the interval $[a, b]$.

Example 5.3
The life of a device used to inspect cracks in aircraft wings is given by X, a continuous random variable assuming all values in the range $x \ge 0$. The pdf of the device's lifetime, in years, is as follows:

$$f(x) = \begin{cases} \frac{1}{2}e^{-x/2}, & x \ge 0 \\ 0, & \text{otherwise} \end{cases}$$

This pdf is shown graphically in Figure 5.3. The random variable X is said to have an exponential distribution with mean 2 years.

The probability that the life of the device is between 2 and 3 years is calculated as

$$P(2 \le X \le 3) = \frac{1}{2} \int_2^3 e^{-x/2} \, dx$$

$$= -e^{-3/2} + e^{-1}$$

$$= -0.223 + 0.368$$

$$= 0.145$$

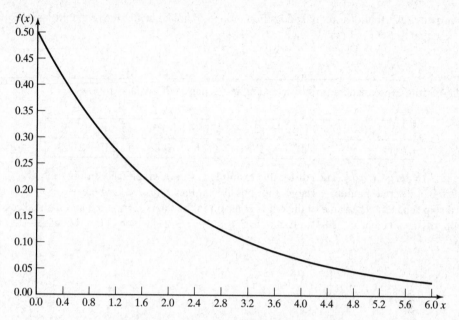

Figure 5.3   pdf for inspection-device life.

5.1.3   Cumulative distribution function

The cumulative distribution function (cdf), denoted by $F(x)$, measures the probability that the random variable X assumes a value less than or equal to x, that is, $F(x) = P(X \le x)$.

If X is discrete, then

$$F(x) = \sum_{\substack{\text{all} \\ x_i \le x}} p(x_i) \tag{5.3}$$

If X is continuous, then

$$F(x) = \int_{-\infty}^{x} f(t)\, dt \tag{5.4}$$

Some properties of the cdf are listed here:

a. F is a nondecreasing function. If $a < b$, then $F(a) \le F(b)$.
b. $\lim_{x \to \infty} F(x) = 1$
c. $\lim_{x \to -\infty} F(x) = 0$

All probability questions about X can be answered in terms of the cdf. For example,

$$P(a < X \le b) = F(b) - F(a) \quad \text{for all } a < b \tag{5.5}$$

For continuous distributions, not only does Equation (5.5) hold, but also the probabilities in Equation (5.2) are equal to $F(b) - F(a)$.

Example 5.4

The die-tossing experiment described in Example 5.2 has a cdf given as follows:

x	$(-\infty, 1)$	$[1, 2)$	$[2, 3)$	$[3, 4)$	$[4, 5)$	$[5, 6)$	$[6, \infty)$
$F(x)$	0	1/21	3/21	6/21	10/21	15/21	21/21

where $[a, b) = \{a \leq x < b\}$. The cdf for this example is shown graphically in Figure 5.4.

If X is a discrete random variable with possible values $x_1, x_2, \ldots$, where $x_1 < x_2 < \ldots$, the cdf is a step function. The value of the cdf is constant in the interval $[x_{i-1}, x_i)$ and then takes a step, or jump, of size $p(x_i)$ at x_i. Thus, in Example 5.4, $p(3) = 3/21$, which is the size of the step when $x = 3$.

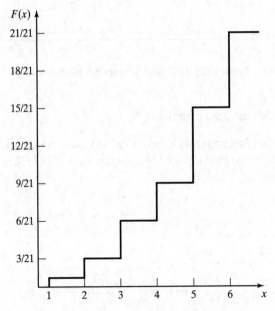

Figure 5.4   cdf for loaded-die example.

Example 5.5

The cdf for the device described in Example 5.3 is given by

$$F(x) = \frac{1}{2} \int_0^x e^{-t/2} dt = 1 - e^{-x/2}$$

The probability that the device will last for less than 2 years is given by

$$P(0 \leq X \leq 2) = F(2) - F(0) = F(2) = 1 - e^{-1} = 0.632$$

The probability that the life of the device is between 2 and 3 years is calculated as

$$P(2 \leq X \leq 3) = F(3) - F(2) = (1 - e^{-3/2}) - (1 - e^{-1})$$
$$= -e^{-3/2} + e^{-1} = -0.223 + 0.368 = 0.145$$

as found in Example 5.3.

5.1.4   Expectation

An important concept in probability theory is that of the expectation of a random variable. If X is a random variable, the expected value of X, denoted by $E(X)$, for discrete and continuous variables, is defined as follows:

$$E(X) = \sum_{\text{all } i} x_i p(x_i) \qquad \text{if } X \text{ is discrete} \tag{5.6}$$

and

$$E(X) = \int_{-\infty}^{\infty} x f(x)\, dx \qquad \text{if } X \text{ is continuous} \tag{5.7}$$

The expected value $E(X)$ of a random variable X is also referred to as the mean, μ, or the first moment of X. The quantity $E(X^n)$, $n \geq 1$, is called the nth moment of X, and is computed as follows:

$$E(X^n) = \sum_{\text{all } i} x_i^n p(x_i) \qquad \text{if } X \text{ is discrete} \tag{5.8}$$

and

$$E(X^n) = \int_{-\infty}^{\infty} x^n f(x)\, dx \qquad \text{if } X \text{ is continuous} \tag{5.9}$$

The variance of a random variable X denoted by $V(X)$ or var(X) or σ^2, is defined by

$$V(X) = E[(X - E[X])^2]$$

A useful identity in computing $V(X)$ is given by

$$V(X) = E(X^2) - [E(X)]^2 \tag{5.10}$$

The mean $E(X)$ is a measure of the central tendency of a random variable. The variance of X measures the expected value of the squared difference between the random variable and its mean. Thus, the variance, $V(X)$, is a measure of the spread or variation of the possible values of X around the mean $E(X)$. The standard deviation, σ, is defined to be the square root of the variance, σ^2. The mean, $E(X)$, and the standard deviation, $\sigma = \sqrt{V(X)}$, are expressed in the same units.

Example 5.6

The mean and variance of the die-tossing experiment described in Example 5.2 are computed as follows:

$$E(X) = 1\left(\frac{1}{21}\right) + 2\left(\frac{2}{21}\right) + \cdots + 6\left(\frac{6}{21}\right) = \frac{91}{21} = 4.33$$

To compute $V(X)$ from Equation (5.10), first compute $E(X^2)$ from Equation (5.8) as follows:

$$E(X^2) = 1^2\left(\frac{1}{21}\right) + 2^2\left(\frac{2}{21}\right) + \cdots + 6^2\left(\frac{6}{21}\right) = 21$$

Thus,

$$V(X) = 21 - \left(\frac{91}{21}\right)^2 = 21 - 18.78 = 2.22$$

and

$$\sigma = \sqrt{V(X)} = 1.49$$

Example 5.7

The mean and variance of the life of the device described in Example 5.3 are computed as follows:

$$E(X) = \frac{1}{2}\int_0^\infty xe^{-x/2}\,dx = -xe^{-x/2}\Big|_0^\infty + \int_0^\infty e^{-x/2}\,dx$$

$$= 0 + \frac{1}{1/2}e^{-x/2}\Big|_0^\infty = 2 \text{ years}$$

To compute $V(X)$ from Equation (5.10), first compute $E(X^2)$ from Equation (5.9) as follows:

$$E(X^2) = \frac{1}{2}\int_0^\infty x^2 e^{-x/2}\,dx$$

Thus,

$$E(X^2) = -x^2 e^{-x/2}\Big|_0^\infty + 2\int_0^\infty xe^{-x/2}\,dx = 8$$

giving

$$V(X) = 8 - 2^2 = 4 \text{ years}^2$$

and

$$\sigma = \sqrt{V(X)} = 2 \text{ years}$$

With a mean life of 2 years and a standard deviation of 2 years, most analysts would conclude that actual lifetimes, X, have a fairly large variability.

5.1.5   The mode

The mode is used in describing several statistical models that appear in this chapter. In the discrete case, the mode is the value of the random variable that occurs most frequently. In the continuous case, the mode is the value at which the pdf is maximized. The mode might not be unique; if the modal value occurs at two values of the random variable, the distribution is said to be *bimodal*.

5.2   Useful Statistical Models

Numerous situations arise in the conduct of a simulation where an investigator may choose to introduce probabilistic events. In Chapter 2, queueing, inventory, and reliability examples were given. In a queueing system, interarrival and service times are often probabilistic. In an inventory model, the time between demands and the lead times (time between placing and receiving an order) can be probabilistic. In a reliability model, the time to failure could be probabilistic. In each of these instances, the simulation analyst desires to generate random events and to use a known statistical model if the underlying distribution can be found. In the following paragraphs, statistical models appropriate to these application areas will be discussed. Additionally, statistical models useful in the case of limited data are mentioned.

5.2.1   Queueing systems

In Chapter 2, examples of waiting-line problems were given. In Chapters 2, 3, and 4, these problems were solved via simulation. In the queueing examples, interarrival- and service-time patterns were given. In these examples, the times between arrivals and the service times were always probabilistic, as is usually the case. However, it is possible to have a constant interarrival time (as in the case of a line moving at a constant speed in the assembly of an automobile), or a constant service time (as in the case of robotized spot welding on the same assembly line). The following example illustrates how probabilistic interarrival times might occur.

Example 5.8
Mechanics arrive at a centralized tool crib as shown in Table 5.1. Attendants check in and check out the requested tools to the mechanics. The collection of data begins at 10:00 A.M. and continues until 20 different interarrival times are recorded. Rather than record the actual time of day, the absolute time from a given origin could have been computed. Thus, the first mechanic could have arrived at time 0, the second mechanic at time 7:13 (7 minutes, 13 seconds), and so on.

Table 5.1   Mechanics' Arrival Data

Arrival Number	Arrival (hour:minutes::seconds)	Interarrival Time (minutes::seconds)
1	10:05::03	—
2	10:12::16	7::13
3	10:15::48	3::32
4	10:24::27	8::39
5	10:32::19	7::52
6	10:35::43	3::24
7	10:39::51	4::08
8	10:40::30	0::39
9	10:41::17	0::47
10	10:44::12	2::55
11	10:45::47	1::35
12	10:50::47	5::00
13	11:00::05	9::18
14	11:04::58	4::53
15	11:06::12	1::14
16	11:11::23	5::11
17	11:16::31	5::08
18	11:17::18	0::47
19	11:21::26	4::08
20	11:24::43	3::17
21	11:31::19	6::36

Example 5.9

Another way of presenting interarrival data is to find the number of arrivals per time period. Here, such arrivals occur over approximately 1.5 hours; it is convenient to look at 10-minute time intervals for the first 20 mechanics. That is, in the first 10-minute time period, one arrival occurred at 10:05::03. In the second time period, two mechanics arrived, and so on. The results are summarized in Table 5.2. This data could then be plotted in a histogram, as shown in Figure 5.5.

The distribution of time between arrivals and the distribution of the number of arrivals per time period are important in the simulation of waiting-line systems. "Arrivals" occur in numerous ways; as machine breakdowns, as jobs coming into a job shop, as units being assembled on a line, as orders to a warehouse, as data packets to a computer system, as calls to a call center, and so on.

Service times could be constant or probabilistic. If service times are completely random, the exponential distribution is often used for simulation purposes; however, there are several other possibilities. It could happen that the service times are constant, but some random variability causes fluctuations in either a positive or a negative way. For example, the time it takes for a lathe to traverse a 10-centimeter shaft should always be the same. However, the material could have slight differences

Table 5.2 Arrivals in Successive Time Periods

Time Period	Number of Arrivals	Time Period	Number of Arrivals
1	1	6	1
2	2	7	3
3	1	8	3
4	3	9	2
5	4	—	—

in hardness or the tool might wear; either event could cause different processing times. In these cases, the normal distribution might describe the service time.

A special case occurs when the phenomenon of interest seems to follow the normal probability distribution, but the random variable is restricted to be greater than or less than a certain value. In this case, the truncated normal distribution can be utilized.

The gamma and Weibull distributions are also used to model interarrival and service times. (Actually, the exponential distribution is a special case of both the gamma and the Weibull distributions.) The differences between the exponential, gamma, and Weibull distributions involve the location of the modes of the pdfs and the shapes of their tails for large and small times. The exponential distribution has its mode at the origin, but the gamma and Weibull distributions have their modes at some point (≥ 0) that is a function of the parameter values selected. The tail of the gamma distribution is long, like an exponential distribution; the tail of the Weibull distribution can decline more rapidly or less rapidly than that of an exponential distribution. In practice, this means that, if there are more large service times than an exponential distribution can account for, a Weibull distribution might provide a better model of these service times.

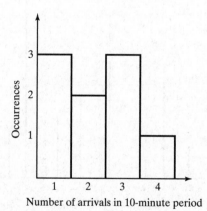

Figure 5.5 Histogram of arrivals per time period.

5.2.2   Inventory and supply-chain systems

In realistic inventory and supply-chain systems, there are at least three random variables: (1) the number of units demanded per order or per time period, (2) the time between demands, and (3) the lead time. (The lead time is defined as the time between the placing of an order for stocking the inventory system and the receipt of that order.) In very simple mathematical models of inventory systems, demand is a constant over time, and lead time is zero, or a constant. However, in most real-world cases, and, hence, in simulation models, demand occurs randomly in time, and the number of units demanded each time a demand occurs is also random, as illustrated by Figure 5.6.

Distributional assumptions for demand and lead time in inventory theory texts are usually based on mathematical tractability, but those assumptions could be invalid in a realistic context. In practice, the lead-time distribution can often be fitted fairly well by a gamma distribution [Hadley and Whitin, 1963]. Unlike analytic models, simulation models can accommodate whatever assumptions appear most reasonable.

The geometric, Poisson, and negative binomial distributions provide a range of distribution shapes that satisfy a variety of demand patterns. The geometric distribution, which is a special case of the negative binomial, has its mode at unity, given that at least one demand has occurred. If demand data are characterized by a long tail, the negative binomial distribution might be appropriate. The Poisson distribution is often used to model demand because it is simple, it is extensively tabulated, and it is well known. The tail of the Poisson distribution is generally shorter than that of the negative binomial, which means that fewer large demands will occur if a Poisson model is used than if a negative binomial distribution is used (assuming that both models have the same mean demand).

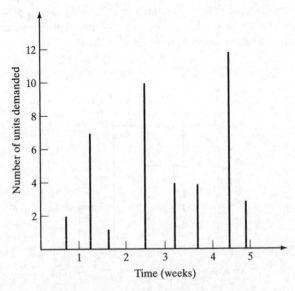

Figure 5.6   Random demands in time.

5.2.3   Reliability and maintainability

Time to failure has been modeled with numerous distributions, including the exponential, gamma, and Weibull. If only random failures occur, the time-to-failure distribution may be modeled as exponential. The gamma distribution arises from modeling standby redundancy, where each component has an exponential time to failure. The Weibull distribution has been extensively used to represent time to failure, and its nature is such that it can be made to approximate many observed phenomena [Hines and Montgomery, 1990]. When there are a number of components in a system and failure is due to the most serious of a large number of defects, or possible defects, the Weibull distribution seems to do particularly well as a model. In situations where most failures are due to wear, the normal distribution might very well be appropriate [Hines and Montgomery, 1990]. The lognormal distribution has been found to be applicable in describing time to failure for some types of components.

5.2.4   Limited data

In many instances, simulations begin before data collection has been completed. There are three distributions that have application to incomplete or limited data. These are the *uniform*, *triangular*, and *beta* distributions. The uniform distribution can be used when an interarrival or service time is known to be random, but no information is immediately available about the distribution [Gordon, 1975]. However, there are those who do not favor using the uniform distribution, calling it the "distribution of maximum ignorance" because it is not necessary to specify more than the continuous interval in which the random variable may occur. The triangular distribution can be used when assumptions are made about the minimum, maximum, and modal values of the random variable. Finally, the beta distribution provides a variety of distributional forms on the unit interval, ones that, with appropriate modification, can be shifted to any desired interval. The uniform distribution is a special case of the beta distribution. Pegden, Shannon, and Sadowski [1995] discuss the subject of limited data in some detail, and we include further discussion in Chapter 9.

5.2.5   Other distributions

Several other distributions may be useful in discrete-system simulation. The Bernoulli and binomial distributions are two discrete distributions which might describe phenomena of interest. The hyperexponential distribution is similar to the exponential distribution, but its greater variability might make it useful in certain instances.

5.3   Discrete Distributions

Discrete random variables are used to describe random phenomena in which only integer values can occur. Numerous examples were given in Section 5.2—for example, demands for inventory items. Four distributions are described in the following subsections.

5.3.1   Bernoulli trials and the Bernoulli distribution

Consider an experiment consisting of n trials, each of which can be a success or a failure. Let $X_j = 1$ if the jth experiment resulted in a success, and let $X_j = 0$ if the jth experiment resulted in a failure.

The n Bernoulli trials are called a *Bernoulli process* if the trials are independent, each trial has only two possible outcomes (success or failure), and the probability of a success remains constant from trial to trial. Thus,

$$p(x_1, x_2, \ldots, x_n) = p_1(x_1) \cdot p_2(x_2) \cdots p_n(x_n)$$

and

$$p_j(x_j) = p(x_j) = \begin{cases} p, & x_j = 1, j = 1, 2, \ldots, n \\ 1 - p = q, & x_j = 0, j = 1, 2, \ldots, n \\ 0, & \text{otherwise} \end{cases} \tag{5.11}$$

For one trial, the distribution given in Equation (5.11) is called the *Bernoulli distribution*. The mean and variance of X_j are calculated as follows:

$$E(X_j) = 0 \cdot q + 1 \cdot p = p$$

and

$$V(X_j) = [(0^2 \cdot q) + (1^2 \cdot p)] - p^2 = p(1 - p)$$

5.3.2   Binomial distribution

The random variable X that denotes the number of successes in n Bernoulli trials has a binomial distribution given by $p(x)$, where

$$p(x) = \begin{cases} \binom{n}{x} p^x q^{n-x}, & x = 0, 1, 2, \ldots, n \\ 0, & \text{otherwise} \end{cases} \tag{5.12}$$

Equation (5.12) is motivated by computing the probability of a particular outcome with all the successes, each denoted by S, occurring in the first x trials, followed by the $n - x$ failures, each denoted by an F—that is,

$$P \, (\overbrace{SSS\ldots\ldots\ldots SS}^{x \text{ of these}} \overbrace{FF\ldots\ldots\ldots FF}^{n-x \text{ of these}}) = p^x q^{n-x}$$

where $q = 1 - p$. There are

$$\binom{n}{x} = \frac{n!}{x!(n - x)!}$$

outcomes having the required number of Ss and Fs. Therefore, Equation (5.12) results. An easy approach to calculating the mean and variance of the binomial distribution is to consider X as a sum of n independent Bernoulli random variables, each with mean p and variance $p(1 - p) = pq$. Then,

$$X = X_1 + X_2 + \cdots + X_n$$

and the mean, $E(X)$, is given by

$$E(X) = p + p + \cdots + p = np \qquad (5.13)$$

and the variance $V(X)$ is given by

$$V(X) = pq + pq + \cdots + pq = npq \qquad (5.14)$$

Example 5.10

A production process manufactures computer chips on the average at 2% nonconforming. Every day, a random sample of size 50 is taken from the process. If the sample contains more than two nonconforming chips, the process will be stopped. Compute the probability that the process is stopped by the sampling scheme.

Consider the sampling process as $n = 50$ Bernoulli trials, each with $p = 0.02$; then the total number of nonconforming chips in the sample, X, would have a binomial distribution given by

$$p(x) = \begin{cases} \dbinom{50}{x} (0.02)^x (0.98)^{50-x}, & x = 0, 1, 2, \ldots, 50 \\ 0, & \text{otherwise} \end{cases}$$

It is much easier to compute the right-hand side of the following identity to compute the probability that more than two nonconforming chips are found in a sample:

$$P(X > 2) = 1 - P(X \le 2)$$

The probability $P(X \le 2)$ is calculated from

$$P(X \le 2) = \sum_{x=0}^{2} \binom{50}{x} (0.02)^x (0.98)^{50-x}$$

$$= (0.98)^{50} + 50(0.02)(0.98)^{49} + 1225(0.02)^2(0.98)^{48}$$

$$\doteq 0.92$$

Thus, the probability that the production process is stopped on any day, based on the sampling process, is approximately 0.08. The mean number of nonconforming chips in a random sample of size 50 is given by

$$E(X) = np = 50(0.02) = 1$$

and the variance is given by

$$V(X) = npq = 50(0.02)(0.98) = 0.98$$

The cdf for the binomial distribution has been tabulated by Banks and Heikes [1984] and others. The tables decrease the effort considerably for computing probabilities such as $P(a < X \leq b)$. Under certain conditions on n and p, both the Poisson distribution and the normal distribution may be used to approximate the binomial distribution [Hines and Montgomery, 1990].

5.3.3   Geometric and Negative Binomial distributions

The geometric distribution is related to a sequence of Bernoulli trials; the random variable of interest, X, is defined to be the number of trials to achieve the first success. The distribution of X is given by

$$p(x) = \begin{cases} q^{x-1}p, & x = 1, 2, \ldots \\ 0, & \text{otherwise} \end{cases} \tag{5.15}$$

The event $\{X = x\}$ occurs when there are $x - 1$ failures followed by a success. Each of the failures has an associated probability of $q = 1 - p$, and each success has probability p. Thus,

$$P(FFF \cdots FS) = q^{x-1}p$$

The mean and variance are given by

$$E(X) = \frac{1}{p} \tag{5.16}$$

and

$$V(X) = \frac{q}{p^2} \tag{5.17}$$

More generally, the negative binomial distribution is the distribution of the number of trials until the kth success, for $k = 1, 2, \ldots$. If Y has a negative binomial distribution with parameters p and k, then the distribution of Y is given by

$$p(y) = \begin{cases} \binom{y-1}{k-1} q^{y-k}p^k, & y = k, k+1, k+2, \ldots \\ 0, & \text{otherwise} \end{cases} \tag{5.18}$$

Because we can think of the negative binomial random variable Y as the sum of k independent geometric random variables, it is easy to see that $E(Y) = k/p$ and $V(X) = kq/p^2$.

Example 5.11

Forty percent of the assembled ink-jet printers are rejected at the inspection station. Find the probability that the first acceptable ink-jet printer is the third one inspected. Considering each inspection as a Bernoulli trial with $q = 0.4$ and $p = 0.6$ yields

$$p(3) = 0.4^2(0.6) = 0.096$$

Thus, in only about 10% of the cases is the first acceptable printer the third one from any arbitrary starting point. To determine the probability that the third printer inspected is the second acceptable printer, we use the negative binomial distribution Equation (5.18),

$$p(3) = \binom{3-1}{2-1}0.4^{3-2}(0.6)^2 = \binom{2}{1}0.4(0.6)^2 = 0.288$$

5.3.4   Poisson distribution

The Poisson distribution describes many random processes quite well and is mathematically quite simple. The Poisson distribution was introduced in 1837 by S. D. Poisson in a book concerning criminal and civil justice matters. (The title of this rather old text is *Recherches sur la probabilité des jugements en matière criminelle et en matière civile*. Evidently, the rumor handed down through generations of probability theory professors concerning the origin of the Poisson distribution is just not true. Rumor has it that the Poisson distribution was first used to model deaths from the kicks of horses in the Prussian Army.)

The Poisson probability mass function is given by

$$p(x) = \begin{cases} \dfrac{e^{-\alpha}\alpha^x}{x!}, & x = 0, 1, \ldots \\ 0, & \text{otherwise} \end{cases} \tag{5.19}$$

where $\alpha > 0$. One of the important properties of the Poisson distribution is that the mean and variance are both equal to α, that is,

$$E(X) = \alpha = V(X)$$

The cumulative distribution function is given by

$$F(x) = \sum_{i=0}^{x} \frac{e^{-\alpha}\alpha^i}{i!} \tag{5.20}$$

The pmf and cdf for a Poisson distribution with $\alpha = 2$ are shown in Figure 5.7. A tabulation of the cdf is given in Table A.4.

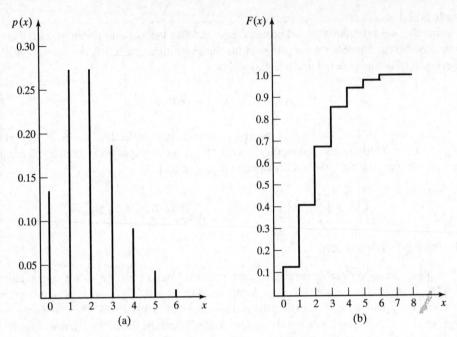

Figure 5.7   Poisson pmf and cdf.

Example 5.12

A computer repair person is "beeped" each time there is a call for service. The number of beeps per hour is known to occur in accordance with a Poisson distribution with a mean of $\alpha = 2$ per hour. The probability of three beeps in the next hour is given by Equation (5.19) with $x = 3$, as follows:

$$p(3) = \frac{e^{-2}2^3}{3!} = \frac{(0.135)(8)}{6} = 0.18$$

This same result can be read from the left side of Figure 5.7 or from Table A.4 by computing

$$F(3) - F(2) = 0.857 - 0.677 = 0.18$$

Example 5.13

In Example 5.12, find the probability of two or more beeps in a 1-hour period.

$$P(2 \text{ or more}) = 1 - p(0) - p(1) = 1 - F(1)$$
$$= 1 - 0.406 = 0.594$$

The cumulative probability, $F(1)$, can be read from the right side of Figure 5.7 or from Table A.4.

Example 5.14

The lead-time demand in an inventory system is the accumulation of demand for an item from the point at which an order is placed until the order is received—that is,

$$L = \sum_{i=1}^{T} D_i \tag{5.21}$$

where L is the lead-time demand, D_i is the demand during the ith time period, and T is the number of time periods during the lead time. Both D_i and T may be random variables.

An inventory manager desires that the probability of a stockout not exceed a certain fraction during the lead time. For example, it may be stated that the probability of a shortage during the lead time not exceed 5%.

If the lead-time demand is Poisson distributed, the determination of the reorder point is greatly facilitated. The reorder point is the level of inventory at which a new order is placed.

Assume that the lead-time demand is Poisson distributed with a mean of $\alpha = 10$ units and that 95% protection from a stockout is desired. Thus, it is desired to find the smallest value of x such that the probability that the lead-time demand not exceed x is greater than or equal to 0.95. Using Equation (5.20) requires finding the smallest x such that

$$F(x) = \sum_{i=0}^{x} \frac{e^{-10}10^i}{i!} \geq 0.95$$

The desired result occurs at $x = 15$, which can be found by using Table A.4 or by computation of $p(0), p(1), \ldots$.

5.4 Continuous Distributions

Continuous random variables can be used to describe random phenomena in which the variable of interest can take on any value in some interval—for example, the time to failure or the length of a rod. Nine distributions are described in the following subsections.

5.4.1 Uniform distribution

A random variable X is uniformly distributed on the interval $[a, b]$ if its pdf is given by

$$f(x) = \begin{cases} \dfrac{1}{b-a}, & a \leq x \leq b \\ 0, & \text{otherwise} \end{cases} \tag{5.22}$$

The cdf is given by

$$F(x) = \begin{cases} 0, & x < a \\ \dfrac{x-a}{b-a}, & a \leq x < b \\ 1, & x \geq b \end{cases} \tag{5.23}$$

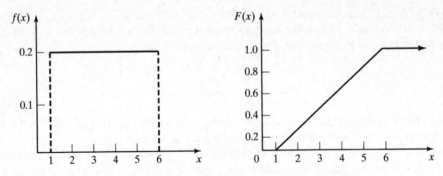

Figure 5.8   pdf and cdf for uniform distribution.

Note that

$$P(x_1 < X < x_2) = F(x_2) - F(x_1) = \frac{x_2 - x_1}{b - a}$$

is proportional to the length of the interval, for all x_1 and x_2 satisfying $a \leq x_1 < x_2 \leq b$. The mean and variance of the distribution are given by

$$E(X) = \frac{a + b}{2} \tag{5.24}$$

and

$$V(X) = \frac{(b - a)^2}{12} \tag{5.25}$$

The pdf and cdf when $a = 1$ and $b = 6$ are shown in Figure 5.8.

The uniform distribution plays a vital role in simulation. Random numbers, uniformly distributed between 0 and 1, provide the means to generate random events. Numerous methods for generating uniformly distributed random numbers have been devised; some will be discussed in Chapter 7. Uniformly distributed random numbers are then used to generate samples of random variates from all other distributions, as will be discussed in Chapter 8.

Example 5.15 _____

A simulation of a warehouse operation is being developed. About every 3 minutes, a call comes for a forklift truck operator to proceed to a certain location. An initial assumption is made that the time between calls (arrivals) is uniformly distributed with a mean of 3 minutes. By Equation (5.25), the uniform distribution with a mean of 3 and the greatest possible variability would have parameter values of $a = 0$ and $b = 6$ minutes. With very limited data (such as a mean of approximately 3 minutes) plus the knowledge that the quantity of interest is variable in a random fashion, the uniform distribution with greatest variance can be assumed, at least until more data are available.

Example 5.16

A bus arrives every 20 minutes at a specified stop beginning at 6:40 A.M. and continuing until 8:40 A.M. A certain passenger does not know the schedule, but arrives randomly (uniformly distributed) between 7:00 A.M. and 7:30 A.M. every morning. What is the probability that the passenger waits more than 5 minutes for a bus?

The passenger has to wait more than 5 minutes only if the arrival time is between 7:00 A.M. and 7:15 A.M. or between 7:20 A.M. and 7:30 A.M. If X is a random variable that denotes the number of minutes past 7:00 A.M. that the passenger arrives, the desired probability is

$$P(0 < X < 15) + P(20 < X < 30)$$

Now, X is a uniform random variable on $(0, 30)$. Therefore, the desired probability is given by

$$F(15) + F(30) - F(20) = \frac{15}{30} + 1 - \frac{20}{30} = \frac{5}{6}$$

5.4.2 Exponential distribution

A random variable X is said to be exponentially distributed with parameter $\lambda > 0$ if its pdf is given by

$$f(x) = \begin{cases} \lambda e^{-\lambda x}, & x \geq 0 \\ 0, & \text{elsewhere} \end{cases} \qquad (5.26)$$

The density function is shown in Figures 5.9 and 5.3. Figure 5.9 also shows the cdf.

The exponential distribution has been used to model interarrival times when arrivals are completely random and to model service times that are highly variable. In these instances, λ is a rate: arrivals per hour or services per minute. The exponential distribution has also been used to model the lifetime of a component that fails catastrophically (instantaneously), such as a light bulb; then λ is the failure rate.

Several different exponential pdfs are shown in Figure 5.10. The value of the intercept on the vertical axis is always equal to the value of λ. Note also that all pdfs eventually intersect. (Why?)

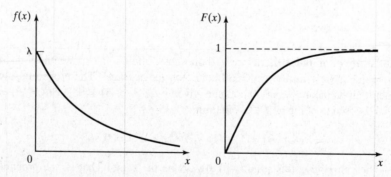

Figure 5.9 Exponential density function and cumulative distribution function.

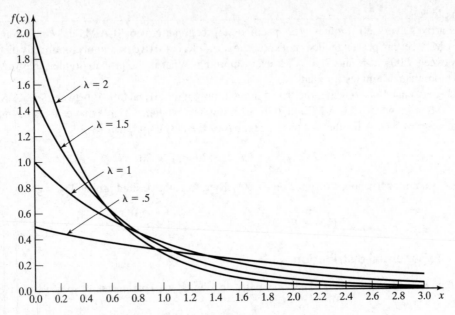

Figure 5.10   pdfs for several exponential distributions.

The exponential distribution has mean and variance given by

$$E(X) = \frac{1}{\lambda} \quad \text{and} \quad V(X) = \frac{1}{\lambda^2} \tag{5.27}$$

Thus, the mean and standard deviation are equal. The cdf can be exhibited by integrating Equation (5.26) to obtain

$$F(x) = \begin{cases} 0, & x < 0 \\ \int_0^x \lambda e^{-\lambda t}\, dt = 1 - e^{-\lambda x}, & x \geq 0 \end{cases} \tag{5.28}$$

Example 5.17

Suppose that the life of an industrial lamp, in thousands of hours, is exponentially distributed with failure rate $\lambda = 1/3$ (one failure every 3000 hours, on the average). The probability that the lamp will last longer than its mean life, 3000 hours, is given by $P(X > 3) = 1 - P(X \leq 3) = 1 - F(3)$. Equation (5.28) is used to compute $F(3)$, obtaining

$$P(X > 3) = 1 - (1 - e^{-3/3}) = e^{-1} = 0.368$$

Regardless of the value of λ, this result will always be the same! That is, the probability that an exponential random variable is greater than its mean is 0.368, for any value of λ.

The probability that the industrial lamp will last between 2000 and 3000 hours is computed as

$$P(2 \leq X \leq 3) = F(3) - F(2)$$

Again, from the cdf given by Equation (5.28),

$$F(3) - F(2) = (1 - e^{-3/3}) - (1 - e^{-2/3})$$
$$= -0.368 + 0.513 = 0.145$$

One of the most important properties of the exponential distribution is that it is "memoryless," which means that, for all $s \geq 0$ and $t \geq 0$,

$$P(X > s + t | X > s) = P(X > t) \tag{5.29}$$

Let X represent the life of a component (a battery, light bulb, computer chip, laser, etc.) and assume that X is exponentially distributed. Equation (5.29) states that the probability that the component lives for at least $s + t$ hours, given that it has survived s hours, is the same as the initial probability that it lives for at least t hours. If the component is alive at time s (if $X > s$), then the distribution of the remaining amount of time that it survives, namely $X - s$, is the same as the original distribution of a new component. That is, the component does not "remember" that it has already been in use for a time s. A used component is as good as new.

That Equation (5.29) holds is shown by examining the conditional probability

$$P(X > s + t | X > s) = \frac{P(X > s + t)}{P(X > s)} \tag{5.30}$$

Equation (5.28) can be used to determine the numerator and denominator of Equation (5.30), yielding

$$P(X > s + t | X > s) = \frac{e^{-\lambda(s+t)}}{e^{-\lambda s}} = e^{-\lambda t}$$
$$= P(X > t)$$

Example 5.18

Find the probability that the industrial lamp in Example 5.17 will last for another 1000 hours, given that it is operating after 2500 hours. This determination can be found using Equations (5.29) and (5.28), as follows:

$$P(X > 3.5 | X > 2.5) = P(X > 1) = e^{-1/3} = 0.717$$

Example 5.18 illustrates the *memoryless* property—namely, that a used component that follows an exponential distribution is as good as a new component. The probability that a new component will have a life greater than 1000 hours is also equal to 0.717. Stated in general, suppose that a component which has a lifetime that follows the exponential distribution with parameter λ is observed and found to be operating at an arbitrary time. Then, the distribution of the remaining lifetime is also exponential with parameter λ. The exponential distribution is the only continuous distribution that has the memoryless property. (The geometric distribution is the only discrete distribution that possesses the memoryless property.)

5.4.3 Gamma distribution

A function used in defining the gamma distribution is the gamma function, which is defined for all $\beta > 0$ as

$$\Gamma(\beta) = \int_0^\infty x^{\beta-1} e^{-x} \, dx \tag{5.31}$$

By integrating Equation (5.31) by parts, it can be shown that

$$\Gamma(\beta) = (\beta - 1)\Gamma(\beta - 1) \tag{5.32}$$

If β is an integer, then, by using $\Gamma(1) = 1$ and applying Equation (5.32), it can be seen that

$$\Gamma(\beta) = (\beta - 1)! \tag{5.33}$$

The gamma function can be thought of as a generalization of the factorial notion to all positive numbers, not just integers.

A random variable X is gamma distributed with parameters β and θ if its pdf is given by

$$f(x) = \begin{cases} \dfrac{\beta\theta}{\Gamma(\beta)} (\beta\theta x)^{\beta-1} e^{-\beta\theta x}, & x > 0 \\ 0, & \text{otherwise} \end{cases} \tag{5.34}$$

β is called the shape parameter, and θ is called the scale parameter. Several gamma distributions for $\theta = 1$ and various values of β are shown in Figure 5.11.

The mean and variance of the gamma distribution are given by

$$E(X) = \frac{1}{\theta} \tag{5.35}$$

and

$$V(X) = \frac{1}{\beta\theta^2} \tag{5.36}$$

The cdf of X is given by

$$F(x) = \begin{cases} 1 - \displaystyle\int_x^\infty \dfrac{\beta\theta}{\Gamma(\beta)} (\beta\theta t)^{\beta-1} e^{-\beta\theta t} \, dt, & x > 0 \\ 0, & x \le 0 \end{cases} \tag{5.37}$$

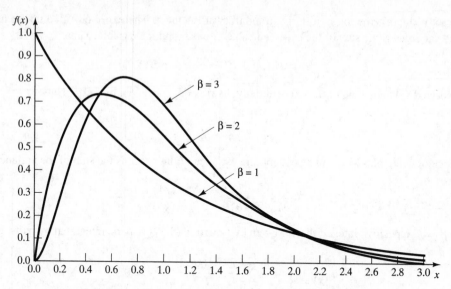

Figure 5.11 pdfs for several gamma distributions with $\theta = 1$.

When β is an integer, the gamma distribution is related to the exponential distribution in the following manner: If the random variable X is the sum of β independent, exponentially distributed random variables, each with parameter $\beta\theta$, then X has a gamma distribution with parameters β and θ. Thus, if

$$X = X_1 + X_2 + \cdots + X_\beta \tag{5.38}$$

where the pdf of X_j is given by

$$g(x_j) = \begin{cases} (\beta\theta)e^{-\beta\theta x_j}, & x \geq 0 \\ 0, & \text{otherwise} \end{cases}$$

and the X_j are mutually independent, then X has the pdf given in Equation (5.34). Note that, when $\beta = 1$, an exponential distribution results. This result follows from Equation (5.38) or from letting $\beta = 1$ in Equation (5.34).

5.4.4 Erlang distribution

The pdf given by Equation (5.34) is often referred to as the Erlang distribution of order (or number of phases) k when $\beta = k$, an integer. Erlang was a Danish telephone engineer who was an early developer of queueing theory. The Erlang distribution could arise in the following context: Consider a series of k stations that must be passed through in order to complete the servicing of a customer. An additional customer cannot enter the first station until the customer in process has negotiated all the stations. Each station has an exponential distribution of service time with parameter $k\theta$. Equations (5.35) and (5.36), which state the mean and variance of a gamma distribution, are valid regardless of the value of β. However, when $\beta = k$, an integer, Equation (5.38) may be used to derive the mean of the distribution in a fairly straightforward manner. The expected value of the sum of random variables is the sum of the expected value of each random variable. Thus,

$$E(X) = E(X_1) + E(X_2) + \cdots + E(X_k)$$

The expected value of each of the exponentially distributed X_j is given by $1/k\theta$. Thus,

$$E(X) = \frac{1}{k\theta} + \frac{1}{k\theta} + \cdots + \frac{1}{k\theta} = \frac{1}{\theta}$$

If the random variables X_j are independent, the variance of their sum is the sum of the variances, or

$$V(X) = \frac{1}{(k\theta)^2} + \frac{1}{(k\theta)^2} + \cdots + \frac{1}{(k\theta)^2} = \frac{1}{k\theta^2}$$

When $\beta = k$, a positive integer, the cdf given by Equation (5.37) may be integrated by parts, giving

$$F(x) = \begin{cases} 1 - \sum_{i=0}^{k-1} \dfrac{e^{-k\theta x}(k\theta x)^i}{i!}, & x > 0 \\ 0, & x \le 0 \end{cases} \tag{5.39}$$

which is the sum of Poisson terms with mean $\alpha = k\theta x$. Tables of the cumulative Poisson distribution may be used to evaluate the cdf when the shape parameter is an integer.

Example 5.19

A college professor of electrical engineering is leaving home for the summer, but would like to have a light burning at all times to discourage burglars. The professor rigs up a device that will hold two light bulbs. The device will switch the current to the second bulb if the first bulb fails. The box in which the light bulbs are packaged says, "Average life 1000 hours, exponentially distributed." The professor will be gone 90 days (2160 hours). What is the probability that a light will be burning when the summer is over and the professor returns?

The probability that the system will operate at least x hours is called the reliability function $R(x)$:

$$R(x) = 1 - F(x)$$

In this case, the total system lifetime is given by Equation (5.38) with $\beta = k = 2$ bulbs and $k\theta = 1/1000$ per hour, so $\theta = 1/2000$ per hour. Thus, $F(2160)$ can be determined from Equation (5.39) as follows:

$$F(2160) = 1 - \sum_{i=0}^{1} \frac{e^{-(2)(1/2000)(2160)}[(2)(1/2000)(2160)]^i}{i!}$$

$$= 1 - e^{-2.16} \sum_{i=0}^{1} \frac{(2.16)^i}{i!} = 0.636$$

Therefore, the chances are about 36% that a light will be burning when the professor returns.

Example 5.20

A medical examination is given in three stages by a physician. Each stage is exponentially distributed with a mean service time of 20 minutes. Find the probability that the exam will take 50 minutes or less. Also, compute the expected length of the exam. In this case, $k = 3$ stages and $k\theta = 1/20$, so that $\theta = 1/60$ per minute. Thus, $F(50)$ can be calculated from Equation (5.39) as follows:

$$F(50) = 1 - \sum_{i=0}^{2} \frac{e^{-(3)(1/60)(50)}[(3)(1/60)(50)]^i}{i!}$$

$$= 1 - \sum_{i=0}^{2} \frac{e^{-5/2}(5/2)^i}{i!}$$

The cumulative Poisson distribution, shown in Table A.4, can be used to calculate that

$$F(50) = 1 - 0.543 = 0.457$$

The probability is 0.457 that the exam will take 50 minutes or less. The expected length of the exam is found from Equation (5.35):

$$E(X) = \frac{1}{\theta} = \frac{1}{1/60} = 60 \text{ minutes}$$

In addition, the variance of X is $V(X) = 1/\beta\theta^2 = 1200$ minutes^2—incidentally, the mode of the Erlang distribution is given by

$$\text{Mode} = \frac{k-1}{k\theta} \tag{5.40}$$

Thus, the modal value in this example is

$$\text{Mode} = \frac{3-1}{3(1/60)} = 40 \text{ minutes}$$

5.4.5 Normal distribution

A random variable X with mean $-\infty < \mu < \infty$ and variance $\sigma^2 > 0$ has a normal distribution if it has the pdf

$$f(x) = \frac{1}{\sigma\sqrt{2\pi}} \exp\left[-\frac{1}{2}\left(\frac{x-\mu}{\sigma}\right)^2\right], \quad -\infty < x < \infty \tag{5.41}$$

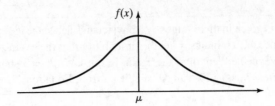

Figure 5.12   pdf of the normal distribution.

The normal distribution is used so often that the notation $X \sim N(\mu, \sigma^2)$ has been adopted by many authors to indicate that the random variable X is normally distributed with mean μ and variance σ^2. The normal pdf is shown in Figure 5.12.

Some of the special properties of the normal distribution are listed here:

a. $\lim_{x \to -\infty} f(x) = 0$ and $\lim_{x \to \infty} f(x) = 0$; the value of $f(x)$ approaches zero as x approaches negative infinity and, similarly, as x approaches positive infinity.

b. $f(\mu - x) = f(\mu + x)$; the pdf is symmetric about μ.

c. The maximum value of the pdf occurs at $x = \mu$; the mean and mode are equal.

The cdf for the normal distribution is given by

$$F(x) = P(X \le x) = \int_{-\infty}^{x} \frac{1}{\sigma\sqrt{2\pi}} \exp\left[-\frac{1}{2}\left(\frac{t-\mu}{\sigma}\right)^2\right] dt \tag{5.42}$$

It is not possible to evaluate Equation (5.42) in closed form. Numerical methods could be used, but it appears that it would be necessary to evaluate the integral for each pair (μ, σ^2). However, a transformation of variables, $z = (t - \mu)/\sigma$, allows the evaluation to be independent of μ and σ. If $X \sim N(\mu, \sigma^2)$, let $Z = (X - \mu)/\sigma$ to obtain

$$F(x) = P(X \le x) = P\left(Z \le \frac{x-\mu}{\sigma}\right)$$

$$= \int_{-\infty}^{(x-\mu)/\sigma} \frac{1}{\sqrt{2\pi}} e^{-z^2/2}\, dz \tag{5.43}$$

$$= \int_{-\infty}^{(x-\mu)/\sigma} \phi(z)\, dz = \Phi\left(\frac{x-\mu}{\sigma}\right)$$

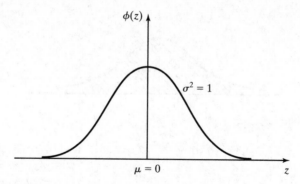

Figure 5.13   pdf of the standard normal distribution.

The pdf

$$\phi(z) = \frac{1}{\sqrt{2\pi}}e^{-z^2/2}, \quad -\infty < z < \infty \tag{5.44}$$

is the pdf of a normal distribution with mean 0 and variance 1. Thus, $Z \sim N(0, 1)$ and it is said that Z has a standard normal distribution. The standard normal distribution is shown in Figure 5.13. The cdf for the standard normal distribution is given by

$$\Phi(z) = \int_{-\infty}^{z} \frac{1}{\sqrt{2\pi}}e^{-t^2/2}\,dt \tag{5.45}$$

Equation (5.45) has been widely tabulated. The probabilities $\Phi(z)$ for $z \geq 0$ are given in Table A.3. Several examples are now given that indicate how Equation (5.43) and Table A.3 are used.

Example 5.21
Suppose that it is known that $X \sim N(50, 9)$. Compute $F(56) = P(X \leq 56)$. Using Equation (5.43), we get

$$F(56) = \Phi\left(\frac{56 - 50}{3}\right) = \Phi(2) = 0.9772$$

from Table A.3. The intuitive interpretation is shown in Figure 5.14. Figure 5.14(a) shows the pdf of $X \sim N(50, 9)$ with the specific value, $x_0 = 56$, marked. The shaded portion is the desired probability. Figure 5.14(b) shows the standard normal distribution or $Z \sim N(0, 1)$ with the value 2 marked; $x_0 = 56$ is 2σ (where $\sigma = 3$) greater than the mean. It is helpful to make both sketches such as those in Figure 5.14 to avoid confusion in figuring out required probabilities.

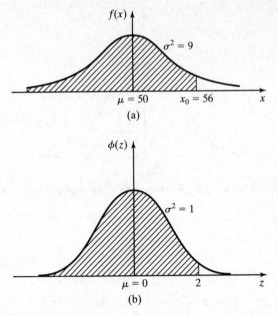

Figure 5.14   Transforming to the standard normal distribution.

Example 5.22

The time in hours required to load an ocean-going vessel, X, is distributed as $N(12, 4)$. The probability that the vessel will be loaded in less than 10 hours is given by $F(10)$, where

$$F(10) = \Phi\left(\frac{10 - 12}{2}\right) = \Phi(-1) = 0.1587$$

The value of $\Phi(-1) = 0.1587$ is looked up in Table A.3 by using the symmetry property of the normal distribution. Note that $\Phi(1) = 0.8413$. The complement of 0.8413, or 0.1587, is contained in the tail, the shaded portion of the standard normal distribution shown in Figure 5.15(a). In Figure 5.15(b), the symmetry property is used to work out the shaded region to be $\Phi(-1) = 1 - \Phi(1) = 0.1587$. [From this logic, it can be seen that $\Phi(2) = 0.9772$ and $\Phi(-2) = 1 - \Phi(2) = 0.0228$. In general, $\Phi(-x) = 1 - \Phi(x)$.]

The probability that 12 or more hours will be required to load the ship can also be discovered by inspection, by using the symmetry property of the normal pdf and the mean as shown by Figure 5.16. The shaded portion of Figure 5.16(a) shows the problem as originally stated [i.e., evaluate $P(X < 12)$]. Now, $P(X > 12) = 1 - F(12)$. The standardized normal in Figure 5.16(b) is used to evaluate $F(12) = \Phi(0) = 0.50$. Thus, $P(X > 12) = 1 - 0.50 = 0.50$. (The shaded portions in both Figure 5.16(a) and (b) contain 0.50 of the area under the normal pdf.)

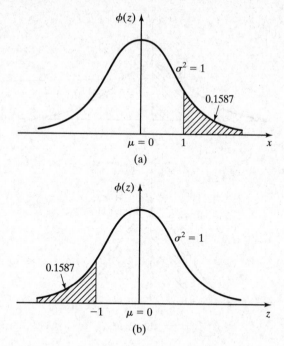

Figure 5.15   Using the symmetry property of the normal distribution.

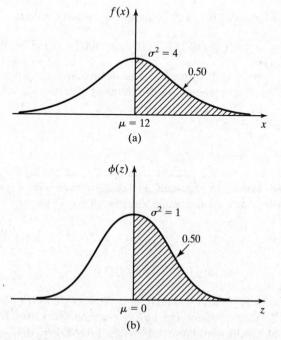

Figure 5.16   Evaluation of probability by inspection.

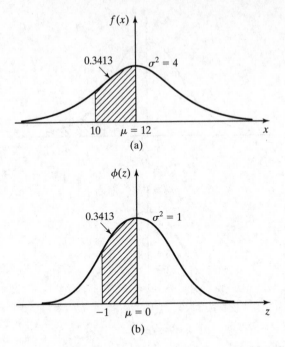

Figure 5.17 Transformation to standard normal for vessel-loading problem.

The probability that between 10 and 12 hours will be required to load a ship is given by

$$P(10 \le X \le 12) = F(12) - F(10) = 0.5000 - 0.1587 = 0.3413$$

using earlier results presented in this example. The desired area is shown in the shaded portion of Figure 5.17(a). The equivalent problem shown in terms of the standardized normal distribution is shown in Figure 5.17(b). The probability statement is $F(12) - F(10) = \Phi(0) - \Phi(-1) = 0.5000 - 0.1587 = 0.3413$, from Table A.3.

Example 5.23 _____

The time to pass through a queue to begin self-service at a cafeteria has been found to be $N(15, 9)$. The probability that an arriving customer waits between 14 and 17 minutes is computed as follows:

$$P(14 \le X \le 17) = F(17) - F(14) = \Phi\left(\frac{17 - 15}{3}\right) - \Phi\left(\frac{14 - 15}{3}\right)$$

$$= \Phi(0.667) - \Phi(-0.333)$$

The shaded area shown in Figure 5.18(a) represents the probability $F(17) - F(14)$. The shaded area shown in Figure 5.18(b) represents the equivalent probability, $\Phi(0.667) - \Phi(-0.333)$, for the standardized normal distribution. From Table A.3, $\Phi(0.667) = 0.7476$. Now, $\Phi(-0.333) =$

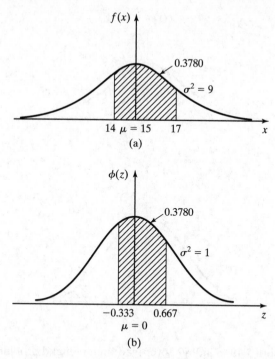

Figure 5.18   Transformation to standard normal for cafeteria problem.

$1 - \Phi(0.333) = 1 - 0.6304 = 0.3696$. Thus, $\Phi(0.667) - \Phi(-0.333) = 0.3780$. The probability is 0.3780 that the customer will pass through the queue in a time between 14 and 17 minutes.

Example 5.24

Lead-time demand, X, for an item is approximated by a normal distribution having mean 25 and variance 9. It is desired to compute the value for lead time that will be exceeded only 5% of the time. Thus, the problem is to find x_0 such that $P(X > x_0) = 0.05$, as shown by the shaded area in Figure 5.19(a). The equivalent problem is shown as the shaded area in Figure 5.19(b). Now,

$$P(X > x_0) = P\left(Z > \frac{x_0 - 25}{3}\right) = 1 - \Phi\left(\frac{x_0 - 25}{3}\right) = 0.05$$

or, equivalently,

$$\Phi\left(\frac{x_0 - 25}{3}\right) = 0.95$$

From Table A.3, it can be seen that $\Phi(1.645) = 0.95$. Thus, x_0 can be found by solving

$$\frac{x_0 - 25}{3} = 1.645$$

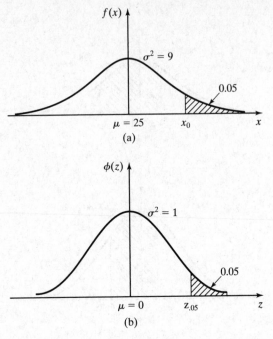

Figure 5.19 Finding x_0 for lead-time-demand problem.

or

$$x_0 = 29.935$$

Therefore, in only 5% of the cases will demand during lead time exceed available inventory if an order to purchase is made when the stock level reaches 30.

5.4.6 Weibull distribution

The random variable X has a Weibull distribution if its pdf has the form

$$f(x) = \begin{cases} \dfrac{\beta}{\alpha} \left(\dfrac{x - \nu}{\alpha} \right)^{\beta-1} \exp\left[-\left(\dfrac{x - \nu}{\alpha} \right)^{\beta} \right], & x \geq \nu \\ 0, & \text{otherwise} \end{cases} \tag{5.46}$$

The three parameters of the Weibull distribution are ν ($-\infty < \nu < \infty$), which is the location parameter; α ($\alpha > 0$), which is the scale parameter; and β ($\beta > 0$), which is the shape parameter. When $\nu = 0$, the Weibull pdf becomes

$$f(x) = \begin{cases} \dfrac{\beta}{\alpha} \left(\dfrac{x}{\alpha} \right)^{\beta-1} \exp\left[-\left(\dfrac{x}{\alpha} \right)^{\beta} \right], & x \geq 0 \\ 0, & \text{otherwise} \end{cases} \tag{5.47}$$

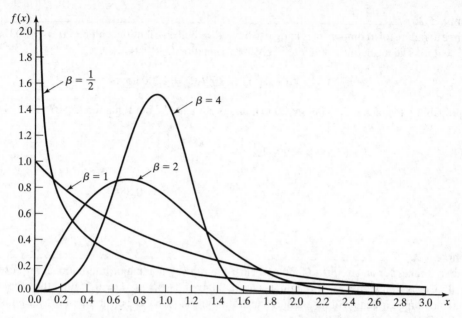

Figure 5.20   Weibull pdfs for $\nu = 0$; $\alpha = 1$; $\beta = \frac{1}{2}, 1, 2, 4$.

Figure 5.20 shows several Weibull densities when $\nu = 0$ and $\alpha = 1$. When $\beta = 1$, the Weibull distribution is reduced to

$$f(x) = \begin{cases} \dfrac{1}{\alpha} e^{-x/\alpha}, & x \geq 0 \\[2mm] 0, & \text{otherwise} \end{cases}$$

which is an exponential distribution with parameter $\lambda = 1/\alpha$.

The mean and variance of the Weibull distribution are given by the following expressions:

$$E(X) = \nu + \alpha \Gamma\left(\frac{1}{\beta} + 1\right) \tag{5.48}$$

$$V(X) = \alpha^2 \left[\Gamma\left(\frac{2}{\beta} + 1\right) - \left[\Gamma\left(\frac{1}{\beta} + 1\right)\right]^2 \right] \tag{5.49}$$

where $\Gamma(\cdot)$ is defined by Equation (5.31). Thus, the location parameter ν has no effect on the variance; however, the mean is increased or decreased by ν. The cdf of the Weibull distribution is given by

$$F(x) = \begin{cases} 0, & x < \nu \\[2mm] 1 - \exp\left[-\left(\dfrac{x - \nu}{\alpha}\right)^{\beta} \right], & x \geq \nu \end{cases} \tag{5.50}$$

Example 5.25

The time to failure for a component is known to have a Weibull distribution with $v = 0, \alpha = 200$ hours, and $\beta = 1/3$. The mean time to failure is given by Equation (5.48) as

$$E(X) = 200\Gamma(3 + 1) = 200(3!) = 1200 \text{ hours}$$

The probability that a unit fails before 2000 hours is computed from Equation (5.50) as

$$F(2000) = 1 - \exp\left[-\left(\frac{2000}{200}\right)^{1/3}\right]$$

$$= 1 - e^{-3\sqrt{10}} = 1 - e^{-2.15} = 0.884$$

Example 5.26

The time it takes for an aircraft to land and clear the runway at a major international airport has a Weibull distribution with $v = 1.34$ minutes, $\alpha = 0.04$ minutes, and $\beta = 0.5$. Find the probability that an incoming airplane will take more than 1.5 minutes to land and clear the runway. In this case $P(X > 1.5)$ is computed as follows:

$$P(X \leq 1.5) = F(1.5)$$

$$= 1 - \exp\left[-\left(\frac{1.5 - 1.34}{0.04}\right)^{0.5}\right]$$

$$= 1 - e^{-2} = 1 - 0.135 = 0.865$$

Therefore, the probability that an aircraft will require more than 1.5 minutes to land and clear the runway is 0.135.

5.4.7 Triangular distribution

A random variable X has a triangular distribution if its pdf is given by

$$f(x) = \begin{cases} \dfrac{2(x - a)}{(b - a)(c - a)}, & a \leq x \leq b \\[2ex] \dfrac{2(c - x)}{(c - b)(c - a)}, & b < x \leq c \\[2ex] 0, & \text{elsewhere} \end{cases} \tag{5.51}$$

where $a \leq b \leq c$. The mode occurs at $x = b$. A triangular pdf is shown in Figure 5.21. The parameters (a, b, c) can be related to other measures, such as the mean and the mode, as follows:

$$E(X) = \frac{a + b + c}{3} \tag{5.52}$$

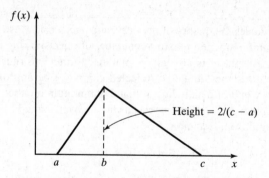

Figure 5.21 pdf of the triangular distribution.

From Equation (5.52) the mode can be determined as

$$\text{Mode} = b = 3E(X) - (a + c) \tag{5.53}$$

Because $a \le b \le c$,

$$\frac{2a + c}{3} \le E(X) \le \frac{a + 2c}{3}$$

The mode is used more often than the mean to characterize the triangular distribution. As is shown in Figure 5.21, its height is $2/(c-a)$ above the x axis. The variance, $V(X)$, of the triangular distribution is left as an exercise for the student. The cdf for the triangular distribution is given by

$$F(x) = \begin{cases} 0, & x \le a \\[2mm] \dfrac{(x-a)^2}{(b-a)(c-a)}, & a < x \le b \\[2mm] 1 - \dfrac{(c-x)^2}{(c-b)(c-a)}, & b < x \le c \\[2mm] 1, & x > c \end{cases} \tag{5.54}$$

Example 5.27
The central processing unit requirements, for programs that will execute, have a triangular distribution with $a = 0.05$ milliseconds, $b = 1.1$ milliseconds, and $c = 6.5$ milliseconds. Find the probability that the CPU requirement for a random program is 2.5 milliseconds or less. The value of $F(2.5)$ is from the portion of the cdf in the interval $[0.05, 1.1]$ plus that portion in the interval $[1.1, 2.5]$. By using Equation (5.54), both portions can be addressed at one time, to yield

$$F(2.5) = 1 - \frac{(6.5 - 2.5)^2}{(6.5 - 0.05)(6.5 - 1.1)} = 0.541$$

Thus, the probability is 0.541 that the CPU requirement is 2.5 milliseconds or less.

Example 5.28

An electronic sensor evaluates the quality of memory chips, rejecting those that fail. Upon demand, the sensor will give the minimum and maximum number of rejects during each hour of production over the past 24 hours. The mean is also given. Without further information, the quality control department has assumed that the number of rejected chips can be approximated by a triangular distribution. The current dump of data indicates that the minimum number of rejected chips during any hour was 0, the maximum was 10, and the mean was 4. Given that $a = 0, c = 10$, and $E(X) = 4$, the value of b can be found from Equation (5.53):

$$b = 3(4) - (0 + 10) = 2$$

The height of the mode is $2/(10 - 0) = 0.2$. Thus, Figure 5.22 can be drawn.

The median is the point at which 0.5 of the area is to the left and 0.5 is to the right. The median in this example is 3.7, also shown on Figure 5.22. Finding the median of the triangular distribution requires an initial location of the value to the left or to the right of the mode. The area to the left of the mode is computed from Equation (5.54) as

$$F(2) = \frac{2^2}{20} = 0.2$$

Thus, the median is between b and c. Setting $F(x) = 0.5$ in Equation (5.54) and solving for $x =$ median yields

$$0.5 = 1 - \frac{(10 - x)^2}{(10)(8)}$$

with

$$x = 3.7$$

This example clearly shows that the mean, mode, and median are not necessarily equal.

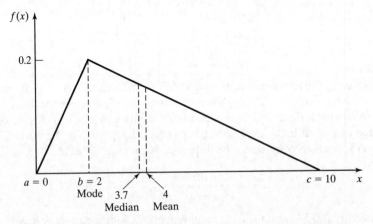

Figure 5.22 Mode, median, and mean for triangular distribution.

5.4.8 Lognormal distribution

A random variable X has a lognormal distribution if its pdf is given by

$$f(x) = \begin{cases} \dfrac{1}{\sqrt{2\pi}\sigma x} \exp\left[-\dfrac{(\ln x - \mu)^2}{2\sigma^2}\right], & x > 0 \\ 0, & \text{otherwise} \end{cases} \tag{5.55}$$

where $\sigma^2 > 0$. The mean and variance of a lognormal random variable are

$$E(X) = e^{\mu + \sigma^2/2} \tag{5.56}$$

$$V(X) = e^{2\mu + \sigma^2}(e^{\sigma^2} - 1) \tag{5.57}$$

Three lognormal pdfs, all having mean 1, but variances 1/2, 1, and 2, are shown in Figure 5.23.

Notice that the parameters μ and σ^2 are not the mean and variance of the lognormal. These parameters come from the fact that when Y has a $N(\mu, \sigma^2)$ distribution then $X = e^Y$ has a lognormal distribution with parameters μ and σ^2. If the mean and variance of the lognormal are known to be μ_L and σ_L^2, respectively, then the parameters μ and σ^2 are given by

$$\mu = \ln\left(\frac{\mu_L^2}{\sqrt{\mu_L^2 + \sigma_L^2}}\right) \tag{5.58}$$

$$\sigma^2 = \ln\left(\frac{\mu_L^2 + \sigma_L^2}{\mu_L^2}\right) \tag{5.59}$$

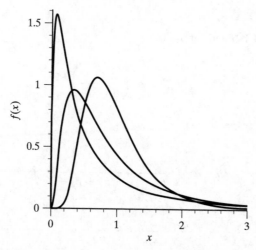

Figure 5.23 pdf of the lognormal distribution.

Example 5.29

The rate of return on a volatile investment is modeled as having a lognormal distribution with mean 20% and standard deviation 5%. Compute the parameters for the lognormal distribution. From the information given, we have $\mu_L = 20$ and $\sigma_L^2 = 5^2$. Thus, from Equations (5.58) and (5.59),

$$\mu = \ln\left(\frac{20^2}{\sqrt{20^2 + 5^2}}\right) \doteq 2.9654$$

$$\sigma^2 = \ln\left(\frac{20^2 + 5^2}{20^2}\right) \doteq 0.06$$

5.4.9 Beta distribution

A random variable X is beta distributed with parameters $\beta_1 > 0$ and $\beta_2 > 0$ if its pdf is given by

$$f(x) = \begin{cases} \dfrac{x^{\beta_1 - 1}(1 - x)^{\beta_2 - 1}}{B(\beta_1, \beta_2)}, & 0 < x < 1 \\ 0, & \text{otherwise} \end{cases} \tag{5.60}$$

where $B(\beta_1, \beta_2) = \Gamma(\beta_1)\Gamma(\beta_2)/\Gamma(\beta_1 + \beta_2)$. The cdf of the beta does not have a closed form, in general.

The beta distribution is very flexible and has a finite range from 0 to 1, as shown in Figure 5.24. In practice, we often need a beta distribution defined on a different range, say (a, b), with $a < b$, rather than $(0, 1)$. This is easily accomplished by defining a new random variable

$$Y = a + (b - a)X$$

The mean and variance of Y are given by

$$a + (b - a)\left(\frac{\beta_1}{\beta_1 + \beta_2}\right) \tag{5.61}$$

and

$$(b - a)^2 \left(\frac{\beta_1 \beta_2}{(\beta_1 + \beta_2)^2(\beta_1 + \beta_2 + 1)}\right) \tag{5.62}$$

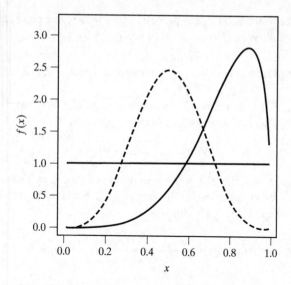

Figure 5.24 pdf for several beta distributions.

5.5 Poisson Process

Consider random events such as the arrival of jobs at a job shop, the arrival of e-mail to a mail server, the arrival of boats to a dock, the arrival of calls to a call center, the breakdown of machines in a large factory, and so on. These events may be described by a counting function $N(t)$ defined for all $t \geq 0$. This counting function will represent the number of events that occurred in $[0, t]$. Time zero is the point at which the observation began, regardless of whether an arrival occurred at that instant. For each interval $[0, t]$, the value $N(t)$ is an observation of a random variable where the only possible values that can be assumed by $N(t)$ are the integers $0, 1, 2, \ldots$.

The counting process, $\{N(t), t \geq 0\}$, is said to be a Poisson process with mean rate λ if the following assumptions are fulfilled:

a. Arrivals occur one at a time.
b. $\{N(t), t \geq 0\}$ has stationary increments: The distribution of the number of arrivals between t and $t + s$ depends only on the length of the interval s, not on the starting point t. Thus, arrivals are completely at random without rush or slack periods.
c. $\{N(t), t \geq 0\}$ has independent increments: The number of arrivals during nonoverlapping time intervals are independent random variables. Thus, a large or small number of arrivals in one time interval has no effect on the number of arrivals in subsequent time intervals. Future arrivals occur completely at random, independent of the number of arrivals in past time intervals.

If arrivals occur according to a Poisson process, meeting the three preceding assumptions, it can be shown that the probability that $N(t)$ is equal to n is given by

$$P[N(t) = n] = \frac{e^{-\lambda t}(\lambda t)^n}{n!} \quad \text{for } t \geq 0 \text{ and } n = 0, 1, 2, \ldots \tag{5.63}$$

Comparing Equation (5.63) to Equation (5.19), it can be seen that $N(t)$ has the Poisson distribution with parameter $\alpha = \lambda t$. Thus, its mean and variance are given by

$$E[N(t)] = \alpha = \lambda t = V[N(t)]$$

For any times s and t such that $s < t$, the assumption of stationary increments implies that the random variable $N(t) - N(s)$, representing the number of arrivals in the interval from s to t, is also Poisson-distributed with mean $\lambda(t - s)$. Thus,

$$P[N(t) - N(s) = n] = \frac{e^{-\lambda(t-s)}[\lambda(t - s)]^n}{n!} \quad \text{for } n = 0, 1, 2, \ldots$$

and

$$E[N(t) - N(s)] = \lambda(t - s) = V[N(t) - N(s)]$$

Now, consider the time at which arrivals occur in a Poisson process. Let the first arrival occur at time A_1, the second occur at time $A_1 + A_2$, and so on, as shown in Figure 5.25. Thus, $A_1, A_2, \ldots$ are successive interarrival times. The first arrival occurs after time t if and only if there are no arrivals in the interval $[0, t]$, so it is seen that

$$\{A_1 > t\} = \{N(t) = 0\}$$

and, therefore,

$$P(A_1 > t) = P[N(t) = 0] = e^{-\lambda t}$$

the last equality following from Equation (5.63). Thus, the probability that the first arrival will occur in $[0, t]$ is given by

$$P(A_1 \leq t) = 1 - e^{-\lambda t}$$

which is the cdf for an exponential distribution with parameter λ. Hence, A_1 is distributed exponentially with mean $E(A_1) = 1/\lambda$. It can also be shown that all interarrival times, $A_1, A_2, \ldots$, are

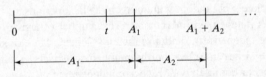

Figure 5.25   Arrival process.

exponentially distributed and independent with mean $1/\lambda$. As an alternative definition of a Poisson process, it can be shown that, if interarrival times are distributed exponentially and independently, then the number of arrivals by time t, say $N(t)$, meets the three previously mentioned assumptions and, therefore, is a Poisson process.

Recall that the exponential distribution is memoryless—that is, the probability of a future arrival in a time interval of length s is independent of the time of the last arrival. The probability of the arrival depends only on the length of the time interval s. Thus, the memoryless property is related to the properties of independent and stationary increments of the Poisson process.

Additional readings concerning the Poisson process may be obtained from many sources, including Parzen [1999], Feller [1968], and Ross [2002].

Example 5.30 _____

The jobs at a machine shop arrive according to a Poisson process with a mean of $\lambda = 2$ jobs per hour. Therefore, the interarrival times are distributed exponentially, with the expected time between arrivals being $E(A) = 1/\lambda = \frac{1}{2}$ hour.

5.5.1   Properties of a Poisson Process

Several properties of the Poisson process, discussed by Ross [2002] and others, are useful in discrete-system simulation. The first of these properties concerns random splitting. Consider a Poisson process $\{N(t), t \geq 0\}$ having rate λ, as represented by the left side of Figure 5.26.

Suppose that, each time an event occurs, it is classified as either a type I or a type II event. Suppose further that each event is classified as a type I event with probability p and type II event with probability $1 - p$, independently of all other events.

Let $N_1(t)$ and $N_2(t)$ be random variables that denote, respectively, the number of type I and type II events occurring in $[0, t]$. Note that $N(t) = N_1(t) + N_2(t)$. It can be shown that $N_1(t)$ and $N_2(t)$ are both Poisson processes having rates λp and $\lambda(1 - p)$, as shown in Figure 5.26. Furthermore, it can be shown that the two processes are independent.

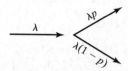

Figure 5.26   Random splitting.

Example 5.31:   Random Splitting _____

Suppose that jobs arrive at a shop in accordance with a Poisson process having rate λ. Suppose further that each arrival is marked "high priority" with probability 1/3 and "low priority" with probability 2/3. Then a type I event would correspond to a high-priority arrival and a type II event would correspond to a low-priority arrival. If $N_1(t)$ and $N_2(t)$ are as just defined, both variables follow the Poisson process, with rates $\lambda/3$ and $2\lambda/3$, respectively.

Example 5.32

The rate in Example 5.31 is $\lambda = 3$ per hour. The probability that no high-priority jobs will arrive in a 2-hour period is given by the Poisson distribution with parameter $\alpha = \lambda pt = 2$. Thus,

$$P(0) = \frac{e^{-2}2^0}{0!} = 0.135$$

Now, consider the opposite situation from random splitting, namely the pooling of two arrival streams. The process of interest is illustrated in Figure 5.27. It can be shown that, if $N_i(t)$ are random variables representing independent Poisson processes with rates λ_i, for $i = 1$ and 2, then $N(t) = N_1(t) + N_2(t)$ is a Poisson process with rate $\lambda_1 + \lambda_2$.

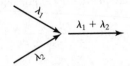

Figure 5.27 Pooled process.

Example 5.33: Pooled Process

A Poisson arrival stream with $\lambda_1 = 10$ arrivals per hour is combined (or pooled) with a Poisson arrival stream with $\lambda_2 = 17$ arrivals per hour. The combined process is a Poisson process with $\lambda = 27$ arrivals per hour.

5.5.2 Nonstationary Poisson Process

If we keep the Poisson Assumptions 1 and 3 (Section 5.5 page 211), but drop Assumption 2 (stationary increments) then we have a *nonstationary Poisson process* (NSPP), which is characterized by $\lambda(t)$, the arrival rate at time t. The NSPP is useful for situations in which the arrival rate varies during the period of interest, including meal times for restaurants, phone calls during business hours, and orders for pizza delivery around 6 P.M.

The key to working with an NSPP is the expected number of arrivals by time t, denoted by

$$\Lambda(t) = \int_0^t \lambda(s)\, ds$$

To be useful as an arrival-rate function, $\lambda(t)$ must be nonnegative and integrable. For a stationary Poisson process with rate λ we have $\Lambda(t) = \lambda t$, as expected.

Let $T_1, T_2, \ldots$ be the arrival times of stationary Poisson process $N(t)$ with $\lambda = 1$, and let $\mathcal{T}_1, \mathcal{T}_2, \ldots$ be the arrival times for an NSPP $\mathcal{N}(t)$ with arrival rate $\lambda(t)$. The fundamental relationship for working with NSPPs is the following:

$$T_i = \Lambda(\mathcal{T}_i)$$
$$\mathcal{T}_i = \Lambda^{-1}(T_i)$$

In other words, an NSPP can be transformed into a stationary Poisson process with arrival rate 1, and a stationary Poisson process with arrival rate 1 can be transformed into an NSPP with rate $\lambda(t)$, and the transformation in both cases is related to $\Lambda(t)$.

Example 5.34

Suppose that arrivals to a post office occur at a rate of 2 per minute from 8 A.M. until 12 P.M., then drop to 1 every 2 minutes until the day ends at 4 P.M. What is the probability distribution of the number of arrivals between 11 A.M. and 2 P.M?

Let time $t = 0$ correspond to 8 A.M. Then this situation could be modeled as an NSPP $\mathcal{N}(t)$ with rate function

$$\lambda(t) = \begin{cases} 2, & 0 \le t < 4 \\ \dfrac{1}{2}, & 4 \le t \le 8 \end{cases}$$

The expected number of arrivals by time t is therefore

$$\Lambda(t) = \begin{cases} 2t, & 0 \le t < 4 \\ \dfrac{t}{2} + 6, & 4 \le t \le 8 \end{cases}$$

Notice that computing the expected number of arrivals for $4 \le t \le 8$ requires that the integration be done in two parts:

$$\Lambda(t) = \int_0^t \lambda(s)\, ds = \int_0^4 2\, ds + \int_4^t \frac{1}{2}\, ds = \frac{t}{2} + 6$$

Since 2 P.M. and 11 A.M. correspond to times 6 and 3, respectively, we have

$$P[\mathcal{N}(6) - \mathcal{N}(3) = k] = P[N(\Lambda(6)) - N(\Lambda(3)) = k]$$

$$= P[N(9) - N(6) = k]$$

$$= \frac{e^{9-6}(9-6)^k}{k!}$$

$$= \frac{e^3(3)^k}{k!}$$

where $N(t)$ is a stationary Poisson process with arrival rate 1.

5.6   Empirical Distributions

An empirical distribution, which may be either discrete or continuous in form, is a distribution whose parameters are the observed values in a sample of data. This is in contrast to parametric distribution families (such as the exponential, normal, or Poisson), which are characterized by specifying a small number of parameters such as the mean and variance. An empirical distribution may be used when it is impossible or unnecessary to establish that a random variable has any particular parametric distribution. One advantage of an empirical distribution is that nothing is assumed beyond the observed values in the sample; however, this is also a disadvantage because the sample might not cover the entire range of possible values.

Example 5.35:   Discrete
Customers at a local restaurant arrive at lunchtime in groups of one to eight people. The number of people per party in the last 300 groups has been observed; the results are summarized in Table 5.3. The relative frequencies appear in Table 5.3 and again in Figure 5.28, which provides a histogram of the data that were gathered. Figure 5.29 provides a cdf of the data. The cdf in Figure 5.29 is called the empirical distribution of the given data.

Table 5.3   Arrivals per Party Distribution

Arrivals per Party	Frequency	Relative Frequency	Cumulative Relative Frequency
1	30	0.10	0.10
2	110	0.37	0.47
3	45	0.15	0.62
4	71	0.24	0.86
5	12	0.04	0.90
6	13	0.04	0.94
7	7	0.02	0.96
8	12	0.04	1.00

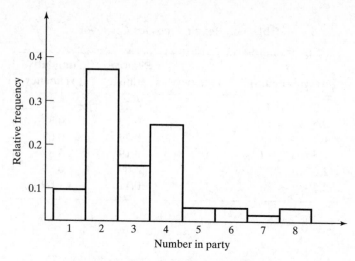

Figure 5.28 Histogram of party size.

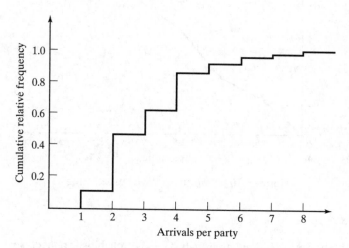

Figure 5.29 Empirical cdf of party size.

Example 5.36: Continuous

The time required to repair a conveyor system that has suffered a failure has been collected for the last 100 instances; the results are shown in Table 5.4. There were 21 instances in which the repair took between 0 and 0.5 hour, and so on. The empirical cdf is shown in Figure 5.30. A piecewise linear curve is formed by the connection of the points of the form $[x, F(x)]$. The points are connected by a straight line. The first connected pair is $(0, 0)$ and $(0.5, 0.21)$; then the points $(0.5, 0.21)$ and $(1.0, 0.33)$ are connected; and so on. More detail on this method is provided in Chapter 8.

Table 5.4   Repair Times for Conveyor

Interval (hours)	Frequency	Relative Frequency	Cumulative Frequency
$0 < x \leq 0.5$	21	0.21	0.21
$0.5 < x \leq 1.0$	12	0.12	0.33
$1.0 < x \leq 1.5$	29	0.29	0.62
$1.5 < x \leq 2.0$	19	0.19	0.81
$2.0 < x \leq 2.5$	8	0.08	0.89
$2.5 < x \leq 3.0$	11	0.11	1.00

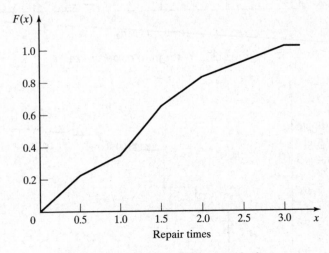

Figure 5.30   Empirical cdf for repair times.

5.7   Summary

In many instances, the world the simulation analyst sees is probabilistic rather than deterministic. The purposes of this chapter were to review several important probability distributions, to familiarize the reader with the notation used in the remainder of the text, and to show applications of the probability distributions in a simulation context.

A major task in simulation is the collection and analysis of input data. One of the first steps in this task is hypothesizing a distributional form for the input data. This is accomplished by comparing the shape of the probability density function or mass function to a histogram of the data and by an understanding that certain physical processes give rise to specific distributions. (Computer software is available to assist in this effort, as will be discussed in Chapter 9.) This chapter was intended to reinforce the properties of various distributions and to give insight into how these distributions arise in practice. In addition, probabilistic models of input data are used in generating random events in a simulation.

Several features that should have made a strong impression on the reader include the differences between discrete, continuous, and empirical distributions; the Poisson process and its properties; and the versatility of the gamma and the Weibull distributions.

REFERENCES

BANKS, J., AND R. G. HEIKES [1984], *Handbook of Tables and Graphs for the Industrial Engineer and Manager*, Reston Publishing, Reston, VA.

DEVORE, J. L. [1999], *Probability and Statistics for Engineers and the Sciences*, 5th ed., Brooks/Cole, Pacific Grove, CA.

FELLER, W. [1968], *An Introduction to Probability Theory and Its Applications*, Vol. I, 3d ed., Wiley, New York.

GORDON, G. [1975], *The Application of GPSS V to Discrete System Simulation*, Prentice-Hall, Englewood Cliffs, NJ.

HADLEY, G., AND T. M. WHITIN [1963], *Analysis of Inventory Systems*, Prentice-Hall, Englewood Cliffs, NJ.

HINES, W. W., AND D. C. MONTGOMERY [1990], *Probability and Statistics in Engineering and Management Science*, 3d ed., Wiley, New York.

LAW, A. M., AND W. D. KELTON [2000], *Simulation Modeling & Analysis*, 3d ed., McGraw-Hill, New York.

PAPOULIS, A. [1990], *Probability and Statistics*, Prentice Hall, Englewood Cliffs, NJ.

PARZEN, E. [1999], *Stochastic Process*, Classics in Applied Mathematics, 24, Society for Industrial & Applied Mathematics, Philadelphia, PA.

PEGDEN, C. D., R. E. SHANNON, AND R. P. SADOWSKI [1995], *Introduction to Simulation Using SIMAN*, 2d ed., McGraw-Hill, New York.

ROSS, S. M. [2002], *Introduction to Probability Models*, 8th ed., Academic Press, New York.

WALPOLE, R. E., AND R. H. MYERS [2002], *Probability and Statistics for Engineers and Scientists*, 7th ed., Prentice Hall, Upper Saddle River, NJ.

EXERCISES

1. A production process manufactures alternators for outboard engines used in recreational boating. On the average, 1% of the alternators will not perform up to the required standards when tested at the engine assembly plant. When a large shipment of alternators is received at the plant, 100 are tested, and, if more than two are nonconforming, the shipment is returned to the alternator manufacturer. What is the probability of returning a shipment?

2. An industrial chemical that will retard the spread of fire in paint has been developed. The local sales representative has estimated, from past experience, that 48% of the sales calls will result in an order.

   (a) What is the probability that the first order will come on the fourth sales call of the day?
   (b) If eight sales calls are made in a day, what is the probability of receiving exactly six orders?
   (c) If four sales calls are made before lunch, what is the probability that one or fewer results in an order?

3. A recent survey indicated that 82% of single women aged 25 years old will be married in their lifetime. Using the binomial distribution, find the probability that two or three women in a sample of twenty will never be married.

4. The Hawks are currently winning 0.55 of their games. There are 5 games in the next two weeks. What is the probability that they will win more games than they lose?

5. Joe Coledge is the third-string quarterback for the University of Lower Alatoona. The probability that Joe gets into any game is 0.40.

   (a) What is the probability that the first game Joe enters is the fourth game of the season?
   (b) What is the probability that Joe plays in no more than two of the first five games?

6. For the random variables X_1 and X_2, which are exponentially distributed with parameter $\lambda = 1$, compute $P(X_1 + X_2 > 2)$.

7. Show that the geometric distribution is memoryless.

8. The number of hurricanes hitting the coast of Florida annually has a Poisson distribution with a mean of 0.8.

   (a) What is the probability that more than two hurricanes will hit the Florida coast in a year?
   (b) What is the probability that exactly one hurricane will hit the coast of Florida in a year?

9. Arrivals at a bank teller's drive-through window are Poisson-distributed at the rate of 1.2 per minute.

   (a) What is the probability of zero arrivals in the next minute?
   (b) What is the probability of zero arrivals in the next 2 minutes?

10. Records indicate that 1.8% of the entering students at a large state university drop out of school by midterm. What is the probability that three or fewer students will drop out of a random group of 200 entering students?

11. Lane Braintwain is quite a popular student. Lane receives, on the average, four phone calls a night (Poisson-distributed). What is the probability that, tomorrow night, the number of calls received will exceed the average by more than one standard deviation?

12. Lead-time demand for condenser units is Poisson-distributed with a mean of 6 units. Prepare a table for the inventory manager that will indicate the order level to achieve protection of the following levels: 50%, 80%, 90%, 95%, 97%, 97.5%, 99%, 99.5%, and 99.9%.

13. A random variable X that has pmf given by $p(x) = 1/(n+1)$ over the range $R_X = \{0, 1, 2, \ldots, n\}$ is said to have a discrete uniform distribution.

    (a) Find the mean and variance of this distribution. *Hint*:

$$\sum_{i=1}^{n} i = \frac{n(n+1)}{2} \text{ and } \sum_{i=1}^{n} i^2 = \frac{n(n+1)(2n+1)}{6}$$

    (b) If $R_X = \{a, a+1, a+2, \ldots, b\}$, compute the mean and variance of X.

14. The lifetime, in years, of a satellite placed in orbit is given by the following pdf:

$$f(x) = \begin{cases} 0.4e^{-0.4x}, & x \geq 0 \\ 0, & \text{otherwise} \end{cases}$$

    (a) What is the probability that this satellite is still "alive" after 5 years?
    (b) What is the probability that the satellite dies between 3 and 6 years from the time it is placed in orbit?

15. A mainframe computer crashes in accordance with a Poisson process, with a mean rate of one crash every 36 hours. Determine the probability that the next crash will occur between 24 and 48 hours after the last crash.

16. (The Poisson distribution can be used to approximate the binomial distribution when n is large and p is small—say, p less than 0.1. In utilizing the Poisson approximation, let $\lambda = np$.) In the production of ball bearings, bubbles or depressions occur, rendering the ball bearing unfit for sale. It has been noted that, on the average, one in every 800 of the ball bearings has one or more of these defects. What is the probability that a random sample of 4000 will yield fewer than three ball bearings with bubbles or depressions?

17. For an exponentially distributed random variable X, find the value of λ that satisfies the following relationship:

$$P(X \leq 3) = 0.9P(X \leq 4)$$

18. Accidents at an industrial site occur one at a time, independently, and completely at random, at a mean rate of one per week. What is the probability that no accidents occur in the next three weeks?

19. A component has an exponential time-to-failure distribution with mean of 10,000 hours.

    (a) The component has already been in operation for its mean life. What is the probability that it will fail by 15,000 hours?
    (b) At 15,000 hours, the component is still in operation. What is the probability that it will operate for another 5000 hours?

20. Suppose that a Die-Hardly-Ever battery has an exponential time-to-failure distribution with a mean of 48 months. At 60 months, the battery is still operating.

   (a) What is the probability that this battery is going to die in the next 12 months?
   (b) What is the probability that the battery dies in an odd year of its life?
   (c) If the battery is operating at 60 months, compute the expected additional months of life.

21. The time to service customers at a bank teller's drive-through window is exponentially distributed with a mean of 50 seconds.

   (a) What is the probability that the two customers in front of an arriving customer will each take less than 60 seconds to complete their transactions?
   (b) What is the probability that the two customers in front will finish their transactions so that an arriving customer can reach the teller's window within 2 minutes?

22. Determine the variance $V(X)$ of the triangular distribution.

23. The daily use of water, in thousands of liters, at the Hardscrabble Tool and Die Works follows a gamma distribution having shape parameter 2 and scale parameter 1/4. What is the probability that the demand exceeds 4000 liters on any given day?

24. When Admiral Byrd went to the North Pole, he wore battery-powered thermal underwear. The batteries failed instantaneously rather than gradually. The batteries had a life that was exponentially distributed, with a mean of 12 days. The trip took 30 days. Admiral Byrd packed three batteries. What is the probability that three batteries would be a number sufficient to keep the Admiral warm?

25. The time intervals between hits on a web page from remote computers are exponentially distributed, with a mean of 15 seconds. Find the probability that the third hit connection occurs after 30 seconds have elapsed.

26. The rail shuttle cars at the Atlanta airport have a dual electrical braking system. A rail car switches to the standby system automatically if the first system fails. If both systems fail, there will be a crash! Assume that the life of a single electrical braking system is exponentially distributed, with a mean of 4000 operating hours. If the systems are inspected every 5000 operating hours, what is the probability that a rail car will not crash before that time?

27. Suppose that cars arriving at a toll booth follow a Poisson process with a mean interarrival time of 15 seconds. What is the probability that up to one minute will elapse until three cars have arrived?

28. Suppose that an average of 30 customers per hour arrive at the Sticky Donut Shop in accordance with a Poisson process. What is the probability that more than 5 minutes will elapse before both of the next two customers walk through the door?

29. Professor Dipsy Doodle gives six problems on each exam. Each problem requires an average of 30 minutes grading time for the entire class of 15 students. The grading time for each problem is exponentially distributed, and the problems are independent of each other.

    (a) What is the probability that the Professor will finish the grading in 2.5 hours or less?
    (b) What is the most likely grading time?
    (c) What is the expected grading time?

30. An aircraft has dual hydraulic systems. The aircraft switches to the standby system automatically if the first system fails. If both systems fail, the plane will crash. Assume that the life of a hydraulic system is exponentially distributed, with a mean of 2000 air hours.

    (a) If the hydraulic systems are inspected every 2500 hours, what is the probability that an aircraft will crash before that time?
    (b) What danger would there be in moving the inspection point to 3000 hours?

31. A random variable X is beta distributed if its pdf is given by

$$f(x) = \begin{cases} \dfrac{(\alpha + \beta + 1)!}{\alpha! \beta!} x^{\alpha}(1 - x)^{\beta}, & 0 < x < 1 \\ 0, & \text{otherwise} \end{cases}$$

Show that the beta distribution becomes the uniform distribution over the unit interval when $\beta_1 = \beta_2 = 1$.

32. Lead time is gamma distributed in 100s of units, with the shape parameter 3 and the scale parameter 1. What is the probability that the lead time exceeds 2 (hundred) units during an upcoming cycle?

33. Lifetime of an inexpensive video card for a PC, in months, denoted by the random variable X, is gamma distributed with $\beta = 4$ and $\theta = 1/16$. What is the probability that the card will last for at least 2 years?

34. Many states have license plates that conform to the following format:

    letter   letter   letter   number   number   number

    The numbers are at random, ranging from 100 to 999.

    (a) What is the probability that the next two plates seen (at random) will have numbers of 500 or higher?
    (b) What is the probability that the sum of the next two plates seen (at random) will have a total of 1000 or higher? [*Hint*: Approximate the discrete uniform distribution with a continuous uniform distribution. The sum of two independent uniform distributions is a triangular distribution.]

35. Let X be a random variable that is normally distributed, with mean 10 and variance 4. Find the values a and b such that $P(a < X < b) = 0.90$ and $|\mu - a| = |\mu - b|$.

36. Given the following distributions:

     Normal $(10, 4)$
     Triangular $(4, 10, 16)$
     Uniform $(4, 16)$

    find the probability that $6 < X < 8$ for each of the distributions.

37. Lead time for an item is approximated by a normal distribution with a mean of 20 days and a variance of 4 days^2. Find values of lead time that will be exceeded only 1%, 5%, and 10% of the time.

38. IQ scores are normally distributed throughout society, with mean 100 and standard deviation 15.

    (a) A person with an IQ of 140 or higher is called a "genius." What proportion of society is in the genius category?
    (b) What proportion of society will miss the genius category by 5 or less points?
    (c) Suppose that an IQ of 110 or higher is required to make it through an accredited college or university. What proportion of society could be eliminated from completing a higher education by having a low IQ score?

39. Three shafts are made and assembled into a linkage. The length of each shaft, in centimeters, is distributed as follows:

     Shaft 1: $N(60, 0.09)$
     Shaft 2: $N(40, 0.05)$
     Shaft 3: $N(50, 0.11)$

    (a) What is the distribution of the length of the linkage?
    (b) What is the probability that the linkage will be longer than 150.2 centimeters?
    (c) The tolerance limits for the assembly are $(149.83, 150.21)$. What proportion of assemblies are within the tolerance limits? [*Hint*: If $\{X_i\}$ are n independent normal random variables, and if X_i has mean μ_i and variance σ_i^2, then the sum

    $$Y = X_1 + X_2 + \cdots + X_n$$

    is normal with mean $\sum_{i=1}^{n} \mu_i$ and variance $\sum_{i=1}^{n} \sigma_i^2$.]

40. The circumferences of battery posts in a nickel-cadmium battery are Weibull-distributed with $v = 3.25$ centimeters, $\alpha = 0.005$ centimeters, and $\beta = 1/3$.

    (a) Find the probability that a battery post chosen at random will have a circumference larger than 3.40 centimeters.
    (b) If battery posts are larger than 3.50 centimeters, they will not go through the hole provided; if they are smaller than 3.30 centimeters, the clamp will not tighten sufficiently. What proportion of posts will have to be scrapped for one of these reasons?

41. The time to failure of a nickel-cadmium battery is Weibull-distributed with parameters $\nu = 0$, $\alpha = 1/2$ years, and $\beta = 1/4$.

    (a) Find the fraction of batteries that are expected to fail prior to 1.5 years.
    (b) What fraction of batteries are expected to last longer than the mean life?
    (c) What fraction of batteries are expected to fail between 1.5 and 2.5 years?

42. Demand for electricity at Gipgip Pig Farm for the merry merry month of May has a triangular distribution, with $a = 100$ kwh and $c = 1800$ kwh. The median kwh is 1425. Compute the modal value of kwh for the month.

43. The time to failure on an electronic subassembly can be modeled by a Weibull distribution whose location parameter is 0, $\alpha = 1000$ hours, and $\beta = 1/2$.

    (a) What is the mean time to failure?
    (b) What fraction of these subassemblies will fail by 3000 hours?

44. The gross weight of three-axle trucks that have been checked at the Hahira Inspection Station on Interstate Highway 85 follows a Weibull distribution with parameters $\nu = 6.8$ tons, $\alpha = 1/2$ ton, and $\beta = 1.5$. Determine the appropriate weight limit such that 0.01 of the trucks will be cited for traveling overweight.

45. The current reading on Sag Revas's gas-mileage indicator is an average of 25.3 miles per gallon. Assume that gas mileage on Sag's car follows a triangular distribution with a minimum value of zero and a maximum value of 50 miles per gallon. What is the value of the median?

46. A postal letter carrier has a route consisting of five segments with the time in minutes to complete each segment being normally distributed, with means and variances as shown:

Tennyson Place	$N(38, 16)$
Windsor Parkway	$N(99, 29)$
Knob Hill Apartments	$N(85, 25)$
Evergreen Drive	$N(73, 20)$
Chastain Shopping Center	$N(52, 12)$

    In addition to the times just mentioned, the letter carrier must organize the mail at the central office, which activity requires a time that is distributed by $N(90, 25)$. The drive to the starting point of the route requires a time that is distributed $N(10, 4)$. The return from the route requires a time that is distributed $N(15, 4)$. The letter carrier then performs administrative tasks with a time that is distributed $N(30, 9)$.

    (a) What is the expected length of the letter carrier's work day?
    (b) Overtime occurs after eight hours of work on a given day. What is the probability that the letter carrier works overtime on any given day?
    (c) What is the probability that the letter carrier works overtime on two or more days in a six-day week?
    (d) What is the probability that the route will be completed within ±24 minutes of eight hours on any given day? [*Hint*: See Exercise 39.]

47. The time to failure of a WD-1 computer chip is known to be Weibull-distributed with parameters $v = 0$, $\alpha = 400$ days, and $\beta = 1/2$. Find the fraction expected to survive 600 days.

48. The TV sets on display at Schocker's Department Store are hooked up such that, when one fails, a model exactly like the one that failed will switch on. Three such units are hooked up in this series arrangement. The lives of these TVs are independent of one another. Each TV has a life which is exponentially distributed, with a mean of 10,000 hours. Find the probability that the combined life of the system is greater than 32,000 hours.

49. High temperature in Biloxi, Mississippi on July 21, denoted by the random variable X, has the following probability density function, where X is in degrees F.

$$f(x) = \begin{cases} \dfrac{2(x - 85)}{119}, & 85 \le x \le 92 \\[2ex] \dfrac{2(102 - x)}{170}, & 92 < x \le 102 \\[2ex] 0, & \text{otherwise} \end{cases}$$

   (a) What is the variance of the temperature $V(X)$? [*Hint*: If you worked Exercise 22, this is quite easy.]
   (b) What is the median temperature?
   (c) What is the modal temperature?

50. The time to failure of Eastinghome light bulbs is Weibull-distributed with $v = 1.8 \times 10^3$ hours, $\alpha = 1/3 \times 10^3$ hours, and $\beta = 1/2$.

   (a) What fraction of bulbs are expected to last longer than the mean lifetime?
   (b) What is the median lifetime of a light bulb?

51. Lead-time demand is gamma distributed in 100s of units, with shape parameter 2 and scale parameter 1/4. What is the probability that the lead time exceeds 4 (hundred) units during an upcoming cycle?

52. Let time $t = 0$ correspond to 6 A.M., and suppose that the arrival rate (in arrivals per hour) of customers to a breakfast restaurant that is open from 6 to 9 A.M. is

$$\lambda(t) = \begin{cases} 30, & 0 \le t < 1 \\ 45, & 1 \le t < 2 \\ 20, & 2 \le t \le 4 \end{cases}$$

Assuming an NSPP model is appropriate, do the following: (a) Derive $\Lambda(t)$. (b) Compute the expected number of arrivals between 6:30 and 8:30 A.M. (c) Compute the probability that there are fewer than 60 arrivals between 6:30 and 8:30 A.M.

6

Queueing Models

Simulation is often used in the analysis of queueing models. In a simple but typical queueing model, shown in Figure 6.1, customers arrive from time to time and join a *queue* (waiting line), are eventually served, and finally leave the system. The term *customer* refers to any type of entity that can be viewed as requesting service from a system. Therefore, many service facilities, production systems, repair and maintenance facilities, communications and computer systems, and transport and material-handling systems can be viewed as queueing systems.

Queueing models, whether solved mathematically or analyzed through simulation, provide the analyst with a powerful tool for designing and evaluating the performance of queueing systems. Typical measures of system performance include server utilization (percentage of time a server is busy), length of waiting lines, and delays of customers. Quite often, when designing or attempting to improve a queueing system, the analyst (or decision maker) is involved in tradeoffs between server utilization and customer satisfaction in terms of line lengths and delays. Queueing theory and simulation analysis are used to predict these measures of system performance as a function of the input parameters. The input parameters include the arrival rate of customers, the service demands of customers, the rate at which a server works, and the number and arrangement of servers. To a certain degree, some of the input parameters are under management's direct control. Consequently, the performance measures could be under their indirect control, provided that the relationship between the performance measures and the input parameters is adequately understood for the given system.

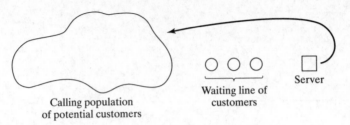

Figure 6.1   Simple queueing model.

For relatively simple systems, these performance measures can be computed mathematically—at great savings in time and expense as compared with the use of a simulation model—but, for realistic models of complex systems, simulation is usually required. Nevertheless, analytically tractable models, although usually requiring many simplifying assumptions, are valuable for rough-cut estimates of system performance. These rough-cut estimates may then be refined by use of a detailed and more realistic simulation model while also providing a way to verify that the simulation model has been programmed correctly (see Chapter 10). Simple models are also useful for developing an understanding of the dynamic behavior of queueing systems and the relationships between various performance measures. This chapter will not develop the mathematical theory of queues but instead will discuss some of the well-known models. For an elementary treatment of queueing theory, the reader is referred to the survey chapters in Hillier and Lieberman [2005] or Winston [2004]. More extensive treatments with a view toward applications are given by Cooper [1990], Gross and Harris [1997], Hall [1991] and Nelson [1995]. The latter two texts especially emphasize engineering and management applications.

This chapter discusses the general characteristics of queues, the meanings and relationships of the important performance measures, estimation of the mean measures of performance from a simulation, the effect of varying the input parameters, and the mathematical solution of a small number of important and basic queueing models.

6.1   Characteristics of Queueing Systems

The key elements of a queueing system are the customers and servers. The term *customer* can refer to people, machines, trucks, mechanics, patients, pallets, airplanes, e-mail, cases, orders, or dirty clothes—anything that arrives at a facility and requires service. The term *server* might refer to receptionists, repair personnel, mechanics, medical personnel, automatic storage and retrieval machines (e.g., cranes), runways at an airport, automatic packers, order pickers, CPUs in a computer, or washing machines—any resource (person, machine, etc.) that provides the requested service. Although the terminology employed will be that of a customer arriving at a service facility, sometimes the server moves to the customer; for example, a repair person moving to a broken machine. This in no way invalidates the models but is merely a matter of terminology. Table 6.1 lists a number of different systems together with a subsystem consisting of arriving customers and one or more servers. The remainder of this section describes the elements of a queueing system in more detail.

Table 6.1   Examples of Queueing Systems

System	Customers	Server(s)
Reception desk	People	Receptionist
Repair facility	Machines	Repair person
Garage	Trucks	Mechanic
Airport security	Passengers	Baggage x-ray
Hospital	Patients	Nurses
Warehouse	Pallets	Fork-lift Truck
Airport	Airplanes	Runway
Production line	Cases	Case-packer
Warehouse	Orders	Order-picker
Road network	Cars	Traffic light
Grocery	Shoppers	Checkout station
Laundry	Dirty linen	Washing machines/dryers
Job shop	Jobs	Machines/workers
Lumberyard	Trucks	Overhead crane
Sawmill	Logs	Saws
Computer	Email	CPU, disk
Telephone	Calls	Exchange
Ticket office	Football fans	Clerk
Mass transit	Riders	Buses, trains

6.1.1   The Calling Population

The population of potential customers, referred to as the *calling population*, may be assumed to be finite or infinite. For example, consider the personal computers of the employees of a small company that are supported by an IT staff of three technicians. When a computer fails, needs new software, etc., it is attended by one of the IT staff. The computers are customers who arrive at the instant they need attention. The IT staff are the servers who provide repairs, software updates, etc. The calling population is finite and consists of the personal computers at the company.

In systems with a large population of potential customers, the calling population is usually assumed to be infinite. For such systems, this assumption is usually innocuous and, furthermore, it might simplify the model. Examples of infinite populations include the potential customers of a restaurant, bank, or other similar service facility and also the personal computers of the employees of a very large company. Even though the actual population could be finite but large, it is generally safe to use infinite population models, provided that the number of customers being served or waiting for service at any given time is a small proportion of the population of potential customers.

The main difference between finite and infinite population models is how the arrival rate is defined. In an infinite population model, the arrival rate (i.e., the average number of arrivals per unit of time) is not affected by the number of customers who have left the calling population and joined the queueing system. When the arrival process is homogeneous over time (e.g., there are no rush hours), the arrival rate is usually assumed to be constant. On the other hand, for finite calling-population models, the arrival rate to the queueing system does depend on the number of customers being served and waiting. To take an extreme case, suppose that the calling population has one member, for example, a corporate jet. When the corporate jet is being serviced by the team of mechanics who are on duty 24 hours per day, the arrival rate is zero, because there are no other potential customers (jets) who can arrive at the service facility (team of mechanics). A more typical example is five hospital patients assigned to a single nurse. When all patients are resting and the nurse is idle, the arrival rate is at its maximum since any of the patients could call the nurse for assistance in the next instant. At those times when all five patients have called for the nurse (four are waiting for the nurse and one is being served) the arrival rate is zero; that is, no arrival is possible until the nurse finishes with a patient, in which case the patient returns to the calling population and becomes a potential arrival. It may seem odd that the arrival rate is at its maximum when all five patients are resting. But the arrival rate is defined as the expected number of arrivals in the next unit of time, so this expectation is largest when all patients could potentially call in the next unit of time.

6.1.2 System Capacity

In many queueing systems, there is a limit to the number of customers that may be in the waiting line or system. For example, an automatic car wash might have room for only 10 cars to wait in line to enter the mechanism. It might be too dangerous (or illegal) for cars to wait in the street. An arriving customer who finds the system full does not enter but returns immediately to the calling population. Some systems, such as in-person concert ticket sales for students, may be considered as having unlimited capacity, since there are no limits on the number of students allowed to wait to purchase tickets. As will be seen later, when a system has limited capacity, a distinction is made between the arrival rate (i.e., the number of arrivals per time unit) and the effective arrival rate (i.e., the number who arrive and enter the system per time unit).

6.1.3 The Arrival Process

The arrival process for infinite-population models is usually characterized in terms of interarrival times of successive customers. Arrivals may occur at scheduled times or at random times. When at random times, the interarrival times are usually characterized by a probability distribution. In addition, customers may arrive one at a time or in batches. The batch may be of constant size or of random size.

The most important model for random arrivals is the Poisson arrival process. If A_n represents the interarrival time between customer $n - 1$ and customer n (A_1 is the actual arrival time of the first customer), then, for a Poisson arrival process, A_n is exponentially distributed with mean $1/\lambda$ time units. The arrival rate is λ customers per time unit. The number of arrivals in a time interval of length t, say $N(t)$, has the Poisson distribution with mean λt customers. For further discussion of the relationship between the Poisson distribution and the exponential distribution, the reader is referred to Section 5.5.

The Poisson arrival process has been employed successfully as a model of the arrival of people to restaurants, drive-in banks, and other service facilities; the arrival of telephone calls to a call center; the arrival of demands, or orders for a service or product; and the arrival of failed components or machines to a repair facility. Typically, the Poisson arrival process is used to describe a large calling population from which customers make independent decisions about when to arrive.

A second important class of arrivals is scheduled arrivals, such as patients to a physician's office or scheduled airline flight arrivals to an airport. In this case it is usually easier to model the positive or negative deviations from the scheduled arrival time, rather than the interarrival times.

A third situation occurs when at least one customer is assumed to always be present in the queue, so that the server is never idle because of a lack of customers. For example, the customers may represent raw material for a product, and sufficient raw material is assumed to be always available.

For finite-population models, the arrival process is characterized in a completely different fashion. Define a customer as *pending* when that customer is outside the queueing system and a member of the potential calling population. For example, a hospital patient is *pending* when they are resting, and becomes *not pending* the instant they call for the nurse. Define a *runtime* of a given customer as the length of time from departure from the queueing system until that customer's next arrival to the queue. Let $A_1^{(i)}, A_2^{(i)}, \ldots$ be the successive runtimes of customer i; let $S_1^{(i)}, S_2^{(i)}, \ldots$ be the corresponding successive service times; and let $W_{Q1}^{(i)}, W_{Q2}^{(i)}, \ldots$ be the corresponding waiting times for service to begin on each visit to the queueing system. Thus, $W_n^{(i)} = W_{Qn}^{(i)} + S_n^{(i)}$ is the total time spent in system by customer i during the nth visit. Figure 6.2 illustrates these concepts for patient 3 in the hospital example. The total arrival process is the superposition of the arrival times of all customers. Figure 6.2 shows the first and second arrival of patient 3, but these two times are not necessarily two successive arrivals to the system. For instance, if it is assumed that all patients are pending at time 0, the first arrival to the system occurs at time $A_1 = \min\{A_1^{(1)}, A_1^{(2)}, A_1^{(3)}, A_1^{(4)}, A_1^{(5)}\}$. If $A_1 = A_1^{(2)}$, then patient 2 is the first arrival (i.e., the first to call for the nurse) after time 0. As discussed earlier, the arrival rate is not constant but is a function of the number of pending customers.

One important application of finite-population models is the machine-repair problem. The machines are the customers, and a runtime is also called *time to failure*. When a machine fails, it *arrives* at the queueing system (the repair facility) and remains there until it is *served* (repaired). Times to failure for a given class of machine have been characterized by the exponential, the Weibull, and the gamma distributions (Chapter 5). Models with an exponential runtime are sometimes analytically tractable; an example is given in Section 6.5. Successive times to failure are usually assumed

Patient 3 Status:

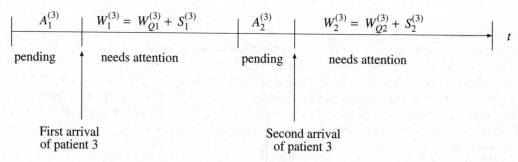

Figure 6.2   Arrival process for a finite-population model.

to be statistically independent, but they could depend on other factors, such as the age of a machine since its last major overhaul.

6.1.4   Queue Behavior and Queue Discipline

Queue behavior refers to the actions of customers while in a queue waiting for service to begin. In some situations, there is a possibility that incoming customers will balk (leave when they see that the line is too long), renege (leave after being in the line when they see that the line is moving too slowly), or jockey (move from one line to another if they think they have chosen a slow line).

Queue discipline refers to the logical ordering of customers in a queue and determines which customer will be chosen for service when a server becomes free. Common queue disciplines include first-in-first-out (FIFO), last-in-first-out (LIFO), service in random order (SIRO), shortest processing time first (SPT), and service according to priority (PR). In a manufacturing system, queue disciplines are sometimes based on due dates and on expected processing time for a given type of job. Notice that a FIFO queue discipline implies that services begin in the same order as arrivals, but that customers could leave the system in a different order because of different-length service times.

6.1.5   Service Times and the Service Mechanism

The service times of successive arrivals are denoted by $S_1, S_2, S_3, \ldots$ They may be constant or of random duration. In the latter case, $\{S_1, S_2, S_3, \ldots\}$ is usually characterized as a sequence of independent and identically distributed random variables. The exponential, Weibull, gamma, lognormal, and truncated normal distributions have all been used successfully as models of service times in different situations. Sometimes services are identically distributed for all customers of a given type or class or priority, whereas customers of different types might have completely different service-time distributions. In addition, in some systems, service times depend upon the time of day or upon the length of the waiting line. For example, servers might work faster than usual when the waiting line is long, thus effectively reducing the service times.

A queueing system consists of a number of service centers and interconnecting queues. Each service center consists of some number of servers, c, working in parallel; that is, upon getting to the head of the line, a customer takes the first available server. Parallel service mechanisms are either single server ($c = 1$), multiple server ($1 < c < \infty$), or unlimited servers ($c = \infty$). A self-service facility is usually characterized as having an unlimited number of servers.

Example 6.1

Consider a discount warehouse where customers may either serve themselves or wait for one of three clerks, then finally leave after paying a single cashier. The system is represented by the flow diagram in Figure 6.3. The subsystem, consisting of queue 2 and service center 2, is shown in more detail in Figure 6.4. Other variations of service mechanisms include batch service (a server serving several customers simultaneously) and a customer requiring several servers simultaneously. In the discount warehouse, a clerk might pick several small orders at the same time, but it may take two of the clerks to handle one heavy item.

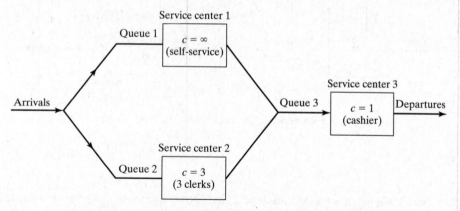

Figure 6.3   Discount warehouse with three service centers.

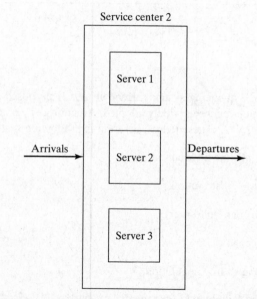

Figure 6.4   Service center 2, with $c = 3$ parallel servers.

Example 6.2

A candy manufacturer has a production line that consists of three machines separated by inventory-in-process buffers. The first machine makes and wraps the individual pieces of candy, the second packs 50 pieces in a box, and the third machine seals and wraps the box. The two inventory buffers have capacities of 1000 boxes each. As illustrated by Figure 6.5, the system is modeled as having three service centers, each center having $c = 1$ server (a machine), with queue capacity constraints between machines. It is assumed that a sufficient supply of raw material is always available at the first queue. Because of the queue capacity constraints, machine 1 shuts down whenever its inventory

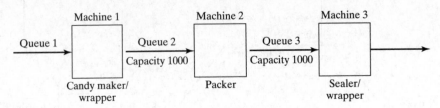

Figure 6.5   Candy-production line.

buffer (queue 2) fills to capacity, and machine 2 shuts down whenever its buffer empties. In brief, the system consists of three single-server queues in series with queue capacity constraints and a continuous arrival stream at the first queue.

6.2   Queueing Notation

Recognizing the diversity of queueing systems, Kendall [1953] proposed a notational system for parallel server systems which has been widely adopted. An abridged version of this convention is based on the format $A/B/c/N/K$. These letters represent the following system characteristics:

A represents the interarrival-time distribution.

B represents the service-time distribution.

c represents the number of parallel servers.

N represents the system capacity.

K represents the size of the calling population.

Common symbols for A and B include M (exponential or Markov), D (constant or deterministic), E_k (Erlang of order k), PH (phase-type), H (hyperexponential), G (arbitrary or general), and GI (general independent).

For example, $M/M/1/\infty/\infty$ indicates a single-server system that has unlimited queue capacity and an infinite population of potential arrivals. The interarrival times and service times are exponentially distributed. When N and K are infinite, they may be dropped from the notation. For example, $M/M/1/\infty/\infty$ is often shortened to $M/M/1$. The nurse attending 5 hospital patients might be represented by $M/M/1/5/5$.

Additional notation used throughout the remainder of this chapter for parallel server systems is listed in Table 6.2. The meanings may vary slightly from system to system. All systems will be assumed to have a FIFO queue discipline.

Table 6.2   Queueing Notation for Parallel Server Systems

P_n	Steady-state probability of having n customers in system
$P_n(t)$	Probability of n customers in system at time t
λ	Arrival rate
λ_e	Effective arrival rate
μ	Service rate of one server
ρ	Server utilization
A_n	Interarrival time between customers $n-1$ and n
S_n	Service time of the nth arriving customer
W_n	Total time spent in system by the nth arriving customer
W_n^Q	Total time spent waiting in queue by customer n
$L(t)$	The number of customers in system at time t
$L_Q(t)$	The number of customers in queue at time t
L	Long-run time-average number of customers in system
L_Q	Long-run time-average number of customers in queue
w	Long-run average time spent in system per customer
w_Q	Long-run average time spent in queue per customer

6.3   Long-Run Measures of Performance of Queueing Systems

The primary long-run measures of performance of queueing systems are the long-run time-average number of customers in the system (L) and in the queue (L_Q), the long-run average time spent in system (w) and in the queue (w_Q) per customer, and the server utilization, or proportion of time that a server is busy (ρ). The term *system* usually refers to the waiting line plus the service mechanism, but, in general, can refer to any subsystem of the queueing system; on the other hand, the term *queue* refers to the waiting line alone. Other measures of performance of interest include the long-run proportion of customers who are delayed in queue longer than t_0 time units, the long-run proportion of customers turned away because of capacity constraints, and the long-run proportion of time the waiting line contains more than k_0 customers.

This section defines the major measures of performance for a general $G/G/c/N/K$ queueing system, discusses their relationships, and shows how they can be estimated from a simulation run. There are two types of estimators: an ordinary sample average, and a time-integrated (or time-weighted) sample average.

6.3.1   Time-Average Number in System *L*

Consider a queueing system over a period of time T, and let $L(t)$ denote the number of customers in the system at time t. A simulation of such a system is shown in Figure 6.6.

Let T_i denote the total time during $[0, T]$ in which the system contained exactly i customers. In Figure 6.6, it is seen that $T_0 = 3, T_1 = 12, T_2 = 4$, and $T_3 = 1$. (The line segments whose lengths total

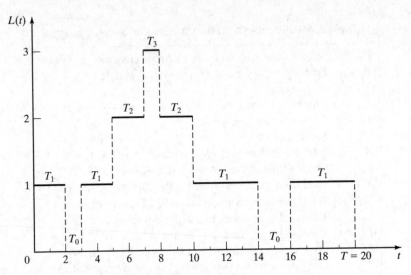

Figure 6.6   Number in system, $L(t)$, at time t.

$T_1 = 12$ are labeled "T_1" in Figure 6.6, etc.) In general, $\sum_{i=0}^{\infty} T_i = T$. The time-weighted-average number in a system is defined by

$$\widehat{L} = \frac{1}{T} \sum_{i=0}^{\infty} i T_i = \sum_{i=0}^{\infty} i \left(\frac{T_i}{T} \right) \tag{6.1}$$

For Figure 6.6, $\widehat{L} = [0(3) + 1(12) + 2(4) + 3(1)]/20 = 23/20 = 1.15$ customers. Notice that T_i/T is the proportion of time the system contains exactly i customers. The estimator $\widehat{L}$ is an example of a time-weighted average.

By considering Figure 6.6, it can be seen that the total area under the function $L(t)$ can be decomposed into rectangles of height i and length T_i. For example, the rectangle of area $3 \times T_3$ has base running from $t = 7$ to $t = 8$ (thus $T_3 = 1$); however, most of the rectangles are broken into parts, such as the rectangle of area $2 \times T_2$ which has part of its base between $t = 5$ and $t = 7$ and the remainder from $t = 8$ to $t = 10$ (thus $T_2 = 2 + 2 = 4$). It follows that the total area is given by $\sum_{i=0}^{\infty} i T_i = \int_0^T L(t)\, dt$, and, therefore, that

$$\widehat{L} = \frac{1}{T} \sum_{i=0}^{\infty} i T_i = \frac{1}{T} \int_0^T L(t)\, dt \tag{6.2}$$

The expressions in Equations (6.1) and (6.2) are always equal for any queueing system, regardless of the number of servers, the queue discipline, or any other special circumstances. Equation (6.2) justifies the terminology *time-integrated average*.

Many queueing systems exhibit a certain kind of long-run stability in terms of their average performance. For such systems, as time T gets large, the observed time-average number in the system $\widehat{L}$ approaches a limiting value, say L, which is called the long-run time-average number in system—that is, with probability 1,

$$\widehat{L} = \frac{1}{T} \int_0^T L(t)\,dt \longrightarrow L \text{ as } T \longrightarrow \infty \tag{6.3}$$

The estimator $\widehat{L}$ is said to be strongly consistent for L. If simulation run length T is sufficiently long, the estimator $\widehat{L}$ becomes arbitrarily close to L. Unfortunately, for $T < \infty$, $\widehat{L}$ depends on the initial conditions at time 0.

Equations (6.2) and (6.3) can be applied to any subsystem of a queueing system as well as to the whole system. If $L_Q(t)$ denotes the number of customers waiting in queue, and T_i^Q denotes the total time during $[0, T]$ in which exactly i customers are waiting in queue, then

$$\widehat{L}_Q = \frac{1}{T} \sum_{i=0}^{\infty} i T_i^Q = \frac{1}{T} \int_0^T L_Q(t)\,dt \longrightarrow L_Q \text{ as } T \longrightarrow \infty \tag{6.4}$$

where $\widehat{L}_Q$ is the observed time-average number of customers waiting in queue from time 0 to time T and L_Q is the long-run time-average number waiting in queue.

Example 6.3

Suppose that Figure 6.6 represents a single-server queue—that is, a $G/G/1/N/K$ queueing system ($N \geq 3, K \geq 3$). Then the number of customers waiting in queue is given by $L_Q(t)$, defined by

$$L_Q(t) = \begin{cases} 0 & \text{if } L(t) = 0 \\ L(t) - 1 & \text{if } L(t) \geq 1 \end{cases}$$

and shown in Figure 6.7. Thus, $T_0^Q = 5 + 10 = 15$, $T_1^Q = 2 + 2 = 4$, and $T_2^Q = 1$. Therefore,

$$\widehat{L}_Q = \frac{0(15) + 1(4) + 2(1)}{20} = 0.3 \text{ customers}$$

6.3.2   Average Time Spent in System Per Customer w

If we simulate a queueing system for some period of time, say, T, then we can record the time each customer spends in the system during $[0, T]$, say $W_1, W_2, \ldots, W_N$, where N is the number of arrivals during $[0, T]$. The average time spent in system per customer, called the *average system time*, is given by the ordinary sample average

$$\widehat{w} = \frac{1}{N} \sum_{i=1}^{N} W_i \tag{6.5}$$

For stable systems, as $N \longrightarrow \infty$,

$$\widehat{w} \longrightarrow w \tag{6.6}$$

with probability 1, where w is called the *long-run average system time*.

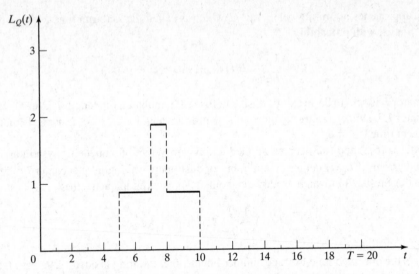

Figure 6.7   Number waiting in queue, $L_Q(t)$, at time t.

If the system under consideration is the queue alone, Equations (6.5) and (6.6) are written as

$$\widehat{w}_Q = \frac{1}{N} \sum_{i=1}^{N} W_i^Q \longrightarrow w_Q \quad \text{as } N \longrightarrow \infty \tag{6.7}$$

where W_i^Q is the total time customer i spends waiting in queue, $\widehat{w}_Q$ is the observed average time spent in queue (called *delay*), and w_Q is the long-run average delay per customer. The estimators $\widehat{w}$ and $\widehat{w}_Q$ are influenced by initial conditions at time 0 and the run length T, analogously to $\widehat{L}$.

Example 6.4
For the system history shown in Figure 6.6, $N = 5$ customers arrive, $W_1 = 2$, and $W_5 = 20 - 16 = 4$, but W_2, W_3, and W_4 cannot be computed unless more is known about the system. Assume that the system has a single server and a FIFO queue discipline. This implies that customers will depart from the system in the same order in which they arrived. Each jump upward of $L(t)$ in Figure 6.6 represents an arrival. Arrivals occur at times 0, 3, 5, 7, and 16. Similarly, departures occur at times 2, 8, 10, and 14. (A departure may or may not have occurred at time 20.) Under these assumptions, it is apparent that $W_2 = 8 - 3 = 5$, $W_3 = 10 - 5 = 5$, $W_4 = 14 - 7 = 7$, and therefore

$$\widehat{w} = \frac{2 + 5 + 5 + 7 + 4}{5} = \frac{23}{5} = 4.6 \text{ time units}$$

Thus, on the average, these customers spent 4.6 time units in the system. As for time spent in the waiting line, it can be computed that $W_1^Q = 0$, $W_2^Q = 0$, $W_3^Q = 8 - 5 = 3$, $W_4^Q = 10 - 7 = 3$, and

$W_5^Q = 0$; thus,

$$\widehat{w}_Q = \frac{0 + 0 + 3 + 3 + 0}{5} = 1.2 \text{ time units}$$

6.3.3   The Conservation Equation: $L = \lambda w$

For the system exhibited in Figure 6.6, there were $N = 5$ arrivals in $T = 20$ time units, and thus the observed arrival rate was $\widehat{\lambda} = N/T = 1/4$ customer per time unit. Recall that $\widehat{L} = 1.15$ and $\widehat{w} = 4.6$; hence, it follows that

$$\widehat{L} = \widehat{\lambda}\widehat{w} \qquad (6.8)$$

This relationship between L, λ, and w is not coincidental; it holds for almost all queueing systems or subsystems regardless of the number of servers, the queue discipline, or any other special circumstances. Allowing $T \longrightarrow \infty$ and $N \longrightarrow \infty$, Equation (6.8) becomes

$$L = \lambda w \qquad (6.9)$$

where $\widehat{\lambda} \longrightarrow \lambda$, and λ is the long-run average arrival rate. Equation (6.9) is called a conservation equation and is usually attributed to Little [1961]. It says that the average number of customers in the system at an arbitrary point in time is equal to the average number of arrivals per time unit, times the average time spent in the system. For Figure 6.6, there is one arrival every 4 time units (on average) and each arrival spends 4.6 time units in the system (on average), so at an arbitrary point in time there will be $(1/4)(4.6) = 1.15$ customers present (on average).

Equation (6.8) can also be derived by reconsidering Figure 6.6 in the following manner: Figure 6.8 shows system history, $L(t)$, exactly as in Figure 6.6, with each customer's time in the system, W_i, represented by a rectangle. This representation again assumes a single-server system with a FIFO queue discipline. The rectangles for the third and fourth customers are in two and three separate pieces, respectively. The ith rectangle has height 1 and length W_i for each $i = 1, 2, \ldots, N$. It follows that the total system time of all customers is given by the total area under the number-in-system function, $L(t)$; that is,

$$\sum_{i=1}^{N} W_i = \int_0^T L(t)\, dt \qquad (6.10)$$

Therefore, by combining Equations (6.2) and (6.5) with $\widehat{\lambda} = N/T$, it follows that

$$\widehat{L} = \frac{1}{T} \int_0^T L(t)\, dt = \frac{N}{T}\frac{1}{N}\sum_{i=1}^{N} W_i = \widehat{\lambda}\widehat{w}$$

which is Little's equation (6.8). The intuitive and informal derivation presented here depended on the single-server FIFO assumptions, but these assumptions are not necessary. In fact, Equation (6.10),

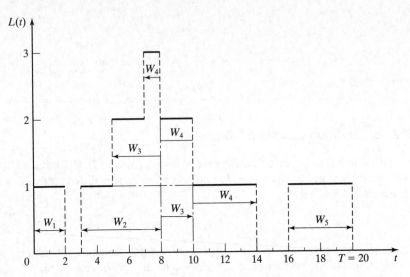

Figure 6.8   System times W_i for single-server FIFO system.

which was the key to the derivation, holds (at least approximately) in great generality, and thus so do Equations (6.8) and (6.9). Exercises 14 and 15 ask the reader to derive Equations (6.10) and (6.8) under different assumptions.

   Technical note: If, as defined in Section 6.3.2, W_i is the system time for customer i during $[0, T]$, then Equation (6.10) and hence Equation (6.8) hold exactly. Some authors choose to define W_i as total system time for customer i; this change will affect the value of W_i only for those customers i who arrive before time T but do not depart until after time T (possibly customer 5 in Figure 6.8). With this change in definition, Equations (6.10) and (6.8) hold only approximately. Nevertheless, as $T \longrightarrow \infty$ and $N \longrightarrow \infty$, the error in Equation (6.8) decreases to zero, and, therefore, the conservation equation (6.9) for long-run measures of performance—namely, $L = \lambda w$—holds exactly.

6.3.4   Server Utilization

Server utilization is defined as the proportion of time that a server is busy. Observed server utilization, denoted by $\widehat{\rho}$, is defined over a specified time interval $[0, T]$. Long-run server utilization is denoted by ρ. For systems that exhibit long-run stability,

$$\widehat{\rho} \longrightarrow \rho \text{ as } T \longrightarrow \infty$$

Example 6.5 _____

Per Figure 6.6 or 6.8, and assuming that the system has a single server, it can be seen that the server utilization is $\widehat{\rho} = (\text{total busy time})/T = (\sum_{i=1}^{\infty} T_i)/T = (T - T_0)/T = 17/20$.

Server utilization in $G/G/1/\infty/\infty$ queues

Consider any single-server queueing system with average arrival rate λ customers per time unit, average service time $E(S) = 1/\mu$ time units, and infinite queue capacity and calling population. Notice that $E(S) = 1/\mu$ implies that, when busy, the server is working at the rate μ customers per time unit, on the average; μ is called the *service rate*. The server alone is a subsystem that can be considered as a queueing system in itself; hence, the conservation Equation (6.9), $L = \lambda w$, can be applied to the server. For stable systems, the average arrival rate to the server, say λ_s, must be identical to the average arrival rate to the system, λ (certainly $\lambda_s \leq \lambda$—customers cannot be served faster than they arrive—but, if $\lambda_s < \lambda$, then the waiting line would tend to grow in length at an average rate of $\lambda - \lambda_s$ customers per time unit, and so we would have an unstable system). For the server subsystem, the average system time is $w = E(S) = \mu^{-1}$. The actual number of customers in the server subsystem is either 0 or 1, as shown in Figure 6.9 for the system represented by Figure 6.6. Hence, the average number in the server subsystem, $\widehat{L}_s$, is given by

$$\widehat{L}_s = \frac{1}{T} \int_0^T (L(t) - L_Q(t)) \, dt = \frac{T - T_0}{T}$$

In this case, $\widehat{L}_s = 17/20 = \widehat{\rho}$. In general, for a single-server queue, the average number of customers being served at an arbitrary point in time is equal to server utilization. As $T \longrightarrow \infty, \widehat{L}_s = \widehat{\rho} \longrightarrow L_s = \rho$. Combining these results into $L = \lambda w$ for the server subsystem yields

$$\rho = \lambda E(S) = \frac{\lambda}{\mu} \tag{6.11}$$

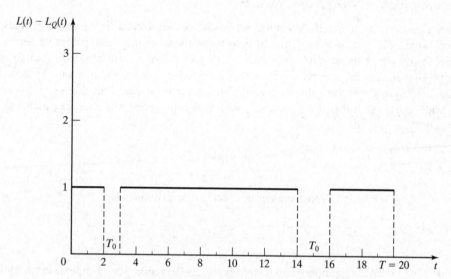

Figure 6.9 Number being served, $L(t) - L_Q(t)$, at time t.

that is, the long-run server utilization in a single-server queue is equal to the average arrival rate divided by the average service rate. For a single-server queue to be stable, the arrival rate λ must be less than the service rate μ:

$$\lambda < \mu$$

or

$$\rho = \frac{\lambda}{\mu} < 1 \qquad (6.12)$$

If the arrival rate is greater than the service rate ($\lambda > \mu$), the server will eventually get further and further behind. After a time, the server will always be busy, and the waiting line will tend to grow in length at an average rate of ($\lambda - \mu$) customers per time unit, because departures will be occurring at rate μ per time unit. For stable single-server systems ($\lambda < \mu$ or $\rho < 1$), long-run measures of performance such as average queue length L_Q (and also L, w, and w_Q) are well defined and have meaning. For unstable systems ($\lambda > \mu$), long-run server utilization is 1, and long-run average queue length is infinite; that is,

$$\frac{1}{T} \int_0^T L_Q(t)\, dt \longrightarrow +\infty \text{ as } T \longrightarrow \infty$$

Similarly, $L = w = w_Q = \infty$. Therefore these long-run measures of performance are meaningless for unstable queues. The quantity λ/μ is also called the *offered load* and is a measure of the workload imposed on the system.

Server utilization in $G/G/c/\infty/\infty$ queues

Consider a queueing system with c identical servers in parallel. If an arriving customer finds more than one server idle, the customer chooses a server without favoring any particular server. (For example, the choice of server might be made at random.) Arrivals occur at rate λ from an infinite calling population, and each server works at rate μ customers per time unit. From Equation (6.9), $L = \lambda w$, applied to the server subsystem alone, an argument similar to the one given for a single server leads to the result that, for systems in statistical equilibrium, the average number of busy servers, say L_s, is given by

$$L_s = \lambda E(S) = \frac{\lambda}{\mu} \qquad (6.13)$$

Clearly, $0 \leq L_s \leq c$. The long-run average server utilization is defined by

$$\rho = \frac{L_s}{c} = \frac{\lambda}{c\mu} \qquad (6.14)$$

and so $0 \leq \rho \leq 1$. The utilization ρ can be interpreted as the proportion of time an arbitrary server is busy in the long run.

The maximum service rate of the $G/G/c/\infty/\infty$ system is $c\mu$, which occurs when all servers are busy. For the system to be stable, the average arrival rate λ must be less than the maximum service rate $c\mu$; that is, the system is stable if and only if

$$\lambda < c\mu \tag{6.15}$$

or, equivalently, if the offered load λ/μ is less than the number of servers c. If $\lambda > c\mu$, then arrivals are occurring, on the average, faster than the system can handle them, all servers will be continuously busy, and the waiting line will grow in length at an average rate of $(\lambda - c\mu)$ customers per time unit. Such a system is unstable, and the long-run performance measures (L, L_Q, w, and w_Q) are again meaningless for such systems.

Notice that Condition (6.15) generalizes Condition (6.12), and the equation for utilization for stable systems, Equation (6.14), generalizes Equation (6.11).

Equations (6.13) and (6.14) can also be applied when some servers work more than others; for example, when customers favor one server over others, or when certain servers serve customers only if all other servers are busy. In this case, the L_s given by Equation (6.13) is still the average number of busy servers, but ρ, as given by Equation (6.14), cannot be applied to an individual server. Instead, ρ must be interpreted as the average utilization of all servers.

Example 6.6

Customers arrive at random to a license bureau at a rate of $\lambda = 50$ customers per hour. Currently, there are 20 clerks, each serving $\mu = 5$ customers per hour on the average. Therefore the long-run, or steady-state, average utilization of a server, given by Equation (6.14), is

$$\rho = \frac{\lambda}{c\mu} = \frac{50}{20(5)} = 0.5$$

and the average number of busy servers is

$$L_s = \frac{\lambda}{\mu} = \frac{50}{5} = 10$$

Thus, in the long run, a typical clerk is busy serving customers only 50% of the time. The office manager asks whether the number of servers can be decreased. By Equation (6.15), it follows that, for the system to be stable, it is necessary for the number of servers to satisfy

$$c > \frac{\lambda}{\mu}$$

or $c > 50/5 = 10$. Thus, possibilities for the manager to consider include $c = 11$, or $c = 12$, or $c = 13, \ldots$. Notice that $c \geq 11$ guarantees long-run stability only in the sense that all servers, when busy, can handle the incoming work load (i.e., $c\mu > \lambda$) on average. The office manager could well desire to have more than the minimum number of servers ($c = 11$) because of other factors, such as customer delays and length of the waiting line. A stable queue can still have very long lines on average.

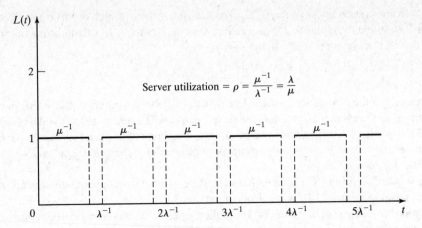

Figure 6.10   Deterministic queue ($D/D/1$).

Server utilization and system performance

As will be illustrated here and in later sections, system performance can vary widely for a given value of utilization, ρ. Consider a $G/G/1/\infty/\infty$ queue, that is, a single-server queue with arrival rate λ, service rate μ, and utilization $\rho = \lambda/\mu < 1$.

At one extreme, consider the $D/D/1$ queue, which has deterministic arrival and service times. Then all interarrival times $\{A_1, A_2, \ldots\}$ are equal to $E(A) = 1/\lambda$, and all service times $\{S_1, S_2, \ldots\}$ are equal to $E(S) = 1/\mu$. Assuming that a customer arrives to an empty system at time 0, the system evolves in a completely deterministic and predictable fashion, as shown in Figure 6.10. Observe that $L = \rho = \lambda/\mu, w = E(S) = \mu^{-1}$, and $L_Q = w_Q = 0$. By varying λ and μ, server utilization can assume any value between 0 and 1, yet there is never any line whatsoever. What, then, causes lines to build, if not a high server utilization? In general, it is the variability of interarrival and service times that causes lines to fluctuate in length.

Example 6.7
Consider a physician who schedules patients every 10 minutes and who spends S_i minutes with the ith patient, where

$$S_i = \begin{cases} 9 \text{ minutes with probability } 0.9 \\ 12 \text{ minutes with probability } 0.1 \end{cases}$$

Thus, arrivals are deterministic ($A_1 = A_2 = \cdots = \lambda^{-1} = 10$) but services are stochastic (or probabilistic), with mean and variance given by

$$E(S_i) = 9(0.9) + 12(0.1) = 9.3 \text{ minutes}$$

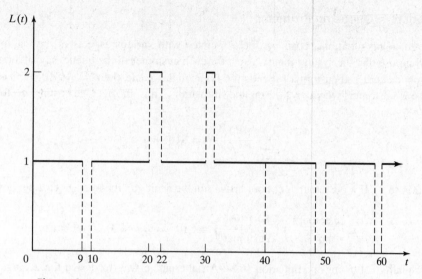

Figure 6.11   Number of patients in the doctor's office at time t.

and

$$V(S_i) = E(S_i^2) - [E(S_i)]^2$$
$$= 9^2(0.9) + 12^2(0.1) - (9.3)^2$$
$$= 0.81 \text{ minutes}^2$$

Here, $\rho = \lambda/\mu = E(S)/E(A) = 9.3/10 = 0.93 < 1$, the system is stable, and the physician will be busy 93% of the time in the long run. In the short run, lines will not build up as long as patients require only 9 minutes of service, but, because of the variability in the service times, 10% of the patients will require 12 minutes, which in turn will cause a temporary line to form.

Suppose the system is simulated with service times, $S_1 = 9, S_2 = 12, S_3 = 9, S_4 = 9, S_5 = 9, \ldots$. Assuming that at time 0 a patient arrived to find the doctor idle and subsequent patients arrived precisely at times 10, 20, 30, $\ldots$, the system evolves as in Figure 6.11. The delays in queue are $W_1^Q = W_2^Q = 0$, $W_3^Q = 22-20 = 2$, $W_4^Q = 31-30 = 1$, $W_5^Q = 0$. The occurrence of a relatively long service time (here $S_2 = 12$) caused a waiting line to form temporarily. In general, because of the variability of the interarrival and service distributions, relatively small interarrival times and relatively large service times occasionally do occur, and these in turn cause lines to lengthen. Conversely, the occurrence of a large interarrival time or a small service time will tend to shorten an existing waiting line. The relationship between utilization, service and interarrival variability, and system performance will be explored in more detail in Section 6.4.

6.3.5   Costs in Queueing Problems

In many queueing situations, costs can be associated with various aspects of the waiting line or servers. Suppose that the system incurs a cost for each customer in the queue, say, at a rate of $10 per hour per customer. If customer j spends W_j^Q hours in the queue, then $\sum_{j=1}^{N}(\$10 \cdot W_j^Q)$ is the total cost of the N customers who arrive during the simulation. Thus, the average cost per customer is

$$\sum_{j=1}^{N} \frac{\$10 \cdot W_j^Q}{N} = \$10 \cdot \widehat{w}_Q$$

by Equation (6.7). If $\widehat{\lambda}$ customers per hour arrive (on the average), the average cost per hour is

$$\left(\widehat{\lambda} \; \frac{\text{customers}}{\text{hour}}\right)\left(\frac{\$10 \cdot \widehat{w}_Q}{\text{customer}}\right) = \$10 \cdot \widehat{\lambda}\widehat{w}_Q = \$10 \cdot \widehat{L}_Q/\text{hour}$$

the last equality following by Equation (6.8). An alternative way to derive the average cost per hour is to consider Equation (6.2). If T_i^Q is the total time over the interval $[0, T]$ that the system contains exactly i customers, then $\$10 \, iT_i^Q$ is the cost incurred by the system during the time exactly i customers are present. Thus, the total cost is $\sum_{i=1}^{\infty}(\$10 \cdot iT_i^Q)$, and the average cost per hour is

$$\sum_{i=1}^{\infty} \frac{\$10 \cdot iT_i^Q}{T} = \$10 \cdot \widehat{L}_Q/ \text{ hour}$$

by Equation (6.2). In these cost expressions, $\widehat{L}_Q$ may be replaced by L_Q (if the long-run number in queue is known), or by L or $\widehat{L}$ (if costs are incurred while the customer is being served in addition to being delayed).

The server may also impose costs on the system. If a group of c parallel servers ($1 \leq c < \infty$) have utilization ρ, and each server imposes a cost of $5 per hour while busy, the total server cost per hour is

$$\$5 \cdot c\rho$$

because $c\rho$ is the average number of busy servers. If server cost is imposed only when the servers are idle, then the server cost per hour would be

$$\$5 \cdot c(1 - \rho)$$

because $c(1 - \rho) = c - c\rho$ is the average number of idle servers. In many problems, two or more of these various costs are combined into a total cost. Such problems are illustrated by Exercises 1, 8, 12, and 19. In most cases, the objective is to minimize total costs (given certain constraints) by varying those parameters that are under management's control, such as the number of servers, the arrival rate, the service rate, and the system capacity.

6.4   Steady-State Behavior of Infinite-Population Markovian Models

This section presents steady-state results for a number of queueing models that can be solved mathematically. For the infinite-population models, the arrivals are assumed to follow a Poisson process with rate λ arrivals per time unit—that is, the interarrival times are assumed to be exponentially distributed with mean $1/\lambda$. Service times may be exponentially distributed (M) or arbitrarily (G). The queue discipline will be FIFO. Because of the exponential distribution assumptions on the arrival process, these models are called *Markovian models*.

A queueing system is said to be in *statistical equilibrium*, or *steady state*, if the probability that the system is in a given state is not time-dependent—that is,

$$P(L(t) = n) = P_n(t) = P_n$$

is independent of time t. Two properties—approaching statistical equilibrium from any starting state, and remaining in statistical equilibrium once it is reached—are characteristic of many stochastic models and, in particular, of all the systems studied in the following subsections. On the other hand, if an analyst were interested in the transient behavior of a queue over a relatively short period of time and were given some specific initial conditions (such as idle and empty), the results to be presented here would be inappropriate. A transient mathematical analysis or, more likely, a simulation model would be the chosen tool of analysis.

The mathematical models whose solutions are shown in the following subsections can be used to obtain approximate results even when the assumptions of the model do not strictly hold. These results may be considered as a rough guide to the behavior of the system. A simulation may then be used for a more refined analysis. However, it should be remembered that a mathematical analysis (when it is applicable) provides the true value of the model parameter (e.g., L), whereas a simulation analysis delivers a statistical estimate (e.g., $\widehat{L}$) of the parameter. On the other hand, for complex systems, a simulation model is often a more faithful representation than a mathematical model.

For the simple models studied here, the steady-state parameter L, the time-average number of customers in the system, can be computed as

$$L = \sum_{n=0}^{\infty} nP_n \tag{6.16}$$

where $\{P_n\}$ are the steady-state probabilities of finding n customers in the system (as defined in Table 6.2). As was discussed in Section 6.3 and was expressed in Equation (6.3), L can also be interpreted as a long-run measure of performance of the system. Once L is given, the other steady-state parameters can be computed readily from Little's equation (6.9) applied to the whole system and to the queue alone:

$$w = \frac{L}{\lambda}$$

$$w_Q = w - \frac{1}{\mu} \tag{6.17}$$

$$L_Q = \lambda w_Q$$

where λ is the arrival rate and μ is the service rate per server.

For the $M/G/c/\infty/\infty$ queues considered in this section to have a statistical equilibrium, a necessary and sufficient condition is that $\lambda/(c\mu) < 1$, where λ is the arrival rate, μ is the service rate of one server, and c is the number of parallel servers. For these unlimited capacity, infinite-calling-population models, it is assumed that the theoretical server utilization, $\rho = \lambda/(c\mu)$, satisfies $\rho < 1$. For models with finite system capacity or finite calling population, the quantity $\lambda/(c\mu)$ may assume any positive value.

6.4.1   Single-Server Queues with Poisson Arrivals and Unlimited Capacity: $M/G/1$

Suppose that service times have mean $1/\mu$ and variance σ^2 and that there is one server. If $\rho = \lambda/\mu < 1$, then the $M/G/1$ queue has a steady-state probability distribution with steady-state characteristics, as given in Table 6.3. In general, there is no simple expression for the steady-state probabilities $P_0, P_1, P_2, \ldots$. When $\lambda < \mu$, the quantity $\rho = \lambda/\mu$ is the server utilization, or long-run proportion of time the server is busy. As can be seen in Table 6.3, $1 - P_0 = \rho$ can also be interpreted as the steady-state probability that the system contains one or more customers. Notice also that $L - L_Q = \rho$ is the time-average number of customers being served.

Table 6.3   Steady-State Parameters of the $M/G/1$ Queue

ρ	$\dfrac{\lambda}{\mu}$
L	$\rho + \dfrac{\lambda^2(1/\mu^2 + \sigma^2)}{2(1-\rho)} = \rho + \dfrac{\rho^2(1+\sigma^2\mu^2)}{2(1-\rho)}$
w	$\dfrac{1}{\mu} + \dfrac{\lambda(1/\mu^2 + \sigma^2)}{2(1-\rho)}$
w_Q	$\dfrac{\lambda(1/\mu^2 + \sigma^2)}{2(1-\rho)}$
L_Q	$\dfrac{\lambda^2(1/\mu^2 + \sigma^2)}{2(1-\rho)} = \dfrac{\rho^2(1+\sigma^2\mu^2)}{2(1-\rho)}$
P_0	$1 - \rho$

Example 6.8

Customers arrive at a walk-in shoe repair shop apparently at random. It is assumed that arrivals occur according to a Poisson process at the rate $\lambda = 1.5$ per hour. Observation over several months has found that shoe repair times by the single worker take an average time of 30 minutes, with a standard deviation of 20 minutes. Thus the mean service time $1/\mu = 1/2$ hour, the service rate is $\mu = 2$ per hour and $\sigma^2 = (20)^2$ minutes^2 = $1/9$ hour^2. The "customers" are the people needing shoe repair, and the appropriate model is the $M/G/1$ queue, because only the mean and variance of service times are known, not their distribution. The proportion of time the worker is busy is $\rho = \lambda/\mu = 1.5/2 = 0.75$, and, by Table 6.3, the steady-state time average number of customers in the shop is

$$L = 0.75 + \frac{(1.5)^2[(0.5)^2 + 1/9]}{2(1 - 0.75)}$$

$$= 0.75 + 1.625 = 2.375 \text{ customers}$$

Thus, an observer who notes the state of the shoe repair system at arbitrary times would find an average of 2.375 customers (over the long run).

A closer look at the formulas in Table 6.3 reveals the source of the waiting lines and delays in an $M/G/1$ queue. For example, L_Q may be rewritten as

$$L_Q = \frac{\rho^2}{2(1 - \rho)} + \frac{\lambda^2 \sigma^2}{2(1 - \rho)}$$

The first term involves only the ratio of the mean arrival rate λ to the mean service rate μ. As shown by the second term, if λ and μ are held constant, the average length of the waiting line (L_Q) depends on the variability, σ^2, of the service times. If two systems have identical mean service times and mean interarrival times, the one with the more variable service times (larger σ^2) will tend to have longer lines on the average. Intuitively, if service times are highly variable, then there is a high probability that a large service time will occur (say, much larger than the mean service time), and, when large service times do occur, there is a higher-than-usual tendency for lines to form and delays of customers to increase. (The reader should not confuse "steady state" with low variability or short lines; a system in steady-state or statistical equilibrium can be highly variable and can have long waiting lines.)

Example 6.9

There are two workers competing for a job. Able claims an average service time that is faster than Baker's, but Baker claims to be more consistent, even if not as fast. The arrivals occur according to a Poisson process at the rate $\lambda = 2$ per hour (1/30 per minute). Able's service statistics are an average service time of 24 minutes with a standard deviation of 20 minutes. Baker's service statistics are an average service time of 25 minutes, but a standard deviation of only 2 minutes. If the average length of the queue is the criterion for hiring, which worker should be hired? For Able, $\lambda = 1/30$ per minute, $1/\mu = 24$ minutes, $\sigma^2 = 20^2 = 400$ minutes^2, $\rho = \lambda/\mu = 24/30 = 4/5$, and the average queue length is computed as

$$L_Q = \frac{(1/30)^2[24^2 + 400]}{2(1 - 4/5)} = 2.711 \text{ customers}$$

For Baker, $\lambda = 1/30$ per minute, $1/\mu = 25$ minutes, $\sigma^2 = 2^2 = 4$ minutes^2, $\rho = 25/30 = 5/6$, and the average queue length is

$$L_Q = \frac{(1/30)^2[25^2 + 4]}{2(1 - 5/6)} = 2.097 \text{ customers}$$

Although working faster on the average, Able's greater service variability results in an average queue length about 30% greater than Baker's. On the basis of average queue length, L_Q, Baker wins. On the other hand, the proportion of arrivals who would find Able idle and thus experience no delay is $P_0 = 1 - \rho = 1/5 = 20\%$, but the proportion who would find Baker idle and thus experience no delay is $P_0 = 1 - \rho = 1/6 = 16.7\%$.

One case of the $M/G/1$ queue that is of special note occurs when service times are exponential, which we describe next.

The $M/M/1$ queue. Suppose that service times in an $M/G/1$ queue are exponentially distributed, with mean $1/\mu$; then the variance as given by Equation (5.27) is $\sigma^2 = 1/\mu^2$. The mean and standard deviation of the exponential distribution are equal, so the $M/M/1$ queue will often be a useful approximate model when service times have standard deviations approximately equal to their means. The steady-state parameters, given in Table 6.4, may be computed by substituting $\sigma^2 = 1/\mu^2$ into the formulas in Table 6.3. Alternatively, L may be computed by Equation (6.16) from the steady-state probabilities P_n given in Table 6.4, and then w, w_Q, and L_Q may be computed from the equation in (6.17). The student can show that the two expressions for each parameter are equivalent by substituting $\rho = \lambda/\mu$ into the right-hand side of each equation in Table 6.4.

Table 6.4   Steady-State Parameters of the $M/M/1$ Queue

$$L \qquad \frac{\lambda}{\mu - \lambda} = \frac{\rho}{1 - \rho}$$

$$w \qquad \frac{1}{\mu - \lambda} = \frac{1}{\mu(1 - \rho)}$$

$$w_Q \qquad \frac{\lambda}{\mu(\mu - \lambda)} = \frac{\rho}{\mu(1 - \rho)}$$

$$L_Q \qquad \frac{\lambda^2}{\mu(\mu - \lambda)} = \frac{\rho^2}{1 - \rho}$$

$$P_n \qquad \left(1 - \frac{\lambda}{\mu}\right)\left(\frac{\lambda}{\mu}\right)^n = (1 - \rho)\rho^n$$

Example 6.10

Suppose that the interarrival times and service times at a single-chair unisex hair-styling shop have been shown to be exponentially distributed. The values of λ and μ are 2 per hour and 3 per hour, respectively—that is, the time between arrivals averages 1/2 hour, exponentially distributed, and the service time averages 20 minutes, also exponentially distributed. The server utilization and the probabilities for 0, 1, 2, 3, and 4 or more customers in the shop are computed as follows:

$$\rho = \frac{\lambda}{\mu} = \frac{2}{3}$$

$$P_0 = 1 - \frac{\lambda}{\mu} = \frac{1}{3}$$

$$P_1 = \left(\frac{1}{3}\right)\left(\frac{2}{3}\right) = \frac{2}{9}$$

$$P_2 = \left(\frac{1}{3}\right)\left(\frac{2}{3}\right)^2 = \frac{4}{27}$$

$$P_3 = \left(\frac{1}{3}\right)\left(\frac{2}{3}\right)^3 = \frac{8}{81}$$

$$P_{\geq 4} = 1 - \sum_{n=0}^{3} P_n = 1 - \frac{1}{3} - \frac{2}{9} - \frac{4}{27} - \frac{8}{81} = \frac{16}{81}$$

From the calculations, the probability that the hair stylist is busy is $1 - P_0 = \rho = 0.67$; thus, the probability that the hair stylist is idle is 0.33. The time-average number of customers in the system is given by Table 6.4 as

$$L = \frac{\lambda}{\mu - \lambda} = \frac{2}{3 - 2} = 2 \text{ customers}$$

The average time an arrival spends in the system can be obtained from Table 6.4 or Equation (6.17) as

$$w = \frac{L}{\lambda} = \frac{2}{2} = 1 \text{ hour}$$

The average time the customer spends in the queue can be obtained from Equation (6.17) as

$$w_Q = w - \frac{1}{\mu} = 1 - \frac{1}{3} = \frac{2}{3} \text{ hour}$$

From Table 6.4, the time-average number in the queue is given by

$$L_Q = \frac{\lambda^2}{\mu(\mu - \lambda)} = \frac{4}{3(1)} = \frac{4}{3} \text{ customers}$$

Finally, notice that multiplying $w = w_Q + 1/\mu$ through by λ and using Little's equation (6.9) yields

$$L = L_Q + \frac{\lambda}{\mu} = \frac{4}{3} + \frac{2}{3} = 2 \text{ customers}$$

Example 6.11

For the $M/M/1$ queue with service rate $\mu = 10$ customers per hour, consider how L and w increase as the arrival rate λ increases from 5 to 8.64 by increments of 20%, and then to $\lambda = 10$.

λ	5.0	6.0	7.2	8.64	10.0
ρ	0.500	0.600	0.720	0.864	1.0
L	1.00	1.50	2.57	6.35	∞
w	0.20	0.25	0.36	0.73	∞

For any $M/G/1$ queue, if $\lambda/\mu \geq 1$, waiting lines tend to continually grow in length; the long-run measures of performance, L, w, w_Q, and L_Q are all infinite ($L = w = w_Q = L_Q = \infty$); and a steady-state probability distribution does not exist. As is shown here for $\lambda < \mu$, if ρ is close to 1, waiting lines and delays will tend to be long. Notice that the increase in average system time w and average number in system L is highly nonlinear as a function of ρ. For example, as λ increases by 20%, L increases first by 50% (from 1.00 to 1.50), then by 71% (to 2.57), and then by 147% (to 6.35).

Example 6.12

If arrivals are occurring at rate $\lambda = 10$ per hour, and management has a choice of two servers, one who works at rate $\mu_1 = 11$ customers per hour and the second at rate $\mu_2 = 12$ customers per hour, the respective utilizations are $\rho_1 = \lambda/\mu_1 = 10/11 = 0.909$ and $\rho_2 = \lambda/\mu_2 = 10/12 = 0.833$. If the $M/M/1$ queue is used as an approximate model, then, with the first server, the average number in the system would be, by Table 6.4,

$$L_1 = \frac{\rho_1}{1 - \rho_1} = 10$$

and, with the second server, the average number in the system would be

$$L_2 = \frac{\rho_2}{1 - \rho_2} = 5$$

Thus, an increase in service rate from 11 to 12 customers per hour, a mere 9.1% increase, would result in a decrease in average number in system from 10 to 5, which is a 50% decrease!

The effect of utilization and service variability

For any $M/G/1$ queue, if lines are too long, they can be reduced by decreasing the server utilization ρ or by decreasing the service time variability σ^2. These remarks hold for almost all queues, not just the $M/G/1$ queue. The utilization factor ρ can be reduced by decreasing the arrival rate λ, by increasing the service rate μ, or by increasing the number of servers, because, in general, $\rho = \lambda/(c\mu)$, where c is the number of parallel servers. The effect of additional servers will be studied in the following subsections.

The squared coefficient of variation (cv) of a positive random variable X is defined as

$$(\text{cv})^2 = \frac{V(X)}{[E(X)]^2}$$

It is a measure of the variability of a distribution. The larger its value, the more variable is the distribution relative to its expected value. For deterministic service times, $V(X) = 0$, so $\text{cv} = 0$. For Erlang service times of order k, $V(X) = 1/(k\mu^2)$ and $E(X) = 1/\mu$, so $\text{cv} = 1/\sqrt{k}$. For exponential service times at service rate μ, the mean service time is $E(X) = 1/\mu$ and the variance is $V(X) = 1/\mu^2$, so $\text{cv} = 1$. If service times have standard deviation greater than their mean (i.e., if $\text{cv} > 1$), then the hyperexponential distribution, which can achieve any desired coefficient of variation greater than 1, provides a good model. One occasion where it arises is given in Exercise 16.

The formula for L_Q for any $M/G/1$ queue can be rewritten in terms of the coefficient of variation by noticing that $(\text{cv})^2 = \sigma^2/(1/\mu)^2 = \sigma^2\mu^2$. Therefore,

$$L_Q = \frac{\rho^2(1 + \sigma^2\mu^2)}{2(1 - \rho)}$$

$$= \frac{\rho^2(1 + (\text{cv})^2)}{2(1 - \rho)}$$

$$= \left(\frac{\rho^2}{1 - \rho}\right)\left(\frac{1 + (\text{cv})^2}{2}\right) \tag{6.18}$$

The first term, $\rho^2/(1 - \rho)$, is L_Q for an $M/M/1$ queue. The second term, $\left(1 + (\text{cv})^2\right)/2$, corrects the $M/M/1$ formula to account for a nonexponential service-time distribution. The formula for w_Q can be obtained from the corresponding $M/M/1$ formula by applying the same correction factor. These factors, $\rho^2/(1 - \rho)$ and $\left(1 + (\text{cv})^2\right)/2$, provide some insight into the relative impact of server utilization and server variability on queue congestion: L_Q explodes as $\rho \to 1$ while increasing linearly in $(\text{cv})^2$ for fixed ρ.

6.4.2 Multiserver Queue: $M/M/c/\infty/\infty$

Suppose that there are c channels operating in parallel. Each of these channels has an independent and identical exponential service-time distribution, with mean $1/\mu$. The arrival process is Poisson with rate λ. Arrivals will join a single queue and enter the first available service channel. The queueing system is shown in Figure 6.12. If the number in system is $n < c$, an arrival will enter an available channel. However, when $n \geq c$, a queue will build if arrivals occur.

The offered load is defined by λ/μ. If $\lambda \geq c\mu$, the arrival rate is greater than or equal to the maximum service rate of the system (the service rate when all servers are busy); thus, the system cannot handle the load put upon it, and therefore it has no statistical equilibrium. If $\lambda > c\mu$, the waiting line grows in length at the rate $\lambda - c\mu$ customers per time unit, on the average. Customers are entering the system at rate λ per time unit but are leaving the system at a maximum rate of $c\mu$ per time unit.

For the $M/M/c$ queue to have statistical equilibrium, the offered load must satisfy $\lambda/\mu < c$, in which case $\lambda/(c\mu) = \rho$, the server utilization. The steady-state parameters are listed in Table 6.5.

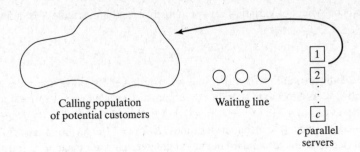

Figure 6.12   Multiserver queueing system.

Table 6.5   Steady-State Parameters for the $M/M/c$ Queue

ρ	$\dfrac{\lambda}{c\mu}$
P_0	$\left\{\left[\displaystyle\sum_{n=0}^{c-1}\dfrac{(\lambda/\mu)^n}{n!}\right]+\left[\left(\dfrac{\lambda}{\mu}\right)^c\left(\dfrac{1}{c!}\right)\left(\dfrac{c\mu}{c\mu-\lambda}\right)\right]\right\}^{-1}$
	$=\left\{\left[\displaystyle\sum_{n=0}^{c-1}\dfrac{(c\rho)^n}{n!}\right]+\left[(c\rho)^c\left(\dfrac{1}{c!}\right)\dfrac{1}{1-\rho}\right]\right\}^{-1}$
$P(L(\infty)\geq c)$	$\dfrac{(\lambda/\mu)^c P_0}{c!(1-\lambda/c\mu)}=\dfrac{(c\rho)^c P_0}{c!(1-\rho)}$
L	$c\rho+\dfrac{(c\rho)^{c+1}P_0}{c(c!)(1-\rho)^2}=c\rho+\dfrac{\rho P(L(\infty)\geq c)}{1-\rho}$
w	$\dfrac{L}{\lambda}$
w_Q	$w-\dfrac{1}{\mu}$
L_Q	$\lambda w_Q=\dfrac{(c\rho)^{c+1}P_0}{c(c!)(1-\rho)^2}=\dfrac{\rho P(L(\infty)\geq c)}{1-\rho}$
$L-L_Q$	$\dfrac{\lambda}{\mu}=c\rho$

Most of the measures of performance can be expressed fairly simply in terms of P_0, the probability that the system is empty, or $\sum_{n=c}^{\infty}P_n$, the probability that all servers are busy, denoted by $P(L(\infty)\geq c)$, where $L(\infty)$ is a random variable representing the number in system in statistical equilibrium (after a very long time). Thus, $P(L(\infty)=n)=P_n, n=0,1,2,\ldots$. The value of P_0 is necessary for computing all the measures of performance, and the equation for P_0 is somewhat more complex than in the previous cases. However, the spreadsheet QueueingTools.xls available at www.bcnn.net performs these calculations for this and all queueing models described in the chapter.

The results in Table 6.5 simplify to those in Table 6.4 when $c = 1$, the case of a single server. Notice that the average number of busy servers, or the average number of customers being served, is given by the simple expression $L - L_Q = \lambda/\mu = c\rho$.

Example 6.13

Many early examples of queueing theory applied to practical problems concerning tool cribs. Attendants manage the tool cribs as mechanics, assumed to be from an infinite calling population, arrive for service. However, this example could as easily be customers arriving to the ticket counter at a movie theater. Assume Poisson arrivals at rate 2 mechanics per minute and exponentially distributed service times with mean 40 seconds.

Now, $\lambda = 2$ per minute, and $\mu = 60/40 = 3/2$ per minute. The offered load is greater than 1:

$$\frac{\lambda}{\mu} = \frac{2}{3/2} = \frac{4}{3} > 1$$

so more than one server is needed if the system is to have a statistical equilibrium. The requirement for steady state is that $c > \lambda/\mu = 4/3$. Thus at least $c = 2$ attendants are needed. The quantity 4/3 is the expected number of busy servers, and for $c \geq 2$, $\rho = 4/(3c)$ is the long-run proportion of time each server is busy. (What would happen if there were only $c = 1$ server?)

Let there be $c = 2$ attendants. First, P_0 is calculated as

$$P_0 = \left\{ \sum_{n=0}^{1} \frac{(4/3)^n}{n!} + \left(\frac{4}{3}\right)^2 \left(\frac{1}{2!}\right) \left[\frac{2(3/2)}{2(3/2) - 2} \right] \right\}^{-1}$$

$$= \left\{ 1 + \frac{4}{3} + \left(\frac{16}{9}\right)\left(\frac{1}{2}\right)(3) \right\}^{-1} = \left(\frac{15}{3}\right)^{-1} = \frac{1}{5} = 0.2$$

Next, the probability that all servers are busy is computed as

$$P(L(\infty) \geq 2) = \frac{(4/3)^2}{2!(1 - 2/3)} \left(\frac{1}{5}\right) = \left(\frac{8}{3}\right)\left(\frac{1}{5}\right) = \frac{8}{15} = 0.533$$

Thus, the time-average length of the waiting line of mechanics is

$$L_Q = \frac{(2/3)(8/15)}{1 - 2/3} = 1.07 \text{ mechanics}$$

and the time-average number in system is given by

$$L = L_Q + \frac{\lambda}{\mu} = \frac{16}{15} + \frac{4}{3} = \frac{12}{5} = 2.4 \text{ mechanics}$$

From Little's relationships, the average time a mechanic spends at the tool crib is

$$w = \frac{L}{\lambda} = \frac{2.4}{2} = 1.2 \text{ minutes}$$

and the average time spent waiting for an attendant is

$$w_Q = w - \frac{1}{\mu} = 1.2 - \frac{2}{3} = 0.533 \text{ minute}$$

An Approximation for the $M/G/c/\infty$ Queue

Recall that formulas for L_Q and w_Q for the $M/G/1$ queue can be obtained from the corresponding $M/M/1$ formulas by multiplying them by the correction factor $\left(1 + (\text{cv})^2\right)/2$, as in Equation (6.18). *Approximate* formulas for the $M/G/c$ queue can be obtained by applying the same correction factor to the $M/M/c$ formulas for L_Q and w_Q (no exact formula exists for $1 < c < \infty$). The nearer the cv is to 1, the better the approximation.

Example 6.14

Recall Example 6.13. Suppose that the service times for the mechanics at the tool crib are not exponentially distributed, but are known to have a standard deviation of 30 seconds. Then we have an $M/G/c$ model, rather than an $M/M/c$. The mean service time is 40 seconds, so the coefficient of variation of the service time is

$$\text{cv} = \frac{30}{40} = \frac{3}{4} < 1$$

Therefore, the accuracy of L_Q and w_Q can be improved by the correction factor

$$\frac{1 + (\text{cv})^2}{2} = \frac{1 + (3/4)^2}{2} = \frac{25}{32} = 0.78$$

For example, when there are $c = 2$ attendants,

$$L_Q = (0.78)(1.07) = 0.83 \text{ mechanics}$$

Notice that, because the coefficient of variation of the service time is less than 1, the congestion in the system, as measured by L_Q, is less than in the corresponding $M/M/2$ model.

The correction factor applies only to the formulas for L_Q and w_Q. Little's formula can then be used to calculate L and w. Unfortunately, there is no general method for correcting the steady-state probabilities, P_n.

When the Number of Servers is Infinite ($M/G/\infty/\infty$)

There are at least three situations in which it is appropriate to treat the number of servers as infinite:

1. when each customer is its own server—in other words, in a self-service system;
2. when service capacity far exceeds service demand, as in a so-called ample-server system; and
3. when we want to know how many servers are required so that customers will rarely be delayed.

The steady-state parameters for the $M/G/\infty$ queue are listed in Table 6.6. In the table, λ is the arrival rate of the Poisson arrival process, and $1/\mu$ is the expected service time of the general service-time distribution (including exponential, constant, or any other).

Table 6.6 Steady-State Parameters for the $M/G/\infty$ Queue

P_0	$e^{-\lambda/\mu}$
w	$\dfrac{1}{\mu}$
w_Q	0
L	$\dfrac{\lambda}{\mu}$
L_Q	0
P_n	$\dfrac{e^{-\lambda/\mu}(\lambda/\mu)^n}{n!}, n = 0, 1, \ldots$

Example 6.15

Prior to introducing their new, subscriber-only, online computer information service, The Connection must plan their system capacity in terms of the number of users that can be logged on simultaneously. If the service is successful, customers are expected to log on at a rate of $\lambda = 500$ per hour, according to a Poisson process, and stay connected for an average of $1/\mu = 180$ minutes (or 3 hours). In the real system, there will be an upper limit on simultaneous users, but, for planning purposes, The Connection can pretend that the number of simultaneous users is infinite. An $M/G/\infty$ model of the system implies that the expected number of simultaneous users is $L = \lambda/\mu = 500(3) = 1500$, so a capacity greater than 1500 is certainly required. To ensure providing adequate capacity 95% of the time, The Connection could allow the number of simultaneous users to be the smallest value c such that

$$P(L(\infty) \leq c) = \sum_{n=0}^{c} P_n = \sum_{n=0}^{c} \frac{e^{-1500}(1500)^n}{n!} \geq 0.95$$

The capacity $c = 1564$ simultaneous users satisfies this requirement.

6.4.3   Multiserver Queues with Poisson Arrivals and Limited Capacity: $M/M/c/N/\infty$

Suppose that service times are exponentially distributed at rate μ, that there are c servers, and that the total system capacity is $N \geq c$ customers. If an arrival occurs when the system is full, that arrival is turned away and does not enter the system. As in the preceding section, suppose that arrivals occur randomly according to a Poisson process with rate λ arrivals per time unit. For any values of λ and μ such that $\rho \neq 1$, the $M/M/c/N$ queue has a statistical equilibrium with steady-state characteristics as given in Table 6.7 (formulas for the case $\rho = 1$ can be found in Hillier and Lieberman [2005]).

Table 6.7   Steady-State Parameters for the $M/M/c/N$ Queue

(N = System Capacity, $a = \lambda/\mu$, $\rho = \lambda/(c\mu)$)

$$P_0 \quad \left[1 + \sum_{n=1}^{c} \frac{a^n}{n!} + \frac{a^c}{c!} \sum_{n=c+1}^{N} \rho^{n-c} \right]^{-1}$$

$$P_N \quad \frac{a^N}{c! c^{N-c}} P_0$$

$$L_Q \quad \frac{P_0 a^c \rho}{c!(1-\rho)^2} \left[1 - \rho^{N-c} - (N-c)\rho^{N-c}(1-\rho) \right]$$

$$\lambda_e \quad \lambda(1 - P_N)$$

$$w_Q \quad \frac{L_Q}{\lambda_e}$$

$$w \quad w_Q + \frac{1}{\mu}$$

$$L \quad \lambda_e w$$

The effective arrival rate λ_e is defined as the mean number of arrivals per time unit who enter and remain in the system. For all systems, $\lambda_e \leq \lambda$; for the unlimited-capacity systems, $\lambda_e = \lambda$; but, for systems such as the present one, which turn customers away when full, $\lambda_e < \lambda$. The effective arrival rate is computed by

$$\lambda_e = \lambda(1 - P_N)$$

because $1 - P_N$ is the probability that a customer, upon arrival, will find space and be able to enter the system. When one is using Little's equations (6.17) to compute mean time spent in system w and in queue w_Q, λ must be replaced by λ_e.

Example 6.16

The unisex hair-styling shop described in Example 6.10 can hold only three customers: one in service, and two waiting. Additional customers are turned away when the system is full. The offered load is as previously determined, namely $\lambda/\mu = 2/3$.

To calculate the performance measures, first compute P_0:

$$P_0 = \left[1 + \frac{2}{3} + \frac{2}{3}\sum_{n=2}^{3}\left(\frac{2}{3}\right)^{n-1}\right]^{-1} = 0.415$$

The probability that there are three customers in the system (the system is full) is

$$P_N = P_3 = \frac{(2/3)^3}{1!1^2}P_0 = \frac{8}{65} = 0.123$$

Then, the average length of the queue (customers waiting for a haircut) is given by

$$L_Q = \frac{(27/65)(2/3)(2/3)}{(1-2/3)^2}\left[1 - (2/3)^2 - 2(2/3)^2(1 - 2/3)\right] = 0.431 \text{ customer}$$

Now, the effective arrival rate λ_e is given by

$$\lambda_e = 2\left(1 - \frac{8}{65}\right) = \frac{114}{65} = 1.754 \text{ customers per hour}$$

Therefore, from Little's equation, the expected time spent waiting in queue is

$$w_Q = \frac{L_Q}{\lambda_e} = \frac{28}{114} = 0.246 \text{ hour}$$

and the expected total time in the shop is

$$w = w_Q + \frac{1}{\mu} = \frac{66}{114} = 0.579 \text{ hour}$$

One last application of Little's equation gives the expected number of customers in the shop (in queue and getting a haircut) as

$$L = \lambda_e w = \frac{66}{65} = 1.015 \text{ customers}$$

Notice that $1 - P_0 = 0.585$ is the average number of customers being served or, equivalently, the probability that the single server is busy. Thus, the server utilization, or proportion of time the server is busy in the long run, is given by

$$1 - P_0 = \frac{\lambda_e}{\mu} = 0.585$$

The reader should compare these results to those of the unisex hair-styling shop before the capacity constraint was placed on the system. Specifically, in systems with limited capacity, the offered load λ/μ can assume any positive value and no longer equals the server utilization $\rho = \lambda_e/\mu$. Notice that server utilization decreases from 67% to 58.5% when the system imposes a capacity constraint.

6.5 Steady-State Behavior of Finite-Population Models ($M/M/c/K/K$)

In many practical problems, the assumption of an infinite calling population leads to invalid results because the calling population is, in fact, small. When the calling population is small, the presence of one or more customers in the system has a strong effect on the distribution of future arrivals, and the use of an infinite-population model can be misleading. Typical examples include a small group of machines that break down from time to time and require repair, or a small group of patients who are the responsibility of a staff of nurses. In the extreme case, if all the machines are broken, no new "arrivals" (breakdowns) of machines can occur; similarly, if all the patients have requested assistance, no arrival is possible. Contrast this to the infinite-population models, in which the arrival rate λ of customers to the system is assumed to be independent of the state of the system.

Consider a finite-calling-population model with K customers. The time between the end of one service visit and the next call for service for each member of the population is assumed to be exponentially distributed with mean $1/\lambda$ time units; service times are also exponentially distributed, with mean $1/\mu$ time units; there are c parallel servers, and system capacity is K, so that all arrivals remain for service. Such a system is depicted in Figure 6.13.

The steady-state parameters for this model are listed in Table 6.8. An electronic spreadsheet or a symbolic calculation program is useful for evaluating these complex formulas. For example, Figure 6.14 is a procedure written for the symbolic calculation program MATLAB to calculate the steady-state probabilities for the $M/M/c/K/K$ queue. The spreadsheet `QueueingTools.xls` available at `www.bcnn.net` also performs these calculations.

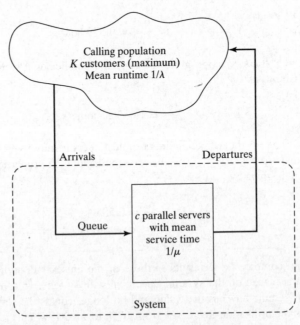

Figure 6.13 Finite-population queueing model.

Table 6.8 Steady-State Parameters for the $M/M/c/K/K$ Queue

$$P_0 \quad \left[\sum_{n=0}^{c-1} \binom{K}{n} \left(\frac{\lambda}{\mu}\right)^n + \sum_{n=c}^{K} \frac{K!}{(K-n)!c!c^{n-c}} \left(\frac{\lambda}{\mu}\right)^n \right]^{-1}$$

$$P_n \quad \begin{cases} \binom{K}{n} \left(\dfrac{\lambda}{\mu}\right)^n P_0, & n = 0, 1, \ldots, c-1 \\[3mm] \dfrac{K!}{(K-n)!c!c^{n-c}} \left(\dfrac{\lambda}{\mu}\right)^n P_0, & n = c, c+1, \ldots, K \end{cases}$$

$$L \quad \sum_{n=0}^{K} n P_n$$

$$L_Q \quad \sum_{n=c+1}^{K} (n-c) P_n$$

$$\lambda_e \quad \sum_{n=0}^{K} (K-n)\lambda P_n$$

$$w \quad L/\lambda_e$$

$$w_Q \quad L_Q/\lambda_e$$

$$\rho \quad \frac{L - L_Q}{c} = \frac{\lambda_e}{c\mu}$$

The effective arrival rate λ_e has several valid interpretations:

λ_e = long-run effective arrival rate of customers to the queue

= long-run effective arrival rate of customers entering service

= long-run rate at which customers exit from service

= long-run rate at which customers enter the calling population
(and begin a new runtime)

= long-run rate at which customers exit from the calling population

Example 6.17

There are two workers who are responsible for 10 milling machines. The machines run on the average for 20 minutes, then require an average 5-minute service period, both times exponentially distributed. Therefore, $\lambda = 1/20$ and $\mu = 1/5$. Compute the various measures of performance for this system.

All of the performance measures depend on P_0, which is

$$\left[\sum_{n=0}^{2-1} \binom{10}{n} \left(\frac{5}{20}\right)^n + \sum_{n=c}^{10} \frac{10!}{(10-n)!2!2^{n-2}} \left(\frac{5}{20}\right)^n \right]^{-1} = 0.065$$

```
p = zeros(K+1,1);
% Note:
%    p(1) = lim_t->infty Pr{N(t)=0}
%    p(n+1) = lim_t->infty Pr{N(t)=n}, for n=1,2,...,K-1
%    p(K+1) = lim_t->infty Pr{N(t)=K}
crho = lambda/mu;
Kfac = factorial(K);
cfac = factorial(c);
% get p(1)
for n=0:c-1
    p(1)=p(1) + (Kfac/(factorial(n)*factorial(K-n)))*(crho^n);
end
for n=c:K
    p(1)=p(1) + (Kfac/((c^(n-c))*factorial(K-n)*cfac))*(crho^n);
end
p(1)=1/p(1);
% get p(n+1), 0 < n < c
for n=1:c-1
    p(n+1)=p(1) * (Kfac/(factorial(n)*factorial(K-n)))*(crho^n);
end
% get p(n+1), c <= n <= K
for n=c:K
    p(n+1)=p(1) * (Kfac/((c^(n-c))*factorial(K-n)*cfac))*(crho^n);
end
% return probability vector
mmcKK = p;
```

Figure 6.14   MATLAB procedure to calculate P_n for the $M/M/c/K/K$ queue.

From P_0, we can obtain the other P_n, from which we can compute the average number of machines waiting for service,

$$L_Q = \sum_{n=3}^{10} (n-2)P_n = 1.46 \text{ machines}$$

the effective arrival rate,

$$\lambda_e = \sum_{n=0}^{10} (10-n)\left(\frac{1}{20}\right)P_n = 0.342 \text{ machines/minute}$$

and the average waiting time in the queue,

$$w_Q = L_Q/\lambda_e = 4.27 \text{ minutes}$$

Similarly, we can compute the expected number of machines being serviced or waiting to be serviced,

$$L = \sum_{n=0}^{10} nP_n = 3.17 \text{ machines}$$

The average number of machines being serviced is given by

$$L - L_Q = 3.17 - 1.46 = 1.71 \text{ machines}$$

Each machine must be running, waiting to be serviced, or in service, so the average number of running machines is given by

$$K - L = 10 - 3.17 = 6.83 \text{ machines}$$

A question frequently asked is this: What will happen if the number of servers is increased or decreased? If the number of workers in this example increases to three ($c = 3$), then the time-average number of running machines increases to

$$K - L = 7.74 \text{ machines}$$

an increase of 0.91 machine, on the average.

Conversely, what happens if the number of servers decreases to one? Then the time-average number of running machines decreases to

$$K - L = 3.98 \text{ machines}$$

The decrease from two servers to one has resulted in a drop of nearly three machines running, on the average. Exercise 17 asks the reader to examine the effect on server utilization of adding or deleting one server.

Example 6.17 illustrates several general relationships that have been found to hold for almost all queues. If the number of servers is decreased, then delays, server utilization, and the probability of an arrival having to wait to begin service all increase.

6.6   Networks of Queues

In this chapter, we have emphasized the study of single queues of the $G/G/c/N/K$ type. However, many systems are naturally modeled as networks of single queues in which customers departing from one queue may be routed to another. Example 6.1 (see, in particular, Figure 6.3) and Example 6.2 (see Figure 6.5) are illustrations.

The study of mathematical models of networks of queues is beyond the scope of this chapter; see, for instance, Gross and Harris [1997], Nelson [1995], and Kleinrock [1976]. However, a few fundamental principles are very useful for rough-cut modeling, perhaps prior to a simulation study. The following results assume a stable system with an infinite calling population and no limit on system capacity:

1. Provided that no customers are created or destroyed in the queue, then the departure rate out of a queue is the same as the arrival rate into the queue, over the long run.
2. If customers arrive to queue i at rate λ_i, and a fraction $0 \le p_{ij} \le 1$ of them are routed to queue j upon departure, then the arrival rate from queue i to queue j is $\lambda_i p_{ij}$, over the long run.
3. The overall arrival rate into queue j, λ_j, is the sum of the arrival rate from all sources. If customers arrive from outside the network at rate a_j, then

$$\lambda_j = a_j + \sum_{\text{all } i} \lambda_i p_{ij}$$

4. If queue j has $c_j < \infty$ parallel servers, each working at rate μ_j, then the long-run utilization of each server is

$$\rho_j = \frac{\lambda_j}{c_j \mu_j}$$

and $\rho_j < 1$ is required for the queue to be stable.
5. If, for each queue j, arrivals from outside the network form a Poisson process with rate a_j, and if there are c_j identical servers delivering exponentially distributed service times with mean $1/\mu_j$ (where c_j may be ∞), then, in steady state, queue j behaves like an $M/M/c_j$ queue with arrival rate $\lambda_j = a_j + \sum_{\text{all } i} \lambda_i p_{ij}$.

Example 6.18

Consider again the discount store described in Example 6.1 and shown in Figure 6.3. Suppose that customers arrive at the rate of 80 per hour and that, of those arrivals, 40% choose self-service; then, the arrival rate to service center 1 is $\lambda_1 = (80)(0.40) = 32$ per hour, and the arrival rate to service center 2 is $\lambda_2 = (80)(0.6) = 48$ per hour. Suppose that each of the $c_2 = 3$ clerks at service center 2 works at the rate $\mu_2 = 20$ customers per hour. Then the long-run utilization of the clerks is

$$\rho_2 = \frac{48}{(3)(20)} = 0.8$$

All customers must see the cashier at service center 3. The overall arrival rate to service center 3 is $\lambda_3 = \lambda_1 + \lambda_2 = 80$ per hour, regardless of the service rate at service center 1, because, over the long run, the departure rate out of each service center must be equal to the arrival rate into it. If the cashier works at rate $\mu_3 = 90$ per hour, then the utilization of the cashier is

$$\rho_3 = \frac{80}{90} = 0.89$$

6.7   Rough-cut Modeling: An Illustration

In this section we show how the tools described in this chapter can be used to do a rough-cut analysis prior to undertaking a more detailed simulation. Rough-cut modeling is useful in a number of ways: In some cases, the results of the rough-cut analysis are so compelling that there is no longer a need for the detailed simulation, saving a great deal of time and money. More typically, the rough-cut model provides the analyst with a better understanding of the system to be modeled—a kind of dress rehearsal prior to constructing a detailed and often complex simulation model. Further, the performance measures obtained at the rough-cut stage provide a sanity check on the simulation outputs, potentially averting erroneous conclusions due simply to mistakes in programming the simulation.

Example 6.19 _____

At a driver's license branch office the manager receives many complaints about the long delays to renew licenses. To obtain quantitative support for more staff she brings in an operations analyst. A diagram of the facility is shown in Figure 6.15.

The branch is open from 8 A.M. to 4 P.M., with an historical average of 464 drivers per day being processed. All arrivals must first check in with one of two clerks. The manager has some time-study data collected on the clerks over several days and finds an average check-in time of 2 minutes with a standard deviation of about 0.4 minutes. After check in, 15% of the drivers need to take a written test that lasts approximately 20 minutes; good data are available on test time because the exams are time stamped when they are issued and returned. These multiple-choice tests are graded nearly instantly by an optical scanner. All arrivals must wait to have their picture taken and their license produced; this station can process about 60 drivers per hour, but the individual times are highly variable as some pictures have to be retaken.

The branch manager wants to know the relative impact on customer delay of adding a check-in clerk or adding a new photo station, and whether either of these changes will impact the number of chairs needed for drivers taking the written test (the 20 chairs she currently has have always been adequate).

This is an ideal application for simulation, but the analyst hired by the manager feels a quick queueing approximation could provide useful insight, as well as a check on the simulation model.

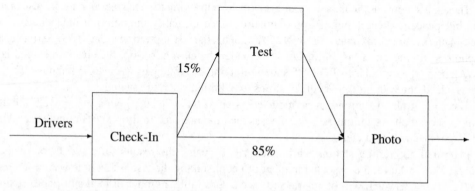

Figure 6.15   Customer flow in the driver's license branch office.

To do a rough-cut analysis, the analyst represents the license branch as a network of queues, with the check-in clerks being queue 1 (with $c_1 = 2$ servers, each working at rate $\mu_1 = 30$ drivers per hour $= 0.5$ drivers per minute), the written test station being queue 2 (with $c_2 = \infty$ servers, because any number of people can be taking the written test simultaneously, and mean service time $1/\mu_2 = 20$ minutes), and with the photo station being queue 3 (with $c_3 = 1$ server working at rate $\mu_3 = 1$ driver per minute). The analyst chooses to ignore the optical scanning of the tests because it happens so quickly. Although the branch manager says that arrivals are heavier around noon, for the purpose of a rough-cut analysis the analyst models the arrival rate as $464/8 = 58$ arrivals per hour throughout the day.

The arrival rates to each queue are as follows:

$$\lambda_1 = a_1 + \sum_{i=1}^{3} p_{i1}\lambda_i = 58 \text{ drivers per hour}$$

$$\lambda_2 = a_2 + \sum_{i=1}^{3} p_{i2}\lambda_i = (0.15)\lambda_1 \text{ drivers per hour}$$

$$\lambda_3 = a_3 + \sum_{i=1}^{3} p_{i3}\lambda_i = (1)\lambda_2 + (0.85)\lambda_1 \text{ drivers per hour}$$

Notice that arrivals from outside the network occur only at queue 1, so $a_1 = 50$ and $a_2 = a_3 = 0$. Solving this system of equations gives $\lambda_1 = \lambda_3 = 58$ and $\lambda_2 = 8.7$. The analyst chooses to work in minutes, and so converts these to $\lambda_1 = \lambda_3 = 0.97$ per minute and $\lambda_2 = 0.15$ per minute.

The analyst approximates the arrival process as Poisson—which is justified by the large population of potential customers who make (largely) independent decisions about when to arrive—and the service times at each queue as exponentially distributed. The service-time approximation is reasonable for queue 3, since the manager believes the picture-taking times are quite variable, and harmless for queue 2 because the results for infinite-server $M/G/\infty$ queues depend only on the service rate, not the distribution. However, the coefficient of variation of the check-in time is $cv = 0.4/2 = 0.2 < 1$, so the analyst decides to also apply the correction $(1 + cv^2)/2 = 0.52$ to the check-in results.

The analyst now has the necessary information to approximate each station in the license branch as an independent queue (another approximation, since in reality they are not independent). The check-in clerks are approximated as an $M/G/c_1$ queue, the testing station as an $M/G/\infty$ queue, and the photo station as an $M/M/c_3$ queue. Thus, under the current set-up, the check-in station is an $M/G/2$; using the formulas in Table 6.5 and multiplying by 0.52 gives $w_Q = 16.5$ minutes. If a third clerk is added, the waiting time in queue drops to 0.4 minute, a huge savings.

The current photo station can be modeled as an $M/M/1$ queue, giving $w_Q = 32.3$ minutes; adding a second photo station ($M/M/2$) causes the time in queue to drop to 0.3 minutes, a savings even larger than adding a clerk.

If desired, the testing station can be analyzed by using the results for an $M/G/\infty$ queue in Table 6.6. For instance, the expected number of people taking the test at any time is $L = \lambda_2/\mu_2 = 0.15/(1/20) = 3$, which will be unaffected by changes in the number of clerks or photo stations, since they do not change λ_2 or μ_2.

Should the manager expect to see precisely these results if either or both of the changes are implemented? No, because the model is rough in a number of ways: The queueing results are steady state, while the actual license facility is open only 8 hours per day, and does not have a constant arrival rate. The system was simplified by leaving out test scoring, employee breaks, etc. However, these results do provide a compelling case for the addition of one or more staff, and provide a baseline against which to check a detailed simulation model that includes all of the relevant features.

6.8   Summary

Queueing models have found widespread use in the analysis of service facilities, production and material-handling systems, telephone and communications systems, and many other situations where congestion or competition for scarce resources can occur. This chapter has introduced the basic concepts of queueing models and shown how simulation, and in some cases a mathematical analysis, can be used to estimate the performance measures of a system.

A simulation may be used to generate one or more artificial histories of a complex system. This simulation-generated data may, in turn, be used to estimate desired performance measures of the system. Commonly used performance measures, including L, L_Q, w, w_Q, ρ, and λ_e, were introduced, and formulas were given for their estimation from data.

When simulating any system that evolves over time, the analyst must decide whether transient behavior or steady-state performance is to be studied. Simple formulas exist for the steady-state behavior of some queues, but estimating steady-state performance measures from simulation-generated data requires recognizing and dealing with the possibly deleterious effect of the initial conditions on the estimators of steady-state performance. These estimators could be severely biased (either high or low) if the initial conditions are unrepresentative of steady state or if the simulation run length is too short. These estimation problems are discussed at greater length in Chapter 11.

Whether the analyst is interested in transient or in steady-state performance of a system, it should be recognized that the estimates obtained from a simulation of a stochastic queue are exactly that— estimates. Every such estimate contains random error, and a proper statistical analysis is required to assess the accuracy of the estimate. Methods for conducting such a statistical analysis are discussed in Chapters 11 and 12.

In the last three sections of this chapter, it was shown that a number of simple models can be solved mathematically. Although the assumptions behind such models might not be met exactly in a practical application, these models can still be useful in providing a rough estimate of a performance measure. In many cases, models with exponentially distributed interarrival and service times will provide a conservative estimate of system behavior. For example, if the model predicts that average waiting time w will be 12.7 minutes, then average waiting time in the real system is likely to be less than 12.7 minutes. The conservative nature of exponential models arises because (a) performance measures, such as w and L, are generally increasing functions of the variance of interarrival times and service times (recall the $M/G/1$ queue), and (b) the exponential distribution is fairly highly variable, having its standard deviation always equal to its mean. Thus, if the arrival process or service mechanism of the real system is less variable than exponentially distributed interarrival or service times, it is likely that the average number in the system, L, and the average time spent in system, w, will be less than what is predicted by the exponential model. Of course, if the interarrival and service times are *more* variable than exponential random variables, then the M/M queueing models could underestimate congestion.

An important application of mathematical queueing models is determining the minimum number of servers needed at a work station or service center. Quite often, if the arrival rate λ and the service rate μ are known or can be estimated, then the simple inequality $\lambda/(c\mu) < 1$ can be used to provide an initial estimate for the number of servers, c, at a work station. For a large system with many work stations, it could be quite time consuming to have to simulate every possibility $(c_1, c_2, \dots)$ for the number of servers, c_i, at work station i. Thus, rough estimates from a bit of analysis could save a great deal of computer time and analysts' time.

Finally, the qualitative behavior of the simple exponential models of queueing carries over to more complex systems. In general, it is the variability of service times and the variability of the arrival process that causes waiting lines to build up and congestion to occur. For most systems, if the arrival rate increases, or if the service rate decreases, or if the variance of service times or interarrival times increases, then the system will become more congested. Congestion can be decreased by adding more servers or by reducing the mean and variability of service times. Simple queueing models can be a great aid in quantifying these relationships and in evaluating alternative system designs.

REFERENCES

COOPER, R. B. [1990], *Introduction to Queueing Theory*, 3d ed., George Washington University, Washington, DC.

DESCLOUX, A. [1962], *Delay Tables for Finite- and Infinite-Source Systems*, McGraw-Hill, New York.

GROSS, D., AND C. HARRIS [1997], *Fundamentals of Queueing Theory*, 3d ed., Wiley, New York.

HALL, R. W. [1991], *Queueing Methods: For Services and Manufacturing*, Prentice Hall, Englewood Cliffs, NJ.

HILLIER, F. S., AND G. J. LIEBERMAN [2005], *Introduction to Operations Research*, 8th ed., McGraw-Hill, New York.

KENDALL, D. G. [1953], "Stochastic Processes Occurring in the Theory of Queues and Their Analysis by the Method of Imbedded Markov Chains," *Annals of Mathematical Statistics*, Vol. 24, pp. 338–354.

KLEINROCK, L. [1976], *Queueing Systems, Vol. 2: Computer Applications*, Wiley, New York.

LITTLE, J. D. C. [1961], "A Proof for the Queueing Formula $L = \lambda w$," *Operations Research*, Vol. 16, pp. 651–665.

NELSON, B. L. [1995], *Stochastic Modeling: Analysis & Simulation*, Dover Publications, Mineola, NY.

WINSTON, W. L. [2004], *Operations Research: Applications and Algorithms*, 4th Edition, Duxbury Press, Pacific Grove, CA.

EXERCISES

1. A tool crib has exponential interarrival and service times and serves a very large group of mechanics. The mean time between arrivals is 4 minutes. It takes 3 minutes on the average for a tool-crib attendant to service a mechanic. The attendant is paid $10 per hour and the mechanic is paid $15 per hour. Would it be advisable to have a second tool-crib attendant?

2. A two-runway (one runway for landing, one runway for taking off) airport is being designed for propeller-driven aircraft. The time to land an airplane is known to be exponentially distributed, with a mean of 1.5 minutes. If airplane arrivals are assumed to occur at random, what arrival rate can be tolerated if the average wait in the sky is not to exceed 3 minutes?

3. The Port of Trop can service only one ship at a time. However, there is mooring space for three more ships. Trop is a favorite port of call, but, if no mooring space is available, the ships have to go to the Port of Poop. An average of seven ships arrive each week, according to a Poisson process. The Port of Trop has the capacity to handle an average of eight ships a week, with service times exponentially distributed. What is the expected number of ships waiting or in service at the Port of Trop?

4. At Metropolis City Hall, two workers "pull strings" (make deals) every day. Strings arrive to be pulled on an average of one every 10 minutes throughout the day. It takes an average of 15 minutes to pull a string. Both times between arrivals and service times are exponentially distributed. What is the probability that there are no strings to be pulled in the system at a random point in time? What is the expected number of strings waiting to be pulled? What is the probability that both string-pullers are busy? What is the effect on performance if a third string-puller, working at the same speed as the first two, is added to the system?

5. At Tony and Cleo's bakery, one kind of birthday cake is offered. It takes 15 minutes to decorate this particular cake, and the job is performed by one particular baker. In fact, it is all this baker does. What mean time between arrivals (exponentially distributed) can be accepted if the mean length of the queue for decorating is not to exceed five cakes?

6. Patients arrive for a physical examination according to a Poisson process at the rate 1 per hour. The physical examination requires three stages, each one independently exponentially distributed, with a service time of 15 minutes. A patient must go through all three stages before the next patient is admitted to the treatment facility. Compute the average number of delayed patients, L_Q, for this system. [*Hint*: The variance of the sum of independent random variables is the sum of the variance.]

7. Suppose that mechanics arrive randomly at a tool crib according to a Poisson process with rate $\lambda = 10$ per hour. It is known that the single tool clerk serves a mechanic in 4 minutes on the average, with a standard deviation of approximately 2 minutes. Suppose that mechanics make $15.00 per hour. Estimate the steady-state average cost per hour of mechanics waiting for tools.

8. Arrivals to an airport are all directed to the same runway. At a certain time of the day, these arrivals form a Poisson process with rate 30 per hour. The time to land an aircraft is a constant 90 seconds. Determine L_Q, w_Q, L, and w for this airport. If a delayed aircraft burns $5000 worth of fuel per hour on the average, determine the average cost per aircraft of delay in waiting to land.

9. A machine shop repairs small electric motors, which arrive according to a Poisson process at the rate of 12 per week (5-day, 40-hour workweek). An analysis of past data indicates that engines can be repaired, on the average, in 2.5 hours, with a variance of 1 hour^2. How many working hours should a customer expect to leave a motor at the repair shop (not knowing the status of the system)? If the variance of the repair time could be controlled, what variance would reduce the expected waiting time to 6.5 hours?

10. Arrivals to a self-service gasoline pump occur in a Poisson fashion at the rate 12 per hour. Service time has a distribution that averages 4 minutes, with a standard deviation of one and a third minutes. What is the expected number of vehicles in the system?

11. Classic Car Care has one worker who washes cars in a four-step method—soap, rinse, dry, vacuum. The time to complete each step is exponentially distributed, with mean 9 minutes. Every car goes through every step before another car begins the process. On the average, one car every 45 minutes arrives for a wash job, according to a Poisson process. What is the average time a car waits to begin the wash job? What is the average number of cars in the car wash system? What is the average time required to wash a car?

12. A room has 10 cotton spinning looms. Once the looms are set up, they run automatically. The setup time is exponentially distributed, with mean 10 minutes. The machines run for an average of 40 minutes, also exponentially distributed. Loom operators are paid $10 an hour, and looms not running incur a cost of $40 an hour. How many loom operators should be employed to minimize the total cost of the loom room? If the objective becomes "on the average, no loom should wait more than 1 minute for an operator," how many people should be employed? How many operators should be employed to ensure that, on average, at least 7.5 looms are running at all times?

13. Consider the following information for a finite calling population problem with exponentially distributed runtimes and service times:

$$K = 10$$

$$\frac{1}{\mu} = 15$$

$$\frac{1}{\lambda} = 82$$

$$c = 2$$

Compute L_Q and w_Q. Find the value of λ such that $L_Q = L/2$.

14. Suppose that Figure 6.6 represents the number in system for a last-in-first-out (LIFO) single-server system. Customers are not preempted (i.e., kicked out of service), but, upon service completion, the most recent arrival next begins service. For this LIFO system, apportion the total area under $L(t)$ to each individual customer, as was done in Figure 6.8 for the FIFO system. Using the figure, show that Equations (6.10) and (6.8) hold for the single-server LIFO system.

15. Repeat Exercise 14, but assuming that

    (a) Figure 6.6 represents a FIFO system with $c = 2$ servers.
    (b) Figure 6.6 represents a LIFO system with $c = 2$ servers.

16. Consider an $M/G/1$ queue with the following type of service distribution: Customers request one of two types of service, in the proportions p and $1 - p$. Type-i service is exponentially distributed at rate μ_i, $i = 1, 2$. Let X_i denote a type-i service time and X an arbitrary service time. Then $E(X_i) = 1/\mu_i$, $V(X_i) = 1/\mu_i^2$ and

$$X = \begin{cases} X_1 & \text{with probability } p \\ X_2 & \text{with probability } (1 - p) \end{cases}$$

The random variable X is said to have a hyperexponential distribution with parameters (μ_1, μ_2, p).

    (a) Show that $E(X) = p/\mu_1 + (1 - p)/\mu_2$ and $E(X^2) = 2p/\mu_1^2 + 2(1 - p)/\mu_2^2$.
    (b) Use $V(X) = E(X^2) - [E(X)]^2$ to show $V(X) = 2p/\mu_1^2 + 2(1-p)/\mu_2^2 - [p/\mu_1 + (1-p)/\mu_2]^2$.
    (c) For any hyperexponential random variable, if $\mu_1 \neq \mu_2$ and $0 < p < 1$, show that its coefficient of variation is greater than 1; that is, $(\text{cv})^2 = V(X)/[E(X)]^2 > 1$. Thus, the hyperexponential distribution provides a family of statistical models for service times that are more variable than exponentially distributed service times. [*Hint*: The algebraic expression for $(\text{cv})^2$, by using parts (a) and (b), can be manipulated into the form $(\text{cv})^2 = 2p(1 - p)(1/\mu_1 - 1/\mu_2)^2/[E(X)]^2 + 1$.]
    (d) Many choices of μ_1, μ_2, and p lead to the same overall mean $E(X)$ and $(\text{cv})^2$. If a distribution with mean $E(X) = 1$ and coefficient of variation $\text{cv} = 2$ is desired, find values of μ_1, μ_2, and p to achieve this. [*Hint*: Choose $p = 1/4$ arbitrarily; then solve the following equations for μ_1 and μ_2.]

$$\frac{1}{4\mu_1} + \frac{3}{4\mu_2} = 1$$

$$\frac{3}{8}\left(\frac{1}{\mu_1} - \frac{1}{\mu_2}\right)^2 + 1 = 4$$

17. In Example 6.17, compare the systems with $c = 1$, $c = 2$, and $c = 3$ servers on the basis of server utilization ρ (the proportion of time a typical server is busy).

18. In Example 6.17, increase the number of machines by 2, then compare the systems with $c = 1$, $c = 2$, and $c = 3$ servers on the basis of server utilization ρ (the proportion of time a typical server is busy).

19. A small lumberyard is supplied by a fleet of 10 trucks. One overhead crane is available to unload the long logs from the trucks. It takes an average of 1 hour to unload a truck. After unloading, a truck takes an average of 3 hours to get the next load of logs and return to the lumberyard.

(a) Certain distributional assumptions are needed to analyze this problem with the models of this chapter. State them and discuss their reasonableness.

(b) With one crane, what is the average number of trucks waiting to be unloaded? On average, how many trucks arrive at the yard each hour? What percentage of trucks upon arrival find the crane busy? Is this the same as the long-run proportion of time the crane is busy?

(c) Suppose that a second crane is installed at the lumberyard. Answer the same questions as in part (b). Make a chart comparing one crane to two cranes.

(d) If the value of the logs brought to the yard is approximately $200 per truck load and if long-run crane costs are $50 per hour per crane (whether busy or idle), compute an optimal number of cranes on the basis of cost per hour.

(e) In addition to the costs in part (d), if management decides to consider the cost of idle trucks and drivers, what is the optimal number of cranes? A truck and its driver together are estimated to cost approximately $40 per hour and are considered to be idle when they are waiting for a crane.

20. A tool crib with one attendant serves a group of 10 mechanics. Mechanics work for an exponentially distributed amount of time with mean 20 minutes, then go to the crib to request a special tool. Service times by the attendant are exponentially distributed, with mean 3 minutes. If the attendant is paid $6 per hour and the mechanic is paid $10 per hour, would it be advisable.to have a second attendant? [*Hint*: Compare with Exercise 1.]

21. This problem is based on Case 8.1 in Nelson [1995]. A large consumer shopping mall is to be constructed. During busy times, on average, the arrival rate of cars is expected to be 1000 per hour, and studies at other malls suggest that customers will spend 3 hours shopping. The mall designers would like to have sufficient parking so that there are enough spaces 99.9% of the time. How many spaces should they have? [*Hint*: Model the system as an $M/G/\infty$ queue, where the spaces are servers, and find out how many spaces are adequate with probability 0.999.]

22. In Example 6.18, suppose that the overall arrival rate is expected to increase to 160 per hour. If the service rates do not change, how many clerks will be needed at service centers 2 and 3, just to keep up with the customer load?

23. A small copy shop has a self-service copier. Currently there is room for only 4 people to line up for the machine (including the person using the machine); when there are more than 4 people, the additional people must line up outside the shop. The owners would like to avoid having people line up outside the shop, as much as possible. For that reason, they are thinking about adding a second self-service copier. Self-service customers have been observed to arrive at the rate of 24 per hour, and they use the machine for 2 minutes, on average. Assess the impact of adding another copier. Carefully state any assumptions or approximations you make.

24. A self-service car wash has 4 washing stalls. When in a stall, a customer may choose from among three options: rinse only; wash and rinse; and wash, rinse, and wax. Each option has a fixed time to complete: rinse only, 3 minutes; wash and rinse, 7 minutes; wash, rinse, and wax, 12 minutes. The owners have observed that 20% of customers choose rinse only; 70% wash and rinse; and 10% wash, rinse, and wax. There are no scheduled appointments, and customers arrive at a rate of about 34 cars per hour. There is room for only 3 cars to wait in the parking lot, so, currently, many customers are lost. The owners wants to know how much more business they will do if they add another stall. Adding a stall will take away one space in the parking lot.

    Develop a queueing model of the system. Estimate the rate at which customers will be lost in the current and proposed system. Carefully state any assumptions or approximations you make.

25. Find examples of queueing models used in real applications. A good source is the journal *Interfaces*.

26. Study the effect of *pooling servers* (having multiple servers draw from a single queue, rather than each having its own queue) by comparing the performance measures for two $M/M/1$ queues, each with arrival rate λ and service rate μ, to an $M/M/2$ queue with arrival rate 2λ and service rate μ for each server.

27. A repair and inspection facility consists of two stations: a repair station with two technicians, and an inspection station with 1 inspector. Each repair technician works at the rate of 3 items per hour; the inspector can inspect 8 items per hour. Approximately 10% of all items fail inspection and are sent back to the repair station. (This percentage holds even for items that have been repaired two or more times.) If items arrive at the rate of 5 per hour, what is the long-run expected delay that items experience at each of the two stations, assuming a Poisson arrival process and exponentially distributed service times? What is the maximum arrival rate that the system can handle without adding personnel?

28. Parts arrive to the first stage of an assembly system at a rate of 50 per hour, then proceed through four additional stages before the product is complete. At each stage there is a 2% scrap rate, which means that 2% of the products at each stage are declared defective and discarded.

    (a) What is the arrival rate in parts per hour to the fifth assembly station?
    (b) If the mean assembly times for the five stations are different, does is make sense to order the stations from smallest to largest processing time, or vice versa? Why?

29. A call center answers questions about financial tracking, personal productivity, and contact management software. Calls arrive at a rate of 1 per minute, and historically 25% of calls are about financial products, 34% for productivity products, and 41% for contact management products. It takes 5 minutes on average to answer a caller's question. The number of customers who can be connected at any one time is essentially unlimited, but each product line has its own operators (2 for financial, 2 for productivity, and 3 for contact management). If an appropriate operator is available then the call is immediately routed to the operator; otherwise, the caller is placed in a hold queue. The company is hoping to reduce the total number of operators they need

by cross-training operators so that they can answer calls for any product line. Since the operators will not be experts across all products, this is expected to increase the time to process a call by about 10%. How many cross-trained operators are needed to provide service at the same level as the current system? Answer the question by approximating the current and proposed system.

Part III

Random Numbers

7

Random-Number Generation

Random numbers are a necessary basic ingredient in the simulation of almost all discrete systems. Most computer languages have a subroutine, object, or function that will generate a random number. Random numbers are used to generate event times and other random variables in simulation languages. In this chapter, the generation of random numbers and their subsequent testing for randomness is described. Chapter 8 shows how random numbers are used to generate a random variable with any desired probability distribution.

7.1 Properties of Random Numbers

A sequence of random numbers, $R_1, R_2, \ldots$, must have two important statistical properties: *uniformity* and *independence*. Each random number R_i must be an independent sample drawn from a continuous uniform distribution between zero and 1—that is, the pdf is given by

$$f(x) = \begin{cases} 1, & 0 \le x \le 1 \\ 0, & \text{otherwise} \end{cases}$$

This density function is shown in Figure 7.1. The expected value of each R_i is given by

$$E(R) = \int_0^1 x\,dx = \left.\frac{x^2}{2}\right|_0^1 = \frac{1}{2}$$

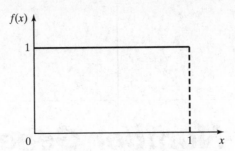

Figure 7.1   pdf for random numbers.

and the variance is given by

$$V(R) = \int_0^1 x^2\,dx - [E(R)]^2 = \left.\frac{x^3}{3}\right|_0^1 - \left(\frac{1}{2}\right)^2 = \frac{1}{3} - \frac{1}{4} = \frac{1}{12}$$

Some consequences of the uniformity and independence properties are the following:

1. If the interval [0, 1] is divided into n classes, or subintervals of equal length, the expected number of observations in each interval is N/n, where N is the total number of observations.
2. The probability of observing a value in a particular interval is independent of the previous values drawn.

In the next section we describe desirable properties of methods that produce random numbers to drive simulations.

7.2   Generation of Pseudo-Random Numbers

Notice that the title of this section has the word "pseudo" in it. "Pseudo" means false, so false random numbers are being generated! In this instance, "pseudo" is used to imply that the very act of generating random numbers by a known method removes the potential for true randomness. If the method is known, the set of random numbers can be repeated. Then an argument can be made that the numbers are not truly random. The goal of any generation scheme, however, is to produce a sequence of numbers between 0 and 1 that simulates, or imitates, the ideal properties of uniform distribution and independence as closely as possible.

To be sure, in the generation of pseudo-random numbers, certain problems or errors can occur. These errors, or departures from ideal randomness, are all related to the properties stated previously in Section 7.1. Some examples of such departures include the following:

1. The generated numbers might not be uniformly distributed.
2. The generated numbers might be discrete-valued instead of continuous-valued.
3. The mean of the generated numbers might be too high or too low.
4. The variance of the generated numbers might be too high or too low.

5. There might be dependence as measured, for instance, by autocorrelation.

Departures from uniformity and independence for a particular generation scheme often can be detected by tests such as those described in Section 7.4. If departures are detected, the generation scheme should be dropped in favor of an acceptable generator. Generators that pass the tests in Section 7.4, and tests even more stringent, have been developed; thus, there is no excuse for using a generator that has been found to be defective.

Usually, random numbers are generated by a digital computer as part of the simulation. There are numerous methods that can be used to generate the values. Before we describe some of these methods, there are several important considerations that we should mention:

1. The method should be fast. Individual computations are inexpensive, but simulation could require many millions of random numbers. The total cost can be managed by selecting a computationally efficient method of random-number generation.
2. The method should be portable to different computers—and, ideally, to different programming languages. This is desirable so that the simulation program will produce the same results wherever it is executed.
3. The method should have a sufficiently long cycle. The cycle length, or period, represents the length of the random number sequence before previous numbers begin to repeat themselves in an earlier order. Thus, if 10,000 events are to be generated, the period should be many times that long.
   A special case of cycling is degenerating. A method degenerates when the same random numbers appear repeatedly. Such an occurrence is certainly unacceptable. This can happen rapidly with some methods.
4. The random numbers should be repeatable. Given the starting point (or conditions) it should be possible to generate the same set of random numbers, completely independent of the system that is being simulated. This is helpful for debugging purposes and facilitates comparisons between systems (see Chapter 12). For the same reasons, it should be possible to easily specify different starting points, widely separated, within the sequence.
5. Most important, the generated random numbers should closely approximate the ideal statistical properties of uniformity and independence.

Inventing techniques that seem to generate random numbers is easy; inventing techniques that really do produce sequences that appear to be independent, uniformly distributed random numbers is incredibly difficult. There is now a vast literature and rich theory on the topic, and many hours of testing have been devoted to establishing the properties of various generators. Even when a technique is known to be theoretically sound, it is seldom easy to implement it in a way that will be fast and portable. The goal of this chapter is to make the reader aware of the central issues in random-number generation, to enhance understanding, and to show some of the techniques that are used by those working in this area.

7.3   Techniques for Generating Random Numbers

The linear congruential method of Section 7.3.1 is the most widely used technique for generating random numbers, so we describe it in detail. We also report on an extension of this method that yields sequences with a longer period. Many other methods have been proposed, and they are reviewed in Bratley, Fox, and Schrage [1996], Law [2007], and Ripley [1987].

7.3.1 Linear Congruential Method

The linear congruential method, initially proposed by Lehmer [1951], first produces a sequence of integers, $X_1, X_2, \ldots$ between zero and $m - 1$ by following a recursive relationship:

$$X_{i+1} = (aX_i + c) \bmod m, \quad i = 0, 1, 2, \ldots \tag{7.1}$$

The initial value X_0 is called the seed, a is called the multiplier, c is the increment, and m is the modulus. If $c \neq 0$ in Equation (7.1), then the form is called the *mixed congruential method*. When $c = 0$, the form is known as the *multiplicative congruential method*. The selection of the values for a, c, m, and X_0 drastically affects the statistical properties and the cycle length. Variations of Equation (7.1) are quite common in the computer generation of random numbers.

Notice that random integers are being generated rather than random numbers. These random integers should appear to be uniformly distributed on the integers zero to m. Random numbers R_i between zero and 1 can then be generated by setting

$$R_i = \frac{X_i}{m}, \quad i = 1, 2, \ldots \tag{7.2}$$

An example will illustrate how this technique operates.

Example 7.1

Use the linear congruential method to generate a sequence of random numbers with $X_0 = 27, a = 17$, $c = 43$, and $m = 100$.

The sequence of seeds X_i and subsequent random numbers R_i, is computed as follows:

$$X_0 = 27$$

$$X_1 = (17 \cdot 27 + 43) \bmod 100 = 502 \bmod 100 = 2$$

$$R_1 = \frac{2}{100} = 0.02$$

$$X_2 = (17 \cdot 2 + 43) \bmod 100 = 77 \bmod 100 = 77$$

$$R_2 = \frac{77}{100} = 0.77$$

$$X_3 = (17 \cdot 77 + 43) \bmod 100 = 1352 \bmod 100 = 52$$

$$R_3 = \frac{52}{100} = 0.52$$

$$\vdots$$

Recall that $a = b \bmod m$ provided that $b - a$ is divisible by m with no remainder; otherwise, the mod operation returns the remainder of b/m. Thus, $X_1 = 502 \bmod 100 = 2$ because 502/100 equals 5 with a remainder of 2.

The ultimate test of the linear congruential method, as of any generation scheme, is how closely the generated numbers $R_1, R_2, \ldots$ approximate uniformity and independence. There are, however, several secondary properties that must be considered. These include *maximum density* and *maximum period*.

First, notice that the numbers generated from Equation (7.2) assume values only from the set $I = \{0, 1/m, 2/m, \ldots, (m-1)/m\}$, because each X_i is an integer in the set $\{0, 1, 2, \ldots, m-1\}$. Thus, each R_i is discrete on I, instead of continuous on the interval [0, 1]. This approximation appears to be of little consequence if the modulus m is a very large integer. (Values such as $m = 2^{31} - 1$ and $m = 2^{48}$ are in common use in generators appearing in many simulation languages.) By maximum density is meant that the values assumed by $R_i, i = 1, 2, \ldots$, leave no large gaps on [0, 1].

Second, to help achieve maximum density, and to avoid cycling in practical applications (i.e., recurrence of the same sequence of generated numbers), the generator should have the largest possible period. Maximal period can be achieved by the proper choice of a, c, m, and X_0 [Law 2007].

- For m a power of 2, say $m = 2^b$, and $c \neq 0$, the longest possible period is $P = m = 2^b$, which is achieved whenever c is relatively prime to m (that is, the greatest common factor of c and m is 1) and $a = 1 + 4k$, where k is an integer.
- For m a power of 2, say $m = 2^b$, and $c = 0$, the longest possible period is $P = m/4 = 2^{b-2}$, which is achieved if the seed X_0 is odd and if the multiplier a is given by $a = 3 + 8k$ or $a = 5 + 8k$, for some $k = 0, 1, \ldots$.
- For m a prime number and $c = 0$, the longest possible period is $P = m - 1$, which is achieved whenever the multiplier a has the property that the smallest integer k such that $a^k - 1$ is divisible by m is $k = m - 1$.

Example 7.2

Using the multiplicative congruential method, find the period of the generator for $a = 13$, $m = 2^6 = 64$, and $X_0 = 1, 2, 3$, and 4. The solution is given in Table 7.1. When the seed is 1 or 3, the sequence has period 16. However, a period of length 8 is achieved when the seed is 2 and a period of length 4 occurs when the seed is 4.

In Example 7.2, $m = 2^6 = 64$ and $c = 0$. The maximal period is therefore $P = m/4 = 16$. Notice that this period is achieved by using odd seeds, $X_0 = 1$ and $X_0 = 3$; even seeds, $X_0 = 2$ and $X_0 = 4$, yield the periods 8 and 4, respectively, both less than the maximum. Notice that $a = 13$ is of the form $5 + 8k$ with $k = 1$, as is required to achieve maximal period.

When $X_0 = 1$, the generated sequence assumes values from the set $\{1, 5, 9, 13, \ldots, 53, 57, 61\}$. The gaps in the sequence of generated random numbers R_i are quite large (i.e., the gap is $5/64 - 1/64$ or 0.0625). Such a gap gives rise to concern about the density of the generated sequence.

The generator in Example 7.2 is not viable for any application—its period is too short and its density is insufficient. However, the example shows the importance of properly choosing a, c, m, and X_0.

Table 7.1 Period Determination Using Various Seeds

i	X_i	X_i	X_i	X_i
0	1	2	3	4
1	13	26	39	52
2	41	18	59	36
3	21	42	63	20
4	17	34	51	4
5	29	58	23	
6	57	50	43	
7	37	10	47	
8	33	2	35	
9	45		7	
10	9		27	
11	53		31	
12	49		19	
13	61		55	
14	25		11	
15	5		15	
16	1		3	

Speed and efficiency in using the generator on a digital computer is also a selection consideration. Speed and efficiency are aided by use of a modulus, m, which is either a power of 2 or close to a power of 2. Since most digital computers use a binary representation of numbers, the modulo, or remaindering, operation of Equation (7.1) can be conducted efficiently when the modulo is a power of 2 (i.e., $m = 2^b$). After ordinary arithmetic yields a value for $aX_i + c$, X_{i+1} is obtained by dropping the leftmost binary digits in $aX_i + c$ and then using only the b rightmost binary digits. The following example illustrates, by analogy, this operation using $m = 10^b$, because most human beings think in decimal representation.

Example 7.3 ———————————————————————————————————————
Let $X_0 = 63$, $a = 19$, $c = 0$, and $m = 10^2 = 100$, and generate a sequence of random integers using Equation (7.1).

$$X_0 = 63$$

$$X_1 = (19)(63) \bmod 100 = 1197 \bmod 100 = 97$$

$$X_2 = (19)(97) \bmod 100 = 1843 \bmod 100 = 43$$

$$X_3 = (19)(43) \bmod 100 = 817 \bmod 100 = 17$$

$$\vdots$$

When m is a power of 10, say $m = 10^b$, the modulo operation is accomplished by saving the b rightmost (decimal) digits. By analogy, the modulo operation is most efficient for binary computers when $m = 2^b$ for some $b > 0$.

Example 7.4

The last example in this section is in actual use. It has been extensively tested [Learmonth and Lewis, 1973; Lewis *et al.*, 1969]. The values for a, c, and m have been selected to ensure that the characteristics desired in a generator are most likely to be achieved.

Let $a = 7^5 = 16,807$, $c = 0$, and $m = 2^{31} - 1 = 2,147,483,647$ (a prime number). These choices satisfy the conditions that ensure a period of $P = m - 1$ (well over 2 billion). Further, specify the seed $X_0 = 123,457$. The first few numbers generated are as follows:

$$X_1 = 7^5(123,457) \bmod (2^{31} - 1) = 2,074,941,799 \bmod (2^{31} - 1)$$

$$X_1 = 2,074,941,799$$

$$R_1 = \frac{X_1}{2^{31}} = 0.9662$$

$$X_2 = 7^5(2,074,941,799) \bmod (2^{31} - 1) = 559,872,160$$

$$R_2 = \frac{X_2}{2^{31}} = 0.2607$$

$$X_3 = 7^5(559,872,160) \bmod (2^{31} - 1) = 1,645,535,613$$

$$R_3 = \frac{X_3}{2^{31}} = 0.7662$$

$$\vdots$$

Notice that this method divides by $m + 1$ instead of m; however, for such a large value of m, the effect is negligible.

7.3.2 Combined Linear Congruential Generators

As computing power has increased, the complexity of the systems that we are able to simulate has also increased. A random-number generator with period $2^{31} - 1 \approx 2 \times 10^9$, such as the popular generator described in Example 7.4, is no longer adequate for all applications. Examples include the simulation of highly reliable systems, in which hundreds of thousands of elementary events must be simulated in order to observe even a single failure event, and the simulation of complex computer networks, in which thousands of users are executing hundreds of programs. An area of current research is deriving generators with substantially longer periods.

One fruitful approach is to combine two or more multiplicative congruential generators in such a way that the combined generator has good statistical properties and a longer period. The following result from L'Ecuyer [1988] suggests how this can be done:

If $W_{i,1}, W_{i,2}, \ldots, W_{i,k}$ are any independent, discrete-valued random variables (not necessarily identically distributed), but one of them, say $W_{i,1}$, is uniformly distributed on the integers from 0 to $m_1 - 2$, then

$$W_i = \left(\sum_{j=1}^{k} W_{i,j} \right) \bmod m_1 - 1$$

is uniformly distributed on the integers from 0 to $m_1 - 2$.

To see how this result can be used to form combined generators, let $X_{i,1}, X_{i,2}, \ldots, X_{i,k}$ be the ith output from k different multiplicative congruential generators, where the jth generator has prime modulus m_j and the multiplier a_j is chosen so that the period is $m_j - 1$. Then the jth generator is producing integers $X_{i,j}$ that are approximately uniformly distributed on the integers from 1 to $m_j - 1$, and $W_{i,j} = X_{i,j} - 1$ is approximately uniformly distributed on the integers from 0 to $m_j - 2$. L'Ecuyer therefore suggests combined generators of the form

$$X_i = \left(\sum_{j=1}^{k} (-1)^{j-1} X_{i,j} \right) \bmod m_1 - 1$$

with

$$R_i = \begin{cases} \dfrac{X_i}{m_1}, & X_i > 0 \\[2mm] \dfrac{m_1 - 1}{m_1}, & X_i = 0 \end{cases}$$

Notice that the "$(-1)^{j-1}$" coefficient implicitly performs the subtraction $X_{i,1} - 1$; for example, if $k = 2$ then $(-1)^0 (X_{i,1} - 1) - (-1)^1 (X_{i,2} - 1) = \sum_{j=1}^{2} (-1)^{j-1} X_{i,j}$.

The maximum possible period for such a generator is

$$P = \frac{(m_1 - 1)(m_2 - 1) \cdots (m_k - 1)}{2^{k-1}}$$

which is achieved by the generator described in the next example.

Example 7.5

For 32-bit computers, L'Ecuyer [1988] suggests combining $k = 2$ generators with $a_1 = 40{,}014$, $m_1 = 2{,}147{,}483{,}563$, $a_2 = 40{,}692$, and $m_2 = 2{,}147{,}483{,}399$. This leads to the following algorithm:

1. Select seed $X_{1,0}$ in the range $[1, \ 2{,}147{,}483{,}562]$ for the first generator, and seed $X_{2,0}$ in the range $[1, \ 2{,}147{,}483{,}398]$ for the second.
   Set $j = 0$.
2. Evaluate each individual generator.

$$X_{1,j+1} = 40{,}014 \, X_{1,j} \bmod 2{,}147{,}483{,}563$$

$$X_{2,j+1} = 40{,}692 \, X_{2,j} \bmod 2{,}147{,}483{,}399$$

3. Set.

$$X_{j+1} = (X_{1,j+1} - X_{2,j+1}) \bmod 2,147,483,562$$

4. Return.

$$R_{j+1} = \begin{cases} \dfrac{X_{j+1}}{2,147,483,563}, & X_{j+1} > 0 \\[2ex] \dfrac{2,147,483,562}{2,147,483,563}, & X_{j+1} = 0 \end{cases}$$

5. Set $j = j + 1$ and go to Step 2.

This combined generator has period $(m_1 - 1)(m_2 - 1)/2 \approx 2 \times 10^{18}$. Perhaps surprisingly, even such a long period might not be adequate for all applications. See L'Ecuyer [1996, 1999] and L'Ecuyer *et al.* [2002] for combined generators with periods as long as $2^{191} \approx 3 \times 10^{57}$. One of these generators has been translated into Visual Basic for Applications and can be found in the spreadsheet `RandomNumberTools.xls` at `www.bcnn.net`.

7.3.3   Random-Number Streams

The initial seed for a linear congruential random-number generator (initial seeds, in the case of a combined linear congruential generator) is the integer value X_0 that initializes the random-number sequence. Since the sequence of integers $X_0, X_1, \dots, X_P, X_0, X_1, \dots$ produced by a generator repeats, any value in the sequence could be used to "seed" the generator.

For a linear congruential generator, a *random-number stream* is nothing more than a convenient way to refer to a starting seed taken from the sequence $X_0, X_1, \dots, X_P$ (for a combined generator, the stream is a reference to the starting seeds for all of the basic generators); typically these starting seeds are far apart in the sequence. For instance, if the streams are b values apart, then stream i could be defined by starting seed

$$S_i = X_{b(i-1)}$$

for $i = 1, 2, \dots, \lfloor P/b \rfloor$. Values of $b = 100,000$ were common in older generators, but values as large as $b = 10^{37}$ are in use in modern combined linear congruential generators. (See, for instance, L'Ecuyer *et al.* [2002] for the implementation of such a generator.) Thus, a single random-number generator with k streams acts like k distinct virtual random-number generators, provided that the current value of seed for each stream is maintained. Exercise 21 illustrates one way to create streams that are widely separated in the random-number sequence.

In Chapter 12, we will consider the problem of comparing two or more alternative systems via simulation, and we will show that there are advantages to dedicating portions of the pseudo-random number sequence to the same purpose in each of the simulated systems. For instance, in comparing the efficiency of several queueing systems, a fairer comparison will be achieved if all of the simulated systems experience exactly the same sequence of customer arrivals. Such synchronization can be achieved by assigning a specific stream to generate arrivals in each of the queueing simulations. If the starting seeds for these streams are spaced far enough apart, then this has the same effect as having a distinct random-number generator whose only purpose is to generate customer arrivals.

7.4 Tests for Random Numbers

The desirable properties of random numbers—uniformity and independence—were discussed in Section 7.1. To check on whether these desirable properties have been achieved, a number of tests can be performed. (Fortunately, the appropriate tests have already been conducted for most commercial simulation software.) The tests can be placed in two categories, according to the properties of interest: uniformity and independence. A brief description of two types of tests is given in this section:

1. *Frequency test.* Uses the Kolmogorov–Smirnov or the chi-square test to compare the distribution of the set of numbers generated to a uniform distribution.
2. *Autocorrelation test.* Tests the correlation between numbers and compares the sample correlation to the desired correlation, zero.

In testing for uniformity, the hypotheses are as follows:

$$H_0: \quad R_i \sim \text{Uniform}[0, 1]$$
$$H_1: \quad R_i \not\sim \text{Uniform}[0, 1]$$

The null hypothesis, H_0, reads that the numbers are distributed uniformly on the interval $[0, 1]$. Failure to reject the null hypothesis means that evidence of nonuniformity has not been detected by this test. This does not imply that further testing of the generator for uniformity is unnecessary.

In testing for independence, the hypotheses are as follows:

$$H_0: \quad R_i \sim \text{independently}$$
$$H_1: \quad R_i \not\sim \text{independently}$$

This null hypothesis, H_0, reads that the numbers are independent. Failure to reject the null hypothesis means that evidence of dependence has not been detected by this test. This does not imply that further testing of the generator for independence is unnecessary.

For each test, a level of significance α must be stated. The level α is the probability of rejecting the null hypothesis when the null hypothesis is true:

$$\alpha = P(\text{reject } H_0 | H_0 \text{ true})$$

The decision maker sets the value of α for any test. Frequently, α is set to 0.01 or 0.05.

If several tests are conducted on the same set of numbers, the probability of rejecting the null hypothesis on at least one test, by chance alone (i.e., making a Type I (α) error), increases. Say that $\alpha = 0.05$ and that five different tests are conducted on a sequence of numbers. The probability of rejecting the null hypothesis on at least one test, by chance alone, could be as large as 0.25.

Similarly, if one test is conducted on many sets of numbers from a generator, the probability of rejecting the null hypothesis on at least one test by chance alone (i.e., making a Type I (α) error), increases as more sets of numbers are tested. For instance, if 100 sets of numbers were subjected to the test, with $\alpha = 0.05$, it would be expected that five of those tests would be rejected by chance alone. If the number of rejections in 100 tests is close to 100α, then there is no compelling reason to discard the generator. The concept discussed in this and the preceding paragraph is discussed further at the conclusion of Example 7.8.

If one of the well-known simulation languages or random-number generators is used, it is probably unnecessary to apply the tests just mentioned or those described in Sections 7.4.1 and 7.4.2. However, random-number generators frequently are added to software that is not specifically developed for simulation, such as spreadsheet programs, symbolic/numerical calculators, and programming languages. If the generator that is at hand is not explicitly known or documented, then the tests in this chapter should be applied to many samples of numbers from the generator. Some additional tests that are commonly used, but are not covered here, are Good's serial test for sampling numbers [1953, 1967], the median-spectrum test [Cox and Lewis, 1966; Durbin, 1967], the runs test [Law 2007], and a variance heterogeneity test [Cox and Lewis, 1966]. Even if a set of numbers passes all the tests, there is no guarantee of randomness; it is always possible that some underlying pattern has gone undetected.

In this book, we emphasize empirical tests that are applied to actual sequences of numbers produced by a generator. Because of the extremely long period of modern pseudo-random-number generators, as described in Section 7.3.2, it is no longer possible to apply these tests to a significant portion of the period of such generators. The tests can be used as a check if one encounters a generator with completely unknown properties (perhaps one that is undocumented and buried deep in a software package), but they cannot be used to establish the quality of a generator throughout its period. Fortunately, there are also families of theoretical tests that evaluate the choices for m, a, and c without actually generating any numbers, the most common being the spectral test. Many of these tests assess how k-tuples of random numbers fill up a k-dimensional unit cube. These tests are beyond the scope of this book; see, for instance, Ripley [1987].

In the examples of tests that follow, the hypotheses are not restated. The hypotheses are as indicated in the foregoing paragraphs. Although few simulation analysts will need to perform these tests, every simulation user should be aware of the qualities of a good random-number generator.

7.4.1 Frequency Tests

A basic test that should always be performed to validate a new generator is the test of uniformity. Two different methods of testing are the Kolmogorov–Smirnov and the chi-square test. Both of these tests measure the degree of agreement between the distribution of a sample of generated random numbers and the theoretical uniform distribution. Both tests are based on the null hypothesis of no significant difference between the sample distribution and the theoretical distribution.

1. *The Kolmogorov–Smirnov test.* This test compares the continuous cdf $F(x)$ of the uniform distribution with the empirical cdf $S_N(x)$ of the sample of N observations. By definition,

$$F(x) = x, \quad 0 \leq x \leq 1$$

If the sample from the random-number generator is $R_1, R_2, \ldots, R_N$, then the empirical cdf $S_N(x)$ is defined by

$$S_N(x) = \frac{\text{number of } R_1, R_2, \ldots, R_N \text{ which are } \leq x}{N}$$

As N becomes larger, $S_N(x)$ should become a better approximation to $F(x)$, provided that the null hypothesis is true.

In Section 5.6, empirical distributions were described. The cdf of an empirical distribution is a step function with jumps at each observed value. This behavior was illustrated by Example 5.35.

The Kolmogorov–Smirnov test is based on the largest absolute deviation between $F(x)$ and $S_N(x)$ over the range of the random variable—that is, it is based on the statistic

$$D = \max |F(x) - S_N(x)| \tag{7.3}$$

The sampling distribution of D is known; it is tabulated as a function of N in Table A.8. For testing against a uniform cdf, the test procedure follows these steps:

Step 1.   Rank the data from smallest to largest. Let $R_{(i)}$ denote the ith smallest observation, so that

$$R_{(1)} \leq R_{(2)} \leq \cdots \leq R_{(N)}$$

Step 2.   Compute

$$D^+ = \max_{1 \leq i \leq N} \left\{ \frac{i}{N} - R_{(i)} \right\}$$

$$D^- = \max_{1 \leq i \leq N} \left\{ R_{(i)} - \frac{i-1}{N} \right\}$$

Step 3.   Compute $D = \max(D^+, D^-)$.

Step 4.   Locate the critical value D_α in Table A.8 for the specified significance level α and the given sample size N.

Step 5.   If the sample statistic D is greater than the critical value D_α, the null hypothesis that the data are a sample from a uniform distribution is rejected. If $D \leq D_\alpha$, conclude that no difference has been detected between the true distribution of $\{R_1, R_2, \ldots, R_N\}$ and the uniform distribution.

Example 7.6
Suppose that the five numbers 0.44, 0.81, 0.14, 0.05, 0.93 were generated, and it is desired to perform a test for uniformity by using the Kolmogorov–Smirnov test with the level of significance $\alpha = 0.05$. First, the numbers must be sorted from smallest to largest. The calculations can be facilitated by use of Table 7.2. The top row lists the numbers from smallest $(R_{(1)})$ to largest $(R_{(5)})$. The computations for D^+, namely $i/N - R_{(i)}$, and for D^-, namely $R_{(i)} - (i-1)/N$, are easily accomplished by using Table 7.2. The statistics are computed as $D^+ = 0.26$ and $D^- = 0.21$. Therefore, $D = \max\{0.26, 0.21\} = 0.26$.

Table 7.2 Calculations for Kolmogorov–Smirnov Test

$R_{(i)}$	0.05	0.14	0.44	0.81	0.93
i/N	0.20	0.40	0.60	0.80	1.00
$i/N - R_{(i)}$	0.15	0.26	0.16	—	0.07
$R_{(i)} - (i-1)/N$	0.05	—	0.04	0.21	0.13

The critical value of D, obtained from Table A.8 for $\alpha = 0.05$ and $N = 5$, is 0.565. Since the computed value, 0.26, is less than the tabulated critical value, 0.565, the hypothesis that the distribution of the generated numbers is the uniform distribution is not rejected.

The calculations in Table 7.2 are illustrated in Figure 7.2, where the empirical cdf $S_N(x)$ is compared to the uniform cdf $F(x)$. It can be seen that D^+ is the largest deviation of $S_N(x)$ above $F(x)$, and that D^- is the largest deviation of $S_N(x)$ below $F(x)$. For example, at $R_{(3)}$, the value of D^+ is given by $3/5 - R_{(3)} = 0.60 - 0.44 = 0.16$, and that of D^- is given by $R_{(3)} - 2/5 = 0.44 - 0.40 = 0.04$. Although the test statistic D is defined by Equation (7.3) as the maximum deviation over all x, it can be seen from Figure 7.2 that the maximum deviation will always occur at one of the jump points $R_{(1)}, R_{(2)}, \ldots, R_{(N)}$; thus, the deviation at other values of x need not be considered.

2. *The chi-square test.* The chi-square test uses the sample statistic

$$\chi_0^2 = \sum_{i=1}^{n} \frac{(O_i - E_i)^2}{E_i}$$

where O_i is the observed number in the ith class, E_i is the expected number in the ith class, and n is the number of classes. For the uniform distribution E_i, the expected number in each class is given by

$$E_i = \frac{N}{n}$$

for equally spaced classes, where N is the total number of observations. It can be shown that the sampling distribution of χ_0^2 is approximately the chi-square distribution with $n-1$ degrees of freedom.

Example 7.7 _____

Use the chi-square test with $\alpha = 0.05$ to test for whether the data shown next are uniformly distributed. Table 7.3 contains the essential computations. The test uses $n = 10$ intervals of equal length, namely $[0, 0.1), [0.1, 0.2), \ldots, [0.9, 1.0)$. The value of χ_0^2 is 3.4. This is compared with the critical value $\chi_{0.05,9}^2 = 16.9$ from Table A.6. Since χ_0^2 is much smaller than the tabulated value of $\chi_{0.05,9}^2$, the null hypothesis of a uniform distribution is not rejected.

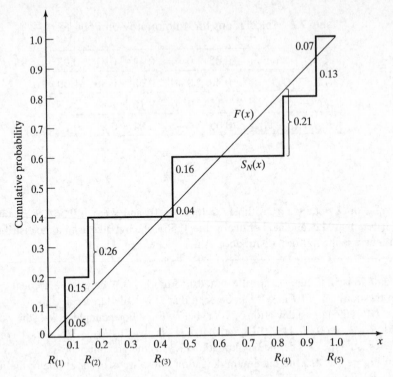

Figure 7.2   Comparison of $F(x)$ and $S_N(x)$.

0.34	0.90	0.25	0.89	0.87	0.44	0.12	0.21	0.46	0.67
0.83	0.76	0.79	0.64	0.70	0.81	0.94	0.74	0.22	0.74
0.96	0.99	0.77	0.67	0.56	0.41	0.52	0.73	0.99	0.02
0.47	0.30	0.17	0.82	0.56	0.05	0.45	0.31	0.78	0.05
0.79	0.71	0.23	0.19	0.82	0.93	0.65	0.37	0.39	0.42
0.99	0.17	0.99	0.46	0.05	0.66	0.10	0.42	0.18	0.49
0.37	0.51	0.54	0.01	0.81	0.28	0.69	0.34	0.75	0.49
0.72	0.43	0.56	0.97	0.30	0.94	0.96	0.58	0.73	0.05
0.06	0.39	0.84	0.24	0.40	0.64	0.40	0.19	0.79	0.62
0.18	0.26	0.97	0.88	0.64	0.47	0.60	0.11	0.29	0.78

Different authors have offered considerations concerning the application of the chi-square test. In the application to a data set the size of that in Example 7.7, the considerations do not apply—that is, if 100 values are in the sample and from 5 to 10 intervals of equal length are used, the test will be acceptable. In general, it is recommended that n and N be chosen so that each $E_i \geq 5$.

Both the Kolmogorov–Smirnov test and the chi-square test are acceptable for testing the uniformity of a sample of data, provided that the sample size is large. However, the Kolmogorov–Smirnov

Table 7.3 Computations for Chi-Square Test

Interval	O_i	E_i	$O_i - E_i$	$(O_i - E_i)^2$	$\dfrac{(O_i - E_i)^2}{E_i}$
1	8	10	−2	4	0.4
2	8	10	−2	4	0.4
3	10	10	0	0	0.0
4	9	10	−1	1	0.1
5	12	10	2	4	0.4
6	8	10	−2	4	0.4
7	10	10	0	0	0.0
8	14	10	4	16	1.6
9	10	10	0	0	0.0
10	11	10	1	1	0.1
	100	100	0		3.4

test is the more powerful of the two and is recommended. Furthermore, the Kolmogorov–Smirnov test can be applied to small sample sizes, whereas the chi-square is valid only for large samples, say $N \geq 50$.

Imagine a set of 100 numbers that are being tested for independence, one where the first 10 values are in the range 0.01–0.10, the second 10 values are in the range 0.11–0.20, and so on. This set of numbers would pass the frequency tests with ease, but the ordering of the numbers produced by the generator would not be random. The test in the next section of this chapter is concerned with the independence of random numbers that are generated.

7.4.2 Tests for Autocorrelation

The tests for autocorrelation are concerned with the dependence between numbers in a sequence. As an example, consider the following sequence of numbers, read from left to right:

0.12	0.01	0.23	0.28	0.89	0.31	0.64	0.28	0.83	0.93
0.99	0.15	0.33	0.35	0.91	0.41	0.60	0.27	0.75	0.88
0.68	0.49	0.05	0.43	0.95	0.58	0.19	0.36	0.69	0.87

From a visual inspection, these numbers appear random, and they would probably pass all the tests presented to this point. However, an examination of the 5th, 10th, 15th (every five numbers beginning with the fifth), and so on, indicates a very large number in that position. Now, 30 numbers is a rather small sample size on which to reject a random number generator, but the notion is that numbers in the sequence might be related. In this particular section, a method for discovering whether such a relationship exists is described. The relationship would not have to be all high numbers. It is possible to have all low numbers in the locations being examined, or the numbers could alternate from very high to very low.

The test to be described shortly requires the computation of the autocorrelation between every ℓ numbers (ℓ is also known as the *lag*), starting with the ith number. Thus, the autocorrelation $\rho_{i\ell}$

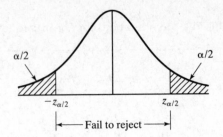

Figure 7.3   Failure to reject hypothesis.

between the following numbers would be of interest: $R_i, R_{i+\ell}, R_{i+2\ell}, \ldots, R_{i+(M+1)\ell}$. The value M is the largest integer such that $i + (M+1)\ell \leq N$, where N is the total number of values in the sequence. (Thus, a subsequence of length $M + 2$ is being tested.)

A nonzero autocorrelation implies a lack of independence, so the following two-tailed test is appropriate:

$$H_0: \quad \rho_{i\ell} = 0$$
$$H_1: \quad \rho_{i\ell} \neq 0$$

For large values of M, the distribution of the estimator of $\rho_{i\ell}$, denoted $\widehat{\rho}_{i\ell}$, is approximately normal if the values $R_i, R_{i+\ell}, R_{i+2\ell}, \ldots, R_{i+(M+1)\ell}$ are uncorrelated. Then the test statistic can be formed as follows:

$$Z_0 = \frac{\widehat{\rho}_{i\ell}}{\sigma_{\widehat{\rho}_{i\ell}}}$$

which is distributed normally with a mean of zero and a variance of 1, under the assumption of independence, for large M.

The formula for $\widehat{\rho}_{i\ell}$, in a slightly different form, and the standard deviation of the estimator $\sigma_{\widehat{\rho}_{i\ell}}$, are given by Schmidt and Taylor [1970] as follows:

$$\widehat{\rho}_{i\ell} = \frac{1}{M+1} \left[\sum_{k=0}^{M} R_{i+k\ell} R_{i+(k+1)\ell} \right] - 0.25$$

and

$$\sigma_{\widehat{\rho}_{i\ell}} = \frac{\sqrt{13M + 7}}{12(M+1)}$$

After computing Z_0, do not reject the null hypothesis of independence if $-z_{\alpha/2} \leq Z_0 \leq z_{\alpha/2}$ where α is the level of significance and $z_{\alpha/2}$ is obtained from Table A.3. Figure 7.3 illustrates this test.

If $\rho_{i\ell} > 0$, the subsequence is said to exhibit positive autocorrelation. In this case, successive values at lag ℓ have a higher probability than expected of being close in value (i.e., high random numbers in the subsequence followed by high, and low followed by low). On the other hand, if $\rho_{i\ell} < 0$, the subsequence is exhibiting negative autocorrelation, which means that low random numbers tend to be followed by high ones, and vice versa. The desired property, independence (which implies zero autocorrelation), means that there is no discernible relationship of the nature discussed here between successive random numbers at lag ℓ.

Example 7.8

Test for whether the 3rd, 8th, 13th, and so on, numbers in the sequence at the beginning of this section are autocorrelated using $\alpha = 0.05$. Here, $i = 3$ (beginning with the third number), $\ell = 5$ (every five numbers), $N = 30$ (30 numbers in the sequence), and $M = 4$ (largest integer such that $3 + (M + 1)5 \le 30$). Then,

$$\widehat{\rho}_{35} = \frac{1}{4+1}[(0.23)(0.28) + (0.28)(0.33) + (0.33)(0.27) + (0.27)(0.05)$$

$$+(0.05)(0.36)] - 0.25$$

$$= -0.1945$$

and

$$\sigma_{\widehat{\rho}_{35}} = \frac{\sqrt{13(4) + 7}}{12(4 + 1)} = 0.1280$$

Then, the test statistic assumes the value

$$Z_0 = -\frac{0.1945}{0.1280} = -1.516$$

Now, the critical value from Table A.3 is

$$z_{0.025} = 1.96$$

Therefore, the hypothesis of independence cannot be rejected on the basis of this test.

It can be observed that this test is not very sensitive for small values of M, particularly when the numbers being tested are on the low side. Imagine what would happen if each of the entries in the foregoing computation of $\widehat{\rho}_{i\ell}$ were equal to zero. Then $\widehat{\rho}_{i\ell}$ would be equal to -0.25 and the calculated Z would have the value of -1.95, not quite enough to reject the hypothesis of independence.

There are many sequences that can be formed in a set of data, given a large value of N. For example, beginning with the first number in the sequence, possibilities include (a) the sequence of all numbers, (b) the sequence formed from the first, third, fifth, ..., numbers, (c) the sequence formed from the first, fourth, ..., numbers, and so on. If $\alpha = 0.05$, there is a probability of 0.05 of rejecting a true hypothesis. If 10 independent sequences are examined, the probability of finding no significant autocorrelation, by chance alone, is $(0.95)^{10}$ or 0.60. Thus, 40% of the time significant autocorrelation would be detected when it does not exist. If α is 0.10 and 10 tests are conducted, there is a 65% chance of finding autocorrelation by chance alone. In conclusion, in "fishing" for autocorrelation by performing numerous tests, autocorrelation might eventually be detected, perhaps by chance alone, even when there is no autocorrelation present.

7.5   Summary

This chapter described the generation of random numbers and the subsequent testing of the generated numbers for uniformity and independence. Random numbers are used to generate random variates, the subject of Chapter 8.

Of the many types of random-number generators available, ones based on the linear congruential method are the most widely used, but they are being replaced by combined linear congruential generators. Of the many types of statistical tests that are used in testing random-number generators, two different types are described: one testing for uniformity and one testing for independence.

The simulation analyst might never work directly with a random-number generator or with the testing of random numbers from a generator. Most computers and simulation languages have methods that generate a random number, or streams of random numbers, for the asking. But even generators that have been used for years, some of which are still in use, have been found to be inadequate. So this chapter calls the simulation analyst's attention to such possibilities, with a warning to investigate and confirm that the generator has been tested thoroughly. Some researchers have attained sophisticated expertise in developing methods for generating and testing random numbers and the subsequent application of these methods. This chapter provides only a basic introduction to the subject matter; more depth and breadth are required for the reader to become a specialist in the area. An important foundational reference is Knuth [1998]; see also the reviews in Bratley, Fox, and Schrage [1996], Law [2007], L'Ecuyer [1998], and Ripley [1987].

One final caution is due. Even if generated numbers pass all the tests (those covered in this chapter and those mentioned in the chapter), some underlying pattern might have gone undetected without the generator's having been rejected as faulty. However, the generators available in widely used simulation languages have been extensively tested and validated.

REFERENCES

BRATLEY, P., B. L. FOX, AND L. E. SCHRAGE [1996], *A Guide to Simulation*, 2d ed., Springer-Verlag, New York.

COX, D. R., AND P. A. W. LEWIS [1966], *The Statistical Analysis of Series of Events*, Methuen, London.

DURBIN, J. [1967], "Tests of Serial Independence Based on the Cumulated Periodogram," *Bulletin of the International Institute of Statistics*, vol. 42, pp. 1039–1049.

GOOD, I. J. [1953], "The Serial Test for Sampling Numbers and Other Tests of Randomness," *Proceedings of the Cambridge Philosophical Society*, Vol. 49, pp. 276–284.

GOOD, I. J. [1967], "The Generalized Serial Test and the Binary Expansion of 4," *Journal of the Royal Statistical Society*, Ser. A, Vol. 30, No. 1, pp. 102–107.

KNUTH, D. W. [1998], *The Art of Computer Programming: Vol. 2, Semi-numerical Algorithms*, 2d ed., Addison–Wesley, Reading, MA.

LAW, A. M. [2007], *Simulation Modeling and Analysis*, 4th ed., McGraw-Hill, New York.

LEARMONTH, G. P., AND P. A. W. LEWIS [1973], "Statistical Tests of Some Widely Used and Recently Proposed Uniform Random Number Generators," *Proceedings of the Conference on Computer Science and Statistics: Seventh Annual Symposium on the Interface*, Western Publishing, North Hollywood, CA, pp. 163–171.

L'ECUYER, P. [1988], "Efficient and Portable Combined Random Number Generators," *Communications of the ACM*, Vol. 31, pp. 742–749, 774.

L'ECUYER, P. [1996], "Combined Multiple Recursive Random Number Generators," *Operations Research*, Vol. 44, pp. 816–822.

L'ECUYER, P. [1998], "Random Number Generation," Chapter 4 in *Handbook of Simulation*, J. Banks, ed., pp. 93–137. Wiley, New York.

L'ECUYER, P. [1999], "Good Parameters and Implementations for Combined Multiple Recursive Random Number Generators," *Operations Research*, Vol. 47, pp. 159–164.

L'ECUYER, P., R. SIMARD, E. J. CHEN, AND W. D. KELTON [2002], "An Object-Oriented Random-Number Package with Many Long Streams and Substreams," *Operations Research*, Vol. 50, pp. 1073–1075.

LEHMER, D. H. [1951], "Mathematical Methods in large-Scale Computing Units," *Proceedings of the Second Symposium on Large-Scale Digital Calculating Machinery*, Harvard University Press, Cambridge, MA, pp. 141–146.

LEWIS, P. A. W., A. S. GOODMAN, AND J. M. MILLER [1969], "A Pseudo-Random Number Generator for the System/360," *IBM Systems Journal*, Vol. 8, pp. 136–145.

RIPLEY, B. D. [1987], *Stochastic Simulation*, Wiley, New York.

SCHMIDT, J. W., and R. E. TAYLOR [1970], *Simulation and Analysis of Industrial Systems*, Irwin, Homewood, IL.

EXERCISES

1. Describe a procedure to physically generate random numbers on the interval [0, 1] with 2-digit accuracy. [*Hint*: Consider drawing something out of a hat.]

2. List applications, other than systems simulation, for pseudo-random numbers—for example, video gambling games.

3. How could random numbers that are uniform on the interval [0, 1] be transformed into random numbers that are uniform on the interval [−11, 17]? Transformations to more general distributions are described in Chapter 8.

4. Use the linear congruential method to generate a sequence of three two-digit random integers and corresponding random numbers. Let $X_0 = 27$, $a = 8$, $c = 47$, and $m = 100$.

5. Do we encounter a problem in the previous exercise if $X_0 = 0$?

6. Use the multiplicative congruential method to generate a sequence of four three-digit random integers and corresponding random numbers. Let $X_0 = 117$, $a = 43$, and $m = 1000$.

7. The sequence of numbers 0.54, 0.73, 0.98, 0.11, and 0.68 has been generated. Use the Kolmogorov–Smirnov test with $\alpha = 0.05$ to learn whether the hypothesis that the numbers are uniformly distributed on the interval [0, 1] can be rejected.

8. Reverse the 100 two-digit random numbers in Example 7.7 to get a new set of random numbers. Thus, the first random number in the new set will be 0.43. Use the chi-square test, with $\alpha = 0.05$, to learn whether the hypothesis that the numbers are uniformly distributed on the interval $[0, 1]$ can be rejected.

9. Figure out whether these linear congruential generators can achieve a maximum period; also, state restrictions on X_0 to obtain this period:

   (a) the mixed congruential method with

   $$a = 2, 814, 749, 767, 109$$
   $$c = 59, 482, 661, 568, 307$$
   $$m = 2^{48}$$

   (b) the multiplicative congruential generator with

   $$a = 69, 069$$
   $$c = 0$$
   $$m = 2^{32}$$

   (c) the mixed congruential generator with

   $$a = 4951$$
   $$c = 247$$
   $$m = 256$$

   (d) the multiplicative congruential generator with

   $$a = 6507$$
   $$c = 0$$
   $$m = 1024$$

10. Use the mixed congruential method to generate a sequence of three two-digit random integers and corresponding random numbers with $X_0 = 37$, $a = 7$, $c = 29$, and $m = 100$.

11. Use the mixed congruential method to generate a sequence of three two-digit random integers between 0 and 24 and corresponding random numbers with $X_0 = 13$, $a = 9$, and $c = 35$.

12. Write a computer program that will generate four-digit random numbers, using the multiplicative congruential method. Allow the user to input values of X_0, a, c, and m.

13. If $X_0 = 3579$ in Exercise 9(c), generate the first random number in the sequence. Compute the random number to four-place accuracy.

14. Investigate the random-number generator in a spreadsheet program on a computer to which you have access. In many spreadsheets, random numbers are generated by a function called RAND or @RAND.

    (a) Check the user's manual to see whether it describes how the random numbers are generated.
    (b) Write macros to conduct each of the tests described in this chapter. Generate 100 sets of random numbers, each set containing 1000 random numbers. Perform each test on each set of random numbers. Draw conclusions.

15. Consider the multiplicative congruential generator under the following circumstances:

    (a) $X_0 = 7, a = 11, m = 16$
    (b) $X_0 = 8, a = 11, m = 16$
    (c) $X_0 = 7, a = 7, m = 16$
    (d) $X_0 = 8, a = 7, m = 16$

    Generate enough values in each case to complete a cycle. What inferences can be drawn? Is maximum period achieved?

16. L'Ecuyer [1988] provides a generator that combines three multiplicative generators, with $a_1 = 157$, $m_1 = 32,363$, $a_2 = 146$, $m_2 = 31,727$, $a_3 = 142$, and $m_3 = 31,657$. The period of this generator is approximately 8×10^{12}. Generate 5 random numbers with the combined generator, using the initial seeds $X_{1,0} = 100$, $X_{2,0} = 300$ and $X_{3,0} = 500$, for the individual generators.

17. Apply the tests described in this chapter to the generator given in the previous exercise.

18. Use the principles described in this chapter to develop your own linear congruential random-number generator.

19. Use the principles described in this chapter to develop your own combined linear congruential random-number generator.

20. Test the following sequence of numbers for uniformity and independence, using procedures you learned in this chapter: 0.594, 0.928, 0.515, 0.055, 0.507, 0.351, 0.262, 0.797, 0.788, 0.442, 0.097, 0.798, 0.227, 0.127, 0.474, 0.825, 0.007, 0.182, 0.929, 0.852.

21. In some applications, it is useful to be able to quickly skip ahead in a pseudo-random number sequence without actually generating all of the intermediate values.

    (a) For a linear congruential generator with $c = 0$, show that $X_{i+n} = (a^n X_i) \bmod m$.
    (b) Next, show that $(a^n X_i) \bmod m = (a^n \bmod m)X_i \bmod m$ (this result is useful because $a^n \bmod m$ can be precomputed, making it easy to skip ahead n random numbers from any point in the sequence).
    (c) In Example 7.3, use this result to compute X_5, starting with $X_0 = 63$. Check your answer by computing X_5 in the usual way.

22. Producing good pseudo-random-number generators is much more difficult than it seems. For instance, mathematical genius John Von Neumann proposed the middle-square method. To generate, say, six-digit pseudo-random numbers by the middle-square method, begin with a six-digit integer as the seed. Square the number and extract the middle six digits as the next number in the sequence. Continue the process by repeatedly squaring the current number and extracting the middle six digits as the next number. To obtain numbers between 0 and 1, a decimal point can be placed in front of the six digits. Try a few examples to show how this method can fail spectacularly. (Von Neumann was aware of these problems.)

8

Random-Variate Generation

This chapter deals with procedures for sampling from a variety of widely used continuous and discrete distributions. Previous discussions and examples indicated the usefulness of probability distributions in modeling activities that are generally unpredictable or uncertain. For example, interarrival times and service times at queues and demands for a product are quite often unpredictable in nature, at least to a certain extent. Usually, such variables are modeled as random variables with some specified probability distribution, and standard statistical procedures exist for estimating the parameters of the hypothesized distribution and for testing the validity of the assumed statistical model. Such procedures are discussed in Chapter 9.

In this chapter, it is assumed that a distribution has been completely specified, and ways are sought to generate samples from this distribution to be used as input to a simulation model. The purpose of the chapter is to explain and illustrate some widely used techniques for generating random variates, not to give a state-of-the-art survey of the most efficient techniques. In practice, most simulation modelers will use either existing routines available in programming libraries or the routines built into a simulation language. However, some programming languages do not have built-in routines for all of the regularly used distributions, and some computer installations do not have random-variate-generation libraries; in such cases the modeler must construct an acceptable routine. Even though the chance of this happening is small, it is nevertheless worthwhile to understand how random-variate generation occurs.

This chapter discusses the inverse-transform technique and, more briefly, the acceptance-rejection technique and special properties. Another technique, the composition method, is discussed by Devroye [1986], Dagpunar [1988], and Law [2007]. All the techniques in this chapter assume that a source of uniform [0,1] random numbers, $R_1, R_2, \ldots$ is readily available, where each R_i has pdf

$$f_R(x) = \begin{cases} 1, & 0 \le x \le 1 \\ 0, & \text{otherwise} \end{cases}$$

and cdf

$$F_R(x) = \begin{cases} 0, & x < 0 \\ x, & 0 \le x \le 1 \\ 1, & x > 1 \end{cases}$$

Throughout this chapter R and $R_1, R_2, \ldots$ represent random numbers uniformly distributed on [0,1] and generated by one of the techniques in Chapter 7 or taken from a random number table, such as Table A.1. Many of the algorithms in this chapter have been translated into Visual Basic for Applications and can be found in the spreadsheet `RandomNumberTools.xls` at `www.bcnn.net`.

8.1   Inverse-Transform Technique

The inverse-transform technique can be used to sample from the exponential, uniform, Weibull, and triangular distributions, and from empirical distributions. Additionally, it is the underlying principle for sampling from a wide variety of discrete distributions. The technique will be explained in detail for the exponential distribution and then applied to other distributions. Computationally, it is the most straightforward, but not always the most efficient, technique.

8.1.1   Exponential Distribution

The exponential distribution, discussed in Section 5.4.2, has the probability density function (pdf)

$$f(x) = \begin{cases} \lambda e^{-\lambda x}, & x \ge 0 \\ 0, & x < 0 \end{cases}$$

and the cumulative distribution function (cdf)

$$F(x) = \int_{-\infty}^{x} f(t)\, dt = \begin{cases} 1 - e^{-\lambda x}, & x \ge 0 \\ 0, & x < 0 \end{cases}$$

The parameter λ can be interpreted as the mean number of occurrences per time unit. For example, if interarrival times $X_1, X_2, X_3, \ldots$ had an exponential distribution with rate λ, then λ could be interpreted as the mean number of arrivals per time unit, or the arrival rate. Notice that, for any i,

$$E(X_i) = \frac{1}{\lambda}$$

and so $1/\lambda$ is the mean interarrival time. The goal here is to develop a procedure for generating values $X_1, X_2, X_3, \ldots$ that have an exponential distribution.

The inverse-transform technique can be utilized, at least in principle, for any distribution, but it is most useful when the cdf, $F(x)$, is of a form so simple that its inverse, F^{-1}, can be computed easily. (The notation F^{-1} denotes the solution of the equation $r = F(x)$ for x; it does not denote $1/F$.) One step-by-step procedure for the inverse-transform technique, illustrated by the exponential distribution, is shown here:

Step 1. Compute the cdf of the desired random variable X.

For the exponential distribution, the cdf is $F(x) = 1 - e^{-\lambda x}, x \geq 0$.

Step 2. Set $F(X) = R$ on the range of X.

For the exponential distribution, it becomes $1 - e^{-\lambda X} = R$ on the range $X \geq 0$.

X is a random variable (with the exponential distribution in this case), so $1 - e^{-\lambda X}$ is also a random variable, here called R. As will be shown later, R has a uniform distribution over the interval $[0, 1]$.

Step 3. Solve the equation $F(X) = R$ for X in terms of R.

For the exponential distribution, the solution proceeds as follows:

$$1 - e^{-\lambda X} = R$$
$$e^{-\lambda X} = 1 - R$$
$$-\lambda X = \ln(1 - R)$$
$$X = -\frac{1}{\lambda} \ln(1 - R) \tag{8.1}$$

Equation (8.1) is called a random-variate generator for the exponential distribution. In general, Equation (8.1) is written as $X = F^{-1}(R)$. Generating a sequence of values is accomplished through Step 4.

Step 4. Generate (as needed) uniform random numbers $R_1, R_2, R_3, \ldots$ and compute the desired random variates by

$$X_i = F^{-1}(R_i)$$

For the exponential case, $F^{-1}(R) = (-1/\lambda) \ln(1 - R)$ by Equation (8.1), so

$$X_i = -\frac{1}{\lambda} \ln(1 - R_i) \tag{8.2}$$

for $i = 1, 2, 3, \ldots$. One simplification that is usually employed in Equation (8.2) is to replace $1 - R_i$ by R_i to yield

$$X_i = -\frac{1}{\lambda} \ln R_i \tag{8.3}$$

This alternative is justified by the fact that both R_i and $1 - R_i$ are uniformly distributed on $[0,1]$.

Example 8.1

Table 8.1 gives a sequence of random numbers from Table A.1 and the computed exponential vari-ates, X_i, given by Equation (8.2) with the value $\lambda = 1$. Figure 8.1(a) is a histogram of 200 values, $R_1, R_2, \ldots, R_{200}$ from the uniform distribution, and Figure 8.1(b) is a histogram of the 200 values, $X_1, X_2, \ldots, X_{200}$, computed by Equation (8.2). Compare these empirical histograms with the theo-retical density functions in Figure 8.1(c) and (d). As illustrated here, a histogram is an estimate of the underlying density function. (This fact is used in Chapter 9 as a way to identify distributions.)

Table 8.1   Generation of Exponential Variates X_i with Mean 1, given Random Numbers R_i

i	1	2	3	4	5
R_i	0.1306	0.0422	0.6597	0.7965	0.7696
X_i	0.1400	0.0431	1.078	1.592	1.468

Figure 8.2 gives a graphical interpretation of the inverse-transform technique. The cdf shown is $F(x) = 1 - e^{-x}$, an exponential distribution with rate $\lambda = 1$. To generate a value X_1 with cdf $F(x)$, a random number R_1 between 0 and 1 is generated, then a horizontal line is drawn from R_1 to the graph of the cdf, then a vertical line is dropped to the x axis to obtain X_1, the desired result. Notice the inverse relation between R_1 and X_1, namely

$$R_1 = 1 - e^{-X_1}$$

and

$$X_1 = -\ln(1 - R_1)$$

In general, the relation is written as

$$R_1 = F(X_1)$$

and

$$X_1 = F^{-1}(R_1)$$

Why does the random variable X_1 generated by this procedure have the desired distribution? Pick a value x_0 and compute the cumulative probability

$$P(X_1 \le x_0) = P(R_1 \le F(x_0)) = F(x_0) \tag{8.4}$$

To see the first equality in Equation (8.4), refer to Figure 8.2, where the fixed numbers x_0 and $F(x_0)$ are drawn on their respective axes. It can be seen that $X_1 \le x_0$ when and only when $R_1 \le F(x_0)$. Since $0 \le F(x_0) \le 1$, the second equality in Equation (8.4) follows immediately from the fact that R_1 is uniformly distributed on [0,1]. Equation (8.4) shows that the cdf of X_1 is F; hence, X_1 has the desired distribution.

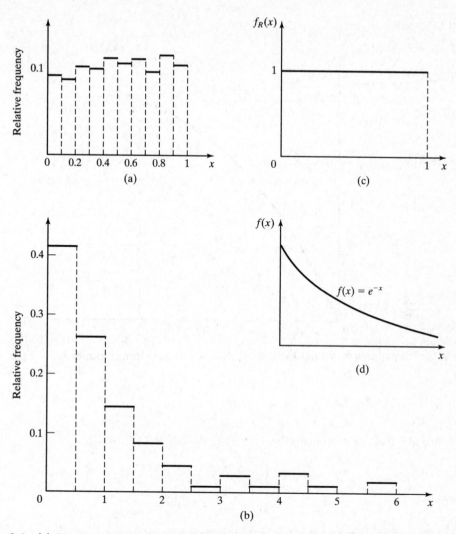

Figure 8.1   (a) Empirical histogram of 200 uniform random numbers; (b) empirical histogram of 200 exponential variates; (c) theoretical uniform density on [0,1]; (d) theoretical exponential density with mean 1.

8.1.2   Uniform Distribution

Consider a random variable X that is uniformly distributed on the interval $[a, b]$. A reasonable guess for generating X is given by

$$X = a + (b - a)R \qquad (8.5)$$

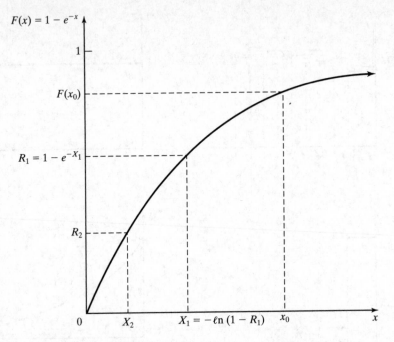

Figure 8.2   Graphical view of the inverse-transform technique.

(Recall that R is always a random number on $[0, 1]$.) The pdf of X is given by

$$f(x) = \begin{cases} \dfrac{1}{b-a}, & a \le x \le b \\ 0, & \text{otherwise} \end{cases}$$

The derivation of Equation (8.5) follows Steps 1 through 3 of Section 8.1.1:

Step 1. The cdf is given by

$$F(x) = \begin{cases} 0, & x < a \\ \dfrac{x-a}{b-a}, & a \le x \le b \\ 1, & x > b \end{cases}$$

Step 2. Set $F(X) = (X - a)/(b - a) = R$.

Step 3. Solving for X in terms of R yields $X = a + (b - a)R$, which agrees with Equation (8.5).

8.1.3 Weibull Distribution

The Weibull distribution was introduced in Section 5.4.6 as a model for *time to failure* for machines or electronic components. When the location parameter ν is set to 0, its pdf is given by Equation (5.47):

$$f(x) = \begin{cases} \dfrac{\beta}{\alpha^{\beta}} x^{\beta-1} e^{-(x/\alpha)^{\beta}}, & x \geq 0 \\ 0, & \text{otherwise} \end{cases}$$

where $\alpha > 0$ and $\beta > 0$ are the scale and shape parameters of the distribution. To generate a Weibull variate, follow Steps 1 through 3 of Section 8.1.1:

Step 1. The cdf is given by $F(X) = 1 - e^{-(x/\alpha)^{\beta}}$, $x \geq 0$.

Step 2. Set $F(X) = 1 - e^{-(X/\alpha)^{\beta}} = R$.

Step 3. Solving for X in terms of R yields

$$X = \alpha[-\ln(1-R)]^{1/\beta} \tag{8.6}$$

The derivation of Equation (8.6) is left as Exercise 10 for the reader. By comparing Equations (8.6) and (8.1), it can be seen that, if X is a Weibull variate, then X^{β} is an exponential variate with mean α^{β}. Conversely, if Y is an exponential variate with mean μ, then $Y^{1/\beta}$ is a Weibull variate with shape parameter β and scale parameter $\alpha = \mu^{1/\beta}$.

8.1.4 Triangular Distribution

Consider a random variable X that has pdf

$$f(x) = \begin{cases} x, & 0 \leq x \leq 1 \\ 2 - x, & 1 < x \leq 2 \\ 0, & \text{otherwise} \end{cases}$$

as shown in Figure 8.3. This distribution is called a *triangular distribution* with endpoints $(0, 2)$ and mode at 1. Its cdf is given by

$$F(x) = \begin{cases} 0, & x \leq 0 \\ \dfrac{x^2}{2}, & 0 < x \leq 1 \\ 1 - \dfrac{(2-x)^2}{2}, & 1 < x \leq 2 \\ 1, & x > 2 \end{cases}$$

For $0 \leq X \leq 1$,

$$R = \frac{X^2}{2} \tag{8.7}$$

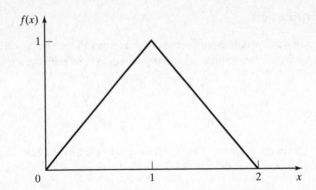

Figure 8.3 Density function for a particular triangular distribution.

and for $1 \le X \le 2$,

$$R = 1 - \frac{(2-X)^2}{2} \tag{8.8}$$

In Equation (8.7), $0 \le X \le 1$ implies that $0 \le R \le \frac{1}{2}$, in which case $X = \sqrt{2R}$. By Equation (8.8), $1 \le X \le 2$ implies that $\frac{1}{2} \le R \le 1$, in which case $X = 2 - \sqrt{2(1-R)}$. Thus, X is generated by

$$X = \begin{cases} \sqrt{2R}, & 0 \le R \le \frac{1}{2} \\ 2 - \sqrt{2(1-R)}, & \frac{1}{2} < R \le 1 \end{cases} \tag{8.9}$$

Exercises 2, 3, and 4 give the student practice in dealing with other triangular distributions. Notice that, if the pdf and cdf of the random variable X come in parts (i.e., require different formulas over different parts of the range of X), then the application of the inverse-transform technique for generating X will result in separate formulas over different parts of the range of R, as in Equation (8.9). A general form of the triangular distribution was discussed in Section 5.4.7.

8.1.5 Empirical Continuous Distributions

If the modeler has been unable to find a theoretical distribution that provides a good model for the input data, then it may be necessary to use the empirical distribution of the data. One possibility is to simply resample the observed data itself. This is known as using the *empirical distribution*, and it makes particularly good sense when the input process is known to take on a finite number of values. See Section 8.1.7 for an example of this type of situation and for a method for generating random inputs.

On the other hand, if the data are drawn from what is believed to be a continuous-valued input process, then it makes sense to interpolate between the observed data points to fill in the gaps. This section describes a method for defining and generating data from a continuous empirical distribution.

Example 8.2

Five observations of fire-crew response times (in minutes) to incoming alarms have been collected to be used in a simulation investigating possible alternative staffing and crew-scheduling policies. The data are

$$2.76 \quad 1.83 \quad 0.80 \quad 1.45 \quad 1.24$$

Before collecting more data, it is desired to develop a preliminary simulation model that uses a response-time distribution based on these five observations. Thus, a method for generating random variates from the response-time distribution is needed. Initially, it will be assumed that response times X have a range $0 \leq X \leq c$, where c is unknown, but will be estimated by $\hat{c} = \max\{X_i : i = 1, \ldots, n\} = 2.76$, where $\{X_i, i = 1, \ldots, n\}$ are the raw data and $n = 5$ is the number of observations.

Arrange the data from smallest to largest and let $x_{(1)} \leq x_{(2)} \leq \cdots \leq x_{(n)}$ denote these sorted values. The smallest possible value is believed to be 0, so define $x_{(0)} = 0$. Assign the probability $1/n = 1/5$ to each interval $x_{(i-1)} < x \leq x_{(i)}$, as shown in Table 8.2. The resulting empirical cdf, $\widehat{F}(x)$, is illustrated in Figure 8.4. The slope of the ith line segment is given by

$$a_i = \frac{x_{(i)} - x_{(i-1)}}{i/n - (i-1)/n} = \frac{x_{(i)} - x_{(i-1)}}{1/n}$$

The inverse cdf is calculated by

$$X = \widehat{F}^{-1}(R) = x_{(i-1)} + a_i \left(R - \frac{(i-1)}{n} \right) \tag{8.10}$$

when $(i-1)/n < R \leq i/n$.

For example, if a random number $R_1 = 0.71$ is generated, then R_1 is seen to lie in the fourth interval (between $3/5 = 0.60$ and $4/5 = 0.80$); so, by Equation (8.10),

$$X_1 = x_{(4-1)} + a_4(R_1 - (4-1)/n)$$

$$= 1.45 + 1.90(0.71 - 0.60) = 1.66$$

The reader is referred to Figure 8.4 for a graphical view of the generation procedure.

Table 8.2   Summary of Fire-Crew Response-Time Data

i	Interval $x_{(i-1)} < x \leq x_{(i)}$	Probability $1/n$	Cumulative Probability, i/n	Slope a_i
1	$0.0 \ < x \leq 0.80$	0.2	0.2	4.00
2	$0.80 < x \leq 1.24$	0.2	0.4	2.20
3	$1.24 < x \leq 1.45$	0.2	0.6	1.05
4	$1.45 < x \leq 1.83$	0.2	0.8	1.90
5	$1.83 < x \leq 2.76$	0.2	1.0	4.65

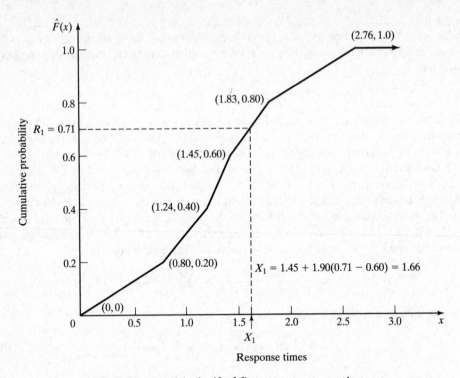

Figure 8.4   Empirical cdf of fire-crew response times.

In Example 8.2, each data point was represented in the empirical cdf. If a large sample of data is available (and sample sizes from several hundred to tens of thousands are possible with automated data collection), then it might be more convenient (and computationally efficient) to first summarize the data into a frequency distribution with a much smaller number of intervals and then fit a continuous empirical cdf to the frequency distribution. Only a slight generalization of Equation (8.10) is required to accomplish this. Now the slope of the ith line segment is given by:

$$a_i = \frac{x_{(i)} - x_{(i-1)}}{c_i - c_{i-1}}$$

where c_i is the cumulative probability of the first i intervals of the frequency distribution and $x_{(i-1)} < x \leq x_{(i)}$ is the ith interval. The inverse cdf is calculated as

$$X = \widehat{F}^{-1}(R) = x_{(i-1)} + a_i \left(R - c_{i-1}\right) \tag{8.11}$$

when $c_{i-1} < R \leq c_i$.

Example 8.3

Suppose that 100 machine repair times have been collected. The data are summarized in Table 8.3 in terms of the number of observations in various intervals. For example, there were 31 observations between 0 and 0.5 hour, 10 between 0.5 and 1 hour, and so on. Suppose it is known that all repairs take at least 15 minutes, so that $X \geq 0.25$ hours always. Then we set $x_{(0)} = 0.25$, as shown in Table 8.3 and Figure 8.5.

Table 8.3   Summary of Repair-Time Data

i	Interval (Hours)	Frequency	Relative Frequency	Cumulative Frequency, c_i	Slope a_i
1	$0.25 \leq x \leq 0.5$	31	0.31	0.31	0.81
2	$0.5 < x \leq 1.0$	10	0.10	0.41	5.0
3	$1.0 < x \leq 1.5$	25	0.25	0.66	2.0
4	$1.5 < x \leq 2.0$	34	0.34	1.00	1.47

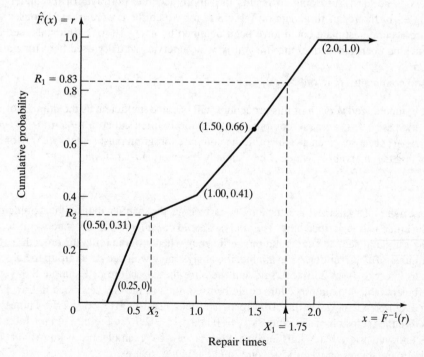

Figure 8.5   Generating variates from the empirical distribution function for repair-time data $(X \geq 0.25)$.

For example, suppose the first random number generated is $R_1 = 0.83$. Then, since R_1 is between $c_3 = 0.66$ and $c_4 = 1.00$,

$$X_1 = x_{(4-1)} + a_4(R_1 - c_{4-1}) = 1.5 + 1.47(0.83 - 0.66) = 1.75 \qquad (8.12)$$

As another illustration, suppose that $R_2 = 0.33$. Since $c_1 = 0.31 < R_2 \leq 0.41 = c_2$,

$$X_2 = x_{(1)} + a_2(R_2 - c_1)$$
$$= 0.5 + 5.0(0.33 - 0.31)$$
$$= 0.6$$

The point $(R_2 = 0.33, X_2 = 0.6)$ is also shown in Figure 8.5.

Now reconsider the data of Table 8.3. The data are restricted in the range $0.25 \leq X \leq 2.0$, but the underlying distribution might have a wider range. This provides one important reason for attempting to find a theoretical probability distribution (such as the gamma or Weibull) for the data: that these distributions do allow a wider range—namely, $0 \leq X < \infty$. On the other hand, an empirical distribution adheres closely to what is present in the data itself, and the data are often the best source of information available.

When data are summarized in terms of frequency intervals, it is recommended that relatively short intervals be used, for doing so results in a more accurate portrayal of the underlying cdf. For example, for the repair-time data of Table 8.3, for which there were $n = 100$ observations, a much more accurate estimate could have been obtained by using 10 to 20 intervals, certainly not an excessive number, rather than the four fairly wide intervals actually used here for purposes of illustration.

Several comments are in order:

1. A computerized version of this procedure will become inefficient as the number of intervals n increases. A systematic computerized version is often called a *table-lookup generation scheme*, because, given a value of R, the computer program must search an array of c_i values to find the interval i in which R lies, namely the interval i satisfying

$$c_{i-1} < R \leq c_i$$

The more intervals there are, the longer, on average, the search will take if it is implemented in the crude way described here. The analyst should consider this tradeoff between accuracy of the estimating cdf and computational efficiency when programming the procedure. If a large number of observations are available, the analyst may well decide to group the observations into intervals from 20 to 50, say, and then use the procedure of Example 8.3—or a more efficient table-lookup procedure could be used, such as the one described in Law [2007].

2. In Example 8.2, it was assumed that response times X satisfied $0 \leq X \leq 2.76$. This assumption led to the inclusion of the points $x_{(0)} = 0$ and $x_{(5)} = 2.76$ in Figure 8.4 and Table 8.2. If it is known a priori that X falls in some other range—for example, if it is known that response times are always between 15 seconds and 3 minutes, that is,

$$0.25 \leq X \leq 3.0$$

—then the points $x_{(0)} = 0.25$ and $x_{(6)} = 3.0$ would be used to estimate the empirical cdf of response times. Notice that, because of inclusion of the new point $x_{(6)}$, there are now six intervals instead of five, and each interval is assigned probability $1/6 = 0.167$. Exercise 12 illustrates the use of these additional assumptions.

8.1.6   Continuous Distributions without a Closed-Form Inverse

A number of useful continuous distributions do not have a closed form expression for their cdf or its inverse; examples include the normal, gamma, and beta distributions. For this reason, it is often stated that the inverse-transform technique for random-variate generation is not available for these distributions. It can, in effect, become available if we are willing to *approximate* the inverse cdf, or numerically integrate and search the cdf. Although this approach sounds inaccurate, notice that even a closed-form inverse requires approximation to evaluate it on a computer. For example, generating exponentially distributed random variates via the inverse cdf $X = F^{-1}(R) = -\ln(1 - R)/\lambda$ requires a numerical approximation for the logarithm function. Thus, there is no essential difference between using an approximate inverse cdf and approximately evaluating a closed-form inverse. The problem with using approximate inverse cdfs is that some of them are computationally slow to evaluate.

To illustrate the idea, consider a simple approximation to the inverse cdf of the standard normal distribution, proposed by Schmeiser [1979]:

$$X = F^{-1}(R) \approx \frac{R^{0.135} - (1 - R)^{0.135}}{0.1975}$$

This approximation gives at least one-decimal-place accuracy for $0.0013499 \leq R \leq 0.9986501$. Table 8.4 compares the approximation with exact values (to four decimal places) obtained by numerical integration for several values of R. Much more accurate approximations exist that are only slightly more complicated. A good source of these approximations for a number of distributions is Bratley, Fox, and Schrage [1996].

Table 8.4 Comparison of Approximate Inverse with Exact Values (to four decimal places) for the Standard Normal Distribution

R	*Approximate Inverse*	*Exact Inverse*
0.01	−2.3263	−2.3373
0.10	−1.2816	−1.2813
0.25	−0.6745	−0.6713
0.50	0.0000	0.0000
0.75	0.6745	0.6713
0.90	1.2816	1.2813
0.99	2.3263	2.3373

8.1.7   Discrete Distributions

All discrete distributions can be generated via the inverse-transform technique, either numerically through a table-lookup procedure or, in some cases, algebraically, the final generation scheme being in terms of a formula. Other techniques are sometimes used for certain distributions, such as the convolution technique for the binomial distribution. Some of these methods are discussed in later sections. This subsection gives examples covering both empirical distributions and two of the standard discrete distributions, the (discrete) uniform and the geometric. Highly efficient table-lookup procedures for these and other distributions are found in Bratley, Fox, and Schrage [1996] and in Ripley [1987].

Example 8.4:   An Empirical Discrete Distribution _____

At the end of any day, the number of shipments on the loading dock of the IHW Company is either 0, 1, or 2, with observed relative frequency of occurrence of 0.50, 0.30, and 0.20, respectively. Internal consultants have been asked to develop a model to improve the efficiency of the loading and hauling operations; as part of this model, they will need to be able to generate values X to represent the number of shipments on the loading dock at the end of each day. The consultants decide to model X as a discrete random variable with the distribution given in Table 8.5 and shown in Figure 8.6.

Table 8.5   Distribution of Number of Shipments X

x	$p(x)$	$F(x)$
0	0.50	0.50
1	0.30	0.80
2	0.20	1.00

The probability mass function (pmf), $p(x)$, is given by

$$p(0) = P(X = 0) = 0.50$$

$$p(1) = P(X = 1) = 0.30$$

$$p(2) = P(X = 2) = 0.20$$

and the cdf, $F(x) = P(X \leq x)$, is given by

$$F(x) = \begin{cases} 0 & x < 0 \\ 0.5 & 0 \leq x < 1 \\ 0.8 & 1 \leq x < 2 \\ 1.0 & 2 \leq x \end{cases}$$

Recall that the cdf of a discrete random variable always consists of horizontal line segments with jumps of size $p(x)$ at those points, x, that the random variable can assume. For example, in

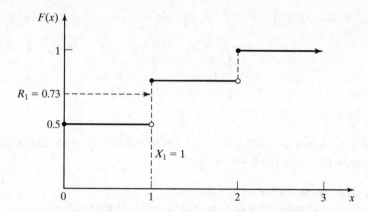

Figure 8.6   Cdf of number of shipments X.

Figure 8.6, there is a jump of size $p(0) = 0.5$ at $x = 0$, of size $p(1) = 0.3$ at $x = 1$, and of size $p(2) = 0.2$ at $x = 2$.

For generating discrete random variables, the inverse-transform technique becomes a table-lookup procedure, but, unlike in the case of continuous variables, interpolation is not required. To illustrate the procedure, suppose that $R_1 = 0.73$ is generated. Graphically, as illustrated in Figure 8.6, first locate $R_1 = 0.73$ on the vertical axis, next draw a horizontal line segment until it hits a "jump" in the cdf, and then drop a perpendicular to the horizontal axis to get the generated variate. Here $R_1 = 0.73$ is transformed to $X_1 = 1$. This procedure is analogous to the procedure used for empirical continuous distributions in Section 8.1.5 and illustrated in Figure 8.4, except that the final step, linear interpolation, is eliminated.

The table-lookup procedure is facilitated by construction of a table such as Table 8.6. When $R_1 = 0.73$ is generated, first find the interval in which R_1 lies. In general, for $R = R_1$, if

$$F(x_{i-1}) = r_{i-1} < R \leq r_i = F(x_i) \tag{8.13}$$

then set $X_1 = x_i$. Here, $r_0 = 0$; $x_0 = -\infty$; $x_1, x_2, \ldots, x_n$ are the possible values of the random variable; and $r_k = p(x_1) + \cdots + p(x_k)$, $k = 1, 2, \ldots, n$. For this example, $n = 3$, $x_1 = 0$, $x_2 = 1$, and $x_3 = 2$; hence, $r_1 = 0.5$, $r_2 = 0.8$, and $r_3 = 1.0$. (Notice that $r_n = 1.0$ in all cases.)

Table 8.6   Table for Generating the Discrete Variate X

i	Input, r_i	Output, x_i
1	0.50	0
2	0.80	1
3	1.00	2

Since $r_1 = 0.5 < R_1 = 0.73 \leq r_2 = 0.8$, set $X_1 = x_2 = 1$. The generation scheme is summarized as follows:

$$X = \begin{cases} 0, & R \leq 0.5 \\ 1, & 0.5 < R \leq 0.8 \\ 2, & 0.8 < R \leq 1.0 \end{cases}$$

Example 8.4 illustrated the table-lookup procedure; the next example illustrates an algebraic approach that can be used for certain distributions.

Example 8.5: A Discrete Uniform Distribution _____

Consider the discrete uniform distribution on $\{1, 2, \ldots, k\}$ with pmf and cdf given by

$$p(x) = \frac{1}{k}, \quad x = 1, 2, \ldots, k$$

and

$$F(x) = \begin{cases} 0, & x < 1 \\ \dfrac{1}{k}, & 1 \leq x < 2 \\ \dfrac{2}{k}, & 2 \leq x < 3 \\ \vdots & \vdots \\ \dfrac{k-1}{k}, & k-1 \leq x < k \\ 1, & k \leq x \end{cases}$$

Let $x_i = i$ and $r_i = p(1) + \cdots + p(x_i) = F(x_i) = i/k$ for $i = 1, 2, \ldots, k$. Then, from Inequality (8.13), it can be seen that, if the generated random number R satisfies

$$r_{i-1} = \frac{i-1}{k} < R \leq r_i = \frac{i}{k} \tag{8.14}$$

then X is generated by setting $X = i$. Now Inequality (8.14) can be solved for i:

$$i - 1 < Rk \leq i$$

$$Rk \leq i < Rk + 1 \tag{8.15}$$

Let $\lceil y \rceil$ denote the smallest integer $\geq y$. For example, $\lceil 7.82 \rceil = 8$, $\lceil 5.13 \rceil = 6$, and $\lceil -1.32 \rceil = -1$. For $y \geq 0$, $\lceil y \rceil$ is a function that rounds up. This notation and Inequality (8.15) yield a formula for generating X, namely

$$X = \lceil Rk \rceil \tag{8.16}$$

For example, consider the generating of a random variate X that is uniformly distributed on $\{1, 2, \ldots, 10\}$. The variate X might represent the number of pallets to be loaded onto a truck. Using Table A.1 as a source of random numbers R and using Equation (8.16) with $k = 10$ yields

$$R_1 = 0.78 \quad X_1 = \lceil 7.8 \rceil = 8$$

$$R_2 = 0.03 \quad X_2 = \lceil 0.3 \rceil = 1$$

$$R_3 = 0.23 \quad X_3 = \lceil 2.3 \rceil = 3$$

$$R_4 = 0.97 \quad X_4 = \lceil 9.7 \rceil = 10$$

The procedure discussed here can be modified to generate a discrete uniform random variate with any range consisting of consecutive integers. Exercise 13 asks the student to devise a procedure for one such case.

Example 8.6: The Geometric Distribution _____

Consider the geometric distribution with pmf

$$p(x) = p(1-p)^x, \quad x = 0, 1, 2, \ldots$$

where $0 < p < 1$. Its cdf is given by

$$F(x) = \sum_{j=0}^{x} p(1-p)^j$$

$$= \frac{p\{1 - (1-p)^{x+1}\}}{1 - (1-p)}$$

$$= 1 - (1-p)^{x+1}$$

for $x = 0, 1, 2, \ldots$ Using the inverse-transform technique [i.e., Inequality (8.13)], recall that a geometric random variable X will assume the value x whenever

$$F(x-1) = 1 - (1-p)^x < R \le 1 - (1-p)^{x+1} = F(x) \tag{8.17}$$

where R is a generated random number assumed $0 < R < 1$. Solving Inequality (8.17) for x proceeds as follows:

$$(1-p)^{x+1} \le 1 - R < (1-p)^x$$

$$(x+1)\ln(1-p) \le \ln(1-R) < x\ln(1-p)$$

But $1 - p < 1$ implies that $\ln(1-p) < 0$, so that

$$\frac{\ln(1-R)}{\ln(1-p)} - 1 \le x < \frac{\ln(1-R)}{\ln(1-p)} \tag{8.18}$$

Thus, $X = x$ for that integer value of x satisfying Inequality (8.18). In brief, and using the round-up function $\lceil \cdot \rceil$, we have

$$X = \left\lceil \frac{\ln(1 - R)}{\ln(1 - p)} - 1 \right\rceil \tag{8.19}$$

Since p is a fixed parameter, let $\beta = -1/\ln(1 - p)$. Then $\beta > 0$ and, by Equation (8.19), $X = \lceil -\beta \ln(1 - R) - 1 \rceil$. By Equation (8.1), $-\beta \ln(1 - R)$ is an exponentially distributed random variable with mean β; so one way of generating a geometric variate with parameter p is to generate (by any method) an exponential variate with parameter $\beta^{-1} = -\ln(1 - p)$, subtract one, and round up.

Occasionally, there is needed a geometric variate X that can assume values $\{q, q + 1, q + 2, \dots\}$ with pmf $p(x) = p(1-p)^{x-q}$ $(x = q, q+1, \dots)$. Such a variate X can be generated, via Equation (8.19), by

$$X = q + \left\lceil \frac{\ln(1 - R)}{\ln(1 - p)} - 1 \right\rceil \tag{8.20}$$

One of the most common cases is $q = 1$.

Example 8.7

Generate three values from a geometric distribution on the range $\{X \geq 1\}$ with mean 2. Such a geometric distribution has pmf $p(x) = p(1 - p)^{x-1}$ $(x = 1, 2, \dots)$ with mean $1/p = 2$, or $p = 1/2$. Thus, X can be generated by Equation (8.20) with $q = 1, p = 1/2$, and $1/\ln(1 - p) = -1.443$. Using Table A.1, $R_1 = 0.932$, $R_2 = 0.105$, and $R_3 = 0.687$ yields

$$X_1 = 1 + \lceil -1.443 \ln(1 - 0.932) - 1 \rceil$$

$$= 1 + \lceil 3.878 - 1 \rceil = 4$$

$$X_2 = 1 + \lceil -1.443 \ln(1 - 0.105) - 1 \rceil = 1$$

$$X_3 = 1 + \lceil -1.443 \ln(1 - 0.687) - 1 \rceil = 2$$

Exercise 15 deals with an application of the geometric distribution.

8.2   Acceptance-Rejection Technique

Suppose that an analyst needed to devise a method for generating random variates, X, uniformly distributed between 1/4 and 1. One way to proceed would be to follow these steps:

Step 1.   Generate a random number R.

Step 2a.   If $R \geq 1/4$, accept $X = R$, then go to Step 3.

Step 2b.   If $R < 1/4$, reject R, and return to Step 1.

Step 3.   If another uniform random variate on [1/4, 1] is needed, repeat the procedure beginning at Step 1. If not, stop.

Each time Step 1 is executed, a new random number R must be generated. Step 2a is an "acceptance" and Step 2b is a "rejection" in this acceptance-rejection technique. To summarize the technique, random variates R with some distribution (here uniform on $[0, 1]$) are generated until some condition (here $R > 1/4$) is satisfied. When the condition is finally satisfied, the desired random variate X (here uniform on $[1/4, 1]$) can be computed ($X = R$). This procedure can be shown to be correct by recognizing that the accepted values of R are conditioned values; that is, R itself does not have the desired distribution, but R conditioned on the event $\{R \geq 1/4\}$ does have the desired distribution. To show this, take $1/4 \leq a < b \leq 1$; then

$$P(a < R \leq b | 1/4 \leq R \leq 1) = \frac{P(a < R \leq b)}{P(1/4 \leq R \leq 1)} = \frac{b - a}{3/4} \tag{8.21}$$

which is the correct probability for a uniform distribution on $[1/4, 1]$. Equation (8.21) says that the probability distribution of R, given that R is between $1/4$ and 1 (all other values of R are thrown out), is the desired distribution. Therefore, if $1/4 \leq R \leq 1$, set $X = R$.

The efficiency of an acceptance-rejection technique depends heavily on being able to minimize the number of rejections. In this example, the probability of a rejection is $P(R < 1/4) = 1/4$, so that the number of rejections is a geometrically distributed random variable with probability of "success" being $p = 3/4$ and mean number of rejections $(1/p - 1) = 4/3 - 1 = 1/3$. (Example 8.6 discussed the geometric distribution.) The mean number of random numbers R required to generate one variate X is one more than the number of rejections; hence, it is $4/3 = 1.33$. In other words, to generate 1000 values of X would require approximately 1333 random numbers R.

In the present situation, an alternative procedure exists for generating a uniform variate on $[1/4, 1]$—namely, Equation (8.5), which reduces to $X = 1/4 + (3/4)R$. Whether the acceptance-rejection technique or an alternative procedure, such as the inverse-transform technique [Equation (8.5)], is the more efficient depends on several considerations. The computer being used, the skills of the programmer, and the relative inefficiency of generating the additional (rejected) random numbers needed by acceptance-rejection should be compared to the computations required by the alternative procedure. In practice, concern with generation efficiency is left to specialists who conduct extensive tests comparing alternative methods (i.e., until a simulation model begins to require excessive computer runtime due to the generator being used).

For the uniform distribution on $[1/4, 1]$, the inverse-transform technique of Equation (8.5) is undoubtedly much easier to apply and more efficient than the acceptance-rejection technique. The main purpose of this example was to explain and motivate the basic concept of the acceptance-rejection technique. However, for some important distributions, such as the normal, gamma and beta, the inverse cdf does not exist in closed form and therefore the inverse-transform technique is difficult. These more advanced techniques are summarized by Bratley, Fox, and Schrage [1996], and Law [2007].

In the following subsections, the acceptance-rejection technique is illustrated for the generation of random variates for the Poisson, nonstationary Poisson, and gamma distributions.

8.2.1   Poisson Distribution

A Poisson random variable N with mean $\alpha > 0$ has pmf

$$p(n) = P(N = n) = \frac{e^{-\alpha}\alpha^n}{n!}, \quad n = 0, 1, 2, \ldots$$

More important, however, is that N can be interpreted as the number of arrivals from a Poisson arrival process in one unit of time. Recall from Section 5.5 that the interarrival times, $A_1, A_2, \ldots$ of successive customers are exponentially distributed with rate α (i.e., α is the mean number of arrivals per unit time); in addition, an exponential variate can be generated by Equation (8.3). Thus, there is a relationship between the (discrete) Poisson distribution and the (continuous) exponential distribution:

$$N = n \tag{8.22}$$

if and only if

$$A_1 + A_2 + \cdots + A_n \leq 1 < A_1 + \cdots + A_n + A_{n+1} \tag{8.23}$$

Equation (8.22), $N = n$, says there were exactly n arrivals during one unit of time; but Inequality (8.23) says that the nth arrival occurred before time 1 while the $(n+1)$st arrival occurred after time 1. Clearly, these two statements are equivalent. Proceed now by generating exponential interarrival times until some arrival, say $n + 1$, occurs after time 1; then set $N = n$.

For efficient generation purposes, Inequality (8.23) is usually simplified by first using Equation (8.3), $A_i = (-1/\alpha) \ln R_i$, to obtain

$$\sum_{i=1}^{n} -\frac{1}{\alpha} \ln R_i \leq 1 < \sum_{i=1}^{n+1} -\frac{1}{\alpha} \ln R_i$$

Next multiply through by $-\alpha$, which reverses the sign of the inequality, and use the fact that a sum of logarithms is the logarithm of a product, to get

$$\ln \prod_{i=1}^{n} R_i = \sum_{i=1}^{n} \ln R_i \geq -\alpha > \sum_{i=1}^{n+1} \ln R_i = \ln \prod_{i=1}^{n+1} R_i$$

Finally, use the relation $e^{\ln x} = x$ for any number x to obtain

$$\prod_{i=1}^{n} R_i \geq e^{-\alpha} > \prod_{i=1}^{n+1} R_i \tag{8.24}$$

which is equivalent to Inequality (8.23). The procedure for generating a Poisson random variate, N, is given by the following steps:

Step 1.   Set $n = 0, P = 1$.

Step 2. Generate a random number R_{n+1}, and replace P by $P \cdot R_{n+1}$.

Step 3. If $P < e^{-\alpha}$, then accept $N = n$. Otherwise, reject the current n, increase n by one, and return to step 2.

Notice that, upon completion of Step 2, P is equal to the rightmost expression in Inequality (8.24). The basic idea of a rejection technique is again exhibited; if $P \geq e^{-\alpha}$ in Step 3, then n is rejected and the generation process must proceed through at least one more trial.

How many random numbers will be required, on the average, to generate one Poisson variate N? If $N = n$, then $n + 1$ random numbers are required, so the average number is given by

$$E(N + 1) = \alpha + 1$$

which is quite large if the mean α of the Poisson distribution is large.

Example 8.8

Generate three Poisson variates with mean $\alpha = 0.2$. First, compute $e^{-\alpha} = e^{-0.2} = 0.8187$. Next, get a sequence of random numbers R from Table A.1 and follow the previously described Steps 1 to 3:

Step 1. Set $n = 0, P = 1$.

Step 2. $R_1 = 0.4357, P = 1 \cdot R_1 = 0.4357$.

Step 3. Since $P = 0.4357 < e^{-\alpha} = 0.8187$, accept $N = 0$.

Step 1–3. ($R_1 = 0.4146$ leads to $N = 0$.)

Step 1. Set $n = 0, P = 1$.

Step 2. $R_1 = 0.8353, P = 1 \cdot R_1 = 0.8353$.

Step 3. Since $P \geq e^{-\alpha}$, reject $n = 0$ and return to Step 2 with $n = 1$.

Step 2. $R_2 = 0.9952, P = R_1 R_2 = 0.8313$.

Step 3. Since $P \geq e^{-\alpha}$, reject $n = 1$ and return to Step 2 with $n = 2$.

Step 2. $R_3 = 0.8004, P = R_1 R_2 R_3 = 0.6654$.

Step 3. Since $P < e^{-\alpha}$, accept $N = 2$.

The calculations required for the generation of these three Poisson random variates are summarized as follows:

n	R_{n+1}	P	*Accept/Reject*	*Result*
0	0.4357	0.4357	$P < e^{-\alpha}$ (accept)	$N = 0$
0	0.4146	0.4146	$P < e^{-\alpha}$ (accept)	$N = 0$
0	0.8353	0.8353	$P \geq e^{-\alpha}$ (reject)	
1	0.9952	0.8313	$P \geq e^{-\alpha}$ (reject)	
2	0.8004	0.6654	$P < e^{-\alpha}$ (accept)	$N = 2$

It took five random numbers R to generate three Poisson variates here ($N = 0, N = 0$, and $N = 2$), but in the long run, to generate, say, 1000 Poisson variates with mean $\alpha = 0.2$, it would require approximately $1000(\alpha + 1)$ or 1200 random numbers.

Example 8.9

Buses arrive at the bus stop at Peachtree and North Avenue according to a Poisson process with a mean of one bus per 15 minutes. Generate a random variate N which represents the number of arriving buses during a 1-hour time slot. Now, N is Poisson distributed with a mean of four buses per hour. First, compute $e^{-\alpha} = e^{-4} = 0.0183$. Using a sequence of 12 random numbers from Table A.1 yields the following summarized results:

n	R_{n+1}	P	Accept/Reject	Result
0	0.4357	0.4357	$P \geq e^{-\alpha}$ (reject)	
1	0.4146	0.1806	$P \geq e^{-\alpha}$ (reject)	
2	0.8353	0.1508	$P \geq e^{-\alpha}$ (reject)	
3	0.9952	0.1502	$P \geq e^{-\alpha}$ (reject)	
4	0.8004	0.1202	$P \geq e^{-\alpha}$ (reject)	
5	0.7945	0.0955	$P \geq e^{-\alpha}$ (reject)	
6	0.1530	0.0146	$P < e^{-\alpha}$ (accept)	$N = 6$

It is immediately seen that a larger value of α (here $\alpha = 4$) usually requires more random numbers; if 1000 Poisson variates were desired, approximately $1000(\alpha + 1) = 5000$ random numbers would be required.

When α is large, say, $\alpha \geq 15$, the rejection technique outlined here becomes quite expensive, but fortunately an approximate technique based on the normal distribution works quite well. When the mean α is large, then

$$Z = \frac{N - \alpha}{\sqrt{\alpha}}$$

is approximately normally distributed with mean zero and variance 1; this observation suggests an approximate technique. First, generate a standard normal variate Z, by Equation (8.28) in Section 8.3.1, then generate the desired Poisson variate N by

$$N = \lceil \alpha + \sqrt{\alpha}Z - 0.5 \rceil \tag{8.25}$$

where $\lceil \cdot \rceil$ is the round-up function described in Section 8.1.7. (If $\alpha + \sqrt{\alpha}Z - 0.5 < 0$, then set $N = 0$.) The "0.5" used in the formula makes the round-up function become a "round to the nearest integer" function. Equation (8.25) is not an acceptance-rejection technique, but, when used as an alternative to the acceptance-rejection method, it provides a fairly efficient and accurate method for generating Poisson variates with a large mean.

8.2.2 Nonstationary Poisson Process

Another type of acceptance-rejection method (which is called *thinning*) can be used to generate interarrival times from a nonstationary Poisson process (NSPP) with arrival rate $\lambda(t), 0 \leq t \leq T$. An NSPP is an arrival process with an arrival rate that varies with time; see Section 5.5.2.

   Consider, for instance, the arrival-rate function given in Table 8.7 that changes every hour. The idea behind thinning is to generate a stationary Poisson arrival process at the fastest rate (1/5 customer per minute in the example), but "accept" or admit only a portion of the arrivals, thinning out just enough to get the desired time-varying rate. Next we give the generic algorithm, which generates T_i as the time of the ith arrival. Remember that, in a stationary Poisson arrival process, the times between arrivals are exponentially distributed.

Step 1. Let $\lambda^* = \max_{0 \leq t \leq T} \lambda(t)$ be the maximum of the arrival rate function and set $t = 0$ and $i = 1$.

Step 2. Generate E from the exponential distribution with rate λ^* and let $t = t + E$ (this is the arrival time of the stationary Poisson process).

Step 3. Generate random number R from the $U(0, 1)$ distribution. If $R \leq \lambda(t)/\lambda^*$ then $T_i = t$ and $i = i + 1$.

Step 4. Go to Step 2.

   The thinning algorithm can be inefficient if there are large differences between the typical and the maximum arrival rate. However, thinning has the advantage that it works for any integrable arrival rate function, not just a piecewise-constant function, as in this example.

Table 8.7 Arrival Rate for NSPP Example

t (min)	Mean Time between Arrivals (min)	Arrival Rate $\lambda(t)$ (arrivals/min)
0	15	1/15
60	12	1/12
120	7	1/7
180	5	1/5
240	8	1/8
300	10	1/10
360	15	1/15
420	20	1/20
480	20	1/20

Example 8.10 _____

For the arrival-rate function in Table 8.7, generate the first two arrival times.

Step 1. $\lambda^* = \max_{0 \leq t \leq T} \lambda(t) = 1/5, t = 0$ and $i = 1$.

Step 2. For random number $R = 0.2130, E = -5\ln(0.213) = 13.13$ and $t = 0 + 13.13 = 13.13$.

Step 3. Generate $R = 0.8830$. Since $R = 0.8830 \nleq \lambda(13.13)/\lambda^* = (1/15)/(1/5) = 1/3$, do not generate the arrival.

Step 4. Go to Step 2.

Step 2. For random number $R = 0.5530$, $E = -5\ln(0.553) = 2.96$ and $t = 13.13 + 2.96 = 16.09$.

Step 3. Generate $R = 0.0240$. Since $R = 0.0240 \leq \lambda(16.09)/\lambda^* = (1/15)/(1/5) = 1/3$, set $T_1 = t = 16.09$ and $i = i + 1 = 2$.

Step 4. Go to Step 2.

Step 2. For random number $R = 0.0001$, $E = -5\ln(0.0001) = 46.05$ and $t = 16.09 + 46.05 = 62.14$.

Step 3. Generate $R = 0.1443$. Since $R = 0.1443 \leq \lambda(62.14)/\lambda^* = (1/12)/(1/5) = 5/12$, set $T_2 = t = 62.14$ and $i = i + 1 = 3$.

Step 4. Go to Step 2.

8.2.3   Gamma Distribution

Several acceptance-rejection techniques for generating gamma random variates have been developed. (See Bratley, Fox, and Schrage [1996] and Law [2007].) One of the more efficient is by Cheng [1977]; the mean number of trials is between 1.13 and 1.47 for any value of the shape parameter $\beta \geq 1$.

    If the shape parameter β is an integer, say, $\beta = k$, one possibility is to use the convolution technique in Example 8.12, because the Erlang distribution is a special case of the more general gamma distribution. On the other hand, the acceptance-rejection technique described here would be a highly efficient method for the Erlang distribution, especially if $\beta = k$ were large. The routine generates gamma random variates with scale parameter θ and shape parameter β—that is, with mean $1/\theta$ and variance $1/\beta\theta^2$. The steps are as follows:

Step 1. Compute $a = 1/(2\beta - 1)^{1/2}$, $b = \beta - \ln 4$.

Step 2. Generate R_1 and R_2. Set $V = R_1/(1 - R_1)$.

Step 3. Compute $X = \beta V^a$.

Step 4a. If $X > b + (\beta a + 1)\ln(V) - \ln(R_1^2 R_2)$, reject X and return to Step 2.

Step 4b. If $X \leq b + (\beta a + 1)\ln(V) - \ln(R_1^2 R_2)$, use X as the desired variate.

    The generated variates from Step 4b will have mean and variance both equal to β. If it is desired to have mean $1/\theta$ and variance $1/\beta\theta^2$, as in Section 5.4.3, then include Step 5.

(Step 5. Replace X by $X/(\beta\theta)$.)

The basic idea of all acceptance-rejection methods is again illustrated here, but the proof of this example is beyond the scope of this book. In Step 3, $X = \beta V^a = \beta[R_1/(1 - R_1)]^a$ is not gamma distributed, but rejection of certain values of X in Step 4a guarantees that the accepted values in Step 4b do have the gamma distribution.

Example 8.11
Downtimes for a high-production candy-making machine have been found to be gamma distributed with mean 2.2 minutes and variance 2.10 minutes². Thus, $1/\theta = 2.2$ and $1/\beta\theta^2 = 2.10$, which together imply that $\beta = 2.30$ and $\theta = 0.4545$.

Step 1. $a = 0.53, b = 0.91$.

Step 2. Generate $R_1 = 0.832, R_2 = 0.021$. Set $V = 0.832/(1 - 0.832) = 4.952$.

Step 3. Compute $X = 2.3(4.952)^{0.53} = 5.37$.

Step 4. $X = 5.37 > 0.91 + [2.3(0.53) + 1]\ln(4.952) - \ln[(0.832)^2 0.021] = 8.68$, so reject X and return to Step 2.

Step 2. Generate $R_1 = 0.434, R_2 = 0.716$. Set $V = 0.434/(1 - 0.434) = 0.767$.

Step 3. Compute $X = 2.3(0.767)^{0.53} = 2.00$.

Step 4. Since $X = 2.00 \leq 0.91 + [2.3(0.53) + 1]\ln(0.767) - \ln[(0.434)^2 0.716] = 2.32$, accept X.

Step 5. Divide X by $\beta\theta = 1.045$ to get $X = 1.91$.

This example took two trials (i.e., one rejection) to generate an acceptable gamma-distributed random variate, but, on the average, to generate, say, 1000 gamma variates, the method will require between 1130 and 1470 trials, or equivalently, between 2260 and 2940 random numbers. The method is somewhat cumbersome for hand calculations, but is easy to program on the computer and is one of the most efficient gamma generators known.

8.3 Special Properties

"Special properties" are just as the name implies: They are variate-generation techniques that are based on features of a particular family of probability distributions, rather than being general-purpose techniques like the inverse-transform or acceptance-rejection techniques.

8.3.1   Direct Transformation for the Normal and Lognormal Distributions

Many methods have been developed for generating normally distributed random variates. The inverse-transform technique cannot easily be applied, however, because the inverse cdf cannot be written in closed form. The standard normal cdf is given by

$$\Phi(x) = \int_{-\infty}^{x} \frac{1}{\sqrt{2\pi}} e^{-t^2/2} \, dt, \quad -\infty < x < \infty$$

This section describes an intuitively appealing direct transformation that produces an independent pair of standard normal variates with mean zero and variance 1. The method is due to Box and Muller [1958]. Although not as efficient as many more modern techniques, it is easy to program in a scientific language, such as FORTRAN, C, C++, Visual Basic, or Java. We then show how to transform a standard normal variate into a normal variate with mean μ and variance σ^2. Once we have a method (this or any other) for generating X from an $N(\mu, \sigma^2)$ distribution, then we can generate a lognormal random variate Y with parameters μ and σ^2 by using the direct transformation $Y = e^X$. (Recall that μ and σ^2 are *not* the mean and variance of the lognormal; see Equations (5.58) and (5.59).)

Consider two standard normal random variables, Z_1 and Z_2, plotted as a point in the plane as shown in Figure 8.7 and represented in polar coordinates as

$$\begin{aligned} Z_1 &= B \cos \theta \\ Z_2 &= B \sin \theta \end{aligned} \tag{8.26}$$

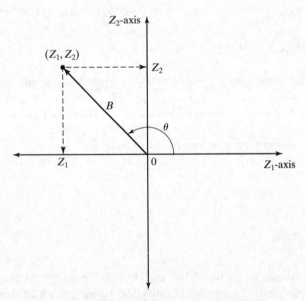

Figure 8.7   Polar representation of a pair of standard normal variables.

It is known that $B^2 = Z_1^2 + Z_2^2$ has the chi-square distribution with 2 degrees of freedom, which is equivalent to an exponential distribution with mean 2. Thus, the radius B can be generated by use of Equation (8.3):

$$B = (-2 \ln R)^{1/2} \qquad\qquad (8.27)$$

By the symmetry of the normal distribution, it seems reasonable to suppose, and indeed it is the case, that the angle is uniformly distributed between 0 and 2π radians. In addition, the radius B and the angle θ are mutually independent. Combining Equations (8.26) and (8.27) gives a direct method for generating two independent standard normal variates, Z_1 and Z_2, from two independent random numbers, R_1 and R_2:

$$\begin{aligned} Z_1 &= (-2 \ln R_1)^{1/2} \cos(2\pi R_2) \\ Z_2 &= (-2 \ln R_1)^{1/2} \sin(2\pi R_2) \end{aligned} \qquad\qquad (8.28)$$

To illustrate the generation scheme, consider Equation (8.28) with $R_1 = 0.1758$ and $R_2 = 0.1489$. Two standard normal random variates are generated as follows:

$$Z_1 = [-2 \ln(0.1758)]^{1/2} \cos(2\pi 0.1489) = 1.11$$

$$Z_2 = [-2 \ln(0.1758)]^{1/2} \sin(2\pi 0.1489) = 1.50$$

To obtain normal variates X_i with mean μ and variance σ^2, we then apply the transformation

$$X_i = \mu + \sigma Z_i \qquad\qquad (8.29)$$

to the standard normal variates. For example, to transform the two standard normal variates into normal variates with mean $\mu = 10$ and variance $\sigma^2 = 4$, we compute

$$X_1 = 10 + 2(1.11) = 12.22$$

$$X_2 = 10 + 2(1.50) = 13.00$$

Recall that to obtain a lognormal variate Y from a normal variate X we set $Y = e^X$. Also recall that the mean and variance of the lognormal Y is not the same as the normal X; see Equations (5.58) and (5.59) for the relationship.

8.3.2   Convolution Method

The probability distribution of a sum of two or more independent random variables is called a *convolution* of the distributions of the original variables. The convolution method thus refers to adding together two or more random variables to obtain a new random variable with the desired distribution. This technique can be applied to obtain Erlang variates and binomial variates. What is important is not the cdf of the desired random variable, but rather its relation to other variates more easily generated.

Example 8.12:   Erlang Distribution

As was discussed in Section 5.4.4, an Erlang random variable X with parameters (k, θ) can be shown to be the sum of k independent exponential random variables, $X_i, i = 1, 2, \ldots, k$, each having mean $1/k\theta$—that is,

$$X = \sum_{i=1}^{k} X_i$$

The convolution approach is to generate $X_1, X_2, \ldots, X_k$, then sum them to get X. In the case of the Erlang, each X_i can be generated by Equation (8.3) with $1/\lambda = 1/k\theta$. Therefore, an Erlang variate can be generated by

$$X = \sum_{i=1}^{k} -\frac{1}{k\theta} \ln R_i$$

$$= -\frac{1}{k\theta} \ln \left(\prod_{i=1}^{k} R_i \right) \tag{8.30}$$

It is more efficient computationally to multiply all the random numbers first and then to compute only one logarithm.

Example 8.13

Trucks arrive at a large warehouse in a completely random fashion that is modeled as a Poisson process with arrival rate $\lambda = 10$ trucks per hour. The guard at the entrance sends trucks alternately to the north and south docks. An analyst has developed a model to study the loading/unloading process at the south docks and needs a model of the arrival process at the south docks alone. An interarrival time X between successive truck arrivals at the south docks is equal to the sum of two interarrival times at the entrance and thus it is the sum of two exponential random variables, each having mean 0.1 hour, or 6 minutes. Thus, X has the Erlang distribution with $k = 2$ and mean $1/\theta = 2/\lambda = 0.2$ hour. To generate the variate X, first obtain $k = 2$ random numbers from Table A.1, say $R_1 = 0.937$ and $R_2 = 0.217$. Then, by Equation (8.30),

$$X = -0.1 \ln[(0.937)(0.217)]$$

$$= 0.159 \text{ hour } = 9.56 \text{ minutes}$$

In general, Equation (8.30) implies that k uniform random numbers are needed for each Erlang variate generated. If k is large, it is more efficient to generate Erlang variates by other techniques, such as one of the many acceptance-rejection techniques for the gamma distribution given in Section 8.2.3, or by Bratley, Fox and Schrage [1996] and Law [2007].

8.3.3 More Special Properties

There are many relationships among probability distributions that can be exploited for random-variate generation. The convolution method in the Section 8.3.2 is one example. Another particularly useful example is the relationship between the beta distribution and the gamma distribution.

Suppose that X_1 has a gamma distribution with shape parameter β_1 and scale parameter $\theta_1 = 1/\beta_1$, while X_2 has a gamma distribution with shape parameter β_2 and scale parameter $\theta_2 = 1/\beta_2$, and that these two random variables are independent. Then

$$Y = \frac{X_1}{X_1 + X_2}$$

has a beta distribution with parameters β_1 and β_2 on the interval $[0, 1]$. If, instead, we want Y to be defined on the interval $[a, b]$, then set

$$Y = a + (b - a)\left(\frac{X_1}{X_1 + X_2}\right)$$

Thus, using the acceptance-rejection technique for gamma variates defined in the previous section, we can generate beta variates, with two gamma variates required for each beta.

Although this method of beta generation is convenient, there are faster methods based on acceptance-rejection ideas. See, for instance, Devroye [1986] or Dagpunar [1988].

8.4 Summary

The basic principles of random-variate generation via the inverse-transform technique, the acceptance-rejection technique, and special properties have been introduced and illustrated by examples. Methods for generating many of the important continuous and discrete distributions, plus all empirical distributions, have been given. See Schmeiser [1980] for an excellent survey; for advanced treatments the reader is referred to Devroye [1986] or Dagpunar [1988].

REFERENCES

BRATLEY, P., B. L. FOX, AND L. E. SCHRAGE [1996], *A Guide to Simulation*, 2d ed., Springer-Verlag, New York.

BOX, G. E. P., AND M. F. MULLER [1958], "A Note on the Generation of Random Normal Deviates," *Annals of Mathematical Statistics*, Vol. 29, pp. 610–611.

CHENG, R. C. H. [1977], "The Generation of Gamma Variables with Nonintegral Shape Parameter," *Applied Statistics*, Vol. 26, No. 1, pp. 71–75.

DAGPUNAR, J. [1988], *Principles of Random Variate Generation*, Clarendon Press, Oxford.

DEVROYE, L. [1986], *Non-Uniform Random Variate Generation*, Springer-Verlag, New York.

LAW, A. M. [2007], *Simulation Modeling and Analysis*, 4th ed., McGraw-Hill, New York.

RIPLEY, B. D. [1987], *Stochastic Simulation*, Wiley, New York.

SCHMEISER, B. W. [1979], "Approximations to the Inverse Cumulative Normal Function for Use on Hand Calculators," *Applied Statistics*, Vol. 28, pp. 175–176.

SCHMEISER, B. W. [1980], "Random Variate Generation: A Survey," in *Simulation with Discrete Models: A State of the Art View*, T. I. Ören, C. M. Shub, and P. F. Roth, eds., IEEE, NY, pp. 79–104.

EXERCISES

Many of the exercises in this section ask the reader to develop a generation scheme for a given distribution and generate values. The results should be implemented in a programming language or spreadsheet so that large numbers of variates can be generated.

1. Develop a random-variate generator for a random variable X with the pdf

$$f(x) = \begin{cases} e^{2x}, & -\infty < x \le 0 \\ e^{-2x}, & 0 < x < \infty \end{cases}$$

2. Develop a generation scheme for the triangular distribution with pdf

$$f(x) = \begin{cases} \dfrac{1}{2}(x - 2), & 2 \le x \le 3 \\ \dfrac{1}{2}\left(2 - \dfrac{x}{3}\right), & 3 < x \le 6 \\ 0, & \text{otherwise} \end{cases}$$

Generate 1000 values of the random variate, compute the sample mean, and compare it to the true mean of the distribution.

3. Develop a generator for a triangular distribution with range (1, 10) and mode at $x = 4$. Generate 1000 values of the random variate, compute the sample mean, and compare it to the true mean of the distribution.

4. Develop a generator for a triangular distribution with range (1, 10) and a mean of 4. Generate 1000 values of the random variate, compute the sample mean, and compare it to the true mean of the distribution.

5. Given the following cdf for a continuous variable with range from -3 to 4, develop a generator for the variable, generate 1000 values, and plot a histogram.

$$F(x) = \begin{cases} 0, & x \le -3 \\ \dfrac{1}{2} + \dfrac{x}{6}, & -3 < x \le 0 \\ \dfrac{1}{2} + \dfrac{x^2}{32}, & 0 < x \le 4 \\ 1, & x > 4 \end{cases}$$

6. Given the cdf $F(x) = x^4/16$ on $0 \le x \le 2$, develop a generator for this distribution. Generate 1000 values of the random variate, compute the sample mean, and compare it to the true mean of the distribution.

7. Given the pdf $f(x) = x^2/9$ on $0 \le x \le 3$, develop a generator for this distribution. Generate 1000 values of the random variate, compute the sample mean, and compare it to the true mean of the distribution.

8. Develop a generator for a random variable whose pdf is

$$f(x) = \begin{cases} \dfrac{1}{3}, & 0 \le x \le 2 \\ \dfrac{1}{24}, & 2 < x \le 10 \\ 0, & \text{otherwise} \end{cases}$$

Generate 1000 values and plot a histogram.

9. The cdf of a discrete random variable X is given by

$$F(x) = \frac{x(x+1)(2x+1)}{n(n+1)(2n+1)}, \quad x = 1, 2, \dots, n$$

When $n = 4$, generate three values of X, using $R_1 = 0.83$, $R_2 = 0.24$, and $R_3 = 0.57$.

10. Times to failure for an automated production process have been found to be randomly distributed with a Weibull distribution with parameters $\beta = 2$ and $\alpha = 10$. Derive Equation (8.6), and then use it to generate five values from this Weibull distribution, using five random numbers taken from Table A.1.

11. Data have been collected on service times at a drive-in bank window at the Shady Lane National Bank. This data are summarized into intervals as follows:

Interval (Seconds)	Frequency
15–30	10
30–45	20
45–60	25
60–90	35
90–120	30
120–180	20
180–300	10

Set up a table like Table 8.3 for generating service times by the table-lookup method, and generate five values of service time, using four-digit random numbers from Table A.1.

12. In Example 8.2, assume that fire-crew response times satisfy $0.25 \leq x \leq 3$. Modify Table 8.2 to accommodate this assumption. Then generate five values of response time, using four-digit uniform random numbers from Table A.1.

13. For a preliminary version of a simulation model, the number of pallets X to be loaded onto a truck at a loading dock was assumed to be uniformly distributed between 8 and 24. Devise a method for generating X, assuming that the loads on successive trucks are independent. Use the technique of Example 8.5 for discrete uniform distributions. Finally, generate loads for 10 successive trucks by using four-digit random numbers from Table A.1.

14. Develop a method for generating values from a negative binomial distribution with parameters p and k, as described in Section 5.3. Generate 3 values when $p = 0.8$ and $k = 2$. [*Hint*: Think about the definition of the negative binomial as the number of Bernoulli trials until the kth success.]

15. The weekly demand X for a slow-moving item has been found to be approximated well by a geometric distribution on the range $\{0, 1, 2, \ldots\}$ with mean weekly demand of 2.5 items. Generate 10 values of X demand per week, using random numbers from Table A.1. [*Hint*: For a geometric distribution on the range $\{q, q+1, \ldots\}$ with parameter p, the mean is $1/p + q - 1$.]

16. In Exercise 15, suppose that the demand has been found to have a Poisson distribution with mean 2.5 items per week. Generate 10 values of X demand per week, using random numbers from Table A.1. Discuss the differences between the geometric and the Poisson distributions.

17. Lead times have been found to be exponentially distributed with mean 3.7 days. Generate five random lead times from this distribution.

18. Regular maintenance of a production routine has been found to vary and has been modeled as a normally distributed random variable with mean 33 minutes and variance 4 minutes^2. Generate five random maintenance times from the given distribution.

19. A machine is taken out of production either if it fails or after 5 hours, whichever comes first. By running similar machines until failure, it has been found that time to failure X has the Weibull distribution with $\alpha = 8$, $\beta = 0.75$, and $\nu = 0$ (refer to Sections 5.4.6 and 8.1.3). Thus, the time until the machine is taken out of production can be represented as $Y = \min(X, 5)$. Develop a step-by-step procedure for generating Y.

20. The time until a component is taken out of service is uniformly distributed on from 0 to 8 hours. Two such independent components are put in series, and the whole system goes down when one of the components goes down. If X_i ($i = 1, 2$) represents the component runtimes, then $Y = \min(X_1, X_2)$ represents the system lifetime. Devise two distinct ways to generate Y. [*Hint*: One way is relatively straightforward. For a second method, first compute the cdf of $Y : F_Y(y) = P(Y \leq y) = 1 - P(Y > y)$, for $0 \leq y \leq 8$. Use the equivalence $\{Y > y\} = \{X_1 > y$ and $X_2 > y\}$ and the independence of X_1, and X_2. After finding $F_Y(y)$, proceed with the inverse-transform technique.]

21. In Exercise 20, component lifetimes are exponentially distributed, one with mean 2 hours and the other with mean 6 hours. Rework Exercise 20 under this new assumption. Discuss the relative efficiency of the two generation schemes devised.

22. Develop a technique for generating a binomial random variable X via the convolution technique. [*Hint*: X can be represented as the number of successes in n independent Bernoulli trials, each success having probability p. Thus, $X = \sum_{i=1}^{n} X_i$, where $P(X_i = 1) = p$ and $P(X_i = 0) = 1 - p$.]

23. Develop an acceptance-rejection technique for generating a geometric random variable X with parameter p on the range $\{0, 1, 2, \ldots\}$. [*Hint*: X can be thought of as the number of failures before the first success occurs in a sequence of independent Bernoulli trials.]

24. Write a computer routine to generate standard normal variates by the exact method discussed in this chapter. Use it to generate 10,000 values. Compare the true probability, $\Phi(z)$, that a value lies in $(-\infty, z)$ to the actual observed relative frequency that values were $\leq z$, for $z = -4, -3, -2, -1, 0, 1, 2, 3$, and 4.

25. Write a computer routine to generate gamma variates with shape parameter β and scale parameter θ. Generate 1000 values with $\beta = 2.5$ and $\theta = 0.2$ and compare the true mean, $1/\theta = 5$, to the sample mean.

26. Write a computer routine to generate 2000 values from one of the variates in Exercises 1 to 23. Make a histogram of the 2000 values and compare it to the theoretical density function (or probability mass function for discrete random variables).

27. Many spreadsheet, symbolic-calculation, and statistical-analysis programs have built-in routines for generating random variates from standard distributions. Try to find out what variate-generation methods are used in one of these packages by looking at the documentation. Should you trust a variate generator if the method is not documented?

28. Suppose that, somehow, we have available a source of exponentially distributed random variates with mean 1. Write an algorithm to generate random variates with a triangular distribution by transforming the exponentially distributed random variates. [*Hint*: First, transform to obtain uniformly distributed random variates.]

29. Let time $t = 0$ correspond to 6 A.M., and suppose that the arrival rate (in arrivals per hour) of customers to a breakfast restaurant that is open 6–9 A.M. is

$$\lambda(t) = \begin{cases} 30, & 0 \leq t < 1 \\ 45, & 1 \leq t < 2 \\ 20, & 2 \leq t \leq 4 \end{cases}$$

Derive a thinning algorithm for generating arrivals from this NSPP and generate the first 100 arrival times.

30. Generate 10 values from a beta distribution on the interval $[0, 1]$ with parameters $\beta_1 = 1.47$ and $\beta_2 = 2.16$. Next transform them to be on the interval $[-10, 20]$.

31. Derive a thinning algorithm for the NSPP with rate function given in Example 5.34 and generate the first 600 arrival times.

32. Derive a thinning algorithm for the NSPP with rate function given in Section 9.5 and generate the first 120 arrival times.

33. Develop a generation scheme for the general triangular distribution in Section 5.4.7. Test it by generating samples of 1000 values for different choices of the parameters.

34. Suppose that a discrete distribution can take values $v_1, v_2, \ldots, v_k$. These values are equally likely, but are not $1, 2, \ldots, k$. Come up with a generation scheme that exploits the idea in Example 8.5. [*Hint*: Consider storing the values in an array, then generating a random array index.]

35. Suppose that the arrival rate of hits to a web site is $\lambda(t) = 1000 + 50 \sin(2\pi t/24)$ hits per hour. Derive a thinning algorithm for generating arrivals from this NSPP.

Part IV

Analysis of Simulation Data

9

Input Modeling

Input models provide the driving force for a simulation. In the simulation of a queueing system, typical input models are the distributions of time between arrivals and of service times. For a supply-chain simulation, input models include the distributions of demand and of lead time. For the simulation of a reliability problem, the distribution of time to failure of a component is an example of an input model.

In the examples and exercises in Chapters 2 and 3, the appropriate distributions were specified for you. In real-world simulation applications, however, choosing appropriate distributions for input data is a major task from the standpoint of time and resource requirements. Regardless of the sophistication of the analyst, faulty models of the inputs will lead to outputs whose interpretation could give rise to misleading recommendations.

There are four steps in the development of a useful model of input data:

1. Collect data from the real system of interest. This often requires a substantial time and resource commitment. Unfortunately, in some situations it is not possible to collect data (for example, when time is extremely limited, when the input process does not yet exist, or when laws or rules prohibit the collection of data). When data are not available, expert opinion and knowledge of the process must be used to make educated guesses.

2. Identify a probability distribution to represent the input process. When data are available, this step typically begins with the development of a histogram of the data. Given the histogram and a structural knowledge of the process, a family of distributions is chosen. Fortunately, as was

described in Chapter 5, several well-known distributions often provide good approximations in practice.

3. Choose parameters that determine a specific instance of the distribution family. When data are available, these parameters may be estimated from the data.

4. Evaluate the chosen distribution and the associated parameters for goodness of fit. Goodness of fit may be evaluated informally, via graphical methods, or formally, via statistical tests. The chi-square and the Kolmogorov–Smirnov tests are standard goodness-of-fit tests. If not satisfied that the chosen distribution is a good approximation of the data, then the analyst returns to the second step, chooses a different family of distributions, and repeats the procedure. If several iterations of this procedure fail to yield a fit between an assumed distributional form and the collected data, the empirical form of the distribution may be used, as was described in Section 8.1.5.

Each of these steps is discussed in this chapter. Although software is now widely available to accomplish Steps 2, 3, and 4—including such stand-alone programs as ExpertFit® and Stat::Fit® and such integrated programs as Arena's Input Processor and @Risk's BestFit®—it is still important to understand what the software does, so that it can be used appropriately. Unfortunately, software is not as readily available for input modeling when there is a relationship between two or more variables of interest or when no data are available. These two topics are discussed toward the end of the chapter.

9.1   Data Collection

Problems are found at the end of each chapter, as exercises for the reader, in textbooks about mathematics, physics, chemistry, and other technical subjects. Years and years of working these problems could give the reader the impression that data are readily available. Nothing could be further from the truth. Data collection is one of the biggest tasks in solving a real problem and it is one of the most important and difficult problems in simulation.

The focus of input modeling is on the statistical aspects of fitting probability distributions to data, distributions that will later provide the driving inputs to the simulation. However, before "fitting" can occur there needs to be accurate and relevant input data available (or collectible), and the properties of the data need to be well understood. Too often data are handed to input modeling software, which will always fit *some* probability model to the data, whether it is justified or not. The following examples illustrate common pitfalls of this approach:

Example 9.1:   Stale Data
A simulation study will investigate ways to reduce the time patients spend in a cancer screening clinic. One part of the screening protocol involves a nurse obtaining the patient's history. In a 2002 study of nursing activities, observations of nurses taking patient histories were collected by a consulting firm. The hospital still has this data. Should it be used for input modeling in the current project?

In this case the answer is probably no, because deeper investigation by the engineers on the current project discovered that in 2008 the nurses stopped recording screening notes on a paper chart and began entering them directly into a laptop computer. Therefore, the 2002 data are of a different history-taking process and may significantly misrepresent the current process. If current data cannot be collected, either due to the time required or the expense, then it would be better to use estimates provided by the nurses themselves rather than rely on the out-of-date study. Section 9.6 describes methods for converting expert opinion into probability distributions. The engineers might then check the nurses' estimates by volunteering to have their own histories taken and timing the process to get a few actual data points.

Example 9.2: Unexpected Data

For security purposes people who enter a government building are required to pass through a security screening that includes a metal detector. Before new security protocols are implemented a simulation study will be conducted to assess their impact. To facilitate the study, 1000 observations of the time, in minutes, required to pass through the metal detector station were collected. The average value of these data was about 1/2 minute, with standard deviation almost the same. This suggests that a good input model for the time to pass through the metal detector is the exponential distribution, since its mean and standard deviation are always the same and exponential distributions are often used to model service times in queueing systems.

However, the analyst is uncomfortable with this input model because in her experience it only takes a few seconds to pass through a metal detector. She creates a histogram of the data, shown in Figure 9.1, which makes it immediately apparent that the time to pass through metal detection consists of two distinct distributions. Thinking carefully, she realizes that some people pass through the detector without incident, but others trigger the alarm and have to remove remaining metal

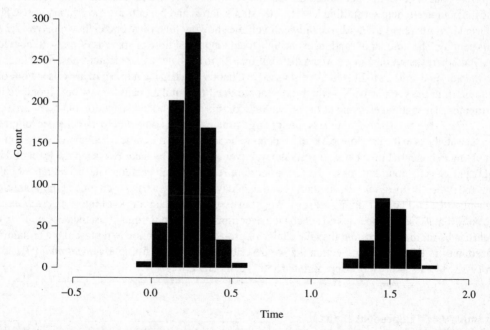

Figure 9.1 Histogram of 1000 observations of time through the metal detector measured in minutes.

objects and try again, in some cases even requiring a personal inspection. Therefore, three distinct input models are required for this data: one for the time to pass through without setting off the detector (the 764 left-most observations in the histogram); one for the time to pass through when the alarm does go off (the 236 right-most observations in the histogram); and a model for the chance that a person sets off the alarm. While an exponential distribution could certainly match the summary statistics from the data (mean and standard deviation), the fitted exponential distribution would have missed a key feature of security screening, almost certainly leading to an inaccurate simulation.

Unfortunately, inspecting the data carefully the analyst spots an additional complication: The histogram in Figure 9.1 shows negative values in the sample (nine total). A negative time to pass through metal detection is clearly impossible, so what should the analyst do? It may be that the observations were recorded correctly but a mistake was made in calculating the times; if that is the case then the analyst may be able to reconstruct the correct values. If not, then the nine negative values should be discarded before fitting the distribution. But should they also be discarded when estimating the probability that someone sets off the alarm? That is, should the estimate be 236/1000 or 236/991? If the analyst can convince herself that the 9 negative values were actually small times that were misrecorded, then she might want to retain them for estimating this probability. If she is unsure, then the simulation should be run with both estimates to see if there is a substantial impact.

Example 9.3:   Time-varying Data

A computer manufacturer provides a customer support center where owners of the product may call with questions or concerns. The call center is staffed from 8 A.M. to 8 P.M. Eastern Time. A simulation study is planned to help set staffing levels, since staff salaries and benefits are the primary costs. The call center has extensive records on daily call volume showing that, on average, they receive 12,277 calls per day. The number of random events in a fixed amount of time or space is often well modeled by a Poisson distributions, so a Poisson distribution is fit to the data (using methods described later in the chapter) and appears to fit well. As discussed in Chapter 6, when the number of arrivals is Poisson, then the time between arrivals—which is what we need to simulate calls—must be exponentially distributed. The natural estimate of the arrival rate for this exponential distribution is 12,277 arrivals per 12-hour day, but since a "day" is a pretty large time unit, it is converted to 17 calls per minute.

The analysis in the previous paragraph seems to be sound, and the choice of a Poisson distribution (implying exponential time between arrivals) was even supported by data. However, the level of data aggregation, a 12-hour day, masked a very important feature of call centers: The arrival rate of calls can, and typically does, vary substantially by time of day, day of week, and even time of the year (e.g., right after the holiday gift-giving season). We often assume that the process of interest is *stationary*, meaning that its basic properties do not change over time. However, if trends exist then accounting for them can be far more important than the choice of probability distribution to represent the remaining uncertainty in the system. In Section 9.5 we describe a method for fitting a *nonstationary Poisson process*, which is likely more appropriate for this call center.

Example 9.4:   Dependent Data

Monthly demands for a particular brand of sports drink by a large retailer have been extracted from an orders database to use in a supply chain simulation. A simple input model for this data is that monthly

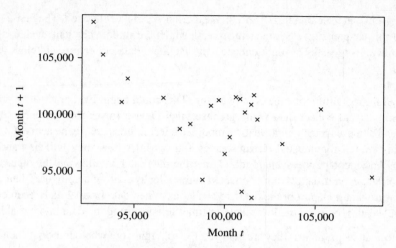

Figure 9.2   Time-lagged demand data.

demands are independent and identically distributed (i.i.d.), and this chapter describes methods for fitting a distribution to such data and evaluating the fit. But it is important to evaluate the "i.i.d." assumption also.

One informal way to check that these data are identically distributed is to plot the demands in time order to look for trends (e.g., demands are decreasing over time) or seasonality (there is a regular cycle of demand, for instance, high demand each summer followed by low demand each winter). Data with trends or seasonality should typically be "de-trended" first (usually by fitting a model to the trend or seasonal pattern) and then a distribution can be fit to any remaining variation.

Independence can be checked via time-lagged scatter plots. Figure 9.2 is a scatter plot of each month's demand for the sports drink plotted against the following month's demand (e.g., January's demand vs. February's demand). Clearly there is a strong negative correlation, which means large demands in one month tend to be followed by low demands in the next month, and vice versa. This feature might be the result of the retailer overstocking in one month, and then ordering less the following month so as to work off the excess inventory. Time lags greater than one month can also be examined. By blindly making the standard assumption of i.i.d. inputs this important phenomenon would be missed, leading to less effective supply chain management. In Section 9.7 we present methods for modeling dependent input data.

As these examples illustrate, simply having data is not enough to support effective input modeling. Data can be out of date or "dirty" (containing errors). Sometimes the effort or cost to transform data into a usable form, or "clean" it of errors, can be as significant as that required to obtain it. Input modeling nearly always requires the analyst to use their judgment as well as to apply appropriate statistical tools. Understanding which input models are the most and least reliable is important for judging the reliability of the conclusions drawn from the simulation study. And since uncertainty about the correctness of the input models can never be entirely eliminated, it is sensible to run the simulation with several plausible input models to see if the conclusions are robust or highly sensitive to the choices.

Many lessons can be learned from an actual experience at data collection. The first five exercises at the end of this chapter suggest some situations in which the student can gain such experience.

The following suggestions might enhance and facilitate data collection, although they are not all-inclusive.

1. A useful expenditure of time is in planning. This could begin by a practice or preobserving session. Try to collect data while preobserving. Devise forms for this purpose. It is very likely that these forms will have to be modified several times before the actual data collection begins. Watch for unusual circumstances and consider how they will be handled. When possible, videotape the system and extract the data later by viewing the tape. Planning is important, even if data will be collected automatically (e.g., via computer data collection), to ensure that the appropriate data are available. When data have already been collected by someone else, be sure to allow plenty of time for converting the data into a usable format.

2. Try to analyze the data as they are being collected. Figure out whether the data being collected are adequate to provide the distributions needed as input to the simulation. Find out whether any data being collected are useless to the simulation. There is no need to collect superfluous data.

3. Try to combine homogeneous data sets. Check data for homogeneity in successive time periods and during the same time period on successive days. For example, check for homogeneity of data from 2:00 P.M. to 3:00 P.M. and 3:00 P.M. to 4:00 P.M., and check to see whether the data are homogeneous for 2:00 P.M. to 3:00 P.M. on Thursday and Friday. When checking for homogeneity, an initial test is to see whether the means of the distributions (the average interarrival times, for example) are the same. The two-sample t test can be used for this purpose. A more thorough analysis would require a test of the equivalence of the distributions, perhaps via a quantile-quantile plot (described later).

4. Be aware of the possibility of data censoring, in which a quantity of interest is not observed in its entirety. This problem most often occurs when the analyst is interested in the time required to complete some process (for example, produce a part, treat a patient, or have a component fail), but the process begins prior to, or finishes after the completion of, the observation period. Censoring can result in especially long process times being left out of the data sample.

5. To discover whether there is a relationship between two variables, build a scatter diagram. Sometimes a look at the scatter diagram will indicate whether there is a relationship between the variables of interest.

6. Consider the possibility that a sequence of observations that appear to be independent actually has autocorrelation. Autocorrelation can exist in successive time periods or for successive customers. A brief introduction to autocorrelation was provided in Section 7.4.2.

7. Keep in mind the difference between input data and output or performance data, and be sure to collect input data. Input data typically represent the uncertain quantities that are largely beyond the control of the system and will not be altered by changes made to improve the system. Output data, on the other hand, represent the performance of the system when

subjected to the inputs, performance that we might be trying to improve. In a queueing simulation, the customer arrival times are usually inputs, whereas the customer delay is an output. Performance data are useful for model validation; however, see Chapter 10.

Again, these are just a few suggestions. As a rule, data collection and analysis must be approached with great care.

9.2   Identifying the Distribution with Data

In this section, we discuss methods for selecting families of input distributions when data are available and are believed to be independent and identically distributed. The specific distribution within a family is specified by estimating its parameters, as described in Section 9.3. Section 9.6 takes up the case in which data are unavailable.

9.2.1   Histograms

A frequency distribution or histogram is useful in identifying the shape of a distribution. A histogram is constructed as follows:

1. Divide the range of the data into intervals. (Intervals are usually of equal width; however, unequal widths may be used if the heights of the frequencies are adjusted.)

2. Label the horizontal axis to conform to the intervals selected.

3. Find the frequency of occurrences within each interval.

4. Label the vertical axis so that the total occurrences can be plotted for each interval.

5. Plot the frequencies on the vertical axis.

The number of class intervals depends on the number of observations and on the amount of scatter or dispersion in the data. Hines *et al.* [2002] state that choosing the number of class intervals approximately equal to the square root of the sample size often works well in practice. If the intervals are too wide, the histogram will be coarse, or blocky, and its shape and other details will not show well. If the intervals are too narrow, the histogram will be ragged and will not smooth the data. Examples of ragged, coarse, and appropriate histograms of the same data are shown in Figure 9.3. Modern data-analysis software often allows the interval sizes to be changed easily and interactively until a good choice is found.

The histogram for continuous data corresponds to the probability density function of a theoretical distribution. If continuous, a line drawn through the center point of each class interval frequency should result in a shape like that of a pdf.

Histograms for discrete data, where there are a large number of data points, should have a cell for each value in the range of the data. However, if there are few data points, it could be necessary to combine adjacent cells to eliminate the ragged appearance of the histogram. If the histogram is associated with discrete data, it should look like a probability mass function.

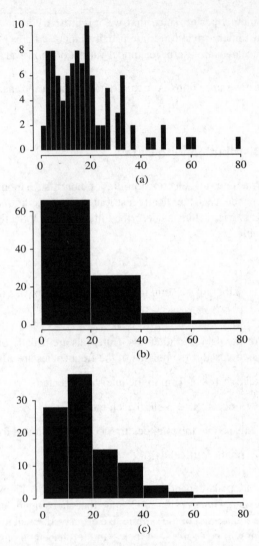

Figure 9.3 Ragged, coarse, and appropriate histograms: (a) original data—too ragged; (b) combining adjacent cells—too coarse; (c) combining adjacent cells—appropriate.

Example 9.5:   Discrete Data

The number of vehicles arriving at the northwest corner of an intersection in a 5-minute period between 7:00 A.M. and 7:05 A.M. was monitored for five workdays over a 20-week period. Table 9.1 shows the resulting data. The first entry in the table indicates that there were 12 5-minute periods during which zero vehicles arrived, 10 periods during which one vehicle arrived, and so on.

Table 9.1   Number of Arrivals in a 5-Minute Period

Arrivals per Period	Frequency	Arrivals per Period	Frequency
0	12	6	7
1	10	7	5
2	19	8	5
3	17	9	3
4	10	10	3
5	8	11	1

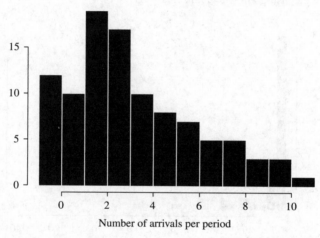

Figure 9.4   Histogram of number of arrivals per period.

The number of automobiles is a discrete variable, and there are ample data, so the histogram may have a cell for each possible value in the range of the data. The resulting histogram is shown in Figure 9.4.

Example 9.6:   Continuous Data _____

Life tests were performed on a random sample of electronic components at 1.5 times the nominal voltage, and their lifetime (or time to failure), in days, was recorded:

79.919	3.081	0.062	1.961	5.845
3.027	6.505	0.021	0.013	0.123
6.769	59.899	1.192	34.760	5.009
18.387	0.141	43.565	24.420	0.433
144.695	2.663	17.967	0.091	9.003
0.941	0.878	3.371	2.157	7.579
0.624	5.380	3.148	7.078	23.960
0.590	1.928	0.300	0.002	0.543
7.004	31.764	1.005	1.147	0.219
3.217	14.382	1.008	2.336	4.562

Lifetime, usually considered a continuous variable, is recorded here to three-decimal-place accuracy. The histogram is prepared by placing the data in class intervals. The range of the data is rather large, from 0.002 day to 144.695 days. However, most of the values (30 of 50) are in the zero-to-5-day range. Using intervals of width three results in Table 9.2. The data of Table 9.2 are then used to prepare the histogram shown in Figure 9.5.

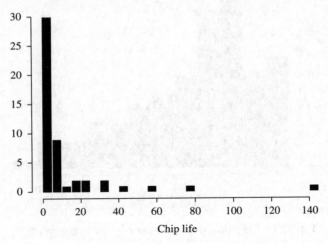

Figure 9.5   Histogram of component life.

9.2.2   Selecting the Family of Distributions

In Chapter 5, some distributions that arise often in simulation were described. Additionally, the shapes of these distributions were displayed. The purpose of preparing a histogram is to infer a known pdf or pmf. A family of distributions is selected on the basis of what might arise in the context being investigated along with the shape of the histogram. Thus, if interarrival-time data have been collected, and the histogram has a shape similar to the pdf in Figure 5.9, the assumption of an exponential distribution would be warranted. Similarly, if measurements of the weights of pallets of

Table 9.2 Electronic Component Data

Component Life (days)	Frequency
$0 \leq x_j < 3$	23
$3 \leq x_j < 6$	10
$6 \leq x_j < 9$	5
$9 \leq x_j < 12$	1
$12 \leq x_j < 15$	1
$15 \leq x_j < 18$	2
$18 \leq x_j < 21$	0
$21 \leq x_j < 24$	1
$24 \leq x_j < 27$	1
$27 \leq x_j < 30$	0
$30 \leq x_j < 33$	1
$33 \leq x_j < 36$	1
$\vdots$	$\vdots$
$42 \leq x_j < 45$	1
$\vdots$	$\vdots$
$57 \leq x_j < 60$	1
$\vdots$	$\vdots$
$78 \leq x_j < 81$	1
$\vdots$	$\vdots$
$144 \leq x_j < 147$	1

freight are being made, and the histogram appears symmetric about the mean with a shape like that shown in Figure 5.12, the assumption of a normal distribution would be warranted.

The exponential, normal, and Poisson distributions are frequently encountered and are not difficult to analyze from a computational standpoint. Although more difficult to analyze, the beta, gamma, and Weibull distributions provide a wide array of shapes and should not be overlooked during modeling of an underlying probabilistic process. Perhaps an exponential distribution was assumed, but it was found not to fit the data. The next step would be to examine where the lack of fit occurred. If the lack of fit was in one of the tails of the distribution, perhaps a gamma or Weibull distribution would fit the data more adequately.

There are literally hundreds of probability distributions that have been created; many were created with some specific physical process in mind. One aid to selecting distributions is to use the physical basis of the distributions as a guide. Here are some examples:

Binomial: Models the number of successes in n trials, when the trials are independent with common success probability p; for example, the number of defective computer chips found in a lot of n chips.

Negative Binomial (includes the geometric distribution): Models the number of trials required to achieve k successes; for example, the number of computer chips that we must inspect to find 4 defective chips.

Poisson: Models the number of independent events that occur in a fixed amount of time or space; for example, the number of customers that arrive to a store during 1 hour, or the number of defects found in 30 square meters of sheet metal.

Normal: Models the distribution of a process that can be thought of as the sum of a number of component processes; for example, a time to assemble a product that is the sum of the times required for each assembly operation. Notice that the normal distribution admits negative values, which could be impossible for process times.

Lognormal: Models the distribution of a process that can be thought of as the product of (meaning to multiply together) a number of component processes; for example, the rate of return on an investment, when interest is compounded, is the product of the returns for a number of periods.

Exponential: Models the time between independent events, or a process time that is memoryless (knowing how much time has passed gives no information about how much additional time will pass before the process is complete); for example, the times between the arrivals from a large population of potential customers who act independently of one another. The exponential is a highly variable distribution; it is sometimes overused, because it often leads to mathematically tractable models. Recall that, if the time between events is exponentially distributed, then the number of events in a fixed period of time is Poisson.

Gamma: An extremely flexible distribution used to model nonnegative random variables. The gamma can be shifted away from 0 by adding a constant.

Beta: An extremely flexible distribution used to model bounded (fixed upper and lower limits) random variables. The beta can be shifted away from 0 by adding a constant and can be given a range larger than [0, 1] by multiplying by a constant.

Erlang: Models processes that can be viewed as the sum of several exponentially distributed processes; for example, a computer network fails when a computer and two backup computers fail, and each has a time to failure that is exponentially distributed. The Erlang is a special case of the gamma.

Weibull: Models the time to failure for components; for example, the time to failure for a disk drive. The exponential is a special case of the Weibull.

Discrete or Continuous Uniform: Models complete uncertainty: All outcomes are equally likely. This distribution often is used inappropriately when there are no data.

Triangular: Models a process for which only the minimum, most likely, and maximum values of the distribution are known; for example, the minimum, most likely, and maximum time required to test a product. This model is often a marked improvement over a uniform distribution.

Empirical: Resamples from the actual data collected; often used when no theoretical distribution seems appropriate.

Do not ignore physical characteristics of the process when selecting distributions. Is the process naturally discrete or continuous valued? Is it bounded or is there no natural bound? This knowledge, which does not depend on data, can help narrow the family of distributions from which to choose. And keep in mind that there is no "true" distribution for any stochastic input process. An input model is an approximation of reality, so the goal is to obtain an approximation that yields useful results from the simulation experiment.

The reader is encouraged to complete Exercises 6 through 11 to learn more about the shapes of the distributions mentioned in this section. Examining the variations in shape as the parameters change is very instructive.

9.2.3   Quantile-Quantile Plots

The construction of histograms, as discussed in Section 9.2.1, and the recognition of a distributional shape, as discussed in Section 9.2.2, are necessary ingredients for selecting a family of distributions to represent a sample of data. However, a histogram is not as useful for evaluating the *fit* of the chosen distribution. When there is a small number of data points, say, 30 or fewer, a histogram can be rather ragged. Further, our perception of the fit depends on the widths of the histogram intervals. But, even if the intervals are well chosen, grouping data into cells makes it difficult to compare a histogram to a continuous probability density function. A quantile-quantile (q-q) plot is a useful tool for evaluating distribution fit, one that does not suffer from these problems.

If X is a random variable with cdf F, then the q-quantile of X is that value γ such that $F(\gamma) = P(X \leq \gamma) = q$, for $0 < q < 1$. When F has an inverse, we write $\gamma = F^{-1}(q)$.

Now let $\{x_i, i = 1, 2, \ldots, n\}$ be a sample of data from X. Order the observations from the smallest to the largest, and denote these as $\{y_j, j = 1, 2, \ldots n\}$, where $y_1 \leq y_2 \leq \cdots \leq y_n$. Let j denote the ranking or order number. Therefore, $j = 1$ for the smallest and $j = n$ for the largest. The q-q plot is based on the fact that y_j is an estimate of the $(j - 1/2)/n$ quantile of X. In other words,

$$y_j \text{ is approximately } F^{-1}\left(\frac{j - \frac{1}{2}}{n}\right)$$

Now, suppose that we have chosen a distribution with cdf F as a possible representation of the distribution of X. If F is a member of an appropriate family of distributions, then a plot of y_j versus $F^{-1}((j - 1/2)/n)$ will be *approximately a straight line*. If F is from an appropriate family of distributions and also has appropriate parameter values, then the line will have slope 1. On the other hand, if the assumed distribution is inappropriate, the points will deviate from a straight line, usually in a systematic manner. The decision about whether to reject some hypothesized model is subjective.

Example 9.7: Normal *Q-Q* Plot _____

A robot is used to install the doors on automobiles along an assembly line. It was thought that the installation times followed a normal distribution. The robot is capable of measuring installation times accurately. A sample of 20 installation times was automatically taken by the robot, with the following results, where the values are in seconds:

99.79	99.56	100.17	100.33
100.26	100.41	99.98	99.83
100.23	100.27	100.02	100.47
99.55	99.62	99.65	99.82
99.96	99.90	100.06	99.85

The sample mean is 99.99 seconds, and the sample variance is $(0.2832)^2$ seconds^2. These values can serve as the parameter estimates for the mean and variance of the normal distribution. The observations are now ordered from smallest to largest as follows:

j	Value	j	Value	j	Value	j	Value
1	99.55	6	99.82	11	99.98	16	100.26
2	99.56	7	99.83	12	100.02	17	100.27
3	99.62	8	99.85	13	100.06	18	100.33
4	99.65	9	99.90	14	100.17	19	100.41
5	99.79	10	99.96	15	100.23	20	100.47

The ordered observations are then plotted versus $F^{-1}((j-1/2)/20)$, for $j = 1, 2, \ldots, 20$, where F is the cdf of the normal distribution with mean 99.99 and variance $(0.2832)^2$, to obtain a *q-q* plot. The plotted values are shown in Figure 9.6, along with a histogram of the data. Notice that it is difficult to tell whether the data are well represented by a normal distribution from looking at the histogram, but the general perception of a straight line is quite clear in the *q-q* plot and supports the hypothesis of a normal distribution.

_____

In the evaluation of the linearity of a *q-q* plot, the following should be considered:

a. The observed values will never fall exactly on a straight line.

b. The ordered values are not independent; they have been ranked. Hence, if one point is above a straight line, it is likely that the next point will also lie above the line. And it is unlikely that the points will be evenly scattered about the line.

c. The variances of the extremes (largest and smallest values) are much higher than the variances in the middle of the plot. Greater discrepancies can be accepted at the extremes. The linearity of the points in the middle of the plot is more important than the linearity at the extremes.

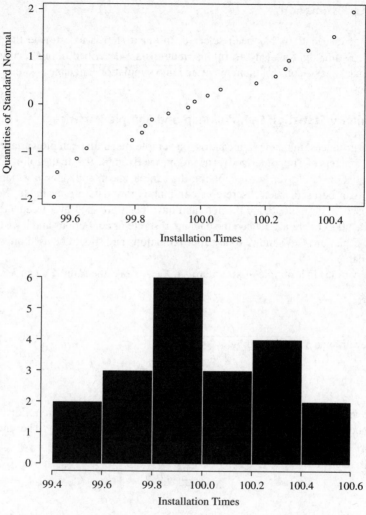

Figure 9.6   A q-q plot and histogram of the installation times.

Modern data-analysis software often includes tools for generating q-q plots, especially for the normal distribution. The q-q plot can also be used to compare two samples of data to see whether they can be represented by the same distribution (that is, that they are homogeneous). If $x_1, x_2, \ldots, x_n$ are a sample of the random variable X, and $z_1, z_2, \ldots, z_n$ are a sample of the random variable Z, then plotting the ordered values of X versus the ordered values of Z will reveal approximately a straight line if both samples are well represented by the same distribution (Chambers, Cleveland, and Tukey [1983]).

9.3   Parameter Estimation

After a family of distributions has been selected, the next step is to estimate the parameters of the distribution. Estimators for many useful distributions are described in this section. In addition, many software packages—some of them integrated into simulation languages—are now available to compute these estimates.

9.3.1   Preliminary Statistics: Sample Mean and Sample Variance

In a number of instances, the sample mean, or the sample mean and sample variance, are used to estimate the parameters of a hypothesized distribution; see Example 9.7. In the following paragraphs, three sets of equations are given for computing the sample mean and sample variance. Equations (9.1) and (9.2) can be used when discrete or continuous raw data are available. Equations (9.3) and (9.4) are used when the data are discrete and have been grouped in a frequency distribution. Equations (9.5) and (9.6) are used when the data are discrete or continuous and have been placed in class intervals. Equations (9.5) and (9.6) are approximations and should be used only when the raw data are unavailable.

If the observations in a sample of size n are $X_1, X_2, \ldots, X_n$, the sample mean $\bar{X}$ is defined by

$$\bar{X} = \frac{\sum_{i=1}^{n} X_i}{n} \tag{9.1}$$

and the sample variance S^2 is defined by

$$S^2 = \frac{\sum_{i=1}^{n} X_i^2 - n\bar{X}^2}{n - 1} \tag{9.2}$$

If the data are discrete and have been grouped in a frequency distribution, Equations (9.1) and (9.2) can be modified to provide for much greater computational efficiency. The sample mean can be computed as

$$\bar{X} = \frac{\sum_{j=1}^{k} f_j X_j}{n} \tag{9.3}$$

and the sample variance as

$$S^2 = \frac{\sum_{j=1}^{k} f_j X_j^2 - n\bar{X}^2}{n - 1} \tag{9.4}$$

where k is the number of distinct values of X and f_j is the observed frequency of the value X_j of X.

Example 9.8:   Grouped Data
The data in Table 9.1 can be analyzed to obtain $n = 100$, $f_1 = 12$, $X_1 = 0$, $f_2 = 10$, $X_2 = 1, \ldots$, $\sum_{j=1}^{k} f_j X_j = 364$, and $\sum_{j=1}^{k} = f_j X_j^2 = 2080$. From Equation (9.3),

$$\bar{X} = \frac{364}{100} = 3.64$$

and, from Equation (9.4),

$$S^2 = \frac{2080 - 100(3.64)^2}{99} = 7.63$$

The sample standard deviation S is just the square root of the sample variance. In this case, $S = \sqrt{7.63} = 2.76$. Equations (9.1) and (9.2) would have yielded exactly the same results for $\bar{X}$ and S^2.

It is preferable to use the raw data, if possible, when the values are continuous. However, data sometimes are received after having been placed in class intervals. Then it is no longer possible to obtain the exact sample mean and variance. In such cases, the sample mean and sample variance are approximated from the following equations:

$$\bar{X} \doteq \frac{\sum_{j=1}^{c} f_j m_j}{n} \tag{9.5}$$

and

$$S^2 \doteq \frac{\sum_{j=1}^{c} f_j m_j^2 - n\bar{X}^2}{n - 1} \tag{9.6}$$

where f_j is the observed frequency in the jth class interval, m_j is the midpoint of the jth interval, and c is the number of class intervals.

Example 9.9:   Continuous Data in Class Intervals

Assume that the raw data on component life shown in Example 9.6 either was discarded or was lost. However, the data shown in Table 9.2 are still available. To approximate values for $\bar{X}$ and S^2, Equations (9.5) and (9.6) are used. The following values are created: $f_1 = 23$, $m_1 = 1.5$, $f_2 = 10$, $m_2 = 4.5, \ldots, \sum_{j=1}^{49} f_j m_j = 614$ and $\sum_{j=1}^{49} f_j m_j^2 = 37{,}226.5$. With $n = 50$, $\bar{X}$ is approximated from Equation (9.5) as

$$\bar{X} \doteq \frac{614}{50} = 12.28$$

Then, S^2 is approximated from Equation (9.6) as

$$S^2 \doteq \frac{37{,}226.5 - 50(12.28)^2}{49} = 605.849$$

and

$$S \doteq 24.614$$

Applying Equations (9.1) and (9.2) to the original data in Example 9.6 results in $\bar{X} = 11.894$ and $S = 24.953$. Thus, when the raw data are either discarded or lost, inaccuracies could result.

Table 9.3 Suggested Estimators for Distributions Often Used in Simulation

Distribution	Parameter(s)	Suggested Estimator(s)
Poisson	α	$\widehat{\alpha} = \bar{X}$
Exponential	λ	$\widehat{\lambda} = \dfrac{1}{\bar{X}}$
Gamma	β, θ	$\widehat{\beta}$ (see Table A.9)
		$\widehat{\theta} = \dfrac{1}{\bar{X}}$
Normal	μ, σ^2	$\widehat{\mu} = \bar{X}$
		$\widehat{\sigma}^2 = S^2$ (unbiased)
Lognormal	μ, σ^2	$\widehat{\mu} = \bar{X}$ (after taking ln of the data)
		$\widehat{\sigma}^2 = S^2$ (after taking ln of the data)
Weibull with $\nu = 0$	α, β	$\widehat{\beta}_0 = \dfrac{\bar{X}}{S}$
		$\widehat{\beta}_j = \widehat{\beta}_{j-1} - \dfrac{f(\widehat{\beta}_{j-1})}{f'(\widehat{\beta}_{j-1})}$
		See Equations (9.11) and (9.14) for $f(\widehat{\beta})$ and $f'(\widehat{\beta})$
		Iterate until convergence
		$\widehat{\alpha} = \left(\dfrac{1}{n} \sum\limits_{i=1}^{n} X_i^{\widehat{\beta}} \right)^{1/\widehat{\beta}}$
Beta	β_1, β_2	$\Psi(\widehat{\beta}_1) + \Psi(\widehat{\beta}_1 - \widehat{\beta}_2) = \ln(G_1)$
		$\Psi(\widehat{\beta}_2) + \Psi(\widehat{\beta}_1 - \widehat{\beta}_2) = \ln(G_2)$
		where Ψ is the digamma function,
		$G_1 = \left(\prod_{i=1}^{n} X_i \right)^{1/n}$ and
		$G_2 = \left(\prod_{i=1}^{n} (1 - X_i) \right)^{1/n}$

9.3.2 Suggested Estimators

Numerical estimates of the distribution parameters are needed to reduce the family of distributions to a specific distribution and to test the resulting hypothesis. Table 9.3 contains suggested estimators for distributions often used in simulation, all of which were described in Chapter 5. Except for an adjustment to remove bias in the estimate of σ^2 for the normal distribution, these estimators are the maximum-likelihood estimators based on the raw data. (If the data are in class intervals, these estimators must be modified.) The reader is referred to Fishman [1973] and Law [2007] for parameter estimates for the uniform, binomial, and negative binomial distributions. The triangular distribution is usually employed when no data are available, with the parameters obtained from educated guesses for the minimum, most likely, and maximum possible values; the uniform distribution may also be used in this way if only minimum and maximum values are available.

Examples of the use of the estimators are given in the following paragraphs. The reader should keep in mind that a parameter is an unknown constant, but the estimator is a statistic (or random variable), because it depends on the sample values. To distinguish the two clearly here, if, say, a parameter is denoted by α, the estimator will be denoted by $\widehat{\alpha}$.

Example 9.10:   Exponential Distribution

We will use the exponential distribution to illustrate the maximum likelihood method for obtaining parameter estimators. Recall from Chapter 5, Equation (5.26), that the density function of the exponential distribution with rate parameter λ is

$$f(x) = \lambda e^{-\lambda x}$$

for $x \geq 0$. If $X_1, X_2, \ldots, X_n$ are i.i.d., and each have this exponential distribution, then their joint distribution is

$$f(x_1, x_2, \ldots, x_n) = \lambda e^{-\lambda x_1} \lambda e^{-\lambda x_2} \cdots \lambda e^{-\lambda x_n} = \lambda^n e^{-\lambda \sum_{i=1}^{n} x_i} \qquad (9.7)$$

The joint distribution represents, roughly, the likelihood of any collection of x_i values.

The maximum likelihood method turns this around in the sense that, if we have a sample of data, say $X_1, X_2, \ldots, X_n$, but λ is not known, then the maximum-likelihood estimator (MLE) is the value of λ that maximizes the likelihood of the sample, as measured by Equation (9.7). Thus, the MLE of λ for the exponential distribution maximizes the likelihood function

$$L(\lambda) = \lambda^n e^{-\lambda \sum_{i=1}^{n} X_i}$$

Finding the MLE is often a calculus exercise of maximizing the likelihood function with respect to the unknown parameters. Because the likelihood function is a product of density functions and it is positive, it is sometimes easier to maximize the log of the likelihood function, which in this case is

$$\ln L(\lambda) = n \ln(\lambda) - \lambda \sum_{i=1}^{n} X_i$$

Taking the derivative with respect to λ and setting it equal to 0 we get

$$\frac{n}{\lambda} - \sum_{i=1}^{n} X_i = 0$$

giving the MLE $\widehat{\lambda} = n / \sum_{i=1}^{n} X_i = 1/\bar{X}$ where $\bar{X}$ is the sample mean of the observations.

For example, assuming that the data in Example 9.6 come from an exponential distribution, the MLE of λ is obtained from $\bar{X}$ as

$$\widehat{\lambda} = \frac{1}{\bar{X}} = \frac{1}{11.894} = 0.084 \text{ per day}$$

Example 9.11:   Weibull Distribution

Suppose that a random sample of size n, $X_1, X_2, \ldots, X_n$, has been taken and that the observations are assumed to come from a Weibull distribution. The likelihood function derived by using the pdf given by Equation (5.47) can be shown to be

$$L(\alpha, \beta) = \frac{\beta^n}{\alpha^{\beta n}} \left[\prod_{i=1}^{n} X_i^{(\beta-1)} \right] \exp\left[-\sum_{i=1}^{n} \left(\frac{X_i}{\alpha} \right)^{\beta} \right] \tag{9.8}$$

The maximum-likelihood estimates are those values of $\widehat{\alpha}$ and $\widehat{\beta}$ that maximize $L(\alpha, \beta)$ or, equivalently, maximize $\ln L(\alpha, \beta)$, denoted by $l(\alpha, \beta)$. The maximum value of $l(\alpha, \beta)$ is obtained by taking the partial derivatives $\partial l(\alpha, \beta)/\partial \alpha$ and $\partial l(\alpha, \beta)/\partial \beta$, setting each to zero, and solving the resulting equations, which, after substitution, become

$$f(\beta) = 0 \tag{9.9}$$

and

$$\alpha = \left(\frac{1}{n} \sum_{i=1}^{n} X_i^{\beta} \right)^{1/\beta} \tag{9.10}$$

where

$$f(\beta) = \frac{n}{\beta} + \sum_{i=1}^{n} \ln X_i - \frac{n \sum_{i=1}^{n} X_i^{\beta} \ln X_i}{\sum_{i=1}^{n} X_i^{\beta}} \tag{9.11}$$

The maximum-likelihood estimates, $\widehat{\alpha}$ and $\widehat{\beta}$, are the solutions of Equations (9.9) and (9.10). First, $\widehat{\beta}$ is found via the iterative procedure explained below. Then $\widehat{\alpha}$ is found from Equation (9.10), with $\beta = \widehat{\beta}$.

Equation (9.9) is nonlinear, so it is necessary to use a numerical-analysis technique to solve it. In Table 9.3, a suggested iterative method for computing $\widehat{\beta}$ is given as

$$\widehat{\beta}_j = \widehat{\beta}_{j-1} - \frac{f(\widehat{\beta}_{j-1})}{f'(\widehat{\beta}_{j-1})} \tag{9.12}$$

Equation (9.12) employs Newton's method in reaching $\widehat{\beta}$, where $\widehat{\beta}_j$ is the jth iteration, beginning with an initial estimate for $\widehat{\beta}_0$, given in Table 9.3, as follows:

$$\widehat{\beta}_0 = \frac{\bar{X}}{S} \tag{9.13}$$

If the initial estimate, $\widehat{\beta}_0$, is sufficiently close to the solution $\widehat{\beta}$, then $\widehat{\beta}_j$ approaches $\widehat{\beta}$ as $j \longrightarrow \infty$. In Newton's method, $\widehat{\beta}$ is approached through increments of size $f(\widehat{\beta}_{j-1})/f'(\widehat{\beta}_{j-1})$. Equation (9.11) is used to compute $f(\widehat{\beta}_{j-1})$ and Equation (9.14) is used to compute $f'(\widehat{\beta}_{j-1})$, as follows:

$$f'(\beta) = -\frac{n}{\beta^2} - \frac{n\sum_{i=1}^{n} X_i^{\beta}(\ln X_i)^2}{\sum_{i=1}^{n} X_i^{\beta}} + \frac{n\left(\sum_{i=1}^{n} X_i^{\beta}\ln X_i\right)^2}{\left(\sum_{i=1}^{n} X_i^{\beta}\right)^2} \qquad (9.14)$$

Equation (9.14) can be derived from Equation (9.11) by differentiating $f(\beta)$ with respect to β. The iterative process continues until $f(\widehat{\beta}_j) \doteq 0$, that is, until $|f(\widehat{\beta}_j)| \leq 0.001$.

Consider the data given in Example 9.6. These data concern the failure of electronic components and look to come from an exponential distribution. In Example 9.10, the parameter $\widehat{\lambda}$ was estimated on the hypothesis that the data were from an exponential distribution. If this hypothesis is rejected, an alternative hypothesis is that the data come from a Weibull distribution. The Weibull distribution is suspected because the data pertain to electronic component failures, which occur suddenly.

Equation (9.13) is used to compute $\widehat{\beta}_0$. For the data in Example 9.6, $n = 50$, $\bar{X} = 11.894$, $\bar{X}^2 = 141.467$, and $\sum_{i=1}^{50} X_i^2 = 37{,}575.850$; so S^2 is found by Equation (9.2) to be

$$S^2 = \frac{37{,}578.850 - 50(141.467)}{49} = 622.650$$

and $S = 24.953$. Thus,

$$\widehat{\beta}_0 = \frac{11.894}{24.953} = 0.477$$

To compute $\widehat{\beta}_1$ by using Equation (9.12) requires the calculation of $f(\widehat{\beta}_0)$ and $f'(\widehat{\beta}_0)$ from Equations (9.11) and (9.14). The following additional values are needed: $\sum_{i=1}^{50} X_i^{\widehat{\beta}_0} = 115.125$, $\sum_{i=1}^{50} \ln X_i = 38.294$, $\sum_{i=1}^{50} X_i^{\widehat{\beta}_0}\ln X_i = 292.629$, and $\sum_{i=1}^{50} X_i^{\widehat{\beta}_0}(\ln X_i)^2 = 1057.781$. Thus,

$$f(\widehat{\beta}_0) = \frac{50}{0.477} + 38.294 - \frac{50(292.629)}{115.125} = 16.024$$

and

$$f'(\widehat{\beta}_0) = \frac{-50}{(0.477)^2} - \frac{50(1057.781)}{115.125} + \frac{50(292.629)^2}{(115.125)^2} = -356.110$$

Then, by Equation (9.12),

$$\widehat{\beta}_1 = 0.477 - \frac{16.024}{-356.110} = 0.522$$

After four iterations, $|f(\widehat{\beta}_3)| \leq 0.001$, at which point $\widehat{\beta} \doteq \widehat{\beta}_4 = 0.525$, is the approximate solution to Equation (9.9). Table 9.4 contains the values needed to complete each iteration.

Now, $\widehat{\alpha}$ can be computed from Equation (9.10) with $\beta = \widehat{\beta} = 0.525$, as follows:

$$\widehat{\alpha} = \left[\frac{130.608}{50}\right]^{1/0.525} = 6.227$$

Table 9.4   Iterative Estimation of Parameters of the Weibull Distribution

j	$\widehat{\beta}_j$	$\sum\limits_{i=1}^{50} X_i^{\widehat{\beta}_j}$	$\sum\limits_{i=1}^{50} X_i^{\widehat{\beta}_j} \ln X_i$	$\sum\limits_{i=1}^{50} X_i^{\widehat{\beta}_j} (\ln X_i)^2$	$f(\widehat{\beta}_j)$	$f'(\widehat{\beta}_j)$	$\widehat{\beta}_{j+1}$
0	0.477	115.125	292.629	1057.781	16.024	−356.110	0.522
1	0.522	129.489	344.713	1254.111	1.008	−313.540	0.525
2	0.525	130.603	348.769	1269.547	0.004	−310.853	0.525
3	0.525	130.608	348.786	1269.614	0.000	−310.841	0.525

If $\widehat{\beta}_0$ is sufficiently close to $\widehat{\beta}$, the procedure converges quickly, usually in four to five iterations. However, if the procedure appears to be diverging, try other initial guesses for $\widehat{\beta}_0$; for example, one-half the initial estimate or twice the initial estimate.

The difficult task of estimating parameters for the Weibull distribution by hand emphasizes the value of having software support for input modeling.

Example 9.12:   Poisson Distribution

Assume that the arrival data in Table 9.1 require analysis. By comparison with Figure 5.7, an examination of Figure 9.4 suggests a Poisson distribution assumption with unknown parameter α. From Table 9.3, the estimator of α is $\bar{X}$, which was found in Example 9.8. Thus, $\widehat{\alpha} = 3.64$. Recall that the true mean and variance are equal for the Poisson distribution. In Example 9.8, the sample variance was estimated as $S^2 = 7.63$. However, it should never be expected that the sample mean and the sample variance will be precisely equal, because each is a random variable.

Example 9.13:   Lognormal Distribution

The rates of return on 10 investments in a portfolio are 18.8, 27.9, 21.0, 6.1, 37.4, 5.0, 22.9, 1.0, 3.1, and 8.3 percent. To estimate the parameters of a lognormal model of these data, we first take the natural log of the data and obtain 2.9, 3.3, 3.0, 1.8, 3.6, 1.6, 3.1, 0, 1.1, and 2.1. Then we set $\widehat{\mu} = \bar{X} = 2.3$ and $\widehat{\sigma}^2 = S^2 = 1.3$.

Example 9.14:   Normal Distribution

The parameters of the normal distribution, μ and σ^2, are estimated by $\bar{X}$ and S^2, as shown in Table 9.3. The q-q plot in Example 9.7 leads to a distributional assumption that the installation times are normal. From Equations (9.1) and (9.2), the data in Example 9.7 yield $\widehat{\mu} = \bar{X} = 99.9865$ and $\widehat{\sigma}^2 = S^2 = (0.2832)^2$ seconds^2.

Example 9.15:   Gamma Distribution

The estimator $\widehat{\beta}$ for the gamma distribution is chosen by referring to Table A.9. Table A.9 requires the computation of the quantity $1/M$, where

$$M = \ln \bar{X} - \frac{1}{n} \sum_{i=1}^{n} \ln X_i \tag{9.15}$$

Also, it can be seen in Table 9.3 that $\widehat{\theta}$ is given by

$$\widehat{\theta} = \frac{1}{\bar{X}}$$ (9.16)

In Chapter 5, it was stated that lead time is often gamma-distributed. Suppose that the lead times (in days) associated with 20 orders have been accurately measured as follows:

Order	Lead Time (days)	Order	Lead Time (days)
1	70.292	11	30.215
2	10.107	12	17.137
3	48.386	13	44.024
4	20.480	14	10.552
5	13.053	15	37.298
6	25.292	16	16.314
7	14.713	17	28.073
8	39.166	18	39.019
9	17.421	19	32.330
10	13.905	20	36.547

To estimate $\widehat{\beta}$ and $\widehat{\theta}$, it is first necessary to compute M from Equation (9.15). Here, $\bar{X}$ is found, from Equation (9.1), to be

$$\bar{X} = \frac{564.32}{20} = 28.22$$

Then

$$\ln \bar{X} = 3.34$$

Next

$$\sum_{i=1}^{20} \ln X_i = 63.99$$

Then

$$M = 3.34 - \frac{63.99}{20} = 0.14$$

and

$$1/M = 7.14$$

By interpolation in Table A.9, $\widehat{\beta} = 3.728$. Finally, Equation (9.16) results in

$$\widehat{\theta} = \frac{1}{28.22} = 0.035$$

Example 9.16:   Beta Distribution

The percentage of customers each month who bring in store coupons must be between 0 and 100 percent. Observations at a store for eight months gave the values 25%, 74%, 20%, 32%, 81%, 47%, 31%, and 8%. To fit a beta distribution to these data, we first need to rescale it to the interval [0, 1] by dividing all the values by 100, which yields 0.25, 0.74, 0.20, 0.32, 0.81, 0.47, 0.31, 0.08.

The maximum-likelihood estimators of the parameters β_1, β_2 solve the system of equations shown in Table 9.3. Such equations can be solved by modern symbolic/numerical calculation programs, such as Maple; a Maple procedure for the beta parameters is shown in Figure 9.7. In this case, the solutions are $\widehat{\beta}_1 = 1.47$ and $\widehat{\beta}_2 = 2.16$.

```
betaMLE := proc(X, n)
           local G1, G2, beta1, beta2, eqns, solns;
           G1 := product(X[i], i=1..n)^(1/n);
           G2 := product(1-X[i],i=1..n)^(1/n);
           eqns := {Psi(beta1) - Psi(beta1 + beta2) = ln(G1),
                    Psi(beta2) - Psi(beta1 + beta2) = ln(G2)};
           solns := fsolve(eqns, {beta1=0..infinity, beta2=0..infinity});
           RETURN(solns);
           end;
```

Figure 9.7   Maple procedure to compute the maximum likelihood estimates for the beta distribution parameters.

9.4   Goodness-of-Fit Tests

Hypothesis testing was discussed in Section 7.4 with respect to testing random numbers. In Section 7.4.1, the Kolmogorov–Smirnov test and the chi-square test were introduced. In this section, these two tests are applied to hypotheses about distributional forms of input data.

Goodness-of-fit tests provide helpful guidance for evaluating the suitability of a potential input model; however, there is no single correct distribution in a real application, so you should not be a slave to the verdict of such a test. It is especially important to understand the effect of sample size. If very little data are available, then a goodness-of-fit test is unlikely to reject *any* candidate distribution; but if a lot of data are available, then a goodness-of-fit test will likely reject *all* candidate distributions. Therefore, failing to reject a candidate distribution should be taken as one piece of evidence in favor of that choice, and rejecting an input model as only one piece of evidence against the choice.

9.4.1   Chi-Square Test

One procedure for testing the hypothesis that a random sample of size n of the random variable X follows a specific distributional form is the chi-square goodness-of-fit test. This test formalizes the intuitive idea of comparing the histogram of the data to the shape of the candidate density or mass function. The test is valid for large sample sizes and for both discrete and continuous distributional assumptions, when parameters are estimated by maximum likelihood. The test procedure begins by arranging the n observations into a set of k class intervals or *cells*. The test statistic is given by

$$\chi_0^2 = \sum_{i=1}^{k} \frac{(O_i - E_i)^2}{E_i} \tag{9.17}$$

where O_i is the observed frequency in the ith class interval and E_i is the expected frequency in that class interval. The expected frequency for each class interval is computed as $E_i = np_i$, where p_i is the theoretical, hypothesized probability associated with the ith class interval.

It can be shown that χ_0^2 approximately follows the chi-square distribution with $k - s - 1$ degrees of freedom, where s represents the number of parameters of the hypothesized distribution estimated by the sample statistics. The hypotheses are the following:

> H_0: The random variable, X, conforms to the distributional assumption with the parameter(s) given by the parameter estimate(s).
> H_1: The random variable X does not conform.

The critical value $\chi_{\alpha,k-s-1}^2$ is found in Table A.6. The null hypothesis H_0 is rejected if $\chi_0^2 > \chi_{\alpha,k-s-1}^2$.

When applying the test, if expected frequencies are too small, χ_0^2 will reflect not only the departure of the observed from the expected frequency, but also the smallness of the expected frequency as well. Although there is no general agreement regarding the minimum size of E_i, values of 3, 4, and 5 have been widely used. In Section 7.4.1, when the chi-square test was discussed, the minimum expected frequency 5 was suggested. If an E_i value is too small, it can be combined with expected frequencies in adjacent class intervals. The corresponding O_i values should also be combined, and k should be reduced by 1 for each cell that is combined.

If the distribution being tested is discrete, each value of the random variable should be a class interval, unless it is necessary to combine adjacent class intervals to meet the minimum-expected cell-frequency requirement. For the discrete case, if combining adjacent cells is not required,

$$p_i = p(x_i) = P(X = x_i)$$

Otherwise, p_i is found by summing the probabilities of appropriate adjacent cells.

Table 9.5   Recommendations for Number of Class Intervals for Continuous Data

Sample Size n	Number of Class Intervals k
20	Do not use the chi-square test
50	5 to 10
100	10 to 20
>100	$\sqrt{n}$ to $n/5$

If the distribution being tested is continuous, the class intervals are given by $[a_{i-1}, a_i)$, where a_{i-1} and a_i are the endpoints of the ith class interval. For the continuous case with assumed pdf $f(x)$, or assumed cdf $F(x)$, p_i can be computed as

$$p_i = \int_{a_{i-1}}^{a_i} f(x)\, dx = F(a_i) - F(a_{i-1})$$

For the discrete case, the number of class intervals is determined by the number of cells resulting after combining adjacent cells as necessary. However, for the continuous case, the number of class intervals must be specified. Although there are no general rules to be followed, the recommendations in Table 9.5 are made to aid in determining the number of class intervals for continuous data.

Example 9.17:   Chi-Square Test Applied to Poisson Assumption

In Example 9.12, the vehicle-arrival data presented in Example 9.5 were analyzed. The histogram of the data, shown in Figure 9.4, appeared to follow a Poisson distribution; hence the parameter $\widehat{\alpha} = 3.64$ was found. Thus, the following hypotheses are formed:

H_0: The random variable is Poisson distributed.
H_1: The random variable is not Poisson distributed.

The pmf for the Poisson distribution was given in Equation (5.19):

$$p(x) = \begin{cases} \dfrac{e^{-\alpha}\alpha^x}{x!}, & x = 0, 1, \ldots \\ 0, & \text{otherwise} \end{cases} \qquad (9.18)$$

For $\widehat{\alpha} = 3.64$, the probabilities associated with various values of x are obtained from Equation (9.18):

$p(0) = 0.026$ $p(6) = 0.085$
$p(1) = 0.096$ $p(7) = 0.044$
$p(2) = 0.174$ $p(8) = 0.020$
$p(3) = 0.211$ $p(9) = 0.008$
$p(4) = 0.192$ $p(10) = 0.003$
$p(5) = 0.140$ $p(\geq 11) = 0.001$

Table 9.6   Chi-Square Goodness-of-Fit Test for Example 9.17

x_i	Observed Frequency O_i	Expected Frequency E_i	$\dfrac{(O_i - E_i)^2}{E_i}$
0	12 ⎫ 22	2.6 ⎫ 12.2	⎫ 7.87
1	10 ⎭	9.6 ⎭	⎭
2	19	17.4	0.15
3	17	21.1	0.80
4	10	19.2	4.41
5	8	14.0	2.57
6	7	8.5	0.26
7	5 ⎫	4.4 ⎫	⎫
8	5 ⎪	2.0 ⎪	⎪
9	3 ⎬ 17	0.8 ⎬ 7.6	⎬ 11.62
10	3 ⎪	0.3 ⎪	⎪
≥11	1 ⎭	0.1 ⎭	⎭
	$\overline{100}$	$\overline{100.0}$	$\overline{27.68}$

From this information, Table 9.6 is constructed. The value of E_1 is given by $np_0 = 100(0.026) = 2.6$. In a similar manner, the remaining E_i values are computed. Since $E_1 = 2.6 < 5$, E_1 and E_2 are combined. In that case, O_1 and O_2 are also combined, and k is reduced by one. The last five class intervals are also combined, for the same reason, and k is further reduced by four.

The calculated χ_0^2 is 27.68. The degrees of freedom for the tabulated value of χ^2 is $k - s - 1 = 7 - 1 - 1 = 5$. Here, $s = 1$, since one parameter, $\widehat{\alpha}$, was estimated from the data. At the 0.05 level of significance, the critical value $\chi_{0.05,5}^2$ is 11.1. Thus, H_0 would be rejected at level of significance 0.05. The analyst, therefore, might want to search for a better-fitting model or use the empirical distribution of the data.

9.4.2   Chi-Square Test with Equal Probabilities

If a continuous distributional assumption is being tested, class intervals that are equal in probability rather than equal in width of interval should be used. This has been recommended by a number of authors [Mann and Wald, 1942; Gumbel, 1943; Law, 2007; Stuart, Ord, and Arnold, 1998]. It should be noted that the procedure is not applicable to data collected in class intervals, where the raw data have been discarded or lost.

Unfortunately, there is no easy method for figuring out the probability associated with each interval that would maximize the power for a test of a given size. (The power of a test is defined as the probability of rejecting a false hypothesis.) However, if using equal probabilities, then $p_i = 1/k$. We recommend

$$E_i = np_i \geq 5$$

so, substituting for p_i yields

$$\frac{n}{k} \geq 5$$

and solving for k yields

$$k \leq \frac{n}{5} \qquad (9.19)$$

Equation (9.19) was used in coming up with the recommendations for maximum number of class intervals in Table 9.5.

   If the assumed distribution is normal, exponential, or Weibull, the method described in this section is straightforward. Example 9.18 indicates how the procedure is accomplished for the exponential distribution. If the assumed distribution is gamma (but not Erlang) or certain other distributions, then the computation of endpoints for class intervals is complex and could require numerical integration of the density function. Statistical-analysis software is very helpful in such cases.

Example 9.18:   Chi-Square Test for Exponential Distribution
In Example 9.10, the failure data presented in Example 9.6 were analyzed. The histogram of the data, shown in Figure 9.5, appeared to follow an exponential distribution, so the parameter $\widehat{\lambda} = 1/\bar{X} = 0.084$ was computed. Thus, the following hypotheses are formed:

   H_0: The random variable is exponentially distributed.
   H_1: The random variable is not exponentially distributed.

   In order to perform the chi-square test with intervals of equal probability, the endpoints of the class intervals must be found. Equation (9.19) indicates that the number of intervals should be less than or equal to $n/5$. Here, $n = 50$, and so $k \leq 10$. In Table 9.5, it is recommended that 5 to 10 class intervals be used. Let $k = 8$; then each interval will have probability $p = 0.125$. The endpoints for each interval are computed from the cdf for the exponential distribution, given in Equation (5.28), as follows:

$$F(a_i) = 1 - e^{-\lambda a_i} \qquad (9.20)$$

where a_i represents the endpoint of the ith interval, $i = 1, 2, \ldots, k$. Since $F(a_i)$ is the cumulative area from zero to a_i, $F(a_i) = ip$, so Equation (9.20) can be written as

$$ip = 1 - e^{-\lambda a_i}$$

or

$$e^{-\lambda a_i} = 1 - ip$$

Taking the logarithm of both sides and solving for a_i gives a general result for the endpoints of k equiprobable intervals for the exponential distribution:

$$a_i = -\frac{1}{\lambda} \ln(1 - ip), \quad i = 0, 1, \ldots, k \qquad (9.21)$$

Table 9.7   Chi-Square Goodness-of-Fit Test for Example 9.18

Class Interval	Observed Frequency O_i	Expected Frequency E_i	$\dfrac{(O_i - E_i)^2}{E_i}$
[0, 1.590)	19	6.25	26.01
[1.590, 3.425)	10	6.25	2.25
[3.425, 5.595)	3	6.25	0.81
[5.595, 8.252)	6	6.25	0.01
[8.252, 11.677)	1	6.25	4.41
[11.677, 16.503)	1	6.25	4.41
[16.503, 24.755)	4	6.25	0.81
[24.755, ∞)	6	6.25	0.01
	$\overline{50}$	$\overline{50}$	$\overline{39.6}$

Regardless of the value of λ, Equation (9.21) will always result in $a_0 = 0$ and $a_k = \infty$. With $\widehat{\lambda} = 0.084$ and $k = 8$, a_1 is computed from Equation (9.21) as

$$a_1 = -\frac{1}{0.084}\ln(1 - 0.125) = 1.590$$

Continued application of Equation (9.21) for $i = 2, 3, \dots , 7$ results in $a_2, \dots , a_7$ as 3.425, 5.595, 8.252, 11.677, 16.503, and 24.755. Since $k = 8$, $a_8 = \infty$. The first interval is [0, 1.590), the second interval is [1.590, 3.425), and so on. The expectation is that 0.125 of the observations will fall in each interval. The observations, the expectations, and the contributions to the calculated value of χ_0^2 are shown in Table 9.7. The calculated value of χ_0^2 is 39.6. The degrees of freedom are given by $k - s - 1 = 8 - 1 - 1 = 6$. At $\alpha = 0.05$, the tabulated value of $\chi_{0.05,6}^2$ is 12.6. Since $\chi_0^2 > \chi_{0.05,6}^2$, the null hypothesis is rejected. (The value of $\chi_{0.01,6}^2$ is 16.8, so the null hypothesis would also be rejected at level of significance $\alpha = 0.01$.)

9.4.3   Kolmogorov–Smirnov Goodness-of-Fit Test

The chi-square goodness-of-fit test can accommodate the estimation of parameters from the data with a resultant decrease in the degrees of freedom (one for each parameter estimated). The chi-square test requires that the data be placed in class intervals; in the case of a continuous distributional assumption, this grouping is arbitrary. Changing the number of classes and the interval width affects the value of the calculated and tabulated chi-square. A hypothesis could be accepted when the data are grouped one way, but rejected when they are grouped another way. Also, the distribution of the chi-square test statistic is known only approximately, and the power of the test is sometimes rather low. As a result of these considerations, goodness-of-fit tests other than the chi-square, are desired. The Kolmogorov–Smirnov test formalizes the idea behind examining a *q-q* plot.

The Kolmogorov–Smirnov test was presented in Section 7.4.1 to test for the uniformity of numbers. Both of these uses fall into the category of testing for goodness of fit. Any continuous distributional assumption can be tested for goodness of fit by using the method from Section 7.4.1.

The Kolmogorov–Smirnov test is particularly useful when sample sizes are small and when no parameters have been estimated from the data. When parameter estimates have been made, the critical values in Table A.8 are biased; in particular, they are too conservative. In this context, "conservative" means that the critical values will be too large, resulting in smaller Type I (α) errors than those specified. The exact value of α can be worked out in some instances, as is discussed at the end of this section.

The Kolmogorov–Smirnov test does not take any special tables when an exponential distribution is assumed. The following example indicates how the test is applied in this instance. (Notice that it is not necessary to estimate the parameter of the distribution in this example, so we may use Table A.8.)

Example 9.19: Kolmogorov–Smirnov Test for Exponential Distribution
Suppose that 50 interarrival times (in minutes) are collected over a 100-minute interval (arranged in order of occurrence):

0.44	0.53	2.04	2.74	2.00	0.30	2.54	0.52	2.02	1.89	1.53	0.21
2.80	0.04	1.35	8.32	2.34	1.95	0.10	1.42	0.46	0.07	1.09	0.76
5.55	3.93	1.07	2.26	2.88	0.67	1.12	0.26	4.57	5.37	0.12	3.19
1.63	1.46	1.08	2.06	0.85	0.83	2.44	1.02	2.24	2.11	3.15	2.90
6.58	0.64										

The null hypothesis and its alternate are formed as follows:

H_0: The interarrival times are exponentially distributed.

H_1: The interarrival times are not exponentially distributed.

The data were collected over the interval from 0 to $T = 100$ minutes. It can be shown that, if the underlying distribution of interarrival times $\{T_1, T_2, \dots\}$ is exponential, the arrival times are uniformly distributed on the interval $[0, T]$. The arrival times $T_1, T_1 + T_2, T_1 + T_2 + T_3, \dots, T_1 + \dots + T_{50}$ are obtained by adding interarrival times. The arrival times are then normalized to a $[0, 1]$ interval so that the Kolmogorov–Smirnov test, as presented in Section 7.4.1, can be applied. On a $[0, 1]$ interval, the points will be $[T_1/T, (T_1 + T_2)/T, \dots, (T_1 + \dots + T_{50})/T]$. The resulting 50 data points are as follows:

0.0044	0.0097	0.0301	0.0575	0.0775	0.0805	0.1059	0.1111	0.1313	0.1502
0.1655	0.1676	0.1956	0.1960	0.2095	0.2927	0.3161	0.3356	0.3366	0.3508
0.3553	0.3561	0.3670	0.3746	0.4300	0.4694	0.4796	0.5027	0.5315	0.5382
0.5494	0.5520	0.5977	0.6514	0.6526	0.6845	0.7008	0.7154	0.7262	0.7468
0.7553	0.7636	0.7880	0.7982	0.8206	0.8417	0.8732	0.9022	0.9680	0.9744

Following the procedure in Example 7.6 produces a D^+ of 0.1054 and a D^- of 0.0080. Therefore, the Kolmogorov–Smirnov statistic is $D = \max(0.1054, 0.0080) = 0.1054$. The critical value of D obtained from Table A.8 for a level of significance of $\alpha = 0.05$ and $n = 50$ is $D_{0.05} = 1.36/\sqrt{n} = 0.1923$; however, $D = 0.1054$, so the hypothesis that the interarrival times are exponentially distributed cannot be rejected.

The Kolmogorov–Smirnov test has been modified so that it can be used in several situations where the parameters are estimated from the data. The computation of the test statistic is the same, but different tables of critical values are used. Different tables of critical values are required for different distributional assumptions. Lilliefors [1967] developed a test for normality. The null hypothesis states that the population is one of the family of normal distributions, without specifying the parameters of the distribution. The interested reader might wish to study Lilliefors' original work; he describes how simulation was used to develop the critical values.

Lilliefors [1969] also modified the critical values of the Kolmogorov–Smirnov test for the exponential distribution. Lilliefors again used random sampling to obtain approximate critical values, but Durbin [1975] subsequently obtained the exact distribution. Connover [1998] gives examples of Kolmogorov–Smirnov tests for the normal and exponential distributions. He also refers to several other Kolmogorov–Smirnov-type tests that might be of interest to the reader.

A test that is similar in spirit to the Kolmogorov–Smirnov test is the *Anderson-Darling test*. Like the Kolmogorov–Smirnov test, the Anderson-Darling test is based on the difference between the empirical cdf and the fitted cdf; unlike the Kolmogorov–Smirnov test, the Anderson-Darling test is based on a more comprehensive measure of difference (not just the maximum difference) and is more sensitive to discrepancies in the tails of the distributions. The critical values for the Anderson-Darling test also depend on the candidate distribution and on whether parameters have been estimated. Fortunately, this test and the Kolmogorov–Smirnov test have been implemented in a number of software packages that support simulation-input modeling.

9.4.4   *p*-Values and "Best Fits"

To apply a goodness-of-fit test, a significance level must be chosen. Recall that the significance level is the probability of falsely rejecting H_0: The random variable conforms to the distributional assumption. The traditional significance levels are $0.1, 0.05$, and 0.01. Prior to the availability of high-speed computing, having a small set of standard values made it possible to produce tables of useful critical values. Now, most statistical software computes critical values as needed, rather than storing them in tables. Thus, the analyst can employ a different level of significance, say, 0.07.

However, rather than require a prespecified significance level, many software packages compute a *p-value* for the test statistic. The *p*-value is the significance level at which one would *just reject* H_0 for the given value of the test statistic. Therefore, a large *p*-value tends to indicate a good fit (we would have to accept a large chance of error in order to reject), while a small *p*-value suggests a poor fit (to accept we would have to insist on almost no risk).

Recall Example 9.17, in which a chi-square test was used to check the Poisson assumption for the vehicle-arrival data. The value of the test statistic was $\chi_0^2 = 27.68$, with 5 degrees of freedom. The *p*-value for this test statistic is 0.00004, meaning that we would reject the hypothesis that the data are Poisson at the 0.00004 significance level. (Recall that we rejected the hypothesis at the 0.05 level; now we know that we would also have to reject it at even lower levels.)

The *p*-value can be viewed as a measure of fit, with larger values being better. This suggests that we could fit every distribution at our disposal, compute a test statistic for each fit, and then choose the distribution that yields the largest p-value. We know of no input modeling software that implements this specific algorithm, but many such packages do include a "best fit" option, in which the software recommends an input model to the user after evaluating all feasible models. The software might also take into account other factors—such as whether the data are discrete or continuous, bounded or unbounded—but, in the end, some summary measure of fit, like the p-value, is used to rank the distributions. There is nothing wrong with this, but there are several things to keep in mind:

1. The software might know nothing about the physical basis of the data, whereas that information can suggest distribution families that are appropriate. (See the list in Section 9.2.2.) Remember that the goal of input modeling is often to fill in gaps or smooth the data, rather than to find an input model that conforms as closely as possible to the given sample.

2. Recall that both the Erlang and the exponential distributions are special cases of the gamma and that the exponential is also a special case of the more flexible Weibull. Automated best-fit procedures tend to choose the more flexible distributions (gamma and Weibull over Erlang and exponential) because the extra flexibility allows closer conformance to the data and a better summary measure of fit. But again, close conformance to the data does not always lead to the most appropriate input model.

3. A summary statistic, like the p-value, is just that, a summary measure. It says little or nothing about where the lack of fit occurs (in the body of the distribution, in the right tail, or in the left tail). A human, using graphical tools, can see where the lack of fit occurs and decide whether or not it is important for the application at hand.

Our recommendation is that automated distribution selection be used as one of several ways to suggest candidate distributions. Always inspect the automatic selection, using graphical methods, and remember that the final choice is yours.

9.5 Fitting a Nonstationary Poisson Process

Fitting a nonstationary Poisson process (NSPP) to arrival data is a difficult problem, in general, because we seldom have knowledge about the appropriate form of the arrival rate function $\lambda(t)$. (See Chapter 5, Section 5.5.2 for the definition of an NSPP). One approach is to choose a very flexible model with lots of parameters and fit it with a method such as maximum likelihood; see Johnson, Lee, and Wilson [1994] for an example of this approach. A second method, and the one we consider here, is to approximate the arrival rate as being constant over some basic interval of time, such as an hour, or a day, or a month, but varying from time interval to time interval. The problem then becomes choosing the basic time interval and estimating the arrival rate within each interval.

Suppose we need to model arrivals over a time period, say $[0, T]$. The approach that we describe is most appropriate when it is possible to observe the time period $[0, T]$ repeatedly and count arrivals. For instance, if the problem involves modeling the arrival of e-mail throughout the business day (8 A.M. to 6 P.M.), and we believe that the arrival rate is approximately constant over half-hour intervals, then we need to be able to count arrivals during half-hour intervals for several days. If it is possible to record actual arrival times, rather than counts, then actual arrival times are clearly better since they can later be grouped into any interval lengths we desire. However, we will assume from here on that only counts are available.

Table 9.8   Monday E-mail Arrival Data for NSPP Example

Time Period	Number of Arrivals			Estimated Arrival Rate (arrivals/hour)
	Day 1	Day 2	Day 3	
8:00–8:30	12	14	10	24
8:30–9:00	23	26	32	54
9:00–9:30	27	19	32	52
9:30–10:00	20	13	12	30

Divide the time period $[0, T]$ into k equal intervals of length $\Delta t = T/k$. For instance, if we are considering a 10-hour business day from 8 A.M. to 6 P.M. and if we allow the rate to change every half hour, then $T = 10$, $k = 20$, and $\Delta t = 1/2$. Over n periods of observation (e.g., n days), let C_{ij} be the number of arrivals that occurred during the ith time interval on the jth period of observation. In our example, C_{23} would be the number of arrivals from 8:30 A.M. to 9 A.M. (second half-hour period) on the third day of observation.

The estimated arrival rate during the ith time period, $(i - 1)\Delta t < t \le i\Delta t$, is then just the average number of arrivals scaled by the length of the time interval:

$$\widehat{\lambda}(t) = \frac{1}{n\Delta t} \sum_{j=1}^{n} C_{ij} \tag{9.22}$$

After the arrival rates for each time interval have been estimated, adjacent intervals whose rates appear to be the same can be combined.

For instance, consider the e-mail arrival counts during the first two hours of the business day on three Mondays, shown in Table 9.8. The estimated arrival rate for 8:30–9:00 is

$$\frac{1}{3(1/2)}(23 + 26 + 32) = 54 \text{ arrivals/hour}$$

After seeing these results, we might consider combining the interval 8:30–9:00 with the interval 9:00–9:30, because the rates are so similar. Note also that the goodness-of-fit tests described in the previous section can be applied to the data from each time interval individually, to check the Poisson approximation.

9.6   Selecting Input Models without Data

Unfortunately, it is often necessary in practice to develop a simulation model—perhaps for demonstration purposes or a preliminary study—before any process data are available. In this case, the modeler must be resourceful in choosing input models and must carefully check the sensitivity of results to the chosen models.

There are a number of ways to obtain information about a process even if data are not available:

Engineering data: Often a product or process has performance ratings provided by the manufacturer (for example, the mean time to failure of a disk drive is 10,000 hours; a laser printer can produce 8 pages/minute; the cutting speed of a tool is 1 cm/second; etc.). Company rules might specify time or production standards. These values provide a starting point for input modeling by fixing a central value.

Expert option: Talk to people who are experienced with the process or similar processes. Often, they can provide optimistic, pessimistic, and most-likely times. They might also be able to say whether the process is nearly constant or highly variable, and they might be able to define the source of variability.

Physical or conventional limitations: Most real processes have physical limits on performance; for example, computer data entry cannot be faster than a person can type. Because of company policies, there could be upper limits on how long a process may take. Do not ignore obvious limits or bounds that narrow the range of the input process.

The nature of the process: The descriptions of the distributions in Section 9.2.2 can be used to justify a particular choice even when no data are available.

When data are not available, the uniform, triangular, and beta distributions are often used as input models. The uniform can be a poor choice, because the upper and lower bounds are rarely just as likely as the central values in real processes. If, in addition to upper and lower bounds, a most-likely value can be given, then the triangular distribution can be used. The triangular distribution places much of its probability near the most-likely value, and much less near the extremes. (See Section 5.4.7.) If a beta distribution is used, then be sure to plot the density function of the selected distribution; the beta can take unusual shapes.

A useful refinement is obtained when a minimum, a maximum, and one or more "breakpoints" can be given. A *breakpoint* is an intermediate value together with a probability of being less than or equal to that value. The following example illustrates how breakpoints are used.

Example 9.20 _____

For a production-planning simulation, the sales volume of various products is required. The sales-person responsible for product XYZ-123 says that no fewer than 1000 units will be sold (because of existing contracts) and no more than 5000 units will be sold (because that is the entire market for the product). Given her experience, she believes that there is a 90% chance of selling more than 2000 units, a 25% chance of selling more than 3500 units, and only a 1% chance of selling more than 4500 units.

Table 9.9 summarizes this information. Notice that the chances of exceeding certain sales goals have been translated into the cumulative probability of being less than or equal to those goals. With the information in this form, the method of Section 8.1.5 can be employed to generate simulation-input data.

Table 9.9 Summary of Sales Information

i	Interval (sales)	Cumulative Frequency c_i
1	$1000 \leq x \leq 2000$	0.10
2	$2000 < x \leq 3500$	0.75
3	$3500 < x \leq 4500$	0.99
4	$4500 < x \leq 5000$	1.00

When input models have been selected without data, it is especially important to test the sensitivity of simulation results to the distribution chosen. Check sensitivity not only to the center of the distribution but also to the variability or limits. Extreme sensitivity of output results to the input model provides a convincing argument against making critical decisions based on the results and in favor of undertaking data collection.

9.7 Multivariate and Time-Series Input Models

In Sections 9.1–9.4, the input random variables were considered to be independent of any other variables within the context of the problem. When this is not the case it is critical that input models be used that account for dependence; otherwise, a highly inaccurate simulation may result.

Example 9.21
A supply-chain simulation includes the lead time and annual demand for industrial robots. An increase in demand results in an increase in lead time: The final assembly of the robots must be made according to the specifications of the purchaser. Therefore, rather than treat lead time and demand as independent random variables, a multivariate input model should be developed.

Example 9.22
A simulation of the web-based trading site of a stock broker includes the time between arrivals of orders to buy and sell. Investors tend to react to what other investors are doing, so these buy and sell orders arrive in bursts. Therefore, rather than treat the time between arrivals as independent random variables, a time-series model should be developed.

We distinguish *multivariate* input models of a fixed, finite number of random variables (such as the two random variables *lead time* and *annual demand* in Example 9.21) from *time-series* input models of a (conceptually infinite) sequence of related random variables (such as the successive times between orders in Example 9.22). We will describe input models appropriate for these examples after reviewing two measures of dependence, the covariance and the correlation.

9.7.1   Covariance and Correlation

Let X_1 and X_2 be two random variables, and let $\mu_i = \mathrm{E}(X_i)$ and $\sigma_i^2 = \mathrm{V}(X_i)$ be the mean and variance of X_i, respectively. The *covariance* and *correlation* are measures of the linear dependence between X_1 and X_2. In other words, the covariance and correlation indicate how well the relationship between X_1 and X_2 is described by the model

$$(X_1 - \mu_1) = \beta(X_2 - \mu_2) + \varepsilon$$

where ε is a random variable with mean 0 that is independent of X_2. If, in fact, $(X_1 - \mu_1) = \beta(X_2 - \mu_2)$, then this model is perfect. On the other hand, if X_1 and X_2 are statistically independent, then $\beta = 0$ and the model is of no value. In general, a positive value of β indicates that X_1 and X_2 tend to be above or below their means together; a negative value of β indicates that they tend to be on opposite sides of their means.

The covariance between X_1 and X_2 is defined to be

$$\mathrm{cov}(X_1, X_2) = \mathrm{E}[(X_1 - \mu_1)(X_2 - \mu_2)] = \mathrm{E}(X_1 X_2) - \mu_1 \mu_2 \tag{9.23}$$

The value $\mathrm{cov}(X_1, X_2) = 0$ implies $\beta = 0$ in our model of dependence, and $\mathrm{cov}(X_1, X_2) < 0 \ (>0)$ implies $\beta < 0 \ (>0)$.

The covariance can take any value between $-\infty$ and ∞. The correlation standardizes the covariance to be between -1 and 1:

$$\rho = \mathrm{corr}(X_1, X_2) = \frac{\mathrm{cov}(X_1, X_2)}{\sigma_1 \sigma_2} \tag{9.24}$$

Again, the value $\mathrm{corr}(X_1, X_2) = 0$ implies $\beta = 0$ in our model, and $\mathrm{corr}(X_1, X_2) < 0 \ (>0)$ implies $\beta < 0 \ (>0)$. The closer ρ is to -1 or 1, the stronger the linear relationship is between X_1 and X_2.

Now, suppose that we have a sequence of random variables $X_1, X_2, X_3, \ldots$ that are identically distributed (implying that they all have the same mean and variance), but could be dependent. We refer to such a sequence as a *time series* and to $\mathrm{cov}(X_t, X_{t+h})$ and $\mathrm{corr}(X_t, X_{t+h})$ as the *lag-h autocovariance* and *lag-h autocorrelation*, respectively. If the value of the autocovariance depends only on h and not on t, then we say that the time series is *covariance stationary*; this concept is discussed further in Chapter 11. For a covariance-stationary time series, we use the shorthand notation

$$\rho_h = \mathrm{corr}(X_t, X_{t+h})$$

for the *lag-h* autocorrelation. Notice that autocorrelation measures the dependence between random variables that are separated by $h - 1$ others in the time series.

9.7.2   Multivariate Input Models

If X_1 and X_2 each are normally distributed, then dependence between them can be modeled by the bivariate normal distribution with parameters $\mu_1, \mu_2, \sigma_1^2, \sigma_2^2$, and $\rho = \mathrm{corr}(X_1, X_2)$. Estimation of μ_1, μ_2, σ_1^2, and σ_2^2 was described in Section 9.3.2. To estimate ρ, suppose that we have n independent and identically distributed pairs $(X_{11}, X_{21}), (X_{12}, X_{22}), \ldots, (X_{1n}, X_{2n})$. Then the sample covariance is

$$\widehat{\text{cov}}(X_1, X_2) = \frac{1}{n-1} \sum_{j=1}^{n} (X_{1j} - \bar{X}_1)(X_{2j} - \bar{X}_2)$$

$$= \frac{1}{n-1} \left(\sum_{j=1}^{n} X_{1j}X_{2j} - n\bar{X}_1\bar{X}_2 \right) \tag{9.25}$$

where $\bar{X}_1$ and $\bar{X}_2$ are the sample means. The correlation is estimated by

$$\widehat{\rho} = \frac{\widehat{\text{cov}}(X_1, X_2)}{\widehat{\sigma}_1 \widehat{\sigma}_2} \tag{9.26}$$

where $\widehat{\sigma}_1$ and $\widehat{\sigma}_2$ are the sample variances.

Example 9.23:   Example 9.21 (cont.)

Let X_1 represent the average lead time (in months) to deliver, and X_2 the annual demand, for industrial robots. The following data are available on demand and lead time for the last ten years:

Lead Time	Demand
6.5	103
4.3	83
6.9	116
6.0	97
6.9	112
6.9	104
5.8	106
7.3	109
4.5	92
6.3	96

Standard calculations give $\bar{X}_1 = 6.14$, $\widehat{\sigma}_1 = 1.02$, $\bar{X}_2 = 101.80$, and $\widehat{\sigma}_2 = 9.93$ as estimates of μ_1, σ_1, μ_2, and σ_2, respectively. To estimate the correlation, we need

$$\sum_{j=1}^{10} X_{1j}X_{2j} = 6328.5$$

Therefore, $\widehat{\text{cov}} = [6328.5 - (10)(6.14)(101.80)]/(10 - 1) = 8.66$, and

$$\widehat{\rho} = \frac{8.66}{(1.02)(9.93)} = 0.86$$

Clearly, lead time and demand are strongly dependent. Before we accept this model, however, lead time and demand should be checked individually to see whether they are represented well by normal distributions. In particular, demand is a discrete-valued quantity, so the continuous normal distribution is at best an approximation.

The following simple algorithm can be used to generate bivariate normal random variables:

Step 1. Generate Z_1 and Z_2, independent standard normal random variables (see Section 8.3.1).

Step 2. Set $X_1 = \mu_1 + \sigma_1 Z_1$

Step 3. Set $X_2 = \mu_2 + \sigma_2 \left(\rho Z_1 + \sqrt{1 - \rho^2} Z_2 \right)$

Obviously, the bivariate normal distribution will not be appropriate for all multivariate-input modeling problems. It can be generalized to the k-variate normal distribution to model the dependence among more than two random variables, but, in many instances, a normal distribution is not appropriate in any form. We provide one method for handling nonnormal distributions in Section 9.7.4. Good references for other models are Johnson [1987] and Nelson and Yamnitsky [1998].

9.7.3   Time-Series Input Models

If $X_1, X_2, X_3, \ldots$ is a sequence of identically distributed, but dependent and covariance-stationary random variables, then there are a number of time-series models that can be used to represent the process. We will describe two models that have the characteristic that the autocorrelations take the form

$$\rho_h = \text{corr}(X_t, X_{t+h}) = \rho^h$$

for $h = 1, 2, \ldots$. Notice that the *lag-h* autocorrelation decreases geometrically as the lag increases, so that observations far apart in time are nearly independent. For one model to be shown below, each X_t is normally distributed; for the other model, each X_t is exponentially distributed. More general time-series input models are described in Section 9.7.4 and in Nelson and Yamnitsky [1998].

AR(1) MODEL

Consider the time-series model

$$X_t = \mu + \phi(X_{t-1} - \mu) + \varepsilon_t \tag{9.27}$$

for $t = 2, 3, \ldots$, where $\varepsilon_2, \varepsilon_3, \ldots$ are independent and identically (normally) distributed with mean 0 and variance σ_ε^2, and $-1 < \phi < 1$. If the initial value X_1 is chosen appropriately (see below), then $X_1, X_2, \ldots$ are all normally distributed with mean μ, variance $\sigma_\varepsilon^2/(1 - \phi^2)$, and

$$\rho_h = \phi^h$$

for $h = 1, 2, \ldots$. This time-series model is called the *autoregressive order-1 model*, or AR(1) for short.

Estimation of the parameter ϕ can be obtained from the fact that

$$\phi = \rho^1 = \text{corr}(X_t, X_{t+1})$$

the lag-1 autocorrelation. Therefore, to estimate ϕ, we first estimate the lag-1 autocovariance by

$$\widehat{\text{cov}}(X_t, X_{t+1}) = \frac{1}{n-1} \sum_{t=1}^{n-1} (X_t - \bar{X})(X_{t+1} - \bar{X})$$

$$\doteq \frac{1}{n-1} \left(\sum_{t=1}^{n-1} X_t X_{t+1} - (n-1)\bar{X}^2 \right) \tag{9.28}$$

and the variance $\sigma^2 = \text{var}(X)$ by the usual estimator $\widehat{\sigma}^2$. Then

$$\widehat{\phi} = \frac{\widehat{\text{cov}}(X_t, X_{t+1})}{\widehat{\sigma}^2}$$

Finally, estimate μ and σ_ε^2 by $\widehat{\mu} = \bar{X}$ and $\widehat{\sigma}_\varepsilon^2 = \widehat{\sigma}^2(1 - \widehat{\phi}^2)$, respectively.

The following algorithm generates a stationary AR(1) time series, given values of the parameters ϕ, μ, and σ_ε^2:

Step 1.   Generate X_1 from the normal distribution with mean μ and variance $\sigma_\varepsilon^2/(1-\phi^2)$. Set $t = 2$.

Step 2.   Generate ε_t from the normal distribution with mean 0 and variance σ_ε^2.

Step 3.   Set $X_t = \mu + \phi(X_{t-1} - \mu) + \varepsilon_t$.

Step 4.   Set $t = t + 1$ and go to Step 2.

EAR(1) MODEL

Consider the time-series model

$$X_t = \begin{cases} \phi X_{t-1}, & \text{with probability } \phi \\ \phi X_{t-1} + \varepsilon_t, & \text{with probability } 1 - \phi \end{cases} \tag{9.29}$$

for $t = 2, 3, \ldots$, where $\varepsilon_2, \varepsilon_3, \ldots$ are independent and identically (exponentially) distributed with mean $1/\lambda$ and $0 \le \phi < 1$. If the initial value X_1 is chosen appropriately (see below), then $X_1, X_2, \ldots$ are all exponentially distributed with mean $1/\lambda$ and

$$\rho_h = \phi^h$$

for $h = 1, 2, \ldots$. This time-series model is called the *exponential autoregressive order-1 model*, or EAR(1) for short. Only autocorrelations greater than 0 can be represented by this model. Estimation of the parameters proceeds as for the AR(1) by setting $\widehat{\phi} = \widehat{\rho}$, the estimated lag-1 autocorrelation, and setting $\widehat{\lambda} = 1/\bar{X}$.

The following algorithm generates a stationary EAR(1) time series, given values of the parameters ϕ and λ:

Step 1. Generate X_1 from the exponential distribution with mean $1/\lambda$. Set $t = 2$.

Step 2. Generate U from the uniform distribution on $[0, 1]$. If $U \leq \phi$, then set

$$X_t = \phi X_{t-1}$$

Otherwise, generate ε_t from the exponential distribution with mean $1/\lambda$ and set

$$X_t = \phi X_{t-1} + \varepsilon_t$$

Step 3. Set $t = t + 1$ and go to Step 2.

Example 9.24:   Example 9.22 (cont.) _____

The stock broker would typically have a large sample of data, but, for the sake of illustration, suppose that the following twenty time gaps between customer buy and sell orders had been recorded (in seconds): 1.95, 1.75, 1.58, 1.42, 1.28, 1.15, 1.04, 0.93, 0.84, 0.75, 0.68, 0.61, 11.98, 10.79, 9.71, 14.02, 12.62, 11.36, 10.22, 9.20. Standard calculations give $\bar{X} = 5.2$ and $\hat{\sigma}^2 = 26.7$. To estimate the lag-1 autocorrelation, we need

$$\sum_{j=1}^{19} X_t X_{t+1} = 924.1$$

Thus, $\widehat{\text{cov}} = [924.1 - (20 - 1)(5.2)^2]/(20 - 1) = 21.6$, and

$$\hat{\rho} = \frac{21.6}{26.7} = 0.8$$

Therefore, we could model the interarrival times as an EAR(1) process with $\hat{\lambda} = 1/5.2 = 0.192$ and $\hat{\phi} = 0.8$, provided that an exponential distribution is a good model for the individual gaps.

_____

9.7.4 The Normal-to-Anything Transformation

The bivariate normal distribution and the AR(1) and EAR(1) time-series models are useful input models that are easy to fit and simulate. However, the marginal distribution is either normal or exponential, which is certainly not the best choice for many applications. Fortunately, we can start with a bivariate normal or AR(1) model and *transform* it to have any marginal distributions we want (including exponential).

Suppose we want to simulate a random variable X with cdf $F(x)$. Let Z be a standard normal random variable (mean 0 and variance 1), and let $\Phi(z)$ be its cdf. Then it can be shown that

$$R = \Phi(Z)$$

is a $U(0, 1)$ random variable. As we learned in Chapter 8, if we have a $U(0, 1)$ random variable, we can get X by using the inverse cdf transformation

$$X = F^{-1}[R] = F^{-1}[\Phi(Z)]$$

We refer this as the *normal to anything transformation*, or NORTA for short.

Of course, if all we want is X, then there is no reason to go to this trouble; we can just generate R directly, using the methods in Chapter 8. But suppose we want a bivariate random vector (X_1, X_2) such that X_1 and X_2 are correlated but their distributions are not normal. Then we can start with a bivariate normal random vector (Z_1, Z_2) and apply the NORTA transformation to obtain

$$X_1 = F_1^{-1}[\Phi(Z_1)] \text{ and } X_2 = F_2^{-1}[\Phi(Z_2)]$$

There is not even a requirement that F_1 and F_2 be from the same distribution family; for instance, F_1 could be an exponential distribution and F_2 a beta distribution.

The same idea applies for time series. If Z_t is generated by an AR(1) with $N(0, 1)$ marginals, then

$$X_t = F^{-1}[\Phi(Z_t)]$$

will be a time-series model with marginal distribution $F(x)$. To ensure that Z_t is $N(0, 1)$, we set $\mu = 0$ and $\sigma_\varepsilon^2 = 1 - \phi^2$ in the AR(1) model.

Although the NORTA method is very general, there are two technical issues that must be addressed to implement it:

1. The NORTA approach requires being able to evaluate that standard normal cdf, $\Phi(z)$, and the inverse cdf of the distributions of interest, $F^{-1}(R)$. There is no closed-form expression for $\Phi(z)$ and no closed-form expression for $F^{-1}(R)$ for many distributions. Therefore, numerical approximations are required. Fortunately, these functions are built into many symbolic calculation and spreadsheet programs, and we give one example next. In addition, Bratley, Fox, and Schrage [1987] contains algorithms for many distributions.
2. The correlation between the standard normal random variables (Z_1, Z_2) is distorted when it passes through the NORTA transformation. To be more specific, if (Z_1, Z_2) have correlation ρ, then in general $X_1 = F_1^{-1}[\Phi(Z_1)]$ and $X_2 = F_2^{-1}[\Phi(Z_2)]$ will have a correlation $\rho_X \neq \rho$. The difference is often small, but not always.

The second issue is more critical, because in input-modeling problems we want to specify the bivariate or lag-1 correlation. Thus, we need to find the bivariate normal correlation ρ that gives us the input correlation ρ_X that we want (recall that we specify the time-series model via the lag-1 correlation, $\rho_X = \text{corr}(X_t, X_{t+1})$). There has been much research on this problem, including Cario and Nelson [1996, 1998] and Biller and Nelson [2003]. Fortunately, it has been shown that ρ_X is a nondecreasing function of ρ, and ρ and ρ_X will always have the same sign. Thus, we can do a relatively simple search based on the following algorithm:

Step 1. Set $\rho = \rho_X$ to start.

Step 2. Generate a large number of bivariate normal pairs (Z_1, Z_2) with correlation ρ, and transform them into (X_1, X_2)s, using the NORTA transformation.

Step 3. Compute the sample correlation between (X_1, X_2), using Equation (9.25), and call it $\widehat{\rho}_T$. If $\widehat{\rho}_T > \rho_X$, then reduce ρ and go to Step 2; if $\widehat{\rho}_T < \rho_X$, then increase ρ and go to Step 2. If $\widehat{\rho}_T \approx \rho_X$ then stop.

Example 9.25

Suppose we need X_1 to have an exponential distribution with mean 1, X_2 to have a beta distribution with $\beta_1 = 1$, $\beta_2 = 1/2$, and the two of them to have correlation $\rho_X = 0.45$. Figure 9.8 shows a procedure in Maple that will estimate the required value of ρ. In the procedure, n is the number of sample pairs used to estimate the correlation. Running this procedure with n set to 1000 gives $\rho = 0.52$.

9.8   Summary

Input-data collection and analysis require major time and resource commitments in a discrete-event simulation project. However, regardless of the validity or sophistication of the simulation model, unreliable inputs can lead to outputs whose subsequent interpretation could result in faulty recommendations.

This chapter discussed four steps in the development of models of input data: collecting the raw data, identifying the underlying statistical distribution, estimating the parameters, and testing for goodness of fit.

Some suggestions were given for facilitating the data-collection step. However, experience, such as that obtained by completing any of Exercises 1 through 5, will increase awareness of the difficulty of problems that can arise in data collection and of the need for planning.

Once the data have been collected, a statistical model should be hypothesized. Constructing a histogram is very useful at this point if sufficient data are available. A distribution based on the underlying process and on the shape of the histogram can usually be selected for further investigation.

The investigation proceeds with the estimation of parameters for the hypothesized distribution. Suggested estimators were given for distributions used often in simulation. In a number of instances, these are functions of the sample mean and sample variance.

The last step in the process is the testing of the distributional hypothesis. The q-q plot is a useful graphical method for assessing fit. The chi-square, Kolmogorov–Smirnov, and Anderson-Darling goodness-of-fit tests can be applied to many distributional assumptions. When a distributional assumption is rejected, another distribution is tried. When all else fails, the empirical distribution could be used in the model.

```
NORTARho := proc(rhoX, n)
local Z1, Z2, Ztemp, X1, X2, R1, R2, rho, rhoT, lower, upper;
randomize(123456);
Z1 := [random[normald[0,1]](n)]:
ZTemp := [random[normald[0,1]](n)]:
Z2 := [0]:
# set up bisection search
rho := rhoX:
if (rhoX < 0) then
   lower := -1:
   upper := 0:
else
   lower := 0:
   upper := 1:
fi:
Z2 := rho*Z1 + sqrt(1-rho^2)*ZTemp:
R1 := statevalf[cdf,normald[0,1]](Z1):
R2 := statevalf[cdf,normald[0,1]](Z2):
X1 := statevalf[icdf,exponential[1,0]](R1):
X2 := statevalf[icdf,beta[1,2]](R2):
rhoT := describe[linearcorrelation](X1, X2);
# do bisection search until 5% relative error
while abs(rhoT - rhoX)/abs(rhoX) > 0.05 do
   if (rhoT > rhoX) then
      upper := rho:
   else
      lower := rho:
   fi:
   rho := evalf((lower + upper)/2):
   Z2 := rho*Z1 + sqrt(1-rho^2)*ZTemp:
   R1 := statevalf[cdf,normald[0,1]](Z1):
   R2 := statevalf[cdf,normald[0,1]](Z2):
   X1 := statevalf[icdf,exponential[1,0]](R1):
   X2 := statevalf[icdf,beta[1,2]](R2):
   rhoT := describe[linearcorrelation](X1, X2);
end do;
RETURN(rho);
end;
```

Figure 9.8 Maple procedure to estimate the bivariate normal correlation required for the NORTA method.

Unfortunately, in some situations, a simulation study must be undertaken when there is not time or resources to collect data on which to base input models. When this happens, the analyst must use any available information—such as manufacturer specifications and expert opinion—to construct the input models. When input models are derived without the benefit of data, it is particularly important to examine the sensitivity of the results to the models chosen.

Many, but not all, input processes can be represented as sequences of independent and identically distributed random variables. When inputs should exhibit dependence, then multivariate-input models are required. The bivariate normal distribution (and more generally the multivariate normal distribution) is often used to represent a finite number of dependent random variables. Time-series models are useful for representing a (conceptually infinite) sequence of dependent inputs. The NORTA transformation facilitates developing multivariate-input models with marginal distributions that are not normal.

REFERENCES

BILLER, B., AND B. L. NELSON [2003], "Modeling and Generating Multivariate Time Series with Arbitrary Marginals Using an Autoregressive Technique," *ACM Transactions on Modeling and Computer Simulation*, Vol. 13, pp. 211–237.

BRATLEY, P., B. L. FOX, AND L. E. SCHRAGE [1987], *A Guide to Simulation*, 2d ed., Springer-Verlag, New York.

CARIO, M. C., AND B. L. NELSON [1996], "Autoregressive to Anything: Time-Series Input Processes for Simulation," *Operations Research Letters*, Vol. 19, pp. 51–58.

CARIO, M. C., AND B. L. NELSON [1998], "Numerical Methods for Fitting and Simulating Autoregressive-to-Anything Processes," *INFORMS Journal on Computing*, Vol. 10, pp. 72–81.

CHOI, S. C., AND R. WETTE [1969], "Maximum Likelihood Estimation of the Parameters of the Gamma Distribution and Their Bias," *Technometrics*, Vol. 11, No. 4, pp. 683–890.

CHAMBERS, J. M., W. S. CLEVELAND, AND P. A. TUKEY [1983], *Graphical Methods for Data Analysis*, CRC Press, Boca Raton, FL.

CONNOVER, W. J. [1998], *Practical Nonparametric Statistics*, 3d ed., Wiley, New York.

DURBIN, J. [1975], "Kolmogorov–Smirnov Tests When Parameters Are Estimated with Applications to Tests of Exponentiality and Tests on Spacings," *Biometrika*, Vol. 65, pp. 5–22.

FISHMAN, G. S. [1973], *Concepts and Methods in Discrete Event Digital Simulation*, Wiley, New York.

GUMBEL, E. J. [1943], "On the Reliability of the Classical Chi-squared Test," *Annals of Mathematical Statistics*, Vol. 14, pp. 253ff.

HINES, W. W., D. C. MONTGOMERY, D. M. GOLDSMAN, AND C. M. BORROR [2002], *Probability and Statistics in Engineering and Management Science*, 4th ed., Wiley, New York.

JOHNSON, M. A., S. LEE, AND J. R. WILSON [1994], "NPPMLE and NPPSIM: Software for Estimating and Simulating Nonhomogeneous Poisson Processes Having Cyclic Behavior," *Operations Research Letters*, Vol. 15, pp. 273–282.

JOHNSON, M. E. [1987], *Multivariate Statistical Simulation*, Wiley, New York.

LAW, A. M. [2007], *Simulation Modeling and Analysis*, 4th ed., McGraw-Hill, New York.

LILLIEFORS, H. W. [1967], "On the Kolmogorov–Smirnov Test for Normality with Mean and Variance Unknown," *Journal of the American Statistical Association*, Vol. 62, pp. 339–402.

LILLIEFORS, H. W. [1969], "On the Kolmogorov–Smirnov Test for the Exponential Distribution with Mean Unknown," *Journal of the American Statistical Association*, Vol. 64, pp. 387–389.

MANN, H. B., AND A. WALD [1942], "On the Choice of the Number of Intervals in the Application of the Chi-squared Test," *Annals of Mathematical Statistics*, Vol. 18, p. 50ff.

NELSON, B. L., AND M. YAMNITSKY [1998], "Input Modeling Tools for Complex Problems," in *Proceedings of the 1998 Winter Simulation Conference*, D. Medeiros, E. Watson, J. Carson, and M. Manivannan, eds., Washington, DC, Dec. 13–16, pp. 105–112.

STUART, A., J. K. ORD, AND E. ARNOLD [1998], *Kendall's Advanced Theory of Statistics*, 6th ed., Vol. 2, Oxford University Press, Oxford.

EXERCISES

1. Go to a small store and record the interarrival- and service-time distributions. If there are several workers, how do the service-time distributions compare to one another? Do service-time distributions need to be constructed for each type of appliance? (Make sure that the management gives permission to perform this study.)

2. Go to a cafeteria and collect data on the distributions of interarrival and service times. The distribution of interarrival times is probably different for each of the three daily meals and might also vary during the meal—that is, the interarrival time distribution for 11:00 A.M. to 12:00 noon could be different from that for 12:00 noon to 1:00 P.M. Define service time as the time from when the customer reaches the point at which the first selection could be made until the time of exiting from the cafeteria line. (Any reasonable modification of this definition is acceptable.) The service-time distribution probably changes for each meal. Can times of the day or days of the week for either distribution be grouped to suit the homogeneity of the data? (Make sure that the management gives permission to perform this study.)

3. Go to a major traffic intersection and record the interarrival-time distributions from each direction. Some arrivals want to go straight, some turn left, some turn right. The interarrival-time distribution varies during the day and by day of the week. Every now and then an accident occurs. Develop an input model for your data.

4. Go to a grocery store and construct the interarrival and service distributions at the checkout counters. These distributions might vary by time of day and by day of week. Also record the number of service channels available at all times. (Make sure that the management gives permission to perform this study.)

5. Go to a laundromat and collect data to develop input models for the number of washers and dryers a customer uses (probably dependent). Also collect data on the arrival rate of customers (probably not stationary). (Make sure that the management gives permission to perform this study.)

6. Prepare four theoretical normal density functions, all on the same figure, each distribution having mean zero, but let the standard deviations be 1/4, 1/2, 1, and 2.

7. On one figure, draw the pdfs of the Erlang distribution where $\theta = 1/2$ and $k = 1, 2, 4$, and 8.

8. On one figure, draw the pdfs of the Erlang distribution where $\theta = 2$ and $k = 1, 2, 4$, and 8.

9. Draw the pmf of the Poisson distribution that results when the parameter α is equal to the following:

    (a) $\alpha = 1/2$
    (b) $\alpha = 1$
    (c) $\alpha = 2$
    (d) $\alpha = 4$

10. On one figure, draw the two exponential pdfs that result when the parameter λ equals 0.6 and 1.2.

11. On one figure, draw the three Weibull pdfs that result when $\nu = 0$, $\alpha = 1/2$, and $\beta = 1, 2$, and 4.

12. The following data are generated randomly from a gamma distribution:

1.691	1.437	8.221	5.976
1.116	4.435	2.345	1.782
3.810	4.589	5.313	10.90
2.649	2.432	1.581	2.432
1.843	2.466	2.833	2.361

    Compute the maximum-likelihood estimators $\widehat{\beta}$ and $\widehat{\theta}$.

13. The following data are generated randomly from a Weibull distribution where $\nu = 0$:

7.936	5.224	3.937	6.513
4.599	7.563	7.172	5.132
5.259	2.759	4.278	2.696
6.212	2.407	1.857	5.002
4.612	2.003	6.908	3.326

    Compute the maximum-likelihood estimators $\widehat{\alpha}$ and $\widehat{\beta}$. (This exercise requires a programmable calculator, a computer, or a lot of patience.)

14. The highway between Atlanta, Georgia and Athens, Georgia has a high incidence of accidents along its 100 kilometers. Public safety officers say that the occurrence of accidents along the

highway is randomly (uniformly) distributed, but the news media say otherwise. The Georgia Department of Public Safety published records for the month of September. These records indicated the point at which 30 accidents involving an injury or death occurred, as follows (the data points representing the distance from the city limits of Atlanta):

88.3	40.7	36.3	27.3	36.8
91.7	67.3	7.0	45.2	23.3
98.8	90.1	17.2	23.7	97.4
32.4	87.8	69.8	62.6	99.7
20.6	73.1	21.6	6.0	45.3
76.6	73.2	27.3	87.6	87.2

Use the Kolmogorov–Smirnov test to discover whether the distribution of location of accidents is uniformly distributed for the month of September.

15. Show that the Kolmogorov–Smirnov test statistic for Example 9.19 is $D = 0.1054$.

16. Records pertaining to the monthly number of job-related injuries at an underground coal mine were being studied by a federal agency. The values for the past 100 months were as follows:

Injuries per Month	Frequency of Occurrence
0	35
1	40
2	13
3	6
4	4
5	1
6	1

(a) Apply the chi-square test to these data to test the hypothesis that the underlying distribution is Poisson. Use the level of significance $\alpha = 0.05$.

(b) Apply the chi-square test to these data to test the hypothesis that the distribution is Poisson with mean 1.0. Again let $\alpha = 0.05$.

(c) What are the differences between parts (a) and (b), and when might each case arise?

17. The time required for 50 different employees to compute and record the number of hours worked during the week was measured, with the following results in minutes:

Employee	Time (minutes)	Employee	Time (minutes)
1	1.88	26	0.04
2	0.54	27	1.49
3	1.90	28	0.66
4	0.15	29	2.03
5	0.02	30	1.00
6	2.81	31	0.39
7	1.50	32	0.34
8	0.53	33	0.01
9	2.62	34	0.10
10	2.67	35	1.10
11	3.53	36	0.24
12	0.53	37	0.26
13	1.80	38	0.45
14	0.79	39	0.17
15	0.21	40	4.29
16	0.80	41	0.80
17	0.26	42	5.50
18	0.63	43	4.91
19	0.36	44	0.35
20	2.03	45	0.36
21	1.42	46	0.90
22	1.28	47	1.03
23	0.82	48	1.73
24	2.16	49	0.38
25	0.05	50	0.48

Use the chi-square test (as in Example 9.18) to test the hypothesis that these service times are exponentially distributed. Let the number of class intervals be $k = 6$. Use the level of significance $\alpha = 0.05$.

18. Studentwiser Beer Company is trying to find out the distribution of the breaking strength of their glass bottles. Fifty bottles are selected at random and tested for breaking strength, with the following results (in pounds per square inch):

218.95	232.75	212.80	231.10	215.95
237.55	235.45	228.25	218.65	212.80
230.35	228.55	216.10	229.75	229.00
199.75	225.10	208.15	213.85	205.45
219.40	208.15	198.40	238.60	219.55
243.10	198.85	224.95	212.20	222.90
218.80	203.35	223.45	213.40	206.05
229.30	239.20	201.25	216.85	207.25
204.85	219.85	226.15	230.35	211.45
227.95	229.30	225.25	201.25	216.10

Using input-modeling software, apply as many tests for normality as are available in the software. If the chi-square test is available, apply it with at least two different choices for the number of intervals. Do all of the tests reach the same conclusion?

19. The Crosstowner was a bus that cut a diagonal path from northeast Atlanta to southwest Atlanta. The time required to complete the route was recorded by the bus operator. The bus runs from Monday through Friday. The times of the last fifty 8:00 A.M. runs, in minutes, are as follows:

92.3	92.8	106.8	108.9	106.6
115.2	94.8	106.4	110.0	90.9
104.6	72.0	86.0	102.4	99.8
87.5	111.4	105.9	90.7	99.2
97.8	88.3	97.5	97.4	93.7
99.7	122.7	100.2	106.5	105.5
80.7	107.9	103.2	116.4	101.7
84.8	101.9	99.1	102.2	102.5
111.7	101.5	95.1	92.8	88.5
74.4	98.9	111.9	96.5	95.9

How are these run times distributed? Develop and test a suitable model.

20. The time required for the transmission of a message (in milliseconds) is sampled electronically at a communications center. The last 50 values in the sample are as follows:

7.936	4.612	2.407	4.278	5.132
4.599	5.224	2.003	1.857	2.696
5.259	7.563	3.937	6.908	5.002
6.212	2.759	7.172	6.513	3.326
8.761	4.502	6.188	2.566	5.515
3.785	3.742	4.682	4.346	5.359
3.535	5.061	4.629	5.298	6.492
3.502	4.266	3.129	1.298	3.454
5.289	6.805	3.827	3.912	2.969
4.646	5.963	3.829	4.404	4.924

How are the transmission times distributed? Develop and test an appropriate model.

21. The time (in minutes) between requests for the hookup of electric service was accurately recorded at the Gotwatts Flash and Flicker Company, with the following results for the last 50 requests:

0.661	4.910	8.989	12.801	20.249
5.124	15.033	58.091	1.543	3.624
13.509	5.745	0.651	0.965	62.146
15.512	2.758	17.602	6.675	11.209
2.731	6.892	16.713	5.692	6.636
2.420	2.984	10.613	3.827	10.244
6.255	27.969	12.107	4.636	7.093
6.892	13.243	12.711	3.411	7.897
12.413	2.169	0.921	1.900	0.315
4.370	0.377	9.063	1.875	0.790

How are the times between requests for service distributed? Develop and test a suitable model.

22. Daily demands for transmission overhaul kits for the D-3 dragline were maintained by Earth Moving Tractor Company, with the following results:

0	2	0	0	0
1	0	1	1	1
0	1	0	0	0
2	0	1	0	1
0	1	0	0	2
1	0	1	0	0
0	0	0	0	0
1	0	1	0	1
0	0	3	0	1
1	0	0	0	0

How are the daily demands distributed? Develop and test an appropriate model.

23. A simulation is to be conducted of a job shop that performs two operations: milling and planing, in that order. It would be possible to collect data about processing times for each operation, then generate random occurrences from each distribution. However, the shop manager says that the times might be related; large milling jobs take lots of planing. Data are collected for the next 25 orders, with the following results in minutes:

Order	Milling Time (minutes)	Planing Time (minutes)	Order	Milling Time (minutes)	Planing Time (minutes)
1	12.3	10.6	14	24.6	16.6
2	20.4	13.9	15	28.5	21.2
3	18.9	14.1	16	11.3	9.9
4	16.5	10.1	17	13.3	10.7
5	8.3	8.4	18	21.0	14.0
6	6.5	8.1	19	19.5	13.0
7	25.2	16.9	20	15.0	11.5
8	17.7	13.7	21	12.6	9.9
9	10.6	10.2	22	14.3	13.2
10	13.7	12.1	23	17.0	12.5
11	26.2	16.0	24	21.2	14.2
12	30.4	18.9	25	28.4	19.1
13	9.9	7.7			

(a) Plot milling time on the horizontal axis and planing time on the vertical axis. Do these data seem dependent?
(b) Compute the sample correlation between milling time and planing time.
(c) Fit a bivariate normal distribution to these data.

24. Write a computer program to compute the maximum-likelihood estimators $(\widehat{\alpha}, \widehat{\beta})$ of the Weibull distribution. Inputs to the program should include the sample size n; the observations $x_1, x_2, \ldots, x_n$; a stopping criterion ϵ (stop when $|f(\widehat{\beta_j})| \leq \epsilon$); and a print option OPT (usually set = 0). Output would be the estimates $\widehat{\alpha}$ and $\widehat{\beta}$. If OPT = 1, additional output would be printed, as in Table 9.4, showing convergence. Make the program as "user friendly" as possible.

25. Examine a computer-software library or simulation-support environment to which you have access. Obtain documentation on data-analysis software that would be useful in solving exercises 7 through 24. Use the software as an aid in solving selected problems.

26. The numbers of patrons staying at a small hotel on 20 successive nights was observed to be 20, 14, 21, 19, 14, 18, 21, 25, 27, 26, 22, 18, 13, 18, 18, 18, 25, 23, 20, 21. Fit both an AR(1) and an EAR(1) model to these data. Decide which model provides a better fit by looking at a histogram of the data.

27. The following data represent the time to perform transactions in a bank, measured in minutes: 0.740, 1.28, 1.46, 2.36, 0.354, 0.750, 0.912, 4.44, 0.114, 3.08, 3.24, 1.10, 1.59, 1.47, 1.17, 1.27, 9.12, 11.5, 2.42, 1.77. Develop an input model for these data.

28. Two types of jobs (A and B) are released to the input buffer of a job shop as orders arrive, and the arrival of orders is uncertain. The following data are available from the last week of production:

Day	Number of Jobs	Number of As
1	83	53
2	93	62
3	112	66
4	65	41
5	78	55

    Develop an input model for the number of new arrivals of each type each day.

29. The following data are available on the processing time at a machine (in minutes): 0.64, 0.59, 1.1, 3.3, 0.54, 0.04, 0.45, 0.25, 4.4, 2.7, 2.4, 1.1, 3.6, 0.61, 0.20, 1.0, 0.27, 1.7, 0.04, 0.34. Develop an input model for the processing time.

30. For the arrival-count data shown here, estimate the arrival-rate function $\lambda(t)$ for an NSPP in arrivals per hour:

Time Period	Number of Arrivals			
	Day 1	Day 2	Day 3	Day 4
8:00–10:00	22	24	20	28
10:00–12:00	23	26	32	30
12:00–2:00	40	33	32	38

31. Using the web, research some of the input-modeling software packages mentioned in this chapter (try www.informs-sim.org as a starting place). What are their features? What distributions do they include?

32. As part of a surgical admitting process orderlies are available to escort patients to the pre-surgery preparation area. Patients are not allowed to go by themselves as a matter of policy. A Management Engineer walked the route briskly and found it took 2 minutes to walk to the prep area. Orderlies say that it sometimes takes as long as 7 minutes to escort a patient to the area, but it usually takes 5 minutes. Construct an input model for time to escort a patient based on the information given.

33. At an insertion/sealing/inspection operation electronic components are inserted into a board, the case is assembled and sealed, and the completed unit is tested. The time to accomplish all of this is expected to be 2 minutes for type A12117c assemblies and 3 minutes for the B33433x assemblies. An average deviation of plus-or-minus 20% around these values is considered the standard allowance for variability. Construct an input model for the insertion/sealing/inspection operation based on the information given.

34. The following data are counts of calls arriving to a call center by hour of the day over five Mondays. Does it appear that the process is nonstationary? If so, then develop an NSPP input model for the arrival of calls between 8 A.M. and 4 P.M.

8–9A.M.	9–10A.M.	10–11A.M.	11–12P.M.	12–1P.M.	1–2P.M.	2–3P.M.	3–4P.M.
27	37	58	68	65	33	37	21
27	35	67	93	74	42	39	21
25	39	58	75	82	34	46	19
18	42	48	75	62	51	49	14
29	42	65	88	70	43	45	22

10

Verification, Calibration, and Validation of Simulation Models

One of the most important and difficult tasks facing a model developer is the verification and validation of the simulation model. The engineers and analysts who use the model outputs to aid in making design recommendations and the managers who make decisions based on these recommendations— justifiably look upon a model with some degree of skepticism about its validity. It is the job of the model developer to work closely with the end users throughout the period of development and validation to reduce this skepticism and to increase the model's credibility.

The goal of the validation process is twofold: (a) to produce a model that represents true system behavior closely enough for the model to be used as a substitute for the actual system for the purpose of experimenting with the system, analyzing system behavior, and predicting system performance; and (b) to increase the credibility of the model to an acceptable level, so that the model will be used by managers and other decision makers.

Validation should not be seen as an isolated set of procedures that follows model development, but rather as an integral part of model development. Conceptually, however, the verification and validation process consists of the following components:

1. Verification is concerned with building the model correctly. It proceeds by the comparison of the conceptual model to the computer representation that implements that conception. It asks the questions: Is the model implemented correctly in the simulation software? Are the input parameters and logical structure of the model represented correctly?
2. Validation is concerned with building the correct model. It attempts to confirm that a model is an accurate representation of the real system. Validation is usually achieved through the calibration of the model, an iterative process of comparing the model to actual system behavior and using the discrepancies between the two, and the insights gained, to improve the model. This process is repeated until model accuracy is judged to be acceptable.

This chapter describes methods that have been recommended and used in the verification and validation process. Most of the methods are informal subjective comparisons; a few are formal statistical procedures. The use of the latter procedures involves issues related to output analysis, the subject of Chapters 11 and 12. Output analysis refers to analysis of the data produced by a simulation and to drawing inferences from these data about the behavior of the real system. To summarize their relationship, validation is the process by which model users gain confidence that output analysis is making valid inferences about the real system under study.

Many articles and chapters in texts have been written on verification and validation. For discussion of the main issues, the reader is referred to Balci [1994, 1998, 2003], Carson [1986, 2002], Gass [1983], Kleijnen [1995], Law and Kelton [2000], Naylor and Finger [1967], Oren [1981], Sargent [2003], Shannon [1975], and van Horn [1969, 1971]. For statistical techniques relevant to various aspects of validation, the reader can obtain the foregoing references plus those by Balci and Sargent [1982a,b; 1984a], Kleijnen [1987], and Schruben [1980]. For case studies in which validation is emphasized, the reader is referred to Carson *et al.* [1981a,b], Gafarian and Walsh [1970], Kleijnen [1993], and Shechter and Lucas [1980]. Bibliographies on validation have been published by Balci and Sargent [1984b] and by Youngblood [1993].

10.1   Model Building, Verification, and Validation

The first step in model building consists of observing the real system and the interactions among its various components and of collecting data on their behavior. But observation alone seldom yields sufficient understanding of system behavior. People familiar with the system, or any subsystem, should be questioned, to take advantage of their special knowledge. Operators, technicians, repair and maintenance personnel, engineers, supervisors, and managers understand certain aspects of the system that might be unfamiliar to others. As model development proceeds, new questions may arise, and the model developers will return to this step to learn more about system structure and behavior.

The second step in model building is the construction of a conceptual model—a collection of assumptions about the components and the structure of the system, plus hypotheses about the values of model input parameters. As is illustrated by Figure 10.1, conceptual validation is the comparison of the real system to the conceptual model.

The third step is the implementation of an operational model, usually by using simulation software, and incorporating the assumptions of the conceptual model into the worldview and concepts of the simulation software.

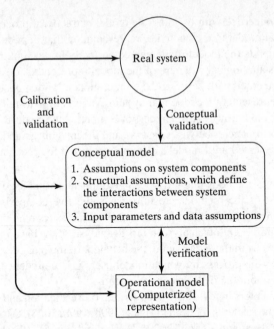

Figure 10.1   Model building, verification, and validation.

In actuality, model building is not a linear process with three steps. Instead, the model builder will return to each of these steps many times while building, verifying, and validating the model. Figure 10.1 depicts the ongoing model building process, in which the need for verification and validation causes continual comparison of the real system to both the conceptual model and the operational model, and induces repeated modification of the model to improve its accuracy.

10.2 Verification of Simulation Models

The purpose of model verification is to assure that the conceptual model is reflected accurately in the operational model. The conceptual model quite often involves some degree of abstraction about system operations or some amount of simplification of actual operations. Verification asks the following question: Is the conceptual model (assumptions about system components and system structure, parameter values, abstractions, and simplifications) accurately represented by the operational model?

Many common-sense suggestions can be used in the verification process:

1. Have the operational model checked by someone other than its developer, preferably an expert in the simulation software being used.
2. Make a flow diagram that includes each logically possible action a system can take when an event occurs, and follow the model logic for each action for each event type. (An example of a logic flow diagram is given in Figures 2.4 and 2.5 for the model of a single-server queue.)

3. Closely examine the model output for reasonableness under a variety of settings of the input parameters. Have the implemented model display a wide variety of output statistics, and examine all of them closely.
4. Have the operational model print the input parameters at the end of the simulation, to be sure that these parameter values have not been changed inadvertently.
5. Make the operational model as self-documenting as possible. Give a precise definition of every variable used and a general description of the purpose of each submodel, procedure (or major section of code), component, or other model subdivision.
6. If the operational model is animated, verify that what is seen in the animation imitates the actual system. Examples of errors that can be observed through animation are automated guided vehicles (AGVs) that pass through one another on a unidirectional path or at an intersection and entities that disappear (unintentionally) during a simulation.
7. The Interactive Run Controller (IRC) or debugger is an essential component of successful simulation model building. Even the best of simulation analysts makes mistakes or commits logical errors when building a model. The IRC assists in finding and correcting those errors in the following ways:
   (a) The simulation can be monitored as it progresses. This can be accomplished by advancing the simulation until a desired time has elapsed, then displaying model information at that time. Another possibility is to advance the simulation until a particular condition is in effect, and then display information.
   (b) Attention can be focused on a particular entity, line of code, or procedure. For instance, every time that an entity enters a specified procedure, the simulation will pause so that information can be gathered. As another example, every time that a specified entity becomes active, the simulation will pause.
   (c) Values of selected model components can be observed. When the simulation has paused, the current value or status of variables, attributes, queues, resources, counters, and so on can be observed.
   (d) The simulation can be temporarily suspended, or paused, not only to view information, but also to reassign values or redirect entities.
8. Graphical interfaces are recommended for accomplishing verification and validation [Bortscheller and Saulnier, 1992]. The graphical representation of the model is essentially a form of self-documentation. It simplifies the task of understanding the model.

These suggestions are basically the same ones any software engineer would follow.

Among these common-sense suggestions, one that is very easily implemented, but quite often overlooked, especially by students who are learning simulation, is a close and thorough examination of model output for reasonableness (suggestion 3). For example, consider a model of a complex network of queues consisting of many service centers in series and parallel configurations. Suppose that the modeler is interested mainly in the response time, defined as the time required for a customer to pass through a designated part of the network. During the verification (and calibration) phase of model development, it is recommended that the program collect and print out many statistics in addition to response times, such as utilizations of servers and time-average number of customers in various subsystems. Examination of the utilization of a server, for example, might reveal that it is unreasonably low (or high), a possible error that could be caused by wrong specification of mean service time, or by a mistake in model logic that sends too few (or too many) customers to this particular server, or by any number of other possible parameter misspecifications or errors in logic.

In a simulation language that automatically collects many standard statistics (average queue lengths, average waiting times, etc.), it takes little or no extra programming effort to display almost all statistics of interest. The effort required can be considerably greater in a general-purpose language such as C, C++, or Java, which do not have statistics-gathering capabilities to aid the programmer.

Two sets of statistics that can give a quick indication of model reasonableness are current contents and total count. These statistics apply to any system having items of some kind flowing through it, whether these items be called customers, transactions, inventory, or vehicles. *Current contents* refers to the number of items in each component of the system at a given time. *Total count* refers to the total number of items that have entered each component of the system by a given time. In some simulation software, these statistics are kept automatically and can be displayed at any point in simulation time. In other simulation software, simple counters might have to be added to the operational model and displayed at appropriate times. If the current contents in some portion of the system are high, this condition indicates that a large number of entities are delayed. If the output is displayed for successively longer simulation run times and the current contents tend to grow in a more or less linear fashion, it is highly likely that a queue is unstable and that the server(s) will fall further behind as time continues. This indicates possibly that the number of servers is too small or that a service time is misspecified. (Unstable queues were discussed in Chapter 6.) On the other hand, if the total count for some subsystem is zero, this indicates that no items entered that subsystem—again, a highly suspect occurrence. Another possibility is that the current count and total count are equal to one. This could indicate that an entity has captured a resource, but never freed that resource. Careful evaluation of these statistics for various run lengths can aid in the detection of mistakes in model logic and data misspecifications. Checking for output reasonableness will usually fail to detect the more subtle errors, but it is one of the quickest ways to discover gross errors. To aid in error detection, it is best for the model developer to forecast a reasonable range for the value of selected output statistics before making a run of the model. Such a forecast reduces the possibility of rationalizing a discrepancy and failing to investigate the cause of unusual output.

For certain models, it is possible to consider more than whether a particular statistic is reasonable. It is possible to compute certain long-run measures of performance. For example, as seen in Chapter 6, the analyst can compute the long-run server utilization for a large number of queueing systems without any special assumptions regarding interarrival or service-time distributions. Typically, the only information needed is the network configuration, plus arrival and service rates. Any measure of performance that can be computed analytically and then compared to its simulated counterpart provides another valuable tool for verification. Presumably, the objective of the simulation is to estimate some measure of performance, such as mean response time, that cannot be computed analytically; but, as illustrated by the formulas in Chapter 6 for a number of special queues ($M/M/1$, $M/G/1$, etc.), all the measures of performance in a queueing system are interrelated. Thus, if a simulation model is predicting one measure (such as utilization) correctly, then confidence in the model's predictive ability for other related measures (such as response time) is increased (even though the exact relation between the two measures is, of course, unknown in general and varies from model to model). Conversely, if a model incorrectly predicts utilization, its prediction of other quantities, such as mean response time, is highly suspect.

Another important way to aid the verification process is the oft-neglected documentation phase. If a model builder writes brief comments in the operational model, plus definitions of all variables and parameters, plus descriptions of each major section of the operational model, it becomes much simpler for someone else, or the model builder at a later date, to verify the model logic. Documentation is also important as a means of clarifying the logic of a model and verifying its completeness.

A more sophisticated technique is the use of a trace. In general, a trace provides detailed, time-stamped simulation output representing the values for a selected set of system states, entity attributes, and model variables whenever an event occurs. That is, at the time an event occurs, the model would write detailed status information to a file for later examination by the model developer to assist in model verification and the detection of modeling errors

Example 10.1

When verifying the operational model (in a general-purpose language such as C, C++ or Java, or most simulation languages) of the single-server queue model of Example 2.5, an analyst made a run over 16 units of time and observed that the time-average length of the waiting line was $\hat{L}_Q = 0.4375$ customer, which is certainly reasonable for a short run of only 16 time units. Nevertheless, the analyst decided that a more detailed verification would be of value.

The trace in Figure 10.2 gives the hypothetical trace output from simulation time CLOCK = 0 to CLOCK = 16 for the simple single-server queue of Example 2.5. This example illustrates how an error can be found with a trace, when no error was apparent from the examination of the summary output statistics (such as $\hat{L}_Q$). Note that, at simulation time CLOCK = 3, the number of customers in the system is NCUST = 1, but the server is idle (STATUS = 0). The source of this error could be incorrect logic, or simply not setting the attribute STATUS to the value 1 (when coding in a general-purpose language or most simulation languages).

In any case, the error must be found and corrected. Note that the less sophisticated practice of examining the summary measures, or output, did not detect the error. By using equation (6.1), the reader can verify that $\hat{L}_Q$ was computed correctly from the data ($\hat{L}_Q$ is the time-average value of NCUST minus STATUS):

$$\hat{L}_Q = \frac{(0-0)3 + (1-0)2 + (0-0)6 + (1-0)1 + (2-1)4}{3+2+6+1+4}$$
$$= \frac{7}{16} = 0.4375$$

as previously mentioned. Thus, the output measure $\hat{L}_Q$ had a reasonable value and was computed correctly from the data, but its value was indeed wrong because the attribute STATUS was not assuming correct values. As is seen from Figure 10.2, a trace yields information on the actual history of the model that is more detailed and informative than the summary measures alone.

Most simulation software has a built-in capability to conduct a trace without the programmer having to do any extensive programming. In addition, a 'print' or 'write' statement can be used to implement a tracing capability in a general-purpose language.

As can be easily imagined, a trace over a large span of simulation time can quickly produce an extremely large amount of output, which would be extremely cumbersome to check in detail for correctness. The purpose of the trace is to verify the correctness of the computer program by making detailed paper-and-pencil calculations. To make this practical, a simulation with a trace is usually

Definition of Variables:
CLOCK   = Simulation clock
EVTYP   = Event type (start, arrival, departure, or stop)
NCUST   = Number of customers in system at time given by CLOCK
STATUS = Status of server (1–busy, 0–idle)

State of System Just After the Named Event Occurs:

CLOCK = 0	EVTYP = 'Start'	NCUST = 0	STATUS = 0
CLOCK = 3	EVTYP = 'Arrival'	NCUST = 1	STATUS = 0
CLOCK = 5	EVTYP = 'Depart'	NCUST = 0	STATUS = 0
CLOCK = 11	EVTYP = 'Arrival'	NCUST = 1	STATUS = 0
CLOCK = 12	EVTYP = 'Arrival'	NCUST = 2	STATUS = 1
CLOCK = 16	EVTYP = 'Depart'	NCUST = 1	STATUS = 1

Figure 10.2   Simulation Trace for Example 2.5.

restricted to a very short period of time. It is desirable, of course, to ensure that each type of event (such as Arrival) occurs at least once, so that its consequences and effect on the model can be checked for accuracy. If an event is especially rare in occurrence, it may be necessary to use artificial data to force it to occur during a simulation of short duration. This is legitimate, as the purpose is to verify that the effect on the system of the rare event is as intended.

Some software allows a selective trace. For example, a trace could be set for specific locations in the model or could be triggered to begin at a specified simulation time. Whenever an entity goes through the designated locations, the simulation software writes a time-stamped message to a trace file. Some simulation software allows for tracing a selected entity; any time the designated entity becomes active, the trace is activated and time-stamped messages are written. This trace is very useful in following one entity through the entire model. Another example of a selective trace is to set it for the occurrence of a particular condition. For example, whenever the queue before a certain resource reaches five or more, turn on the trace. This allows running the simulation until something unusual occurs, then examining the behavior from that point forward in time. Different simulation software packages support tracing to various extents. In practice, it is often implemented by the model developer by adding printed messages at appropriate points into a model.

Of the three classes of techniques—the common-sense techniques, thorough documentation, and traces—it is recommended that the first two always be carried out. Close examination of model output for reasonableness is especially valuable and informative. A generalized trace may provide voluminous data, far more than can be used or examined carefully. A selective trace can provide useful information on key model components and keep the amount of data to a manageable level.

10.3   Calibration and Validation of Models

Calibration and validation, although conceptually distinct, usually are conducted simultaneously by the modeler. Validation is the *overall process* of comparing the model and its behavior to the real system and its behavior. Calibration is the *iterative process* of comparing the model to the real system, making adjustments (or even major changes) to the model, comparing the revised model to reality, making additional adjustments, comparing again, and so on. Figure 10.3 shows the relationship of model calibration to the overall validation process. The comparison of the model to reality is carried out by a variety of tests—some subjective, others objective. *Subjective tests* usually involve people, who are knowledgeable about one or more aspects of the system, making judgments about the model and its output. *Objective tests* always require data on the system's behavior, plus the corresponding data produced by the model. Then one or more statistical tests are performed to compare some aspect of the system data set with the same aspect of the model data set. This iterative process of comparing model with system and then revising both the conceptual and operational models to accommodate any perceived model deficiencies is continued until the model is judged to be sufficiently accurate.

A possible criticism of the calibration phase, were it to stop at this point, is that the model has been validated only for the one data set used—that is, the model has been "fitted" to one data set. One way to alleviate this criticism is to collect a new set of system data (or to reserve a portion of the original system data) to be used at this final stage of validation. That is, after the model has been calibrated by using the original system data set, a "final" validation is conducted, using the second system data set. If unacceptable discrepancies between the model and the real system are discovered in the "final" validation effort, the modeler must return to the calibration phase and modify the model until it becomes acceptable.

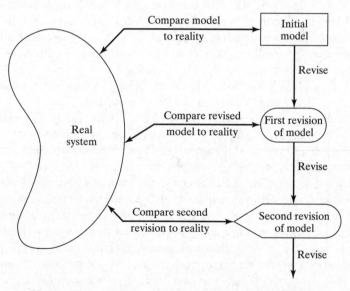

Figure 10.3 Iterative process of calibrating a model.

Validation is not an either/or proposition—no model is ever totally representative of the system under study. In addition, each revision of the model, as pictured in Figure 10.3, involves some cost, time, and effort. The modeler must weigh the possible, but not guaranteed, increase in model accuracy versus the cost of increased validation effort. Usually, the modeler (and model users) have some maximum discrepancy between model predictions and system behavior that would be acceptable. If this level of accuracy cannot be obtained within the budget constraints, either expectations of model accuracy must be lowered, or the model must be abandoned.

As an aid in the validation process, Naylor and Finger [1967] formulated a three-step approach that has been widely followed:

Step 1.　Build a model that has high face validity.

Step 2.　Validate model assumptions.

Step 3.　Compare the model input-output transformations to corresponding input-output transformations for the real system.

The next five subsections investigate these three steps in detail.

10.3.1　Face Validity

The first goal of the simulation modeler is to construct a model that appears reasonable on its face to model users and others who are knowledgeable about the real system being simulated. The potential users of a model should be involved in model construction from its conceptualization to its implementation, to ensure that a high degree of realism is built into the model through reasonable assumptions regarding system structure and through reliable data. Potential users and knowledgeable people can also evaluate model output for reasonableness and can aid in identifying model deficiencies. Thus, the users can be involved in the calibration process as the model is improved iteratively by the insights gained from identification of the initial model deficiencies. Another advantage of user involvement is the increase in the model's perceived validity, or credibility, without which a manager would not be willing to trust simulation results as a basis for decision making.

Sensitivity analysis can also be used to check a model's face validity. The model user is asked whether the model behaves in the expected way when one or more input variables is changed. For example, in most queueing systems, if the arrival rate of customers (or demands for service) were to increase, it would be expected that utilizations of servers, lengths of lines, and delays would tend to increase (although by how much might well be unknown). From experience and from observations on the real system (or similar related systems), the model user and model builder would probably have some notion at least of the direction of change in model output when an input variable is increased or decreased. For most large-scale simulation models, there are many input variables and thus many possible sensitivity tests. The model builder must attempt to choose the most critical input variables for testing if it is too expensive or time consuming to vary all input variables. If real system data are available for at least two settings of the input parameters, objective scientific sensitivity tests can be conducted via appropriate statistical techniques.

10.3.2   Validation of Model Assumptions

Model assumptions fall into two general classes: structural assumptions and data assumptions. *Structural assumptions* involve questions of how the system operates and usually involve simplifications and abstractions of reality. For example, consider the customer queueing and service facility in a bank. Customers can form one line, or there can be an individual line for each teller. If there are many lines, customers could be served strictly on a first-come-first-served basis, or some customers could change lines if one line is moving faster. The number of tellers could be fixed or variable. These structural assumptions should be verified by actual observation during appropriate time periods and by discussions with managers and tellers regarding bank policies and actual implementation of these policies.

Data assumptions should be based on the collection of reliable data and correct statistical analysis of the data. For example, in the bank study previously mentioned, data were collected on

1. interarrival times of customers during several 2-hour periods of peak loading ("rush-hour" traffic);
2. interarrival times during a slack period;
3. service times for commercial accounts;
4. service times for personal accounts.

The reliability of the data was verified by consultation with bank managers, who identified typical rush hours and typical slack times. When combining two or more data sets collected at different times, data reliability can be further enhanced by objective statistical tests for homogeneity of data. (Do two data sets $\{X_i\}$ and $\{Y_i\}$ on service times for personal accounts, collected at two different times, come from the same parent population? If so, the two sets can be combined.) Additional tests might be required, to test for correlation in the data. As soon as the analyst is assured of dealing with a random sample (i.e., correlation is not present), the statistical analysis can begin.

The procedures for analyzing input data from a random sample were discussed in detail in Chapter 9. Whether done manually or by special-purpose software, the analysis consists of three steps:

Step 1.   Identify an appropriate probability distribution.

Step 2.   Estimate the parameters of the hypothesized distribution.

Step 3.   Validate the assumed statistical model by a goodness-of-fit test, such as the chi-square or Kolmogorov–Smirnov test, and by graphical methods.

The use of goodness-of-fit tests is an important part of the validation of data assumptions.

10.3.3   Validating Input-Output Transformations

The ultimate test of a model, and in fact the only objective test of the model as a whole, is the model's ability to predict the future behavior of the real system when the model input data match the real inputs and when a policy implemented in the model is implemented at some point in the system. Furthermore, if the level of some input variables (e.g., the arrival rate of customers to a service facility) were to increase or decrease, the model should accurately predict what would happen in the real system under similar circumstances. In other words, the structure of the model should be accurate enough for the model to make good predictions, not just for one input data set, but for the range of input data sets that are of interest.

In this phase of the validation process, the model is viewed as an input-output transformation—that is, the model accepts values of the input parameters and transforms these inputs into output measures of performance. It is this correspondence that is being validated.

Instead of validating the model input-output transformations by predicting the future, the modeler could use historical data that have been reserved for validation purposes only—that is, if one data set has been used to develop and calibrate the model, it is recommended that a separate data set be used as the final validation test. Thus, accurate "prediction of the past" can replace prediction of the future for the purpose of validating the model.

A model is usually developed with primary interest in a specific set of system responses to be measured under some range of input conditions. For example, in a queueing system, the responses may be server utilization and customer delay, and the range of input conditions (or input variables) may include two or three servers at some station and a choice of scheduling rules. In a production system, the response may be throughput (i.e., production per hour), and the input conditions may be a choice of several machines that run at different speeds, with each machine having its own breakdown and maintenance characteristics.

In any case, the modeler should use the main responses of interest as the primary criteria for validating a model. If the model is used later for a purpose different from its original purpose, the model should be revalidated in terms of the new responses of interest and under the possibly new input conditions.

A necessary condition for the validation of input-output transformations is that some version of the system under study exist, so that system data under at least one set of input conditions can be collected to compare to model predictions. If the system is in the planning stages and no system operating data can be collected, complete input-output validation is not possible. Other types of validation should be conducted, to the extent possible. In some cases, subsystems of the planned system may exist, and a partial input-output validation can be conducted.

Presumably, the model will be used to compare alternative system designs or to investigate system behavior under a range of new input conditions. Assume for now that some version of the system is operating and that the model of the existing system has been validated. What, then, can be said about the validity of the model when different inputs are used? That is, if model inputs are being changed to represent a new system design, or a new way to operate the system, or even hypothesized future conditions, what can be said about the validity of the model with respect to this new but nonexistent proposed system or to the system under new input conditions?

First, the responses of the two models under similar input conditions will be used as the criteria for comparison of the existing system to the proposed system. Validation increases the modeler's confidence that the model of the existing system is accurate. Second, in many cases, the proposed system is a modification of the existing system, and the modeler hopes that confidence in the model of the existing system can be transferred to the model of the new system. This transfer of confidence usually can be justified if the new model is a relatively minor modification of the old model in terms of changes to the operational model (it may be a major change for the actual system). Changes in the operational model ranging from relatively minor to relatively major include the following:

1. minor changes of single numerical parameters, such as the speed of a machine, the arrival rate of customers (with no change in distributional form of interarrival times), the number of servers in a parallel service center, or the mean time to failure or mean time to repair of a machine;

2. minor changes in the form of a statistical distribution, such as the distribution of a service time or a time to failure of a machine;

3. major changes in the logical structure of a subsystem, such as a change in queue discipline for a waiting-line model or a change in the scheduling rule for a job-shop model;

4. major changes involving a different design for the new system, such as a computerized inventory control system replacing an older noncomputerized system, or an automated storage-and-retrieval system replacing a warehouse system in which workers manually pick items using fork trucks.

If the change to the operational model is minor, such as in items 1 or 2, these changes can be carefully verified and output from the new model accepted with considerable confidence. If a sufficiently similar subsystem exists elsewhere, it might be possible to validate the submodel that represents the subsystem and then to integrate this submodel with other validated submodels to build a complete model. In this way, partial validation of the substantial model changes in items 3 and 4 might be possible. Unfortunately, there is no way to validate the input-output transformations of a model of a nonexisting system completely. In any case, within time and budget constraints, the modeler should use as many validation techniques as possible, including input-output validation of subsystem models if operating data can be collected on such subsystems.

Example 10.2 will illustrate some of the techniques that are possible for input-output validation and will discuss the concepts of an input variable, uncontrollable variable, decision variable, output or response variable, and input-output transformation in more detail.

Example 10.2:   The Fifth National Bank of Jaspar

The Fifth National Bank of Jaspar, as shown in Figure 10.4, is planning to expand its drive-in service at the corner of Main Street. Currently, there is one drive-in window serviced by one teller. Only one or two transactions are allowed at the drive-in window, so it was assumed that each service time was a random sample from some underlying population. Service times $\{S_i, i = 1, 2, \ldots, 90\}$ and interarrival times $\{A_i, i = 1, 2, \ldots, 90\}$ were collected for the 90 customers who arrived between 11:00 A.M. and 1:00 P.M. on a Friday. This time slot was selected for data collection after consultation with management and the teller because it was felt to be representative of a typical rush hour.

Data analysis (as outlined in Chapter 9) led to the conclusion that arrivals could be modeled as a Poisson process at a rate of 45 customers per hour and that service times were approximately

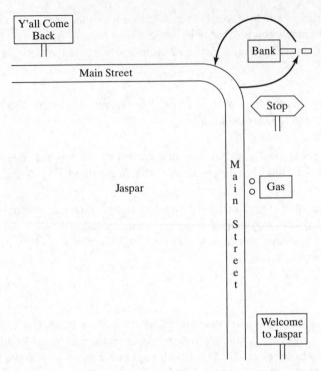

Figure 10.4   Drive-in window at the Fifth National Bank.

normally distributed, with mean 1.1 minutes and standard deviation 0.2 minute. Thus, the model has two input variables:

1. interarrival times, exponentially distributed (i.e., a Poisson arrival process) at rate $\lambda = 45$ per hour;
2. service times, assumed to be $N(1.1, (0.2)^2)$.

Each input variable has a level: the rate $\lambda = 45$ per hour for the interarrival times, and the mean 1.1 minutes and standard deviation 0.2 minute for the service times. The interarrival times are examples of uncontrollable variables (i.e., uncontrollable by management in the real system). The service times are also treated as uncontrollable variables, although the level of the service times might be partially controllable. If the mean service time could be decreased to 0.9 minute by installing a computer terminal, the level of the service-time variable becomes a decision variable or controllable parameter. Setting all decision variables at some level constitutes a policy. For example, the current bank policy is one teller $D_1 = 1$, mean service time $D_2 = 1.1$ minutes, and one line for waiting cars $D_3 = 1$. ($D_1, D_2, \ldots$ are used to denote decision variables.) Decision variables are under management's control; the uncontrollable variables, such as arrival rate and actual arrival times, are not under management's control. The arrival rate might change from time to time, but such change is treated as being due to external factors not under management control.

Table 10.1   Input and Output Variables for Model of Current Bank Operations

Input Variables	*Model Output Variables, Y*
D = decision variables	Variables of primary interest
X = other variables	to management (Y_1, Y_2, Y_3)
	Y_1 = teller's utilization
Poisson arrivals at rate = 45/hour	Y_2 = average delay
$X_{11}, X_{12}, \ldots$	Y_3 = maximum line length
Service times, $N(D_2, 0.2^2)$	Other output variables of
$X_{21}, X_{22}, \ldots$	secondary interest
	Y_4 = observed arrival rate
$D_1 = 1$ (one teller)	Y_5 = average service time
$D_2 = 1.1$ minutes (mean service time)	Y_6 = sample standard deviation of service
$D_3 = 1$ (one line)	times
	Y_7 = average length of waiting line

A model of current bank operations was developed and verified in close consultation with bank management and employees. Model assumptions were validated, as discussed in Section 10.3.2. The resulting model is now viewed as a "black box" that takes all input-variable specifications and transforms them into a set of output or response variables. The output variables consist of all statistics of interest generated by the simulation about the model's behavior. For example, management is interested in the teller's utilization at the drive-in window (percent of time the teller is busy at the window), average delay in minutes of a customer from arrival to beginning of service, and the maximum length of the line during the rush hour. These input and output variables are shown in Figure 10.5 and are listed in Table 10.1, together with some additional output variables. The uncontrollable input variables are denoted by X, the decision variables by D, and the output variables by Y. From the "black box" point of view, the model takes the inputs X and D and produces the outputs Y, namely

$$(X, D) \xrightarrow{f} Y$$

or

$$f(X, D) = Y$$

Here f denotes the transformation that is due to the structure of the model. For the Fifth National Bank study, the exponentially distributed interarrival time generated in the model (by the methods of Chapter 8) between customer $n - 1$ and customer n is denoted by X_{1n}. (Do not confuse X_{1n} with A_n; the latter was an observation made on the real system.) The normally distributed service time generated in the model for customer n is denoted by X_{2n}. The set of decision variables, or policy, is $D = (D_1, D_2, D_3) = (1, 1.1, 1)$ for current operations. The output, or response, variables are denoted by $Y = (Y_1, Y_2, \ldots, Y_7)$ and are defined in Table 10.1.

For validation of the input-output transformations of the bank model to be possible, real system data must be available, comparable to at least some of the model output Y of Table 10.1. The system responses should have been collected during the same time period (from 11:00 A.M. to 1:00 P.M. on

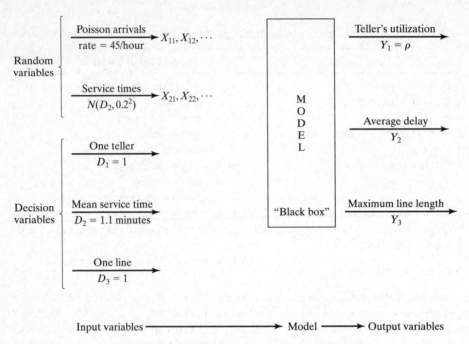

Figure 10.5   Model input-output transformation.

the same Friday) in which the input data $\{A_i, S_i\}$ were collected. This is important because, if system response data were collected on a slower day (say, an arrival rate of 40 per hour), the system responses such as teller utilization Z_1, average delay Z_2, and maximum line length Z_3 would be expected to be lower than the same variables during a time slot when the arrival rate was 45 per hour, as observed. Suppose that the delay of successive customers was measured on the same Friday between 11:00 A.M. and 1:00 P.M. and that the average delay was found to be $Z_2 = 4.3$ minutes. For the purpose of validation, we will consider this to be the true mean value $\mu_0 = 4.3$.

When the model is run with generated random variates X_{1n} and X_{2n}, it is expected that observed values of average delay Y_2 should be close to $Z_2 = 4.3$ minutes. The generated input values X_{1n} and X_{2n} cannot be expected to replicate the actual input values A_n and S_n of the real system exactly, but they are expected to replicate the statistical pattern of the actual inputs. Hence, simulation-generated values of Y_2 are expected to be consistent with the observed system variable $Z_2 = 4.3$ minutes. Now consider how the modeler might test this consistency.

The modeler makes a small number of statistically independent replications of the model. Statistical independence is guaranteed by using nonoverlapping sets of random numbers produced by the random-number generator or by choosing seeds for each replication independently (from a random number table). The results of six independent replications, each of 2 hours duration, are given in Table 10.2.

Observed arrival rate Y_4 and sample average service time Y_5 for each replication of the model are also noted, to be compared with the specified values of 45/hour and 1.1 minutes, respectively. The validation test consists of comparing the system response, namely average delay $Z_2 = 4.3$ minutes, to the model responses Y_2. Formally, a statistical test of the null hypothesis

$$H_0 : E(Y_2) = 4.3 \text{ minutes}$$

versus   (10.1)

$$H_1 : E(Y_2) \neq 4.3 \text{ minutes}$$

is conducted. If H_0 is not rejected, then, on the basis of this test, there is no reason to consider the model invalid. If H_0 is rejected, the current version of the model is rejected, and the modeler is forced to seek ways to improve the model, as illustrated by Figure 10.3. As formulated here, the appropriate statistical test is the t test, which is conducted in the following manner:

Choose a level of significance α and a sample size n. For the bank model, choose

$$\alpha = 0.05, \quad n = 6$$

Compute the sample mean $\bar{Y}_2$ and the sample standard deviation S over the n replications, by using Equations (9.1) and (9.2):

$$\bar{Y}_2 = \frac{1}{n} \sum_{i=1}^{n} Y_{2i} = 2.51 \text{ minutes}$$

and

$$S = \left[\frac{\sum_{i=1}^{n} (Y_{2i} - \bar{Y}_2)^2}{n-1} \right]^{1/2} = 0.82 \text{ minute}$$

where $Y_{2i}, i = 1, \dots, 6$, are as shown in Table 10.2.

Get the critical value of t from Table A.5. For a two-sided test, such as that in equation (10.1), use $t_{\alpha/2,n-1}$; for a one-sided test, use $t_{\alpha,n-1}$ or $-t_{\alpha,n-1}$, as appropriate ($n-1$ being the degrees of freedom). From Table A.5, $t_{0.025,5} = 2.571$ for a two-sided test.

Table 10.2   Results of Six Replications of the Bank Model

Replication	Y_4 (arrivals/hour)	Y_5 (minutes)	$Y_2 =$ Average Delay (minutes)
1	51	1.07	2.79
2	40	1.12	1.12
3	45.5	1.06	2.24
4	50.5	1.10	3.45
5	53	1.09	3.13
6	49	1.07	2.38
Sample mean			2.51
Standard deviation			0.82

Compute the test statistic

$$t_0 = \frac{\bar{Y}_2 - \mu_0}{S/\sqrt{n}} \tag{10.2}$$

where μ_0 is the specified value in the null hypothesis H_0. Here $\mu_0 = 4.3$ minutes, so that

$$t_0 = \frac{2.51 - 4.3}{0.82/\sqrt{6}} = -5.34$$

For the two-sided test, if $|t_0| > t_{\alpha/2,n-1}$, reject H_0. Otherwise, do not reject H_0. [For the one-sided test with $H_1 : E(Y_2) > \mu_0$, reject H_0 if $t > t_{\alpha,n-1}$; with $H_1 : E(Y_2) < \mu_0$, reject H_0 if $t < -t_{\alpha,n-1}$.]

Since $|t| = 5.34 > t_{0.025,5} = 2.571$, reject H_0, and conclude that the model is inadequate in its prediction of average customer delay.

Recall that, in the testing of hypotheses, rejection of the null hypothesis H_0 is a strong conclusion, because

$$P(H_0 \text{ rejected } |H_0 \text{ is true}) = \alpha \tag{10.3}$$

and the level of significance α is chosen small, say $\alpha = 0.05$, as was done here. Equation (10.3) says that the probability is low ($\alpha = 0.05$) of making the error of rejecting H_0 when H_0 is in fact true —that is, the probability is small of declaring the model invalid when it is valid (with respect to the variable being tested). The assumptions justifying a t test are that the observations Y_{2i} are normally and independently distributed. Are these assumptions met in the present case?

1. The ith observation Y_{2i} is the average delay of all drive-in customers who began service during the ith simulation run of 2 hours; thus, by a Central Limit Theorem effect, it is reasonable to assume that each observation Y_{2i} is approximately normally distributed, provided that the number of customers it is based on is not too small.
2. The observations $Y_{2i}, i = 1, \ldots , 6$, are statistically independent by design—that is, by choosing the random-number seeds independently for each replication or by using nonoverlapping streams.
3. The t statistic computed by Equation (10.2) is a robust statistic—that is, it is distributed approximately as the t distribution with $n - 1$ degrees of freedom, even when $Y_{21}, Y_{22}, \ldots$ are not exactly normally distributed, and thus the critical values in Table A.5 can reliably be used.

Now that the model of the Fifth National Bank of Jaspar has been found lacking, what should the modeler do? Upon further investigation, the modeler realized that the model contained two unstated assumptions:

1. When a car arrived to find the window immediately available, the teller began service immediately.
2. There is no delay between one service ending and the next beginning, when a car is waiting.

Assumption 2 was found to be approximately correct, because a service time was considered to begin when the teller actually began service but was not considered to have ended until the car had exited the drive-in window and the next car, if any, had begun service, or the teller saw that the line was empty. On the other hand, assumption 1 was found to be incorrect because the teller had other duties—mainly, serving walk-in customers if no cars were present—and tellers always finished with a previous customer before beginning service on a car. It was found that walk-in customers were always present during rush hour; that the transactions were mostly commercial in nature, taking a considerably longer time than the time required to service drive-up customers; and that, when an arriving car found no other cars at the window, it had to wait until the teller finished with the present walk-in customer. To correct this model inadequacy, the structure of the model was changed to include the additional demand on the teller's time, and data were collected on service times of walk-in customers. Analysis of these data found that they were approximately exponentially distributed with a mean of 3 minutes.

The revised model was run, yielding the results in Table 10.3. A test of the null hypothesis $H_0 : E(Y_2) = 4.3$ minutes was again conducted, according to the procedure previously outlined.

Choose $\alpha = 0.05$ and $n = 6$ (sample size).

Compute $\bar{Y}_2 = 4.78$ minutes, $S = 1.66$ minutes.

Look up, in Table A.5, the critical value $t_{0.25,5} = 2.571$.

Compute the test statistic $t_0 = (\bar{Y}_2 - \mu_0)/(S/\sqrt{n}) = 0.710$.

Since $|t_0| < t_{0.025,5} = 2.571$, do not reject H_0, and thus tentatively accept the model as valid.

Failure to reject H_0 must be considered as a weak conclusion unless the power of the test has been estimated and found to be high (close to 1)—that is, it can be concluded only that the data at hand ($Y_{21}, \ldots , Y_{26}$) were not sufficient to reject the hypothesis $H_0 : \mu_0 = 4.3$ minutes. In other words, this test detects no inconsistency between the sample data ($Y_{21}, \ldots , Y_{26}$) and the specified mean μ_0.

The power of a test is the probability of detecting a departure from $H_0 : \mu = \mu_0$ when, in fact, such a departure exists. In the validation context, the power of the test is the probability of detecting an invalid model. The power may also be expressed as 1 minus the probability of a Type II, or β,

Table 10.3   Results of Six Replications of the Revised Bank Model

Replication	Y_4 (arrivals/hour)	Y_5 (minutes)	$Y_2 = $ Average Delay (minutes)
1	51	1.07	5.37
2	40	1.11	1.98
3	45.5	1.06	5.29
4	50.5	1.09	3.82
5	53	1.08	6.74
6	49	1.08	5.49
Sample mean			4.78
Standard deviation			1.66

error, where $\beta = P(\text{Type II error}) = P(\text{failing to reject } H_0 | H_1 \text{ is true})$ is the probability of accepting the model as valid when it is not valid.

To consider failure to reject H_0 as a strong conclusion, the modeler would want β to be small. Now, β depends on the sample size n and on the true difference between $E(Y_2)$ and $\mu_0 = 4.3$ minutes—that is, on

$$\delta = \frac{|E(Y_2) - \mu_0|}{\sigma}$$

where σ, the population standard deviation of an individual Y_{2i}, is estimated by S. Tables A.10 and A.11 are typical operating-characteristic (OC) curves, which are graphs of the probability of a Type II error $\beta(\delta)$ versus δ for given sample size n. Table A.10 is for a two-sided t test; Table A.11 is for a one-sided t test. Suppose that the modeler would like to reject H_0 (model validity) with probability at least 0.90 if the true mean delay of the model, $E(Y_2)$, differed from the average delay in the system, $\mu_0 = 4.3$ minutes, by 1 minute. Then δ is estimated by

$$\hat{\delta} = \frac{|E(Y_2) - \mu_0|}{S} = \frac{1}{1.66} = 0.60$$

For the two-sided test with $\alpha = 0.05$, use of Table A.10 results in

$$\beta(\hat{\delta}) = \beta(0.6) = 0.75 \text{ for } n = 6$$

To guarantee that $\beta(\hat{\delta}) \leq 0.10$, as was desired by the modeler, Table A.10 reveals that a sample size of approximately $n = 30$ independent replications would be required—that is, for a sample size $n = 6$ and assuming that the population standard deviation is 1.66, the probability of accepting H_0 (model validity), when, in fact, the model is invalid ($|E(Y_2) - \mu_0| = 1$ minute), is $\beta = 0.75$, which is quite high. If a 1-minute difference is critical, and if the modeler wants to control the risk of declaring the model valid when model predictions are as much as 1 minute off, a sample size of $n = 30$ replications is required to achieve a power of 0.9. If this sample size is too high, either a higher β risk (lower power) or a larger difference δ must be considered.

In general, it is always best to control the Type II error, or β error, by specifying a critical difference δ and choosing a sample size by making use of an appropriate OC curve. (Computation of power and use of OC curves for a wide range of tests is discussed in Hines, Montgomery, Goldsman, and Borror [2002].) In summary, in the context of model validation, the Type I error is the rejection of a valid model and is easily controlled by specifying a small level of significance α, say, $\alpha = 0.1$, 0.05, or 0.01. The Type II error is the acceptance of a model as valid when it is invalid. For a fixed sample size n, increasing α will decrease β, the probability of a Type II error. Once α is set, and the critical difference to be detected is selected, the only way to decrease β is to increase the sample size. A Type II error is the more serious of the two types of errors; thus, it is important to design the simulation experiments to control the risk of accepting an invalid model. The two types of error are summarized in Table 10.4, which compares statistical terminology to modeling terminology.

Note that validation is not to be viewed as an either/or proposition, but rather should be viewed in the context of calibrating a model, as conceptually exhibited in Figure 10.3. If the current version of the bank model produces estimates of average delay Y_2 that are not close enough to real system behavior ($\mu_0 = 4.3$ minutes), the source of the discrepancy is sought, and the model is revised in light of this new knowledge. This iterative scheme is repeated until model accuracy is judged adequate.

Table 10.4 Types of Error in Model Validation

Statistical Terminology	Modeling Terminology	Associated Risk
Type I: rejecting H_0 when H_0 is true	Rejecting a valid model	α
Type II: failure to reject H_0 when H_1 is true	Failure to reject an invalid model	β

Philosophically, the hypothesis-testing approach tries to evaluate whether the simulation and the real system are *the same* with respect to some output performance measure or measures. A different, but closely related, approach is to attempt to evaluate whether the simulation and the real-system performance measures are *close enough* by using confidence intervals.

We continue to assume that there is a known output performance measure for the existing system, denoted by μ_0, and an unknown performance measure of the simulation, μ, that we hope is close. The hypothesis-testing formulation tested whether $\mu = \mu_0$; the confidence-interval formulation tries to bound the difference $|\mu - \mu_0|$ to see whether it is $\leq \varepsilon$, a difference that is small enough to allow valid decisions to be based on the simulation. The value of ε is set by the analyst.

Specifically, if Y is the simulation output and $\mu = E(Y)$, then we execute the simulation and form a confidence interval for μ, such as $\bar{Y} \pm t_{\alpha/2,n-1} S/\sqrt{n}$. The determination of whether to accept the model as valid or to refine the model depends on the best-case and worst-case error implied by the confidence interval.

1. Suppose the confidence interval does not contain μ_0. [See Figure 10.6(a).]
   (a) If the best-case error is $> \varepsilon$, then the difference in performance is large enough, even in the best case, to indicate that we need to refine the simulation model.
   (b) If the worst-case error is $\leq \varepsilon$, then we can accept the simulation model as close enough to be considered valid.
   (c) If the best-case error is $\leq \varepsilon$, but the worst-case error is $> \varepsilon$, then additional simulation replications are necessary to shrink the confidence interval until a conclusion can be reached.
2. Suppose the confidence interval does contain μ_0. [See Figure 10.6(b).]
   (a) If either the best-case or worst-case error is $> \varepsilon$, then additional simulation replications are necessary to shrink the confidence interval until a conclusion can be reached.
   (b) If the worst-case error is $\leq \varepsilon$, then we can accept the simulation model as close enough to be considered valid.

In Example 10.2, $\mu_0 = 4.3$ minutes, and "close enough" was $\varepsilon = 1$ minute of expected customer delay. A 95% confidence interval, based on the 6 replications in Table 10.2, is

$$\bar{Y} \pm t_{0.025,5} S/\sqrt{n}$$

$$2.51 \pm 2.571(0.82/\sqrt{6})$$

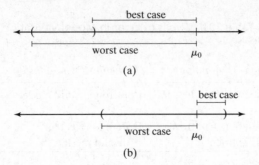

Figure 10.6 Validation of the input-output transformation (a) when the known value falls outside, and (b) when the known value falls inside, the confidence interval.

yielding the interval $[1.65, 3.37]$. As in Figure 10.6(a), $\mu_0 = 4.3$ falls outside the confidence interval. Since in the best case $|3.37 - 4.3| = 0.93 < 1$, but in the worst case $|1.65 - 4.3| = 2.65 > 1$, additional replications are needed to reach a decision.

10.3.4 Input-Output Validation: Using Historical Input Data

When using artificially generated data as input data, as was done to test the validity of the bank models in Section 10.3.3, the modeler expects the model to produce event patterns that are compatible with, but not identical to, the event patterns that occurred in the real system during the period of data collection. Thus, in the bank model, artificial input data $\{X_{1n}, X_{2n}, n = 1, 2, \ldots\}$ for interarrival and service times were generated, and replicates of the output data Y_2 were compared to what was observed in the real system by means of the hypothesis test stated in equation (10.1). An alternative to generating input data is to use the actual historical record, $\{A_n, S_n, n = 1, 2, \ldots\}$, to drive the simulation model and then to compare model output with system data.

To implement this technique for the bank model, the data $A_1, A_2, \ldots$ and $S_1, S_2, \ldots$ would have to be entered into the model into arrays, or stored in a file to be read as the need arose. Just after customer n arrived at time $t_n = \sum_{i=1}^{n} A_i$, customer $n + 1$ would be scheduled on the future event list to arrive at future time $t_n + A_{n+1}$ (without any random numbers being generated). If customer n were to begin service at time t'_n, a service completion would be scheduled to occur at time $t'_n + S_n$ This event scheduling without random-number generation could be implemented quite easily in a general-purpose programming language or most simulation languages by using arrays to store the data or reading the data from a file.

When using this technique, the modeler hopes that the simulation will duplicate as closely as possible the important events that occurred in the real system. In the model of the Fifth National Bank of Jaspar, the arrival times and service durations will exactly duplicate what happened in the real system on that Friday between 11:00 A.M. and 1:00 P.M. If the model is sufficiently accurate, then the delays of customers, lengths of lines, utilizations of servers, and departure times of customers predicted by the model will be close to what actually happened in the real system. It is, of course, the model-builder's and model-user's judgment that determines the level of accuracy required.

To conduct a validation test using historical input data, it is important that all the input data $(A_n, S_n, \ldots)$ and all the system response data, such as average delay (Z_2), be collected during the same time period. Otherwise, the comparison of model responses to system responses, such as the comparison of average delay in the model (Y_2) to that in the system (Z_2), could be misleading. The responses Y_2 and Z_2 depend both on the inputs A_n and S_n and on the structure of the system (or model). Implementation of this technique could be difficult for a large system, because of the need for simultaneous data collection of all input variables and those response variables of primary interest. In some systems, electronic counters and devices are used to ease the data-collection task by automatically recording certain types of data. The following example was based on two simulation models reported in Carson *et al.* [1981a, b], in which simultaneous data collection and the subsequent validation were both completed successfully.

Example 10.3:   The Candy Factory

The production line at the Sweet Lil' Things Candy Factory in Decatur consists of three machines that make, package, and box their famous candy. One machine (the candy maker) makes and wraps individual pieces of candy and sends them by conveyor to the packer. The second machine (the packer) packs the individual pieces into a box. A third machine (the box maker) forms the boxes and supplies them by conveyor to the packer. The system is illustrated in Figure 10.7.

Each machine is subject to random breakdowns due to jams and other causes. These breakdowns cause the conveyor to begin to empty or fill. The conveyors between the two makers and the packer are used as a temporary storage buffer for in-process inventory. In addition to the randomly occurring breakdowns, if the candy conveyor empties, a packer runtime is interrupted and the packer remains idle until more candy is produced. If the box conveyor empties because of a long random breakdown of the box machine, an operator manually places racks of boxes onto the packing machine. If a conveyor fills, the corresponding maker becomes idle. The purpose of the model is to investigate the frequency of those operator interventions that require manual loading of racks of boxes as a function of various combinations of individual machines and lengths of conveyor. Different machines have

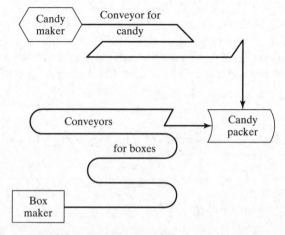

Figure 10.7   Production line at the candy factory.

different production speeds and breakdown characteristics, and longer conveyors can hold more in-process inventory. The goal is to hold operator interventions to an acceptable level while maximizing production. Machine stoppages (whether due to a full or an empty conveyor) cause damage to the product, so this is also a factor in production.

A simulation model of the candy factory was developed, and a validation effort using historical inputs was conducted. Engineers in the candy factory set aside a 4-hour time slot from 7:00 A.M. to 11:00 A.M. to collect data on an existing production line. For each machine—say, machine i—time to failure and downtime duration

$$T_{i1}, D_{i1}, T_{i2}, D_{i2} \ldots$$

were collected. For machine $i (i = 1, 2, 3)$, T_{ij} is the jth runtime (or time to failure), and D_{ij} is the successive downtime. A runtime T_{ij} can be interrupted by a full or empty conveyor (as appropriate), but resumes when conditions are right. Initial system conditions at 7:00 A.M. were recorded so that they could be duplicated in the model as initial conditions at time 0. Additionally, system responses of primary interest—the production level Z_1, and the number Z_2 and time of occurrence Z_3 of operator interventions—were recorded for comparison with model predictions.

The system input data T_{ij} and D_{ij} were fed into the model and used as runtimes and random downtimes. The structure of the model determined the occurrence of shutdowns due to a full or empty conveyor and the occurrence of operator interventions. Model response variables Y_i, $i = 1, 2, 3$ were collected for comparison to the corresponding system response variables Z_i, $i = 1, 2, 3$.

The closeness of model predictions to system performance aided the engineering staff considerably in convincing management of the validity of the model. These results are shown in Table 10.5. A simple display such as Table 10.5 can be quite effective in convincing skeptical engineers and managers of a model's validity—perhaps more effectively than the most sophisticated statistical methods!

With only one set of historical input and output data, only one set of simulated output data can be obtained, and thus no simple statistical tests based on summary measures are possible. However, if K historical input data sets are collected, and K observations $Z_{i1}, Z_{i2}, \ldots, Z_{iK}$ of some system response variable Z_i are collected, such that the output measure Z_{ij} corresponds to the jth input set, an objective statistical test becomes possible. For example, Z_{ij} could be the average delay of all customers who were served during the time the jth input data set was collected. With the K input data sets in hand, the modeler now runs the model K times, once for each input set, and observes the simulated results $W_{i1}, W_{i2}, \ldots, W_{iK}$ corresponding to $Z_{ij}, j = 1, \ldots, K$. Continuing the same example, W_{ij} would be

Table 10.5   Validation of the Candy-Factory Model

Response i	System Z_i	Model Y_i
1. Production level	897,208	883,150
2. Number of operator interventions	3	3
3. Time of occurrence	7:22, 8:41, 10:10	7:24, 8:42, 10:14

Table 10.6 Comparison of System and Model Output Measures for Identical Historical Inputs

Input Data Set	System Output Z_{ij}	Model Output W_{ij}	Observed Difference d_j	Squared Deviation from Mean $(d_j - \bar{d})^2$
1	Z_{i1}	W_{i1}	$d_1 = Z_{i1} - W_{i1}$	$(d_1 - \bar{d})^2$
2	Z_{i2}	W_{i2}	$d_2 = Z_{i2} - W_{i2}$	$(d_2 - \bar{d})^2$
3	Z_{i3}	W_{i3}	$d_3 = Z_{i3} - W_{i3}$	$(d_3 - \bar{d})^2$
.	.	.	.	.
.	.	.	.	.
.	.	.	.	.
K	Z_{iK}	W_{iK}	$d_K = Z_{iK} - W_{iK}$	$(d_K - \bar{d})^2$

$$\bar{d} = \frac{1}{K} \sum_{j=1}^{K} d_j \qquad S_d^2 = \frac{1}{K-1} \sum_{j=1}^{K} (d_j - \bar{d})^2$$

the average delay predicted by the model for the jth input set. The data available for comparison appears as in Table 10.6.

If the K input data sets are fairly homogeneous, it is reasonable to assume that the K observed differences $d_j = Z_{ij} - W_{ij}$, $j = 1, \ldots, K$, are identically distributed. Furthermore, if the collection of the K sets of input data was separated in time—say, on different days—it is reasonable to assume that the K differences $d_1, \ldots, d_K$ are statistically independent and, hence, that the differences $d_1, \ldots, d_K$ constitute a random sample. In many cases, each Z_i and W_i is a sample average over customers, and so (by the Central Limit Theorem) the differences $d_j = Z_{ij} - W_{ij}$ are approximately normally distributed with some mean μ_d and variance σ_d^2. The appropriate statistical test is then a t test of the null hypothesis of no mean difference:

$$H_0 : \mu_d = 0$$

versus the alternative of significant difference:

$$H_1 : \mu_d \neq 0$$

The proper test is a paired t test (Z_{i1} is paired with W_{i1}, each having been produced by the first input data set, and so on). First, compute the sample mean difference, $\bar{d}$, and the sample variance, S_d^2, by the formulas given in Table 10.6. Then, compute the t statistic as

$$t_0 = \frac{\bar{d} - \mu_d}{S_d / \sqrt{K}} \tag{10.4}$$

(with $\mu_d = 0$), and get the critical value $t_{\alpha/2, K-1}$ from Table A.5, where α is the prespecified significance level and $K - 1$ is the number of degrees of freedom. If $|t_0| > t_{\alpha/2, K-1}$, reject the hypothesis H_0 of no mean difference, and conclude that the model is inadequate. If $|t_0| \leq t_{\alpha/2, K-1}$, do not reject H_0, and hence conclude that this test provides no evidence of model inadequacy.

Example 10.4:   The Candy Factory (cont.) _____

Engineers at the Sweet Lil' Things Candy Factory decided to expand the initial validation effort reported in Example 10.3. Electronic devices were installed that could automatically monitor one of the production lines, and the validation effort of Example 10.3 was repeated with $K = 5$ sets of input data. The system and the model were compared on the basis of production level. The results are shown in Table 10.7.

A paired t test was conducted to test $H_0 : \mu_d = 0$, or equivalently, $H_0 : E(Z_1) = E(W_1)$, where Z_1 is the system production level and W_1 is the production level predicted by the simulated model. Let the level of significance be $\alpha = 0.05$. Using the results in Table 10.7, the test statistic, as given by equation (10.4), is

$$t_0 = \frac{\bar{d}}{S_d/\sqrt{K}} = \frac{5343.2}{8705.85/\sqrt{5}} = 1.37$$

From Table A.5, the critical value is $t_{\alpha/2,K-1} = t_{0.025,4} = 2.78$. Since $|t_0| = 1.37 < t_{0.025,4} = 2.78$, the null hypothesis cannot be rejected on the basis of this test—that is, no inconsistency is detected between system response and model predictions in terms of mean production level. If H_0 had been rejected, the modeler would have searched for the cause of the discrepancy and revised the model, in the spirit of Figure 10.3.

Table 10.7   Validation of the Candy-Factory Model (cont.)

Input Data Set j	System Production Z_{1j}	Model Production W_{1j}	Observed Difference d_j	Squared Deviation from Mean $(d_j - \bar{d})^2$
1	897,208	883,150	14,058	7.594×10^7
2	629,126	630,550	$-1,424$	4.580×10^7
3	735,229	741,420	$-6,191$	1.330×10^8
4	797,263	788,230	9,033	1.362×10^7
5	825,430	814,190	11,240	3.4772×10^7
			$\bar{d} = 5,343.2$	$S_d^2 = 7.580 \times 10^7$

10.3.5   Input-Output Validation: Using a Turing Test

In addition to statistical tests, or when no statistical test is readily applicable, personnel knowledgeable about system behavior can be used to compare model output to system output. For example, suppose that five reports of system performance over five different days are prepared, and simulation output data are used to produce five "fake" reports. The 10 reports should all be in exactly the same format and should contain information of the type that managers and engineers have previously seen on the system. The 10 reports are randomly shuffled and given to the engineer, who is asked to decide which reports are fake and which are real. If the engineer identifies a substantial number of the fake reports, the model builder questions the engineer and uses the information gained to improve the model. If the engineer cannot distinguish between fake and real reports with any consistency, the modeler will conclude that this test provides no evidence of model inadequacy. For further discussion and an application to a real simulation, the reader is referred to Schruben [1980]. This type of validation test is commonly called a *Turing test*. Its use as model development proceeds can be a valuable tool in detecting model inadequacies and, eventually, in increasing model credibility as the model is improved and refined.

10.4 Summary

Validation of simulation models is of great importance. Decisions are made on the basis of simulation results; thus, the accuracy of these results should be subject to question and investigation.

Quite often, simulations appear realistic on the surface because simulation models, unlike analytic models, can incorporate any level of detail about the real system. To avoid being "fooled" by this apparent realism, it is best to compare system data to model data and to make the comparison by using a wide variety of techniques, including an objective statistical test, if at all possible.

As discussed by Van Horn [1969, 1971], some of the possible validation techniques, in order of increasing cost-to-value ratios, include

1. Develop models with high face validity by consulting people knowledgeable about system behavior on model structure, model input, and model output. Use any existing knowledge in the form of previous research and studies, observation, and experience.
2. Conduct simple statistical tests of input data for homogeneity, for randomness, and for goodness of fit to assumed distributional forms.
3. Conduct a Turing test. Have knowledgeable people (engineers, managers) compare model output to system output and attempt to detect the difference.
4. Compare model output to system output by means of statistical tests.
5. After model development, collect new system data and repeat techniques 2 to 4.
6. Build the new system (or redesign the old one), conforming to the simulation results, collect data on the new system, and use the data to validate the model (not recommended if this is the only technique used).
7. Do little or no validation. Implement simulation results without validating. (Not recommended.)

It is usually too difficult, too expensive, or too time consuming to use all possible validation techniques for every model that is developed. It is an important part of the model-builder's task to choose those validation techniques most appropriate, both to assure model accuracy and to promote model credibility.

REFERENCES

BALCI, O. [1994], "Validation, Verification and Testing Techniques throughout the Life Cycle of a Simulation Study," *Annals of Operations Research*, Vol. 53, pp. 121–174.

BALCI, O. [1998], "Verification, Validation, and Testing," in *Handbook of Simulation*, J. Banks, ed., John Wiley, New York.

BALCI, O. [2003], "Verification, Validation, and Certification of Modeling and Simulation Applications," in *Proceedings of the 2003 Winter Simulation Conference*, S. Chick, P. J. Sánchez, D. Ferrin, and D. J. Morrice, eds., New Orleans, LA, Dec. 7–10, pp. 150–158.

BALCI, O., AND R. G. SARGENT [1982a], "Some Examples of Simulation Model Validation Using Hypothesis Testing," in *Proceedings of the 1982 Winter Simulation Conference*, H. J. Highland, Y. W. Chao, and O. S. Madrigal, eds., San Diego, CA, Dec. 6–8, pp. 620–629.

BALCI, O., AND R. G. SARGENT [1982b], "Validation of Multivariate Response Models Using Hotelling's Two-Sample T^2 Test," *Simulation*, Vol. 39, No. 6, pp. 185–192.

BALCI, O., AND R. G. SARGENT [1984a], "Validation of Simulation Models via Simultaneous Confidence Intervals," *American Journal of Mathematical Management Sciences*, Vol. 4, Nos. 3 & 4, pp. 375–406.

BALCI, O., AND R. G. SARGENT [1984b], "A Bibliography on the Credibility Assessment and Validation of Simulation and Mathematical Models," *Simuletter*, Vol. 15, No. 3, pp. 15–27.

BORTSCHELLER, B. J., AND E. T. SAULNIER [1992], "Model Reusability in a Graphical Simulation Package," in *Proceedings of the 24th Winter Simulation Conference*, J. J. Swain, D. Goldsman, R. C. Crain, and J. R. Wilson, eds., Arlington, VA, Dec. 13–16, pp. 764–772.

CARSON, J. S., [1986], "Convincing Users of Model's Validity is Challenging Aspect of Modeler's Job," *Industrial Engineering*, June, pp. 76–85.

CARSON, J. S. [2002], "Model Verification and Validation," in *Proceedings of the 34th Winter Simulation Conference*, E. Yücesan, C.-H. Chen, J. L. Snowdon, and J. M. Charnes, eds., San Diego, Dec. 8-11, pp. 52–58.

CARSON, J. S., N. WILSON, D. CARROLL, AND C. H. WYSOWSKI [1981a], "A Discrete Simulation Model of a Cigarette Fabrication Process," *Proceedings of the Twelfth Modeling and Simulation Conference*, University of Pittsburgh, PA., Apr. 30–May 1, pp. 683–689.

CARSON, J. S., N. WILSON, D. CARROLL, AND C. H. WYSOWSKI [1981b], "Simulation of a Filter Rod Manufacturing Process," *Proceedings of the 1981 Winter Simulation Conference*, T. I. Oren, C. M. Delfosse, and C. M. Shub, eds., Atlanta, GA, Dec. 9–11, pp. 535–541.

GAFARIAN, A. V., AND J. E. WALSH [1970], "Methods for Statistical Validation of a Simulation Model for Freeway Traffic near an On-Ramp," *Transportation Research*, Vol. 4, p. 379–384.

GASS, S. I. [1983], "Decision-Aiding Models: Validation, Assessment, and Related Issues for Policy Analysis," *Operations Research*, Vol. 31, No. 4, pp. 601–663.

HINES, W. W., D. C. MONTGOMERY, D. M. GOLDSMAN, AND C. M. BORROR [2002], *Probability and Statistics in Engineering*, 4th ed., Wiley, New York.

KLEIJNEN, J. P. C. [1987], *Statistical Tools for Simulation Practitioners*, Marcel Dekker, New York.

KLEIJNEN, J. P. C. [1993], "Simulation and Optimization in Production Planning: A Case Study," *Decision Support Systems*, Vol. 9, pp. 269–280.

KLEIJNEN, J. P. C. [1995], "Theory and Methodology: Verification and Validation of Simulation Models," *European Journal of Operational Research*, Vol. 82, No. 1, pp. 145–162.

LAW, A. M., AND W. D. KELTON [2000], *Simulation Modeling and Analysis*, 3d ed., McGraw-Hill, New York.

NAYLOR, T. H., AND J. M. FINGER [1967], "Verification of Computer Simulation Models," *Management Science*, Vol. 2, pp. B92–B101.

OREN, T. [1981], "Concepts and Criteria to Assess Acceptability of Simulation Studies: A Frame of Reference," *Communications of the Association for Computing Machinery*, Vol. 24, No. 4, pp. 180–189.

SARGENT, R. G. [2003], "Verification and Validation of Simulation Models," in *Proceedings of the 2003 Winter Simulation Conference*, S. Chick, P. J. Sánchez, D. Ferrin, and D. J. Morrice, eds., New Orleans, LA, Dec. 7–10, pp. 37–48.

SCHECTER, M., AND R. C. LUCAS [1980], "Validating a Large Scale Simulation Model of Wilderness Recreation Travel," *Interfaces*, Vol. 10, pp. 11–18.

SCHRUBEN, L. W. [1980], "Establishing the Credibility of Simulations," *Simulation*, Vol. 34, pp. 101–105.

SHANNON, R. E. [1975], *Systems Simulation: The Art and Science*. Prentice-Hall, Englewood Cliffs, NJ.

VAN HORN, R. L. [1969], "Validation," in *The Design of Computer Simulation Experiments*, T. H. Naylor, ed., pp. 232–235 Duke University Press, Durham, NC.

VAN HORN, R. L. [1971], "Validation of Simulation Results," *Management Science*, Vol. 17, pp. 247–258.

YOUNGBLOOD, S. M. [1993], "Literature Review and Commentary on the Verification, Validation and Accreditation of Models," in *Proceedings of the 1993 Summer Computer Simulation Conference*, J. Schoen, ed., Boston, MA, July 19–21, pp. 10–17.

EXERCISES

1. A simulation model of a job shop was developed to investigate different scheduling rules. To validate the model, the scheduling rule currently used was incorporated into the model and the resulting output was compared against observed system behavior. By searching the previous year's database records, it was estimated that the average number of jobs in the shop was 22.5 on a given day. Seven independent replications of the model were run, each of 30 days' duration, with the following results for average number of jobs in the shop:

$$18.9 \quad 22.0 \quad 19.4 \quad 22.1 \quad 19.8 \quad 21.9 \quad 20.2$$

   (a) Develop and conduct a statistical test to evaluate whether model output is consistent with system behavior. Use the level of significance $\alpha = 0.05$.
   (b) What is the power of this test if a difference of two jobs is viewed as critical? What sample size is needed to guarantee a power of 0.8 or higher? (Use $\alpha = 0.05$.)

2. System data for the job shop of Exercise 1 revealed that the average time spent by a job in the shop was approximately 4 working days. The model made the following predictions, on seven independent replications, for average time spent in the shop:

$$3.70 \quad 4.21 \quad 4.35 \quad 4.13 \quad 3.83 \quad 4.32 \quad 4.05$$

   (a) Is model output consistent with system behavior? Conduct a statistical test, using the level of significance $\alpha = 0.01$.
   (b) If it is important to detect a difference of 0.5 day, what sample size is needed to have a power of 0.90? Interpret your results in terms of model validity or invalidity. (Use $\alpha = 0.01$.)

3. For the job shop of Exercise 1, four sets of input data were collected over four different 10-day periods, together with the average number of jobs in the shop Z_i for each period. The input data were used to drive the simulation model for four runs of 10 days each, and model predictions of average number of jobs in the shop Y_i were collected, with these results:

i	1	2	3	4
Z_i	21.7	19.2	22.8	19.4
Y_i	24.6	21.1	19.7	24.9

   (a) Conduct a statistical test to check the consistency of system output and model output. Use the level of significance $\alpha = 0.05$.
   (b) If a difference of two jobs is viewed as important to detect, what sample size is required to guarantee a probability of at least 0.80 of detecting this difference, if it indeed exists? (Use $\alpha = 0.05$.)

4. Obtain at least two of the papers or reports listed in the References dealing with validation and verification. Write a short essay comparing and contrasting the various philosophies and approaches to the topic of verification and validation.

5. Find several examples of actual simulations reported in the literature in which the authors discuss validation of their model. Is enough detail given to judge the adequacy of the validation effort? If so, compare the reported validation with the criteria set forth in this chapter. Did the authors use any validation technique not discussed in this chapter? [*Hint*: Several potential sources of articles on simulation applications include the journals *Interfaces* and *Simulation*, and the *Winter Simulation Conference Proceedings* at www.informs-cs.org.]

6. Compare and contrast the various simulation-software packages in their capability to aid the modeler in the often arduous task of debugging and verification. [*Hint*: Articles discussing the nature of simulation software packages may be found in the *Winter Simulation Conference Proceedings* at www.informs-cs.org.]

7. (a) Compare validation in simulation to the validation of theories in the physical sciences.
   (b) Compare the issues involved and the techniques available for validation of models of physical systems versus models of social systems.
   (c) Contrast the difficulties, and compare the techniques, in validating a model of a manually operated warehouse with fork trucks and other manually operated vehicles, versus a model of a facility with automated guided vehicles, conveyors, and an automated storage-and-retrieval system.
   (d) Repeat (c) for a model of a production system involving considerable manual labor and human decision making, versus a model of the same production system after it has been automated.

8. Use the confidence-interval approach to assess the validity of the revised bank model in Example 10.2, using the output data in Table 10.3.

11

Estimation of Absolute Performance

Output analysis is the examination of data generated by a simulation. Its purpose is either to predict the performance of a system or to compare the performance of two or more alternative system designs. This chapter deals with estimating *absolute performance*, by which we mean estimating the value of one or more system performance measures; Chapter 12 deals with the comparison of two or more systems, in other words *relative performance*.

The need for statistical output analysis is based on the observation that the output data from a simulation exhibits random variability when random-number generators are used to produce the values of the input variables—that is, two different streams or sequences of random numbers will produce two sets of outputs, which (probably) will differ. If the performance of the system is measured by a parameter θ, the result of a set of simulation experiments will be an estimator $\widehat{\theta}$ of θ. The precision of the estimator $\widehat{\theta}$ can be measured by the standard error of $\widehat{\theta}$ or by the width of a confidence interval for θ. The purpose of the statistical analysis is either to estimate this standard error or confidence interval or to figure out the number of observations required to achieve a standard error or confidence interval of a given size—or both.

Consider a typical output variable Y, the total cost per week of an inventory system; Y should be treated as a random variable with an unknown distribution. A simulation run of length 1 week provides a single sample observation from the population of all possible observations on Y. By increasing the run length, the sample size can be increased to n observations, $Y_1, Y_2, \ldots, Y_n$, based on a run length of n weeks. However, these observations do not constitute a random sample, in the classic sense, because they are not statistically independent. In this case, the inventory on hand at the end of one week is the beginning inventory on hand for the next week, and so the value of Y_i has some influence on the value of Y_{i+1}. Thus, the sequence of random variables $Y_1, Y_2, \ldots, Y_n$ could be *autocorrelated* (i.e., correlated with itself). This autocorrelation, which is a measure of a lack of statistical independence, means that classic methods of statistics, which assume independence, are not directly applicable to the analysis of these output data. The methods must be properly modified and the simulation experiments properly designed for valid inferences to be made.

In addition to the autocorrelation present in most simulation output data, the specification of the initial conditions of the system at time 0 can pose a problem for the simulation analyst and could influence the output data. By "time 0" we mean whatever point in time the beginning of the simulation run represents. For example, the inventory on hand and the number of backorders at time 0 (Monday morning) would most likely influence the value of Y_1, the total cost for week 1. Because of the autocorrelation, these initial conditions would also influence the costs $Y_2, \ldots, Y_n$ for subsequent weeks. The specified initial conditions, if not chosen well, can have an effect on estimation of the steady-state (long-run) performance of a simulation model. For purposes of statistical analysis, the effect of the initial conditions is that the output observations might not be identically distributed and that the initial observations might not be representative of the steady-state behavior of the system.

Section 11.1 distinguishes between two types of simulation—transient versus steady state—and defines commonly used measures of system performance for each type of simulation. Section 11.2 illustrates by example the inherent variability in a stochastic (i.e., probabilistic) discrete-event simulation and thereby demonstrates the need for a statistical analysis of the output. Section 11.3 covers the statistical estimation of performance measures. Section 11.4 discusses the analysis of terminating simulations, and Section 11.5 the analysis of steady-state simulations. To support the calculations in this chapter a spreadsheet called `SimulationTools.xls` is provided at `www.bcnn.net`. `SimulationTools.xls` is a menu-driven application that allows the user to paste in their own data—generated by any simulation application—and perform statistical analyses on it. A detailed user guide is also available on the book web site, and `SimulationTools.xls` has integrated help.

11.1   Types of Simulations with Respect to Output Analysis

In analyzing simulation output data, a distinction is made between terminating or transient simulations and steady-state simulations.

A terminating simulation is one that runs for some duration of time T_E, where E is a specified event (or set of events) that stops the simulation. Such a simulated system opens at time 0 under well-specified initial conditions and closes at the stopping time T_E. The next four examples are terminating simulations.

Example 11.1

The Shady Grove Bank opens at 8:30 A.M. (time 0) with no customers present and 8 of the 11 tellers working (initial conditions) and closes at 4:30 P.M. (time $T_E = 480$ minutes). Here, the event E is merely the fact that the bank has been open for 480 minutes. The simulation analyst is interested in modeling the interaction between customers and tellers over the entire day, including the effect of starting up and closing down at the end of the day.

Example 11.2

Consider the Shady Grove Bank of Example 11.1, but restricted to the period from 11:30 A.M. (time 0) to 1:30 P.M., when it is especially busy. The simulation run length is $T_E = 120$ minutes. The initial conditions at time 0 (11:30 A.M.) could be specified in essentially two ways: (a) the real system could be observed at 11:30 on a number of different days and a distribution of number of customers in system (at 11:30 A.M.) could be estimated, then these data could be used to load the simulation model with customers at time 0; or (b) the model could be simulated from 8:30 A.M. to 11:30 A.M. without collecting output statistics, and the ending conditions at 11:30 A.M. used as initial conditions for the 11:30 A.M. to 1:30 P.M. simulation.

Example 11.3

A communications system consists of several components plus several backup components. It is represented schematically in Figure 11.1. Consider the system over a period of time T_E, until the system fails. The stopping event E is defined by $E = \{A \text{ fails, or } D \text{ fails, or } (B \text{ and } C \text{ both fail})\}$. Initial conditions are that all components are new at time 0.

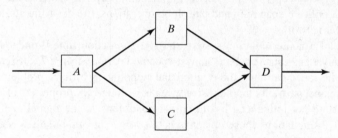

Figure 11.1 Example of a communications system.

Notice that, in the bank model of Example 11.1, the stopping time $T_E = 480$ minutes is known, but in Example 11.3, the stopping time T_E is generally unpredictable in advance; in fact, T_E is probably the output variable of interest, as it represents the total time until the system breaks down. One goal of the simulation might be to estimate $E(T_E)$, the mean time to system failure.

Example 11.4

A manufacturing process runs continuously from Monday mornings until Saturday mornings. The first shift of each work week is used to load inventory buffers and chemical tanks with the components and catalysts needed to make the final product. These components and catalysts are made continually throughout the week, except for the last shift Friday night, which is used for cleanup and maintenance. Thus, most inventory buffers are near empty at the end of the week. During the first shift on Monday, a buffer stock is built up to cover the eventuality of breakdown in some part of the process. It is desired to simulate this system during the first shift (time 0 to time $T_E = 8$ hours) to study various scheduling policies for loading inventory buffers.

In the simulating of a terminating system, the initial conditions of the system at time 0 must be specified, and the stopping time T_E—or, alternatively, the stopping event E—must be well defined. Although it is certainly true that the Shady Grove Bank in Example 11.1 will open again the next day, the simulation analyst has chosen to consider it a terminating system because the object of interest is one day's operation, including start up and close down. On the other hand, if the simulation analyst were interested in some other aspect of the bank's operations, such as the flow of money or operation of automated teller machines, then the system might be considered as a nonterminating one. Similar comments apply to the communications system of Example 11.3. If the failed component were replaced and the system continued to operate, and, most important, if the simulation analyst were interested in studying its long-run behavior, it might be considered as a nonterminating system. In Example 11.3, however, interest is in its short-run behavior, from time 0 until the first system failure at time T_E. Therefore, whether a simulation is considered to be terminating depends on both the objectives of the simulation study and the nature of the system.

Example 11.4 is a terminating system, too. It is also an example of a *transient* (or nonstationary) simulation: the variables of interest are the in-process inventory levels, which are increasing from zero or near zero (at time 0) to full or near full (at time 8 hours).

A *nonterminating system* is a system that runs continuously, or at least over a very long period of time. Examples include assembly lines that shut down infrequently, continuous production systems of many different types, telephone and other communications systems such as the Internet, hospital emergency rooms, police dispatching and patrolling operations, fire departments, and continuously operating computer networks.

A simulation of a nonterminating system starts at simulation time 0 under initial conditions defined by the analyst and runs for some analyst-specified period of time T_E. (Significant problems arise concerning the specification of these initial and stopping conditions, problems that we discuss later.) Usually, the analyst wants to study steady-state, or long-run, properties of the system—that is, properties that are not influenced by the initial conditions of the model at time 0. A steady-state simulation is a simulation whose objective is to study long-run, or steady-state, behavior of a nonterminating system.

The next two examples are steady-state simulations.

Example 11.5

Consider the manufacturing process of Example 11.4, beginning with the second shift, when the complete production process is under way. It is desired to estimate long-run production levels and production efficiencies. For the relatively long period of 13 shifts, this may be considered as a steady-state simulation. To obtain sufficiently precise estimates of production efficiency and other response variables, the analyst could decide to simulate for any length of time, T_E (even longer than 13 shifts). That is, T_E is not determined by the nature of the problem (as it was in terminating simulations); rather, it is set by the analyst as one parameter in the design of the simulation experiment.

Example 11.6

HAL Inc., a large web-based order-processing company, has many customers worldwide. Thus, its large computer system with many servers, workstations, and peripherals runs continuously, 24 hours per day. To handle an increased work load, HAL is considering additional CPUs, memory, and storage devices in various configurations. Although the load on HAL's computers varies throughout the day, management wants the system to be able to accommodate sustained periods of peak load. Furthermore, the time frame in which HAL's business will change in any substantial way is unknown, so there is no fixed planning horizon. Thus, a steady-state simulation at peak-load conditions is appropriate. HAL systems staff develops a simulation model of the existing system with the current peak work load and then explores several possibilities for expanding capacity. HAL is interested in long-run average throughput and utilization of each computer. The stopping time T_E is determined not by the nature of the problem, but rather by the simulation analyst, either arbitrarily or with a certain statistical precision in mind.

11.2 Stochastic Nature of Output Data

Consider one run of a simulation model over a period of time $[0, T_E]$. Since the model is an input-output transformation, as illustrated by Figure 10.5, and since some of the model input variables are random variables, it follows that the model output variables are random variables. Two examples are now given to illustrate the nature of the output data from stochastic simulations and to give a preliminary discussion of several important properties of these data. Do not be concerned if some of these properties and the associated terminology are not entirely clear on a first reading. They will be explained carefully later in the chapter.

Example 11.7: Software Made Personal

Software Made Personal (SMP) customizes software products for clients in two areas: financial tracking and contact management. They have a customer support call center that handles questions for owners of their software from 8 A.M. to 4 P.M. Eastern Time. When a customer calls they use an automated system to select among the two product lines. Each product line has its own operators, and if an appropriate operator is available then the call is immediately routed to the operator; otherwise, the caller is placed in a hold queue. SMP is hoping to reduce the total number of operators they need by cross-training them so that they can answer calls for any product line. This is expected to increase the time to process a call by about 10%. Before considering reducing operators, however, SMP wants to know what their quality of service would be if they simply cross-train the operators they currently have. Using historical data, incoming calls have been modeled as a nonstationary Poisson arrival process and distributions have been fit for operator time to answer questions. Their quality of service measures are caller response time (time from initiating a call until speaking to an operator) and the number of callers on hold. Response-time differences of more than 30 seconds matter.

This is clearly a terminating simulation experiment with time 0 corresponding to 8 A.M. and T_E corresponding to 4 P.M. (another way to terminate the simulation is when the last call that arrives before 4 P.M. is terminated, in which case T_E is a random value). Table 11.1 shows the results from 4 replications of the SMP simulation; each row presents the average number of minutes callers spent on hold, and the average number of callers on hold, during the course of a simulated 8-hour day.

Table 11.1 Results of Four Independent Replications of the SMP Call Center

Replication, r	Average Waiting Time, $\widehat{w}_{Qr}$ (minutes)	Average on Hold, $\widehat{L}_{Qr}$ (people)
1	0.88	0.68
2	5.04	4.18
3	4.13	3.26
4	0.52	0.34

The table illustrates that there is significant variation in these averages from day to day, implying that a substantial number of replications may be required to precisely estimate, say, the mean values $w_Q = E(\widehat{w}_Q)$ and $L_Q = E(\widehat{L}_Q)$. Even with a large number of replications there will still be some estimation error, which is why we want to report not just a single value (the point estimate), but also a measure of its error in the form of a confidence interval or standard error.

These questions are addressed in Section 11.4 for terminating simulations such as Example 11.7. Classic methods of statistics may be used because $\widehat{w}_{Q1}, \widehat{w}_{Q2}, \widehat{w}_{Q3}$, and $\widehat{w}_{Q4}$ constitute a random sample—that is, they are independent and identically distributed. In addition, $w_Q = E(\widehat{w}_{Qr})$ is the parameter being estimated, so each $\widehat{w}_{Qr}$ is an unbiased estimate of the true mean waiting time w_Q. The analysis of Example 11.7 is considered in Example 11.10 of Section 11.4. A survey of statistical methods applicable to terminating simulations is given by Law [1980]. Additional guidance may be found in Alexopoulos and Seila [1998], Kleijnen [1987], Law [2007], and Nelson [2001].

The next example illustrates the effects of correlation and initial conditions on the estimation of long-run mean measures of performance of a system.

Example 11.8

Semiconductor wafer fabrication is an amazingly expensive process involving hundreds of millions of dollars in processing equipment, material handling, and human resources. FastChip, Inc. wants to use simulation to design a new "fab" by evaluating the steady-state mean cycle time (time from wafer release to completion) that they can expect under a particular loading of two distinct products, called C-chip and D-chip. The process of wafer fabrication consists of two basic steps, diffusion and lithography, each of which contains many substeps, and the fabrication process requires multiple passes through these steps. Product will be released in cassettes at the rate of 1 cassette/hour, 7 days a week, 24-hours a day, to achieve a desired throughput of 1 cassette/hour. The hope is that the current design will achieve long-run average cycle times of less than 45 hours for the C-chip and 30 hours for the D-chip.

Figure 11.2 shows the first 30 Chip-C cycle times from one run of the simulation model of the fab. The goal is to estimate the long-run average cycle time as, conceptually, the number of cassettes produced goes to infinity. Notice that the first few cycle times seem lower than the ones that follow; this occurs because the simulation was started with the fab "empty and idle," which is not representative of long-run or steady-state conditions. Also notice that when there is a low cycle time (say, around 40 hours), it tends to be followed by several additional low cycle times. Thus, the cycle time of successive cassettes may be correlated. These features suggest that the cycle-time data

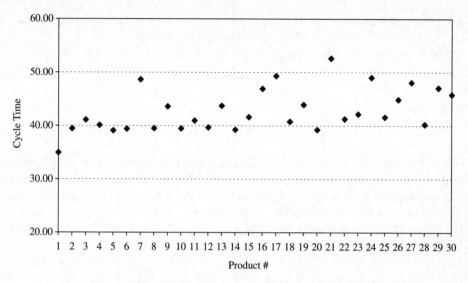

Figure 11.2   First 30 Chip-C cycle times from one simulation run.

are not a random sample, so classic statistical methods may not apply. The analysis of Example 11.8 is considered in Section 11.5.

11.3   Absolute Measures of Performance and Their Estimation

Consider the estimation of a performance parameter θ (or ϕ) of a simulated system. It is desired to have a point estimate and an interval estimate of θ (or ϕ). The length of the interval estimate is a measure of the error in the point estimate. The simulation output data are of the form $\{Y_1, Y_2, \ldots, Y_n\}$ for estimating θ; we refer to such output data as *discrete-time data*, because the index n is discrete valued. The simulation output data are of the form $\{Y(t), 0 \le t \le T_E\}$ for estimating ϕ; we refer to such output data as *continuous-time data*, because the index t is continuous valued. For example, Y_i might be the delay of customer i, or the total cost in week i; $Y(t)$ might be the queue length at time t, or the number of backlogged orders at time t. The parameter θ is an ordinary mean; ϕ will be referred to as a time-weighted mean. Whether we call the performance parameter θ or ϕ does not really matter; we use two different symbols here simply to provide a distinction between ordinary means and time-weighted means. To make the discussions more concrete, we consider output measures of queueing systems as described in Chapter 6.

11.3.1   Point Estimation

The point estimator of θ, based on the data $\{Y_1, \ldots, Y_n\}$, is defined by

$$\widehat{\theta} = \frac{1}{n} \sum_{i=1}^{n} Y_i \tag{11.1}$$

where $\widehat{\theta}$ is a sample mean based on a sample of size n. Computer simulation languages may refer to this as a "discrete-time," "collect," "tally," or "observational" statistic.

The point estimator $\widehat{\theta}$ is said to be unbiased for θ if its expected value is θ; that is, if

$$E(\widehat{\theta}) = \theta \tag{11.2}$$

In general, however,

$$E(\widehat{\theta}) \neq \theta \tag{11.3}$$

and $E(\widehat{\theta}) - \theta$ is called the *bias* in the point estimator $\widehat{\theta}$. It is desirable to have estimators that are unbiased, or, if this is not possible, to have a small bias relative to the magnitude of θ. Examples of estimators of the form of Equation (11.1) include $\widehat{w}$ and $\widehat{w}_Q$ of Equations (6.5) and (6.7), in which case Y_i is the time spent in the (sub)system by customer i.

The point estimator of ϕ, based on the data $\{Y(t), 0 \leq t \leq T_E\}$, where T_E is the simulation run length, is defined by

$$\widehat{\phi} = \frac{1}{T_E} \int_0^{T_E} Y(t)\, dt \tag{11.4}$$

and is called a *time average* of $Y(t)$ over $[0, T_E]$. Simulation languages may refer to this as a "continuous-time," "discrete-change," or "time-persistent" statistic. In general,

$$E(\widehat{\phi}) \neq \phi \tag{11.5}$$

and $\widehat{\phi}$ is said to be biased for ϕ. Again, we would like to obtain unbiased or low-bias estimators. Examples of time averages include $\widehat{L}$ and $\widehat{L}_Q$ of Equations (6.3) and (6.4) and Y_j of Equation (11.4)

Generally, θ and ϕ are regarded as mean measures of performance of the system being simulated. Other measures usually can be put into this common framework. For example, consider estimation of the proportion of days on which sales are lost through an out-of-stock situation. In the simulation, let

$$Y_i = \begin{cases} 1, & \text{if out of stock on day } i \\ 0, & \text{otherwise} \end{cases}$$

With n equal to the total number of days $\widehat{\theta}$, as defined by Equation (11.1), is a point estimator of θ, the proportion of out-of-stock days. For a second example, consider estimation of the proportion of time queue length is greater than 10 customers. If $L_Q(t)$ represents simulated queue length at time t, then (in the simulation) define

$$Y(t) = \begin{cases} 1, & \text{if } L_Q(t) > 10 \\ 0, & \text{otherwise} \end{cases}$$

Then $\widehat{\phi}$, as defined by Equation (11.4), is a point estimator of ϕ, the proportion of time that the queue length is greater than 10 customers. Thus, estimation of proportions or probabilities is a special case of the estimation of means.

A performance measure that does not fit this common framework is a quantile or percentile. Quantiles describe the level of performance that can be delivered with a given probability p. For instance, suppose that Y represents the delay in queue that a customer experiences in a service system, measured in minutes. Then the $p = 0.85$ quantile of Y is the value θ such that

$$\Pr\{Y \le \theta\} = 0.85 \tag{11.6}$$

As a percentage, θ is the $100p$th or 85th percentile of customer delay. Therefore, 85% of all customers will experience a delay of θ minutes or less. Stated differently, a customer has only a 0.15 probability of experiencing a delay of longer than θ minutes. A widely used performance measure is the median, which is the 0.5 quantile or 50th percentile.

The problem of estimating a quantile is the inverse of the problem of estimating a proportion or probability. In estimating a proportion, θ is given and the probability p is to be estimated; however, in estimating a quantile, p is given and θ is to be estimated.

The most intuitive method for estimating a quantile is to form a histogram of the observed values of Y, then find a value $\widehat{\theta}$ such that $100p\%$ of the histogram is to the left of (smaller than) $\widehat{\theta}$. For instance, if we observe $n = 250$ customer delays $\{Y_1, \ldots, Y_{250}\}$, then an estimate of the 85th percentile of delay is a value $\widehat{\theta}$ such that $(0.85)(250) = 212.5 \approx 213$ of the observed values are less than or equal to θ. An obvious estimate is, therefore, to set $\widehat{\theta}$ equal to the 213th smallest value in the sample (this requires sorting the data). When the output is a continuous-time process, such as the queue-length process $\{L_Q(t), 0 \le t \le T_E\}$, then a histogram gives the fraction of time that the process spent at each possible level (queue length, in this example). However, the method for quantile estimation remains the same: Find a value $\widehat{\theta}$ such that $100p\%$ of the histogram is to the left of $\widehat{\theta}$.

11.3.2   Confidence-Interval Estimation

To understand confidence intervals fully, it is important to understand the difference between a *measure of error* and a *measure of risk*. One way to make the difference clear is to contrast a *confidence interval* with a *prediction interval* (which is another useful output-analysis tool).

Both confidence intervals and prediction intervals are based on the premise that the data being produced by the simulation is represented well by a probability model. Suppose that model is the normal distribution with mean θ and variance σ^2, both unknown. To make the example concrete, let Y_i be the average cycle time for parts produced on the ith replication (representing a day of production) of the simulation. Therefore, θ is the mathematical expectation of Y_i, and σ is represents the day-to-day variation of the average cycle time.

Suppose our goal is to estimate θ. If we are planning to be in business for a long time, producing parts day after day, then θ is a relevant parameter, because it is the long-run mean daily cycle time. Our average cycle time will vary from day to day, but over the long run the average of the averages will be close to θ.

The natural estimator for θ is the overall sample mean of n independent replications, $\bar{Y} = \sum_{i=1}^{n} Y_i/n$. But $\bar{Y}$ is not θ, it is an estimate, based on a sample, and it has error. A confidence interval is a measure of that error. Let

$$S^2 = \frac{1}{n-1} \sum_{i=1}^{n} \left(Y_i - \bar{Y}\right)^2$$

be the sample variance across the R replications. The usual confidence interval, which assumes the Y_i are normally distributed, is

$$\bar{Y} \pm t_{\alpha/2, n-1} \frac{S}{\sqrt{n}}$$

where $t_{\alpha/2, R-1}$ is the quantile of the t distribution with $R - 1$ degrees of freedom that cuts off $\alpha/2$ of the area of each tail. (See Table A.5.) We cannot know for certain exactly how far $\bar{Y}$ is from θ, but the confidence interval attempts to bound that error. Unfortunately, the confidence interval itself may be wrong. A confidence level, such as 95%, tells us how much we can trust the interval to actually bound the error between $\bar{Y}$ and θ. The more replications we make, the less error there is in $\bar{Y}$, and our confidence interval reflects that because $t_{\alpha/2, n-1} S/\sqrt{n}$ will tend to get smaller as n increases, converging to 0 as n goes to infinity.

Now suppose we need to make a promise about what the average cycle time will be on a particular day. A good guess is our estimator $\bar{Y}$, but it is unlikely to be exactly right. Even θ itself, which is the center of the distribution, is not likely to be the actual average cycle time on any particular day, because the daily average cycle time varies. A prediction interval, on the other hand, is designed to be wide enough to contain the actual average cycle time on any particular day, with high probability.

The normal-theory prediction interval is

$$\bar{Y} \pm t_{\alpha/2, n-1} S \sqrt{1 + \frac{1}{n}}$$

The length of this interval will not go to 0 as n increases. In fact, in the limit it becomes

$$\theta \pm z_{\alpha/2} \sigma$$

to reflect the fact that, no matter how much we simulate, our daily average cycle time still varies.

In summary, a prediction interval is a measure of risk, and a confidence interval is a measure of error. We can simulate away error by making more and more replications, but we can never simulate away risk, which is an inherent part of the system. We can, however, do a better job of evaluating risk by making more replications.

Example 11.9

Suppose that the overall average of the average cycle time on 120 replications of a manufacturing simulation is 5.80 hours, with a sample standard deviation of 1.60 hours. Since $t_{0.025, 119} = 1.98$, a 95% confidence interval for the long-run expected daily average cycle time is $5.80 \pm 1.98(1.60/\sqrt{120})$ or 5.80 ± 0.29 hours. Thus, our best guess of the long-run average of the daily average cycle times is 5.80 hours, but there could be as much as ± 0.29 hours error in this estimate.

On any particular day, we are 95% confident that the average cycle time for all parts produced on that day will be

$$5.80 \pm 1.98(1.60) \sqrt{1 + \frac{1}{120}}$$

or 5.80 ± 3.18 hours. The ± 3.18 hours reflects the inherent variability in the daily average cycle times and the fact that we want to be 95% confident of covering the actual average cycle time on a particular day, rather than simply covering the long-run average.

A caution: In neither case (prediction interval or confidence interval) can we make statements about the cycle time for *individual* products because we observed the daily average. To make valid statements about the cycle times of individual products, the analysis must be based on individual cycle times.

11.4   Output Analysis for Terminating Simulations

Consider a terminating simulation that runs over a simulated time interval $[0, T_E]$ and results in observations $Y_1, \ldots, Y_n$. The sample size n may be a fixed number, or it may be a random variable (say, the number of observations that occur during time T_E). A common goal in simulation is to estimate

$$\theta = E\left(\frac{1}{n}\sum_{i=1}^{n} Y_i\right)$$

When the output data are of the form $\{Y(t), 0 \le t \le T_E\}$, the goal is to estimate

$$\phi = E\left(\frac{1}{T_E}\int_0^{T_E} Y(t)\,dt\right)$$

The method used in each case is the method of *independent replications*. The simulation is repeated a total of R times, each run using a different random number stream and independently chosen initial conditions (which includes the case that all runs have identical initial conditions). We now address this problem.

11.4.1   Statistical Background

Perhaps the most confusing aspect of simulation output analysis is distinguishing *within-replication* data from *across-replication* data, and understanding the properties and uses of each. The issue can be further confused by the fact that simulation languages often provide only summary measures, like sample means, sample variances, and confidence intervals, rather than all of the raw data. Sometimes these summary measures are *all* that the simulation language provides, without a lot of extra work.

To illustrate the key ideas, think in terms of the simulation of a manufacturing system and two performance measures of that system, the cycle time for parts (time from release into the factory until completion) and the work in process (WIP, the total number of parts in the factory at any time). In computer applications, these two measures could correspond to the response time and the length of the task queue at the CPU; in a service application, they could be the time to fulfill a customer's request and the number of requests on the "to do" list; in a supply-chain application, they could be the order fill time and the inventory level. Similar measures appear in many systems.

Here is the usual set up for something like cycle time: Let Y_{ij} be the cycle time for the jth part produced in the ith replication. If each replication represents two shifts of production, then the

Table 11.2 Within and Across-Replication Cycle-Time Data

Within-Rep Data				*Across-Rep Data*
Y_{11}	Y_{12}	$\cdots$	Y_{1n_1}	$\bar{Y}_{1\cdot}, S_1^2, H_1$
Y_{21}	Y_{22}	$\cdots$	Y_{2n_2}	$\bar{Y}_{2\cdot}, S_2^2, H_2$
$\vdots$	$\vdots$	$\cdots$	$\vdots$	$\vdots$
Y_{R1}	Y_{R2}	$\cdots$	Y_{Rn_R}	$\bar{Y}_{R\cdot}, S_R^2, H_R$
				$\bar{Y}_{\cdot\cdot}, S^2, H$

number of parts produced in each replication might differ. Table 11.2 shows, symbolically, the results of R replications.

The across-replication data are formed by summarizing within-replication data:[1] $\bar{Y}_{i\cdot}$ is the sample mean of the n_i cycle times from the ith replication, S_i^2 is the sample variance of the same data, and

$$H_i = t_{\alpha/2, n_i - 1} \frac{S_i}{\sqrt{n_i}} \tag{11.7}$$

is a confidence-interval half-width based on this dataset.

From the across-replication data, we compute overall statistics, the average of the daily cycle time averages

$$\bar{Y}_{\cdot\cdot} = \frac{1}{R} \sum_{i=1}^{R} \bar{Y}_{i\cdot} \tag{11.8}$$

the sample variance of the daily cycle time averages

$$S^2 = \frac{1}{R-1} \sum_{i=1}^{R} (\bar{Y}_{i\cdot} - \bar{Y}_{\cdot\cdot})^2 \tag{11.9}$$

and finally, the confidence-interval half-width

$$H = t_{\alpha/2, R-1} \frac{S}{\sqrt{R}} \tag{11.10}$$

The quantity $S/\sqrt{R}$ is the standard error, which is sometimes interpreted as the average error in $\bar{Y}_{\cdot\cdot}$ as an estimator of θ. Notice that S^2 is *not* the average of the within-replication sample variances S_i^2; rather, it is the sample variance of the within-replication averages $\bar{Y}_{1\cdot}, \bar{Y}_{2\cdot}, \ldots, \bar{Y}_{R\cdot}$.

---

[1]We use the convention that a dot, as in the subscript $i\cdot$, indicates summation over the indicted subscript; and a bar, as in $\bar{Y}_{i\cdot}$, indicates an average of that subscript.

Table 11.3   Within and Across-Replication WIP Data

Within-Rep Data	Across-Rep Data
$Y_1(t), 0 \leq t \leq T_{E_1}$	$\bar{Y}_1., S_1^2, H_1$
$Y_2(t), 0 \leq t \leq T_{E_2}$	$\bar{Y}_2., S_2^2, H_2$
$\vdots$	$\vdots$
$Y_R(t), 0 \leq t \leq T_{E_R}$	$\bar{Y}_R., S_R^2, H_R$
	$\bar{Y}.., S^2, H$

Within a replication, work in process is a continuous-time output, denoted $Y_i(t)$. The stopping time for the ith replication T_{E_i} could be a random variable, in general; in this example, it is the end of the second shift. Table 11.3 is an abstract representation of the data produced.

The within-replication sample mean and variance are defined appropriately for continuous-time data:

$$\bar{Y}_{i\cdot} = \frac{1}{T_{E_i}} \int_0^{T_{E_i}} Y_i(t)dt \tag{11.11}$$

and

$$S_i^2 = \frac{1}{T_{E_i}} \int_0^{T_{E_i}} \left(Y_i(t) - \bar{Y}_{i\cdot}\right)^2 dt \tag{11.12}$$

A definition for H_i is more problematic, but, to be concrete, take it to be

$$H_i = z_{\alpha/2} \frac{S_i}{\sqrt{T_{E_i}}}. \tag{11.13}$$

It is difficult to conceive of a situation in which H_i is relevant, a topic we discuss later. Although the definitions of the within-replication data change for continuous-time data, the across-replication statistics are unchanged, and this is a critical observation.

Here are the key points that must be understood:

- The overall sample average $\bar{Y}..$ and the individual replication sample averages $\bar{Y}_{i\cdot}$ are always unbiased estimators of the expected daily average cycle time or daily average WIP.
- Across-replication data are independent, since they are based on different random numbers; are identically distributed, since we are running the same model on each replication; and tend to be normally distributed, if they are averages of within-replication data, as they are here. This implies that the confidence interval $\bar{Y}.. \pm H$ is often pretty good.

- Within-replication data, on the other hand, might have none of these properties. The individual cycle times may not be identically distributed, if the first few parts of the day find the system empty; they are almost certainly not independent, because one part follows another; and whether they are normally distributed is difficult to know in advance. For this reason, S_i^2 and H_i, which are computed under the assumption of independent and identically distributed (i.i.d.) data, tend not to be useful, although there are exceptions.
- There are situations in which $\bar{Y}_{..}$ and $\bar{Y}_{i.}$ are valid estimators of the expected cycle time for an individual part or the expected WIP at any point in time, rather than the daily average. (See Section 11.5 on steady-state simulations.) Even when this is the case, the confidence interval $\bar{Y}_{..} \pm H$ is valid, and $\bar{Y}_{i.} \pm H_i$ is not. The difficulty occurs because S_i^2 is a reasonable estimator of the variance of the cycle time, but S_i^2/n_i and S_i^2/T_{E_i} are not good estimators of the $\text{Var}[\bar{Y}_{i.}]$; more on this in Section 11.5.2.

Example 11.10:   Software Made Personal (cont.) _____

Consider Example 11.7, the SMP call center problem, with the data for $R = 4$ replications given in Table 11.1. The four number-on-hold estimates $\widehat{L_Q}_r$ are time averages of the form of Equation (11.11), where $Y_r(t) = L_Q(t)$, the number of customers on hold at time t. Similarly, the four average system times $\widehat{w}_{Q1}, \ldots, \widehat{w}_{Q4}$ are analogous to $\bar{Y}_r$. of Table 11.2, where Y_{ri} is the actual time spent on hold by customer i on replication r.

Suppose that the analyst desires a 95% confidence interval for the mean time spent on hold, w_Q. Using Equation (11.8), compute an overall point estimator

$$\bar{Y}_{..} = \widehat{w}_Q = \frac{0.88 + 5.04 + 4.13 + 0.52}{4} = 2.64$$

Using Equation (11.9), compute its estimated variance:

$$S^2 = \frac{(0.88 - 2.64)^2 + \cdots + (0.52 - 2.64)^2}{4 - 1} = (2.28)^2$$

Thus, the standard error of $\widehat{w}_Q = 2.64$ is estimated by s.e.$(\widehat{w}_Q) = S/\sqrt{4} = 1.14$. Obtain $t_{0.025,3} = 3.18$ from Table A.5, and compute the 95% confidence interval half-width by Equation (11.10) as

$$H = t_{0.025,3}\frac{S}{\sqrt{4}} = (3.18)(1.14) = 3.62$$

giving 2.64 ± 3.62 minutes with 95% confidence. Clearly 4 replications are far too few to make any sort of useful estimate of w_Q, since a negative mean waiting time is impossible. Later in the chapter we return to this problem and determine how many replications are required to get a meaningful estimate.

11.4.2   Confidence Intervals with Specified Precision

By Expression (11.10), the half-length H of a $100(1 - \alpha)\%$ confidence interval for a mean θ, based on the t distribution, is given by

$$H = t_{\alpha/2, R-1} \frac{S}{\sqrt{R}}$$

where S^2 is the sample variance and R is the number of replications. In simulation we seldom need to select R and just accept the confidence interval half-length that results. Instead, we can drive the number of replications to be large enough so that H is small enough to facilitate the decision that the simulation is supposed to support. The acceptable size of H depends on the problem at hand: it might be $\pm\$5000$ in a financial simulation, ± 6 minutes for product cycle time, or ± 1 customer on hold in a call center.

Suppose that an error criterion of ϵ is specified; in other words, it is desired to estimate θ by $\bar{Y}_{..}$ to within $\pm\epsilon$ with high probability—say, at least $1 - \alpha$. Thus, it is desired that a sufficiently large sample size R be taken to satisfy

$$P(|\bar{Y}_{..} - \theta| \leq \epsilon) \geq 1 - \alpha$$

When the sample size R is fixed, no guarantee can be given for the resulting error. But if the sample size can be increased, an error criterion (such as $\epsilon = \$5000$, 6 minutes, or 1 customer) can be specified.

Assume that an initial sample of size R_0 replications has been observed—that is, the simulation analyst initially makes R_0 independent replications. We must have $R_0 \geq 2$, with 10 or more being desirable. The R_0 replications will be used to obtain an initial estimate S_0^2 of the population variance σ^2. To meet the half-length criterion, a sample size R must be chosen such that $R \geq R_0$ and

$$H = t_{\alpha/2, R-1} \frac{S_0}{\sqrt{R}} \leq \epsilon \tag{11.14}$$

Solving for R in Inequality (11.14) shows that R is the smallest integer satisfying $R \geq R_0$ and

$$R \geq \left(\frac{t_{\alpha/2, R-1} S_0}{\epsilon} \right)^2 \tag{11.15}$$

An initial estimate for R is given by

$$R \geq \left(\frac{z_{\alpha/2} S_0}{\epsilon} \right)^2 \tag{11.16}$$

where $z_{\alpha/2}$ is the $100(1 - \alpha/2)$ percentage point of the standard normal distribution from Table A.3. And since $t_{\alpha/2, R-1} \approx z_{\alpha/2}$ for large R, say, $R \geq 50$, the second inequality for R is adequate when R is large. After determining the final sample size R, collect $R - R_0$ additional observations (i.e., make $R - R_0$ additional replications, or start over and make R total replications) and form the $100(1 - \alpha)\%$ confidence interval for θ by

$$\bar{Y}_{..} - t_{\alpha/2, R-1} \frac{S}{\sqrt{R}} \leq \theta \leq \bar{Y}_{..} + t_{\alpha/2, R-1} \frac{S}{\sqrt{R}} \tag{11.17}$$

where $\bar{Y}_{..}$ and S^2 are computed on the basis of all R replications, $\bar{Y}_{..}$ by Equation (11.8), and S^2 by Equation (11.9). The half-length of the confidence interval given by Inequality (11.17) should be approximately ϵ or smaller; however, with the additional $R - R_0$ observations, the variance estimator S^2 could differ somewhat from the initial estimate S_0^2, possibly causing the half-length to be greater than desired. If the confidence interval from Inequality (11.17) is too large, the procedure may be repeated, using Inequality (11.15), to determine an even larger sample size.

Example 11.11

Suppose that it is desired to estimate the mean waiting time in SMP's hold queue in Example 11.7 to within ± 0.5 minutes with 95% confidence. We will illustrate this calculation using the "Get Sample Size" utility of `SimulationTools.xls`. The input dialog box is shown in Figure 11.3 and the results are displayed in Figure 11.4. Notice that "Absolute Precision" means that we directly specify the half width of the confidence interval (± 0.5 minutes in this case); "Relative Precision" specifies the length of the half width as a fraction of the mean.

An initial sample of size $R_0 = 20$ is taken, and the program determines that $R = 175$ is the smallest integer satisfying Inequality (11.16). Therefore, $R - R_0 = 175 - 20 = 155$ additional replications are needed. After obtaining the additional outputs, we would again need to compute the half-width H to ensure that it is as small as is desired. In this case the result—which can also be found at `www.bcnn.net`—was 2.66 ± 0.34 minutes

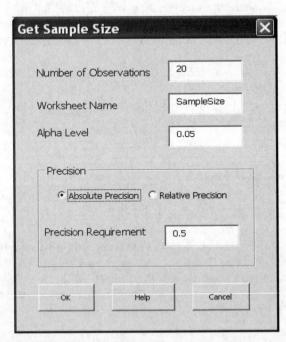

Figure 11.3   Input Dialog for the "Get Sample Size" utility of SimulationTools.xls

Get Sample Size			Completed			Data Column			Minimum Sample Size
					1	0.880897931			175
Number of Observations	20				2	5.04340304			
Alpha	0.05				3	4.132638545			
Precision Type	1				4	0.517418075			
Precision Value	0.5				5	2.584490259			
					6	0.860547939			
					7	1.544612493			
					8	1.762676832			
					9	6.925242515			
					10	2.201491896			
					11	2.915438392			
					12	0.486246064			
					13	1.792947434			
					14	1.244455641			
					15	4.278579049			
					16	13.91696167			
					17	0.235565331			
					18	0.685156864			
					19	1.577792626			
					20	1.070626537			

Figure 11.4   Sample Size Calculation using "Get Sample Size" in SimulationTools.xls

11.4.3   Quantiles

To present the interval estimator for quantiles, it is helpful to review the interval estimator for a mean in the special case when the mean represents a proportion or probability p. In this book, we have chosen to treat a proportion or probability as just a special case of a mean. However, in many statistics texts, probabilities are treated separately.

When the number of independent replications $Y_1, \ldots, Y_R$ is large enough that $t_{\alpha/2, n-1} \doteq z_{\alpha/2}$, the confidence interval for a probability p is often written as

$$\widehat{p} \pm z_{\alpha/2} \sqrt{\frac{\widehat{p}(1 - \widehat{p})}{R - 1}}$$

where $\widehat{p}$ is the sample proportion (algebra shows that this formula for the half-width is precisely equivalent to Equation (11.10) when used in estimating a proportion).

As mentioned in Section 11.3, the quantile-estimation problem is the inverse of the probability-estimation problem: Find θ such that $\Pr\{Y \le \theta\} = p$. Thus, to estimate the p quantile, we find that value $\widehat{\theta}$ such that $100p\%$ of the data in a histogram of Y is to the left of $\widehat{\theta}$, or, stated differently, the npth smallest value of $Y_1, \ldots, Y_R$.

Extending this idea, an approximate $(1 - \alpha)100\%$ confidence interval for θ can be obtained by finding two values: θ_ℓ that cuts off $100p_\ell\%$ of the histogram and θ_u that cuts off $100p_u\%$ of the histogram, where

$$p_\ell = p - z_{\alpha/2}\sqrt{\frac{p(1-p)}{R-1}}$$

$$p_u = p + z_{\alpha/2}\sqrt{\frac{p(1-p)}{R-1}} \tag{11.18}$$

(Recall that we know p.) In terms of sorted values, $\widehat{\theta}_\ell$ is the Rp_ℓ smallest value (rounded down) and $\widehat{\theta}_u$ is the Rp_u smallest value (rounded up), of $Y_1, \ldots, Y_R$.

Example 11.12

Suppose that we want to estimate the 0.75 quantile of average daily waiting time in the hold queue for SMP and form a 95% confidence interval for it, based on $R = 300$ replications. The data may be found in the spreadsheet `Quantile.xls` at www.bcnn.net. The point estimator is $\widehat{\theta} = 3.39$ minutes, because that is the $300 \times 0.75 = 225$th smallest value of the sorted data. The interpretation is that, over the long run, on 75% of the days the average time customers spend on hold will be less than or equal to 3.39 minutes.

To obtain the confidence interval, we first compute

$$p_\ell = p - z_{\alpha/2}\sqrt{\frac{p(1-p)}{R-1}} = 0.75 - 1.96\sqrt{\frac{0.75(0.25)}{299}} = 0.700$$

$$p_u = p + z_{\alpha/2}\sqrt{\frac{p(1-p)}{R-1}} = 0.75 + 1.96\sqrt{\frac{0.75(0.25)}{299}} = 0.799$$

The lower bound of the confidence interval is $\widehat{\theta}_\ell = 3.13$ minutes (the $300 \times p_\ell = 210$th smallest value, rounding down); the upper bound of the confidence interval is $\widehat{\theta}_u = 3.98$ minutes (the $300 \times p_u = 240$th smallest value, rounding up). Thus, with 95% confidence, the point estimate of 3.39 minutes is no more than $\max\{3.98 - 3.39, 3.39 - 3.13\} = 0.59$ minutes off the true 0.75 quantile.

It is important to understand the 3.39 minutes is an estimate of the 0.75 quantile (75th percentile) of daily average customer delay, not an individual customer's delay. Because the call center starts the day empty and idle, and the customer load on the system varies throughout the day, the 0.75 quantile of individual customer waiting time varies throughout the day and cannot be given by a single number.

11.4.4 Estimating Probabilities and Quantiles from Summary Data

Knowing the equation for the confidence interval half-width is important if all that the simulation software provides is $\bar{Y}_{..}$ and H and you need to work out the number of replications required to get a prespecified precision, or if you need to estimate a probability or quantile. You know the number of replications, so the sample standard deviation can be extracted from H by using the formula

$$S = \frac{H\sqrt{R}}{t_{\alpha/2, R-1}}$$

With this information, the method in Section 11.4.2 can be employed.

The more difficult problem is estimating a probability or quantile from summary data. When all we have available is the sample mean and confidence-interval halfwidth (which gives us the sample standard deviation), then one approach is to use a normal-theory approximation for the probabilities or quantiles we desire, specifically

$$\Pr\{\bar{Y}_{i\cdot} \leq c\} \approx \Pr\left\{Z \leq \frac{c - \bar{Y}_{\cdot\cdot}}{S}\right\}$$

and

$$\widehat{\theta} \approx \bar{Y}_{\cdot\cdot} + z_p S$$

The following example illustrates how this is done.

Example 11.13
From 25 replications of a manufacturing simulation, a 90% confidence interval for the daily average WIP is 218 ± 32. What is the probability that the daily average WIP is less than 350? What is the 85th percentile of daily average WIP?

First, we extract the standard deviation:

$$S = \frac{H\sqrt{R}}{t_{0.05,24}} = \frac{32\sqrt{25}}{1.71} = 93$$

Then, we use the normal approximations and Table A.3 to get

$$\Pr\{\bar{Y}_{i\cdot} \leq 350\} \approx \Pr\left\{Z \leq \frac{350 - 218}{93}\right\} = \Pr\{Z \leq 1.42\} = 0.92$$

and

$$\widehat{\theta} \approx \bar{Y}_{\cdot\cdot} + z_{0.85}S = 218 + 1.04(93) = 315 \text{ parts}$$

There are shortcomings to obtaining our probabilities and quantiles this way. The approximation depends heavily on whether the output variable of interest is normally distributed. If the output variable itself is not an average, then this approximation is suspect. Therefore, we expect the approximation to work well for statements about the average daily cycle time, for instance, but very poorly for the cycle time of an individual part.

11.5   Output Analysis for Steady-State Simulations

Consider a single run of a simulation model whose purpose is to estimate a steady-state, or long-run, characteristic of the system. Suppose that the single run produces observations $Y_1, Y_2, \ldots$, which, generally, are samples of an autocorrelated time series. The steady-state (or long-run) measure of performance, θ, is defined by

$$\theta = \lim_{n \longrightarrow \infty} \frac{1}{n} \sum_{i=1}^{n} Y_i \tag{11.19}$$

with probability 1, where the value of θ is independent of the initial conditions. (The phrase "with probability 1" means that essentially all simulations of the model, using different random numbers, will produce a series $Y_i, i = 1, 2, \ldots$ whose sample average converges to θ.) For example, if Y_i was the time customer i spent talking to an operator, then θ would be the long-run average time a customer spends talking to an operator; and, because θ is defined as a limit, it is independent of the call center's conditions at time 0. Similarly, the steady-state performance for a continuous-time output measure $\{Y(t), t \geq 0\}$, such as the number of customers in the call center's hold queue, is defined as

$$\phi = \lim_{T_E \longrightarrow \infty} \frac{1}{T_E} \int_0^{T_E} Y(t) \, dt$$

with probability 1.

Of course, the simulation analyst could decide to stop the simulation after some number of observations, say, n, have been collected; or the simulation analyst could decide to simulate for some length of time T_E that determines n (although n may vary from run to run). The sample size n (or T_E) is a design choice; it is not inherently determined by the nature of the problem. The simulation analyst will choose simulation run length (n or T_E) with several considerations in mind:

1. Any bias in the point estimator that is due to artificial or arbitrary initial conditions. (The bias can be severe if run length is too short, but generally it decreases as run length increases.)
2. The desired precision of the point estimator, as measured by the standard error or confidence interval half-width.
3. Budget constraints on the time available to execute the simulation.

The next subsection discusses initialization bias and the following subsections outline two methods of estimating point-estimator variability. For clarity of presentation, we discuss only estimation of θ from a discrete-time output process. Thus, when discussing one replication (or run), the notation

$$Y_1, Y_2, Y_3, \ldots$$

will be used; if several replications have been made, the output data for replication r will be denoted by

$$Y_{r1}, Y_{r2}, Y_{r3}, \ldots \tag{11.20}$$

11.5.1   Initialization Bias in Steady-State Simulations

There are several methods of reducing the point-estimator bias caused by using artificial and unrealistic initial conditions in a steady-state simulation. The first method is to initialize the simulation in a state that is more representative of long-run conditions. This method is sometimes called *intelligent initialization*. Examples include

1. setting the inventory levels, number of backorders, and number of items on order and their arrival dates in an inventory simulation;
2. placing customers in queue and in service in a queueing simulation;
3. having some components fail or degrade in a reliability simulation.

There are at least two ways to specify the initial conditions intelligently. If the system exists, collect data on it and use these data to specify more nearly typical initial conditions. This method sometimes requires a large data-collection effort. In addition, if the system being modeled does not exist—for example, if it is a variant of an existing system—this method is impossible to implement. Nevertheless, it is recommended that simulation analysts use any available data on existing systems to help initialize the simulation, as this will usually be better than assuming the system to be "completely stocked," "empty and idle," or "brand new" at time 0.

A related idea is to obtain initial conditions from a second model of the system that has been simplified enough to make it mathematically solvable. The queueing models in Chapter 6 are very useful for this purpose. The simplified model can be solved to find long-run expected or most likely conditions—such as the expected number of customers in the queue—and these conditions can be used to initialize the simulation.

A second method to reduce the impact of initial conditions, possibly used in conjunction with the first, is to divide each simulation run into two phases: first, an initialization phase, from time 0 to time T_0, followed by a data-collection phase from time T_0 to the stopping time $T_0 + T_E$; that is, the simulation begins at time 0 under specified initial conditions I_0 and runs for a specified period of time T_0. Data collection on the response variables of interest does not begin until time T_0 and continues until time $T_0 + T_E$. The choice of T_0 is quite important, because the system state at time T_0, denoted by I, should be more nearly representative of steady-state behavior than are the original initial conditions I_0 at time 0. In addition, the length T_E of the data-collection phase should be long enough to guarantee sufficiently precise estimates of steady-state behavior. Notice that the system state I at time T_0 is a random variable and to say that the system has reached an approximate steady state is to say that the probability distribution of the system state at time T_0 is sufficiently close to the steady-state probability distribution as to make the bias in point estimates of response variables negligible. Figure 11.5 illustrates the two phases of a steady-state simulation. The effect of starting a simulation run of a queueing system in the empty and idle state, as well as a useful plot to aid the simulation analyst in choosing an appropriate value for T_0, are given in the following example.

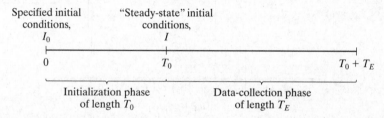

Figure 11.5   Initialization and data collection phases of a steady-state simulation run.

Example 11.14

Consider the FastChip wafer fabrication problem discussed in Example 11.8. Suppose that a total of $R = 10$ independent replications were made, each run long enough so that 250 cycle times were collected for each of Chip-C and Chip-D. We will focus on Chip-C.

Normally we average all data within each replication to obtain a replication average. However, our goal at this stage is to identify the trend in the data due to initialization bias and find out when it dissipates. To do this, we will average corresponding cycle times across replications and plot them (this idea is usually attributed to Welch [1983]). Such averages are known as *ensemble averages*. Specifically, let $Y_{r1}, Y_{r2}, \ldots, Y_{r,250}$ be the 250 Chip-C cycle times, in order of completion, from replication r. For the jth cycle time, define the ensemble average across all R replications to be

$$\bar{Y}_{\cdot j} = \frac{1}{R} \sum_{r=1}^{R} Y_{rj} \tag{11.21}$$

($R = 10$ here). The ensemble averages are plotted as the solid line in Figure 11.6 (the dashed lines will be discussed below). This plot was produced by the `SimulationTools.xls` spreadsheet, and we will refer to it as a *mean plot*.

Notice the gradual upward tread at the beginning. The simulation analyst may suspect that this is due to the downward bias in these estimators, which in turn is due to the fab being empty and idle at time 0. As time becomes larger (so the number of cycle times recorded increases), the effect of the initial conditions on later observations lessens and the observations appear to vary around a common mean. When the simulation analyst feels that this point has been reached, then the data-collection phase begins. Here approximately 120 Chip-C cycle times might be adequate.

Although we have determined a deletion amount of 120 cycle times, the initialization phase is almost always specified in terms of *simulation time* T_0, because we need to consider the initialization phase for each performance measure, and they may differ. In the FastChip example we would also

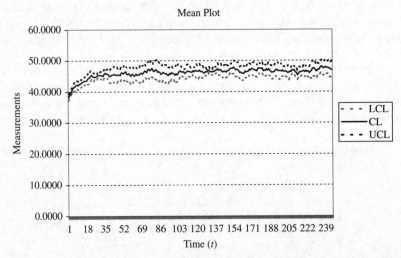

Figure 11.6   Mean Plot Across 10 Replications of 250 Product C Cycle Times

need to examine the cycle times for Chip-D. If we specified the initialization phase as a count for each output, then we would have to track these counts. When the initialization phase is specified by time, on the other hand, then a single event can be scheduled at time T_0, and data only retained beyond that point.

To specify T_0 by time we have to convert the appropriate counts into approximate times and use the largest of these. For instance, for the FastChip simulation 120 Chip-C cycle times corresponds to approximately 200 hours of simulation time, since cassettes are released at 1 per hour and 60% of the releases are Chip-C: 120 cycle times/0.6 releases/hour = 200 hours (we could also increase this somewhat to account for the time it takes the 120th Chip-C release to traverse the fab). This turns out to be larger than the initialization phase for Chip-D, so we set $T_0 = 200$ hours.

Plotting ensemble averages is extra work (although `SimulationTools.xls` makes it easier), so it might be tempting to take a shortcut. For instance, one might plot the output from only replication 1, $Y_{11}, Y_{12}, \ldots, Y_{1n}$, or the cumulative average from a replication 1

$$\bar{Y}_{1\ell} = \frac{1}{\ell} \sum_{j=1}^{\ell} Y_{1j}$$

for $\ell = 1, 2, \ldots, n$. Figure 11.7 shows both. Notice that the raw data from a single replication can be too highly variable to detect the trend, while the cumulative plot is clearly biased low because it retains all of the data from the beginning of the run. Thus, these shortcuts should be avoided and the mean plot used instead.

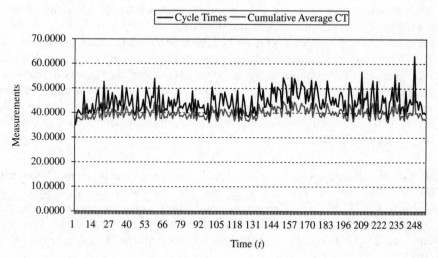

Figure 11.7   Plot of 250 Product C Cycle Times from 1 Replication, Raw and Cumulative Average

Some additional factors to keep in mind when using the mean plot to determine the initialization phase:

1. When first starting to examine the initialization phase, a run length and number of replications will have to be guessed. Ensemble averages, such as Figure 11.6, will reveal a smoother and more precise trend as the number of replications R is increased, and it may have to be increased several times until the trend is clear. It is also possible that the run length guess is not long enough to get past the initialization phase (this would be indicated if the trend in the mean plot does not disappear), so the run length may have to be increased as well.
2. Ensemble averages can also be smoothed by plotting a moving average rather than the original ensemble averages. In a moving average, each plotted point is actually the average of several adjacent ensemble averages. Specifically, the jth plot point would be

$$\tilde{Y}_j = \frac{1}{2m+1} \sum_{i=j-m}^{j+m} \bar{Y}_{\cdot i}$$

for some $m \geq 1$, rather than the original ensemble average $\bar{Y}_{\cdot j}$. The value of m is typically chosen by trial and error until a smooth plot is obtained. The mean plot in Figure 11.6 was obtained with $m = 2$.
3. Since each ensemble average (or moving average) is the sample mean of i.i.d. observations across R replications, a confidence interval based on the t distribution can be placed around each point, as shown by the dashed lines in Figure 11.6, and these intervals can be used to judge whether or not the plot is precise enough to decide that bias has diminished. This is the preferred method to determine a deletion point.
4. Cumulative averages, such as in Figure 11.7, become less variable as more data are averaged. Therefore, it is expected that the left side of the curve will always be less smooth than the right side. Remember that cumulative averages tend to converge more slowly to long-run performance than do ensemble averages, because cumulative averages contain all observations, including the most biased ones from the beginning of the run. For this reason, cumulative averages should be used only if it is not feasible to compute ensemble averages, such as when only a single replication is possible.

There has been no shortage of solutions to the initialization-bias problem. Unfortunately, for every "solution" that works well in some situations, there are other situations in which either it is not applicable or it performs poorly. Important ideas include testing for bias (e.g., Kelton and Law [1983], Schruben [1980], Goldsman, Schruben, and Swain [1994]); modeling the bias (e.g., Snell and Schruben [1985]); and randomly sampling the initial conditions on multiple replications (e.g., Kelton [1989]).

11.5.2   Error Estimation for Steady-State Simulation

If $\{Y_1, \ldots, Y_n\}$ are not statistically independent, then S^2/n, given by Equation (11.9), is a biased estimator of the true variance $V(\widehat{\theta})$. This is almost always the case when $\{Y_1, \ldots, Y_n\}$ is a sequence of output observations from within a single replication. In this situation, $Y_1, Y_2, \ldots$ is an autocorrelated sequence, sometimes called a *time series*.

Suppose that our point estimator for θ is the sample mean $\bar{Y} = \sum_{i=1}^{n} Y_i/n$. A general result[2] from mathematical statistics is that the variance of $\bar{Y}$ is

$$V(\bar{Y}) = \frac{1}{n^2} \sum_{i=1}^{n} \sum_{j=1}^{n} \text{cov}(Y_i, Y_j) \tag{11.22}$$

where $\text{cov}(Y_i, Y_i) = V(Y_i)$. To construct a confidence interval for θ, an estimate of $V(\bar{Y})$ is required. But obtaining an estimate of Equation (11.22) is futile, because each term $\text{cov}(Y_i, Y_j)$ could be different, in general. Fortunately, systems that have a steady state will, if simulated long enough to pass the transient phase, produce an output process that is approximately covariance stationary. Intuitively, stationarity implies that Y_{i+k} depends on Y_{i+1} in the same manner as Y_k depends on Y_1. In particular, the covariance between two random variables in the time series depends only on the number of observations between them, called the *lag*.

For a covariance-stationary time series, $\{Y_1, Y_2, \ldots\}$, define the lag-k autocovariance by

$$\gamma_k = \text{cov}(Y_1, Y_{1+k}) = \text{cov}(Y_i, Y_{i+k}) \tag{11.23}$$

which, by definition of covariance stationarity, is not a function of i. For $k = 0$, γ_0 becomes the population variance σ^2; that is,

$$\gamma_0 = \text{cov}(Y_i, Y_{i+0}) = V(Y_i) = \sigma^2 \tag{11.24}$$

The lag-k autocorrelation is the correlation between any two observations k apart. It is defined by

$$\rho_k = \frac{\gamma_k}{\sigma^2} \tag{11.25}$$

and has the property that

$$-1 \leq \rho_k \leq 1, \quad k = 1, 2, \ldots$$

If a time series is covariance stationary, then Equation (11.22) can be simplified substantially. Algebra shows that

$$V(\bar{Y}) = \frac{\sigma^2}{n} \left[1 + 2 \sum_{k=1}^{n-1} \left(1 - \frac{k}{n} \right) \rho_k \right] \tag{11.26}$$

where ρ_k is the lag-k autocorrelation given by Equation (11.25).

When $\rho_k > 0$ for all k (or most k), the time series is said to be positively autocorrelated. In this case, large observations tend to be followed by large observations, small observations by small ones. Such a series will tend to drift slowly above and then below its mean. Figure 11.8(a) is an example of a stationary time series exhibiting positive autocorrelation. The output data from most queueing simulations are positively autocorrelated.

---

[2]This general result can be derived from the fact that, for two random variables Y_1 and Y_2, $V(Y_1 \pm Y_2) = V(Y_1) + V(Y_2) \pm 2\text{cov}(Y_1, Y_2)$.

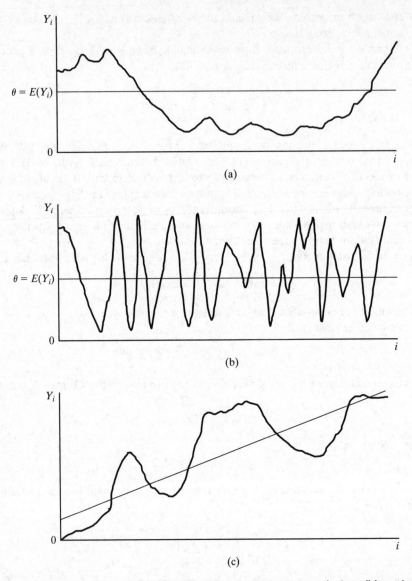

Figure 11.8   (a) Stationary time series Y_i exhibiting positive autocorrelation; (b) stationary time series Y_i exhibiting negative autocorrelation; (c) nonstationary time series with an upward trend.

On the other hand, if some of the $\rho_k < 0$, the series $Y_1, Y_2, \ldots$ will display the characteristics of negative autocorrelation. In this case, large observations tend to be followed by small observations, and vice versa. Figure 11.8(b) is an example of a stationary time series exhibiting negative autocorrelation. The output of certain inventory simulations might be negatively autocorrelated.

Figure 11.8(c) also shows an example of a time series with an upward trend. Such a time series is not stationary; the probability distribution of Y_i is changing with the index i.

Why does autocorrelation make it difficult to estimate $V(\bar{Y})$? Recall that the standard estimator for the variance of a sample mean is S^2/n. By using Equation (11.26), it can be shown [Law, 1977] that the expected value of the variance estimator S^2/n is

$$E\left(\frac{S^2}{n}\right) = BV(\bar{Y}) \tag{11.27}$$

where

$$B = \frac{n/c - 1}{n - 1} \tag{11.28}$$

and c is the quantity in brackets in Equation (11.26). The effect of the autocorrelation on the estimator S^2/n is derived by an examination of Equations (11.26) and (11.28). There are essentially three possibilities:

Case 1

If the Y_i are independent, then $\rho_k = 0$ for $k = 1, 2, 3, \ldots$. Therefore, $c = 1 + 2\sum_{k=1}^{n-1}(1 - k/n)\rho_k = 1$ and Equation (11.26) reduces to the familiar σ^2/n. Notice also that $B = 1$, so S^2/n is an unbiased estimator of $V(\bar{Y})$. The Y_i will always be independent when they are obtained from different replications; that independence is the primary reason that we prefer experiment designs calling for multiple replications.

Case 2

If the autocorrelations ρ_k are primarily positive, then $c = 1 + 2\sum_{k=1}^{n-1}(1 - k/n)\rho_k > 1$, so that $n/c < n$, and hence $B < 1$. Therefore, S^2/n is biased low as an estimator of $V(\bar{Y})$. If this correlation were ignored, the nominal $100(1 - \alpha)\%$ confidence interval given by Expression (11.10) would be too short, and its true confidence coefficient would be less than $1 - \alpha$. The practical effect would be that the simulation analyst would have unjustified confidence in the apparent precision of the point estimator due to the shortness of the confidence interval. If the correlations ρ_k are large, B could be quite small, implying a significant underestimation.

Case 3

If the autocorrelations ρ_k are substantially negative, then $0 \leq c < 1$, and it follows that $B > 1$ and S^2/n is biased high for $V(\bar{Y})$. In other words, the true precision of the point estimator $\bar{Y}$ would be greater than what is indicated by its variance estimator S^2/n, because

$$V(\bar{Y}) < E\left(\frac{S^2}{n}\right)$$

As a result, the nominal $100(1 - \alpha)\%$ confidence interval of Expression (11.10) would have true confidence coefficient greater than $1 - \alpha$. This error is less serious than Case 2, because we are unlikely to make incorrect decisions if our estimate is actually more precise than we think it is.

A simple example demonstrates why we are especially concerned about positive correlation: Suppose you want to know how students on a university campus will vote in an upcoming election. To estimate their preferences, you plan to solicit 100 responses. The standard experiment is to randomly select 100 students to poll; call this experiment A. An alternative is to randomly select 20 students and ask each of them to state their preference 5 times in the same day; call this experiment B. Both experiments obtain 100 responses, but clearly an estimate based on experiment B will be less precise (will have larger variance) than an estimate based on experiment A. Experiment A obtains 100 independent responses, whereas experiment B obtains only 20 independent responses and 80 dependent ones. The five opinions from any one student are perfectly positively correlated (assuming a student names the same candidate all five times). Although this is an extreme example, it illustrates that estimates based on positively correlated data are more variable than estimates based on independent data. Therefore, a confidence interval or other measure of error should account correctly for dependent data, but S^2/n does not.

Two methods for eliminating or reducing the deleterious effects of autocorrelation upon estimation of a mean are given in the following sections. Unfortunately, some simulation languages either use or facilitate the use of S^2/n as an estimator of $V(\bar{Y})$, the variance of the sample mean, in all situations. If used uncritically in a simulation with positively autocorrelated output data, the downward bias in S^2/n and the resulting shortness of a confidence interval for θ will convey the impression of much greater precision than actually exists. When such positive autocorrelation is present in the output data, the true variance of the point estimator $\bar{Y}$ can be many times greater than is indicated by S^2/n.

11.5.3 Replication Method for Steady-State Simulations

If initialization bias in the point estimator has been reduced to a negligible level (through some combination of intelligent initialization and deletion), then the method of independent replications can be used to estimate point-estimator variability and to construct a confidence interval. The basic idea is simple: Make R replications, initializing and deleting from each one the same way.

If, however, significant bias remains in the point estimator and a large number of replications are used to reduce point-estimator variability, the resulting confidence interval can be misleading. This happens because bias is not affected by the number of replications R; it is affected only by deleting more data (i.e., increasing T_0) or extending the length of each run (i.e., increasing T_E). Thus, increasing the number of replications R could produce shorter confidence intervals around the "wrong point." Therefore, it is important to do a thorough job of investigating the initial-condition bias.

If the simulation analyst decides to delete d observations of the total of n observations in a replication, then the point estimator of θ is

$$\bar{Y}_{..}(n, d) = \frac{1}{R} \sum_{r=1}^{R} \frac{1}{n-d} \sum_{j=d+1}^{n} Y_{rj} \tag{11.29}$$

That is, the point estimator is the average of the remaining data. The basic raw output data $\{Y_{rj}, r = 1, \dots, R; j = 1, \dots, n\}$ are exhibited in Table 11.4. For instance, Y_{rj} could be the delay of customer j in queue, or the response time of job j in a job shop, on replication r. The number d of deleted

Table 11.4   Raw Output Data from a Steady-State Simulation

Replication	Observations						Replication Averages
	1	$\cdots$	*d*	*d+1*	$\cdots$	*n*	
1	$Y_{1,1}$	$\cdots$	$Y_{1,d}$	$Y_{1,d+1}$	$\cdots$	$Y_{1,n}$	$\bar{Y}_{1\cdot}(n,d)$
2	$Y_{2,1}$	$\cdots$	$Y_{2,d}$	$Y_{2,d+1}$	$\cdots$	$Y_{2,n}$	$\bar{Y}_{2\cdot}(n,d)$
.	.		.	.		.	.
.	.		.	.		.	.
.	.		.	.		.	.
R	$Y_{R,1}$	$\cdots$	$Y_{R,d}$	$Y_{R,d+1}$	$\cdots$	$Y_{R,n}$	$\bar{Y}_{R\cdot}(n,d)$
	$\bar{Y}_{\cdot 1}$	$\cdots$	$\bar{Y}_{\cdot d}$	$\bar{Y}_{\cdot,d+1}$	$\cdots$	$\bar{Y}_{\cdot n}$	$\bar{Y}_{\cdot\cdot}(n,d)$

observations and the total number of observations n might vary from one replication to the next, in which case replace d by d_r and n by n_r. For simplicity, assume that d and n are constant over replications.

When using the replication method, each replication is regarded as a single sample for the purpose of estimating θ. For replication r, define

$$\bar{Y}_{r\cdot}(n,d) = \frac{1}{n-d} \sum_{j=d+1}^{n} Y_{rj} \tag{11.30}$$

as the sample mean of all (nondeleted) observations in replication r. Because all replications use different random-number streams and all are initialized at time 0 by the same set of initial conditions I_0, the replication averages

$$\bar{Y}_{1\cdot}(n,d), \ldots, \bar{Y}_{R\cdot}(n,d)$$

are independent and identically distributed random variables; that is, they constitute a random sample from some underlying population having unknown mean

$$\theta_{n,d} = \mathrm{E}[\bar{Y}_{r\cdot}(n,d)] \tag{11.31}$$

The overall point estimator, given in Equation (11.29), is also given by

$$\bar{Y}_{\cdot\cdot}(n,d) = \frac{1}{R} \sum_{r=1}^{R} \bar{Y}_{r\cdot}(n,d) \tag{11.32}$$

as can be seen from Table 11.4 or from using Equation (11.21). Thus, it follows that

$$\mathrm{E}[\bar{Y}_{\cdot\cdot}(n,d)] = \theta_{n,d}$$

also. If d and n are chosen sufficiently large, then $\theta_{n,d} \approx \theta$, and $\bar{Y}_{..}(n, d)$ is an approximately unbiased estimator of θ. The bias in $\bar{Y}_{..}(n, d)$ is $\theta_{n,d} - \theta$.

For convenience, when the value of n and d are understood, abbreviate $\bar{Y}_{r.}(n, d)$ (the mean of the undeleted observations from the rth replication) and $\bar{Y}_{..}(n, d)$ [the mean of $\bar{Y}_{1.}(n, d), \ldots, \bar{Y}_{R.}(n, d)$] by $\bar{Y}_{r.}$ and $\bar{Y}_{..}$, respectively. To estimate the standard error of $\bar{Y}_{..}$, first compute the sample variance,

$$S^2 = \frac{1}{R-1}\sum_{r=1}^{R}(\bar{Y}_{r.} - \bar{Y}_{..})^2 = \frac{1}{R-1}\left(\sum_{r=1}^{R}\bar{Y}_{r.}^2 - R\bar{Y}_{..}^2\right) \tag{11.33}$$

The standard error of $\bar{Y}_{..}$ is given by

$$\text{s.e.}(\bar{Y}_{..}) = \frac{S}{\sqrt{R}} \tag{11.34}$$

A $100(1-\alpha)\%$ confidence interval for θ, based on the t distribution, is given by

$$\bar{Y}_{..} - t_{\alpha/2,R-1}\frac{S}{\sqrt{R}} \le \theta \le \bar{Y}_{..} + t_{\alpha/2,R-1}\frac{S}{\sqrt{R}} \tag{11.35}$$

where $t_{\alpha/2,R-1}$ is the $100(1-\alpha/2)$ percentage point of a t distribution with $R-1$ degrees of freedom. This confidence interval is valid only if the bias of $\bar{Y}_{..}$ is approximately zero.

As a rough rule, the length of each replication, beyond the deletion point, should be at least ten times the amount of data deleted. In other words, $(n-d)$ should at least $10d$ (or more generally, T_E should be at least $10T_0$). Given this run length, the number of replications should be as many as time permits, up to about 25 replications. Kelton [1986] established that there is little value in dividing the available time into more than 25 replications, so, if time permits making more than 25 replications of length $T_0 + 10T_0$, then make 25 replications of longer than $T_0 + 10T_0$, instead. Again, these are rough rules that need not be followed slavishly.

In this section we have presented results as if the run length is n observations and d of them are deleted. As discussed earlier, in practice we make a run of length $T_0 + T_E$ time units and delete the first T_0 time units. In this case d becomes the number of observations obtained up to time T_0, and $n-d$ is the number of observations recorded between times T_0 and $T_E + T_0$, both random variables that typically differ from replication to replication. In the following example we will use $(T_0 + T_E, T_0)$ instead of (n, d) in the notation to make this point clear.

Example 11.15

Consider again the FastChip wafer fab simulation of Examples 11.8 and 11.14. Suppose that the simulation analyst decides to make $R = 10$ replications, each of length $T_0 + T_E = 2200$ hours, deleting the first $T_0 = 200$ hours before data collection, begins as determined in Example 11.14. Remember that the raw output data consist of cycle times for Chip-C cassettes, and the output from each replication is the average of all of the Chip-C cassettes that finish between times $T_0 = 200$ hours and $T_0 + T_E = 2200$ hours (approximately $2000 \times 60\% = 1200$ cycle times per replication). The purpose of the simulation is to estimate the long-run average cycle time, denoted w, and to measure the error of the estimate by a 95% confidence interval.

Table 11.5   Data Summary for FastChip Simulation by Replication

Replication r	Sample Mean for Replication r
1	46.86
2	46.09
3	47.64
4	47.43
5	46.94
6	46.43
7	47.11
8	46.56
9	46.73
10	46.80
$\bar{Y}_{..}$	46.86
S	0.46

The replication averages $\bar{Y}_{r.}(T_0 + T_E, T_0), r = 1, 2, \ldots, 10$, are shown in Table 11.5. The point estimator is computed by Equation (11.32) as

$$\bar{Y}_{..}(T_0 + T_E, T_0) = 46.86$$

Its standard error is given by Equation (11.34) as

$$\text{s.e.}(\bar{Y}_{..}T_0 + T_E, T_0)) = \frac{S}{\sqrt{R}} = 0.15$$

and using $\alpha = 0.05$ and $t_{0.025,9} = 2.26$, the 95% confidence interval for long-run mean queue length is given by Inequality (11.35) as

$$46.86 - 2.26(0.15) \leq w \leq 46.86 + 2.26(0.15)$$

or

$$46.52 \leq w \leq 47.20$$

The simulation analyst may conclude with a high degree of confidence that the long-run mean cycle time for Chip-C is between 46.52 and 47.20 hours. The confidence interval computed here as given by Inequality (11.35) should be used with caution, because a key assumption behind its validity is that enough data have been deleted to remove any significant bias due to initial conditions—that is, that T_0 and $T_E + T_0$ are sufficiently large so that the bias $\theta_{T_E+T_0,T_0} - \theta$ is negligible.

11.5.4   Sample Size in Steady-State Simulations

Suppose it is desired to estimate a long-run performance measure θ within $\pm\epsilon$, with confidence $100(1-\alpha)\%$. In a steady-state simulation, a specified precision may be achieved either by increasing the number of replications R or by increasing the run length T_E. The first solution, controlling R, is carried out as given in Section 11.4.2 for terminating simulations.

Example 11.16
Consider the data in Table 11.5 for the FastChip wafer fab simulation as an initial sample of size $R_0 = 10$. The initial estimate of standard deviation is $S_0 = 0.46$ hours. Suppose that it is desired to estimate long-run mean cycle time of Chip-C, w, to within $\epsilon = 0.1$ hours (6 minutes) with 95% confidence. Using $\alpha = 0.05$ in Inequality (11.16) yields an initial estimate of

$$ R \geq \left(\frac{z_{0.025}S_0}{\epsilon} \right)^2 = \left(\frac{(1.96)(0.46)}{0.1} \right)^2 = 81.3 $$

Thus, at least 82 replications will be needed.

An alternative to increasing R is to increase total run length $T_0 + T_E$ within each replication. If the calculations in Section 11.4.2, as illustrated in Example 11.16, indicate that $R - R_0$ additional replications are needed beyond the initial number R_0, then an alternative is to increase run length $(T_0 + T_E)$ in the same proportion (R/R_0) to a new run length $(R/R_0)(T_0 + T_E)$. Thus, additional data will be deleted, from time 0 to time $(R/R_0)T_0$, and more data will be used to compute the point estimates, as illustrated by Figure 11.9. However, the total amount of simulation effort is the same as if we had simply increased the number of replications but maintained the same run length. The advantage of increasing total run length per replication and deleting a fixed proportion $[T_0/(T_0+T_E)]$ of the total run length is that any residual bias in the point estimator should be further reduced by the additional deletion of data at the beginning of the run. A possible disadvantage of the method is that, in order to continue the simulation of all R replications [from time $T_0 + T_E$ to time $(R/R_0)(T_0 + T_E)$],

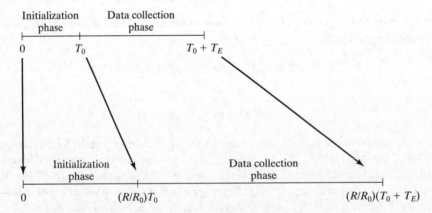

Figure 11.9   Increasing run length to achieve specified accuracy.

it is necessary to have saved the state of the model at time $T_0 + T_E$ and to be able to restart the model and run it for the additional required time. Otherwise, the simulations would have to be rerun from time 0, which could be time consuming for a complex model. Some simulation languages have the capability to save enough information that a replication can be continued from time T_E onward, rather than having to start over from time 0.

Example 11.17 _____

In Example 11.16, suppose that run length was to be increased to achieve the desired error ± 0.1 hours. Since $R/R_0 = 82/10 = 8.2$, the run length should be $(R/R_0)(T_0 + T_E) = 8.2(2200) = 18{,}040$ hours. The data collected from time 0 to time $(R/R_0)T_0 = 8.2(200) = 1640$ hours would be deleted, and the data from time 1640 to time 18,040 used to compute new point estimates and confidence intervals.

_____

11.5.5   Batch Means Method for Steady-State Simulations

One disadvantage of the replication method is that data must be deleted on each replication and, in one sense, deleted data are wasted data, or at least lost information. This suggests that there might be merit in using an experiment design that is based on a single, long replication. The disadvantage of a single-replication design arises when we try to compute the standard error of the sample mean. Since we only have data from within one replication, the data are dependent, and the usual estimator is biased.

The method of batch means attempts to solve this problem by dividing the output data from one replication (after appropriate deletion) into a few large batches and then treating the means of these batches as if they were independent. When the raw output data after deletion form a continuous-time process $\{Y(t), T_0 \leq t \leq T_0 + T_E\}$, such as the length of a queue or the level of inventory, then we form k batches of size $m = T_E/k$ and compute the batch means as

$$\bar{Y}_j = \frac{1}{m} \int_{(j-1)m}^{jm} Y(t + T_0)\, dt$$

for $j = 1, 2, \ldots, k$. In other words, the jth batch mean is just the time-weighted average of the process over the time interval $[T_0 + (j-1)m, T_0 + jm)$.

When the raw output data after deletion form a discrete-time process $\{Y_i, i = d+1, d+2, \ldots, n\}$, such as the customer delays in a queue or the cost per period of an inventory system, then we form k batches of size $m = (n-d)/k$ and compute the batch means as

$$\bar{Y}_j = \frac{1}{m} \sum_{i=(j-1)m+1}^{jm} Y_{i+d}$$

for $j = 1, 2, \ldots, k$ (assuming k divides $n - d$ evenly, otherwise round down to the nearest integer). That is, the batch means are formed as shown here:

$$\underbrace{Y_1, \ldots, Y_d,}_{\text{deleted}} \underbrace{Y_{d+1}, \ldots, Y_{d+m},}_{\bar{Y}_1} \underbrace{Y_{d+m+1}, \ldots, Y_{d+2m},}_{\bar{Y}_2} \ldots, \underbrace{Y_{d+(k-1)m+1}, \ldots, Y_{d+km}}_{\bar{Y}_k}$$

Starting with either continuous-time or discrete-time data, the variance of the sample mean is estimated by

$$\frac{S^2}{k} = \frac{1}{k} \sum_{j=1}^{k} \frac{(\bar{Y}_j - \bar{Y})^2}{k-1} = \frac{\sum_{j=1}^{k} \bar{Y}_j^2 - k\bar{Y}^2}{k(k-1)} \tag{11.36}$$

where $\bar{Y}$ is the overall sample mean of the data after deletion. The batch means $\bar{Y}_1, \bar{Y}_2, \ldots, \bar{Y}_k$ are not independent; however, if the batch size is sufficiently large, successive batch means will be approximately independent, and the variance estimator will be approximately unbiased.

Unfortunately, there is no widely accepted and relatively simple method for choosing an acceptable batch size m (or equivalently, choosing a number of batches k). But there are some general guidelines that can be culled from the research literature:

a. Schmeiser [1982] found that, for a fixed total sample size, there is little benefit from dividing it into more than $k = 30$ batches, even if we could do so and still retain independence between the batch means. Therefore, there is no reason to consider numbers of batches much greater than 30, no matter how much raw data are available. He also found that the performance of the confidence interval, in terms of its width and the variability of its width, is poor for fewer than 10 batches. Therefore, a number of batches between 10 and 30 should be used in most applications.

b. Although there is typically autocorrelation between batch means at all lags, the lag-1 autocorrelation $\rho_1 = \text{corr}(\bar{Y}_j, \bar{Y}_{j+1})$ is usually studied to assess the dependence between batch means. When the lag-1 autocorrelation is nearly 0, then the batch means are treated as independent. This approach is based on the observation that the autocorrelation in many stochastic processes decreases as the lag increases. Therefore, all lag autocorrelations should be smaller, in absolute value, than the lag-1 autocorrelation.

c. The lag-1 autocorrelation between batch means can be estimated (see below). However, the autocorrelation should not be estimated from a small number of batch means (such as the $10 \leq k \leq 30$ recommended above); there is bias in the autocorrelation estimator. Law and Carson [1979] suggest estimating the lag-1 autocorrelation from a large number of batch means based on a smaller batch size, perhaps $100 \leq k \leq 400$. When the autocorrelation between these batch means is approximately 0, then the autocorrelation will be even smaller if we rebatch the data to between 10 and 30 batch means based on a larger batch size. Hypothesis tests for 0 autocorrelation are available, as described next.

d. If the total sample size is to be chosen sequentially, say to attain a specified precision, then it is helpful to allow the batch size and number of batches to grow as the run length increases. It can be shown that a good strategy is to allow the number of batches to increase as the square root of the sample size after first finding a batch size at which the lag-1 autocorrelation is approximately 0. Although we will not discuss this point further, an algorithm based on it can be found in Fishman and Yarberry [1997]; see also Steiger and Wilson [2002] and Lada, Steiger, and Wilson [2006].

Given these insights, we recommend the following general strategy:

1. Obtain output data from a single replication and delete as appropriate. Recall our guideline: collecting at least 10 times as much data as are deleted.
2. Form up to $k = 400$ batches (but at least 100 batches) with the retained data, and compute the batch means. Estimate the sample lag-1 autocorrelation of the batch means as

$$\widehat{\rho}_1 = \frac{\sum_{j=1}^{k-1}(\bar{Y}_j - \bar{Y})(\bar{Y}_{j+1} - \bar{Y})}{\sum_{j=1}^{k}(\bar{Y}_j - \bar{Y})^2}$$

3. Check the correlation to see whether it is sufficiently small.
   (a) If $\widehat{\rho}_1 \leq 0.2$, then rebatch the data into $30 \leq k \leq 40$ batches, and form a confidence interval using $k - 1$ degrees of freedom for the t distribution and Equation (11.36) to estimate the variance of $\bar{Y}$.
   (b) If $\widehat{\rho}_1 > 0.2$, then extend the replication by 50% to 100% and go to Step 2. If it is not possible to extend the replication, then rebatch the data into approximately $k = 10$ batches, and form the confidence interval, using $k - 1$ degrees of freedom for the t distribution and Equation (11.36) to estimate the variance of $\bar{Y}$.
4. As an additional check on the confidence interval, examine the batch means (at the larger or smaller batch size) for independence, using the following test. See, for instance, Alexopoulos and Seila [1998]. Compute the test statistic

$$C = \sqrt{\frac{k^2 - 1}{k - 2}} \left(\widehat{\rho}_1 + \frac{(\bar{Y}_1 - \bar{Y})^2 + (\bar{Y}_k - \bar{Y})^2}{2\sum_{j=1}^{k}(\bar{Y}_j - \bar{Y})^2}\right)$$

If $C < z_\beta$ then accept the independence of the batch means, where β is the Type I error level of the test, such as 0.1, 0.05, 0.01. Otherwise, extend the replication by 50% to 100% and go to Step 2. If it is not possible to extend the replication, then rebatch the data into approximately $k = 10$ batches, and form the confidence interval, using $k - 1$ degrees of freedom for the t distribution and Equation (11.36) to estimate the variance of $\bar{Y}$.

This procedure, including the final check, is conservative in several respects. First, if the lag-1 autocorrelation is substantially negative then we proceed to form the confidence interval anyway. A dominant negative correlation tends to make the confidence interval wider than necessary, which is an error, but not one that will cause us to make incorrect decisions. The requirement that $\widehat{\rho}_1 < 0.2$ at $100 \leq k \leq 400$ batches is pretty stringent and will tend to force us to get more data (and therefore create larger batches) if there is any hint of positive dependence. And finally, the hypothesis test at the end has a probability of β of forcing us to get more data when none are really needed. But this conservatism is by design; the cost of an incorrect decision is typically much greater than the cost of some additional computer run time.

The batch-means approach to confidence-interval estimation is illustrated in the next example.

Example 11.18

Reconsider the FastChip wafer fab simulation of Example 11.8. Suppose that we want to estimate the steady-state mean cycle time for Chip-C, w, by a 95% confidence interval. To illustrate the method of batch means, assume that one run of the model has been made, simulating 5000 cycle times after the deletion point. We then form batch means from $k = 100$ batches of size $m = 50$ and estimate the lag-1 autocorrelation to be $\widehat{\rho}_1 = 0.37 > 0.2$. Thus, we decide to extend the simulation to 10,000 customers after the deletion point, and again we estimate the lag-1 autocorrelation. This estimate, based on $k = 100$ batches of size $m = 100$, is $\widehat{\rho}_1 = 0.15 < 0.2$.

Having passed the correlation check, we rebatch the data into $k = 40$ batches of size $m = 250$. The point estimate is the overall mean

$$\bar{Y} = \frac{1}{10000} \sum_{j=1}^{10000} \bar{Y}_j = 47.00 \text{ hours.}$$

The variance of $\bar{Y}$, computed from the 40 batch means, is

$$\frac{S^2}{k} = \frac{\sum_{j=1}^{40} \bar{Y}_j^2 - 40\bar{Y}^2}{40(39)} = 0.049$$

Thus, a 95% confidence interval is given by

$$\bar{Y} - t_{0.025,39}\sqrt{0.049} \leq w \leq \bar{Y} + t_{0.025,39}\sqrt{0.049}$$

or

$$46.55 = 47.00 - 2.02(0.221) \leq w \leq 47.00 + 2.02(0.221) = 47.45$$

Thus, we assert with 95% confidence that true mean cycle time w is between 46.55 and 47.45 hours. If these results are not sufficiently precise, then the run length should be increased to achieve greater precision.

As a further check on the validity of the confidence interval, we can apply the correlation hypothesis test. To do so, we compute the test statistic from the $k = 40$ batches of size $m = 250$ used to form the confidence interval. This gives

$$C = -0.58 < 1.96 = z_{0.05}$$

which confirms the lack of correlation at the 0.05 significance level. Notice that, at this small number of batches, the estimated lag-1 autocorrelation appears to be slightly negative, illustrating our point about the difficulty of estimating correlation with small numbers of observations.

11.5.6   Steady-State Quantiles

Constructing confidence intervals for quantile estimates in a steady-state simulation can be tricky, especially if the output process of interest is a continuous-time process, such as $L_Q(t)$, the number of customers in queue at time t. In this section, we outline the main issues.

Taking the easier case first, suppose that the output process from a single replication, after appropriate deletion of initial data, is $Y_{d+1}, \ldots, Y_n$. To be concrete, Y_i might be the delay in queue of the ith customer. Then the point estimate of the pth quantile can be obtained as before, either from the histogram of the data or from the sorted values. Of course, only the data after the deletion point are used. Suppose we make R replications and let $\widehat{\theta}_r$ be the quantile estimate from the rth. Then the R quantile estimates, $\widehat{\theta}_1, \ldots, \widehat{\theta}_R$, are independent and identically distributed. Their average is

$$\widehat{\theta}_. = \frac{1}{R} \sum_{i=1}^{R} \widehat{\theta}_i$$

It can be used as the point estimator of θ; and an approximate confidence interval is

$$\widehat{\theta}_. \pm t_{\alpha/2, R-1} \frac{S}{\sqrt{R}}$$

where S^2 is the usual sample variance of $\widehat{\theta}_1, \ldots, \widehat{\theta}_R$.

What if only a single replication is obtained? Then the same reasoning applies if we let $\widehat{\theta}_i$ be the quantile estimate from *within* the ith batch of data. This requires sorting the data or forming a histogram within each batch. If the batches are large enough, then these within-batch quantile estimates will also be approximately i.i.d.

When we have a continuous-time output process, then, in principle, the same methods apply. However, we must be careful not to transform the data in a way that changes the problem. In particular, we cannot first form batch means, as we have done throughout this chapter, and then estimate the quantile from these batch means. The p quantile of the *batch means* of $L_Q(t)$ is *not* the same as the p quantile of $L_Q(t)$ itself. Thus, the quantile point estimate must be formed from the histogram of the raw data, either from each run, if we make replications, or within each batch, if we make a single replication.

11.6   Summary

This chapter emphasized the idea that a stochastic discrete-event simulation is a statistical experiment. Therefore, before sound conclusions can be drawn on the basis of the simulation-generated output data, a proper statistical analysis is required. The purpose of the simulation experiment is to obtain estimates of the performance measures of the system under study. The purpose of the statistical analysis is to acquire some assurance that these estimates are sufficiently precise for the proposed use of the model.

A distinction was made between terminating simulations and steady-state simulations. Steady-state simulation output data are more difficult to analyze, because the simulation analyst must address the problem of initial conditions and the choice of run length. Some suggestions were given regarding these problems, but unfortunately no simple, complete, and satisfactory solution exists. Nevertheless, simulation analysts should be aware of the potential problems, and of the possible solutions—namely, deletion of data and increasing of the run length. More advanced statistical techniques (not discussed in this text) are given in Alexopoulos and Seila [1998], Bratley, Fox, and Schrage [1996], and Law [2007].

The statistical precision of point estimators can be measured by a standard-error estimate or by a confidence interval. The method of independent replications was emphasized. With this method, the simulation analyst generates statistically independent observations, and thus standard statistical methods can be employed. For steady-state simulations, the method of batch means was also discussed.

The main point is that simulation output data contain some amount of random variability; without some assessment of its size, the point estimates cannot be used with any degree of reliability.

REFERENCES

ALEXOPOULOS, C., AND A. F. SEILA [1998], "Output Data Analysis," Chapter 7 in *Handbook of Simulation*, J. Banks, ed., Wiley, New York.

BRATLEY, P., B. L. FOX, AND L. E. SCHRAGE [1996], *A Guide to Simulation*, 2d ed., Springer-Verlag, New York.

FISHMAN, G. S., AND L. S. YARBERRY [1997], "An Implementation of the Batch Means Method," *INFORMS Journal on Computing*, Vol. 9, pp. 296–310.

GOLDSMAN, D., L. SCHRUBEN, AND J. J. SWAIN [1994], "Tests for Transient Means in Simulated Time Series," *Naval Research Logistics*, Vol. 41, pp. 171–187.

KELTON, W. D. [1986], "Replication Splitting and Variance for Simulating Discrete-Parameter Stochastic Processes," *Operations Research Letters*, Vol. 4, pp. 275–279.

KELTON, W. D. [1989], "Random Initialization Methods in Simulation," *IIE Transactions*, Vol. 21, pp. 355–367.

KELTON, W. D., AND A. M. LAW [1983], "A New Approach for Dealing with the Startup Problem in Discrete Event Simulation," *Naval Research Logistics Quarterly*, Vol. 30, pp. 641–658.

KLEIJNEN, J. P. C. [1987], *Statistical Tools for Simulation Practitioners*, Dekker, New York.

LADA, E. K., N. M. Steiger, AND J. R. WILSON [2006], "Performance Evaluation of Recent Procedures for Steady-state Simulation Analysis," *IIE Transactions*, Vol. 38, pp. 711–-727.

LAW, A. M. [1977], "Confidence Intervals in Discrete Event Simulation: A Comparison of Replication and Batch Means," *Naval Research Logistics Quarterly*, Vol. 24, pp. 667–78.

LAW, A. M. [1980], "Statistical Analysis of the Output Data from Terminating Simulations," *Naval Research Logistics Quarterly*, Vol. 27, pp. 131–43.

LAW, A. M. [2007], *Simulation Modeling and Analysis*, 4th ed., McGraw-Hill, New York.

LAW, A. M., AND J. S. CARSON [1979], "A Sequential Procedure for Determining the Length of a Steady-State Simulation," *Operations Research*, Vol. 27, pp. 1011–1025.

NELSON, B. L. [2001], "Statistical Analysis of Simulation Results," Chapter 94 in *Handbook of Industrial Engineering*, 3d ed., G. Salvendy, ed., Wiley, New York.

SCHMEISER, B. [1982], "Batch Size Effects in the Analysis of Simulation Output," *Operations Research*, Vol. 30, pp. 556–568.

SCHRUBEN, L. [1982], "Detecting Initialization Bias in Simulation Output," *Operations Research*, Vol. 30, pp. 569–590.

SNELL, M., AND L. SCHRUBEN [1985], "Weighting Simulation Data to Reduce Initialization Effects," *IIE Transactions*, Vol. 17, pp. 354–363.

STEIGER, N. M., AND J. R. Wilson [2002], "An Improved Batch Means Procedure for Simulation Output Analysis," *Management Science*, Vol. 48, pp. 1569–1586.

WELCH, P. D. [1983], "The Statistical Analysis of Simulation Results," in *The Computer Performance Modeling Handbook*, S. Lavenberg, ed., Academic Press, New York, pp. 268–328.

EXERCISES

1. For each of the systems described below, under what circumstances would it be appropriate to use a terminating simulation versus a steady-state simulation to analyze this system?

    (a) A walk-in medical clinic simulated to determine staffing levels.
    (b) A portfolio of stocks, bonds, and derivative securities simulated to estimate long-run return.
    (c) A fuel injector assembly plant that runs two shifts per day simulated to estimate throughput.
    (d) The United States air traffic control system simulated to evaluate safety.

2. Suppose that the output process from a queueing simulation is $L(t)$, $0 \le t \le T$, the total number in queue at time t. A continuous-time output process can be converted into the sort of discrete-time process $Y_1, Y_2, \ldots$ described in this chapter by first forming $k = T/m$ batch means of size m time units:

$$Y_j = \frac{1}{m} \int_{(j-1)m}^{jm} L(t)\, dt$$

    for $j = 1, 2, \ldots, k$. Ensemble averages of these batch means can be plotted to check for initial-condition bias.

    (a) Show algebraically that the batch mean over $2m$ time units can be obtained by averaging two adjacent batch means over m time units. [*Hint*: This implies that we can start with batch means over rather small time intervals m and build up batch means over longer intervals without reanalyzing all of the data.]
    (b) Simulate an $M/M/1$ queue with $\lambda = 1$ and $\mu = 1.25$ for a run length of 4000 time units, computing batch means of the total number in queue with batch size $m = 4$ time units. Make replications and use a mean plot to determine an appropriate number of batches to delete. Convert this number of batches to delete into a deletion time.

3. Using the results of Exercise 2 above, design and execute an experiment to estimate the steady-state expected number in this $M/M/1$ queue, L, to within ± 0.5 customers with 95% confidence. Check your estimate against the true value of $L = \lambda/(\mu - \lambda)$.

4. In Example 11.7, suppose that management desired a 95% confidence interval on the estimate of mean cycle time and that the error allowed was $\epsilon = 0.05$ hours (3 minutes). Using the same initial data given in Table 11.5, estimate the required total sample size. Although we cut the error in half, by how much did the required sample size increase?

5. Again, simulate an $M/M/1$ queue with $\lambda = 1$ and $\mu = 1.25$, this time recording customer time in system (from arrival to departure) as the performance measure for 4000 customers. Make replications and use a mean plot to determine an appropriate number of customers to delete when starting the system empty, with 4 customers initially in the system, and with 8 customers initially in the system. How does the warmup period change with these different initial conditions? What does this suggest about how to initialize simulations?

6. Consider an (M, L) inventory system, in which the procurement quantity Q is defined by

$$Q = \begin{cases} M - I & \text{if } I < L \\ 0 & \text{if } I \geq L \end{cases}$$

where I is the level of inventory on hand plus on order at the end of a month, M is the maximum inventory level, and L is the reorder point. M and L are under management control, so the pair (M, L) is called the inventory policy. Under certain conditions, the analytical solution of such a model is possible, but not always. Use simulation to investigate an (M, L) inventory system with the following properties: The inventory status is checked at the end of each month. Backordering is allowed at a cost of $4 per item short per month. When an order arrives, it will first be used to relieve the backorder. The lead time is given by a uniform distribution on the interval [0.25, 1.25] months. Let the beginning inventory level stand at 50 units, with no orders outstanding. Let the holding cost be $1 per unit in inventory per month. Assume that the inventory position is reviewed each month. If an order is placed, its cost is $60 + \$5Q$, where $60 is the ordering cost and $5 is the cost of each item. The time between demands is exponentially distributed with a mean of 1/15 month. The sizes of the demands follow this distribution:

Demand	Probability
1	1/2
2	1/4
3	1/8
4	1/8

(a) Make ten independent replications, each of run length 100 months preceded by a 12-month initialization period, for the $(M, L) = (50, 30)$ policy. Estimate long-run mean monthly cost with a 90% confidence interval.

(b) Using the results of part (a), estimate the total number of replications needed to estimate mean monthly cost within $5.

7. Reconsider Exercise 6, except that, if the inventory level at a monthly review is zero or negative, a rush order for Q units is placed. The cost for a rush order is $120 + \$12Q$, where $120 is the ordering cost and $12 is the cost of each item. The lead time for a rush order is given by a uniform distribution on the interval [0.10, 0.25] months.

(a) Make ten independent replications for the (M, L) policy, and estimate long-run mean monthly cost with a 90% confidence interval.

(b) Using the results of part (a), estimate the total number of replications needed to estimate mean monthly cost within $5.

8. Suppose that the items in Exercise 6 are perishable, with a selling price given by the following data:

On the Shelf (months)	Selling Price
0-1	$10
1-2	5
>2	0

Thus, any item that has been on the shelf more than 2 months cannot be sold. The age is measured at the time the demand occurs. If an item is outdated, it is discarded, and the next item is brought forward. Simulate the system for 100 months.

(a) Make ten independent replications for the $(M, L) = (50, 30)$ policy, and estimate long-run mean monthly cost with a 90% confidence interval.
(b) Using the results of part (a), estimate the total number of replications needed to estimate mean monthly cost within $5.

At first, assume that all the items in the beginning inventory are fresh. Is this a good assumption? What effect does this "all-fresh" assumption have on the estimates of long-run mean monthly cost? What can be done to improve these estimates? Carry out a complete analysis.

9. Consider the following inventory system:

(a) Whenever the inventory level falls to or below 10 units, an order is placed. Only one order can be outstanding at a time.
(b) The size of each order is Q. Maintaining an inventory costs $0.50 per day per item in inventory. Placing an order incurs a fixed cost of $10.00.
(c) Lead time is distributed in accordance with a discrete uniform distribution between zero and 5 days.
(d) If a demand occurs during a period when the inventory level is zero, the sale is lost at a cost of $2.00 per unit.
(e) The number of customers each day is given by the following distribution:

Number of Customers per Day	Probability
1	0.23
2	0.41
3	0.22
4	0.14

(f) The demand on the part of each customer is Poisson distributed with a mean of 3 units.
(g) For simplicity, assume that all demands occur at noon and that all orders are placed immediately thereafter.

Assume further that orders are received at 5:00 P.M., or after the demand that occurred on that day. Consider the policy having $Q = 20$. Make ten independent replications, each of length 100 days, and compute a 90% confidence interval for long-run mean daily cost. Investigate the effect of initial inventory level and existence of an outstanding order on the estimate of mean daily cost. Begin with an initial inventory of $Q + 10$ and no outstanding orders.

10. A store selling Mother's Day cards must decide 6 months in advance on the number of cards to stock. Reordering is not allowed. Cards cost $0.45 and sell for $1.25. Any cards not sold by Mother's Day go on sale for $0.50 for 2 weeks. However, sales of the remaining cards is probabilistic in nature according to the following distribution:

    32% of the time, all cards remaining get sold.

    40% of the time, 80% of all cards remaining are sold.

    28% of the time, 60% of all cards remaining are sold.

    Any cards left after 2 weeks are sold for $0.25. The card-shop owner is not sure how many cards can be sold, but thinks it is somewhere (i.e., uniformly distributed) between 200 and 400. Suppose that the card-shop owner decides to order 300 cards. Estimate the expected total profit with an error of at most $5.00. [*Hint*: Make ten initial replications. Use these data to estimate the total sample size needed. Each replication consists of one Mother's Day.]

11. A very large mining operation has decided to control the inventory of high-pressure piping by a "periodic review, order up to M" policy, where M is a target level. The annual demand for this piping is normally distributed, with mean 600 and variance 800. This demand occurs fairly uniformly over the year. The lead time for resupply is Erlang distributed of order $k = 2$ with its mean at 2 months. The cost of each unit is $400. The inventory carrying charge, as a proportion of item cost on an annual basis, is expected to fluctuate normally about the mean 0.25 (simple interest), with a standard deviation of 0.01. The cost of making a review and placing an order is $200, and the cost of a backorder is estimated to be $100 per unit backordered. Suppose that the inventory level is reviewed every 2 months, and let $M = 337$.

    (a) Make ten independent replications, each of run length 100 months, to estimate long-run mean monthly cost by means of a 90% confidence interval.
    (b) Investigate the effects of initial conditions. Calculate an appropriate number of monthly observations to delete, in order to reduce initialization bias to a negligible level.

12. Consider some number, say N, of $M/M/1$ queues in series. The $M/M/1$ queue, described in Section 6.4, has Poisson arrivals at some rate λ customers per hour, exponentially distributed service times with mean $1/\mu$, and a single server. (Recall that "Poisson arrivals" means that interarrival times are exponentially distributed.) By $M/M/1$ queues in series, it is meant that, upon completion of service at a given server, a customer joins a waiting line for the next server. The system can be shown as follows:

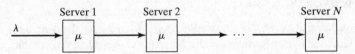

All service times are exponentially distributed with mean $1/\mu$, and the capacity of each waiting line is assumed to be unlimited. Assume that $\lambda = 8$ customers per hour and $1/\mu = 0.1$ hour. The measure of performance is response time, which is defined to be the total time a customer is in the system.

(a) By making appropriate simulation runs, compare the initialization bias for $N = 1$ (i.e., one $M/M/1$ queue) to $N = 2$ (i.e., two $M/M/1$ queues in series). Start each system with all servers idle and no customers present. The purpose of the simulation is to estimate mean response time.

(b) Investigate the initialization bias as a function of N, for $N = 1, 2, 3, 4$, and 5.

(c) Draw some general conclusions concerning initialization bias for "large" queueing systems when at time 0 the system is assumed to be empty and idle.

13. Jobs enter a job shop in random fashion according to a Poisson process at a stationary overall rate, two every 8-hour day. The jobs are of four types. They flow from work station to work station in a fixed order, depending on type, as shown below. The proportions of each type are also shown.

Type	Flow through Stations	Proportion
1	1, 2, 3, 4	0.4
2	1, 3, 4	0.3
3	2, 4, 3	0.2
4	1, 4	0.1

Processing times per job at each station depend on type, but all times are, approximately, normally distributed with mean and s.d. in hours as follows:

	Station			
Type	1	2	3	4
1	(20, 3)	(30, 5)	(75, 4)	(20, 3)
2	(18, 2)		(60, 5)	(10, 1)
3		(20, 2)	(50, 8)	(10, 1)
4	(30, 5)			(15, 2)

Station i will have c_i workers, $i = 1, 2, 3, 4$. Each job occupies one worker at a station for the duration of a processing time. All jobs are processed on a first-in–first-out basis, and all queues for waiting jobs are assumed to have unlimited capacity. Simulate the system for 800 hours, preceded by a 200-hour initialization period. Assume that $c_1 = 8$, $c_2 = 8$, $c_3 = 20$, $c_4 = 7$. Based on $R = 20$ replications, compute a 95% confidence interval for average worker utilization at each of the four stations. Also, compute a 95% confidence interval for mean total response time for each job type, where a total response time is the total time that a job spends in the shop.

14. Change Exercise 13 to give priority at each station to the jobs by type. Type I jobs have priority over type II, type II over type III, and type III over type IV. Use 800 hours as run length, 200 hours as initialization period, and $R = 20$ replications. Compute four 95% confidence intervals for mean total response time by type. Also, run the model without priorities and compute the same confidence intervals. Discuss the trade-offs when using first-in-first-out versus a priority system.

15. Consider a single-server queue with Poisson arrivals at rate $\lambda = 10.82$ per minute and normally distributed service times with mean 5.1 seconds and variance 0.98 seconds^2. It is desired to estimate the mean time in the system for a customer who, upon arrival, finds i other customers in the system; that is, to estimate

$$w_i = E(W|N = i) \quad \text{for } i = 0, 1, 2, \ldots$$

where W is a typical system time and N is the number of customers found by an arrival. For example, w_0 is the mean system time for those customers who find the system empty, w_1 is the mean system time for those customers who find one other customer present upon arrival, and so on. The estimate $\widehat{w}_i$ of w_i will be a sample mean of system times taken over all arrivals who find i in the system. Plot $\widehat{w}_i$ vs i. Hypothesize and attempt to verify a relation between w_i and i.

   (a) Simulate for a 10-hour period with empty and idle initial conditions.
   (b) Simulate for a 10-hour period after an initialization of one hour. Are there observable differences in the results of (a) and (b)?
   (c) Repeat parts (a) and (b) with service times exponentially distributed with mean 5.1 seconds.
   (d) Repeat parts (a) and (b) with deterministic service times equal to 5.1 seconds.
   (e) Find the number of replications needed to estimate $w_0, w_1, \ldots, w_6$ with a standard error for each of at most 3 seconds. Repeat parts (a)–(d), but using this number of replications.

16. At Smalltown U., there is one specialized graphics workstation for student use located across campus from the computer center. At 2:00 A.M. one day, six students arrive at the workstation to complete an assignment. A student uses the workstation for 10 ± 8 minutes, then leaves to go to the computer center to pick up their graphics output. There is a 25% chance that the run will be OK and the student will go to sleep. If it is not OK, the student returns to the workstation and waits until it becomes free. The round trip from workstation to computer center and back takes 30 ± 5 minutes. The computer becomes inaccessible at 5:00 A.M. Estimate the probability p that at least five of the six students will finish their assignment in the 3-hour period. First, make $R = 10$ replications, and compute a 95% confidence interval for p. Next, work out the number of replications needed to estimate p within $\pm.02$, and make this number of replications. Recompute the 95% confidence interval for p.

17. Four workers are spaced evenly along a conveyor belt. Items needing processing arrive according to a Poisson process at the rate of 2 per minute. Processing time is exponentially distributed, with mean 1.6 minutes. If a worker becomes idle, then he or she takes the first item to come by on the conveyor. If a worker is busy when an item comes by, that item moves down the conveyor to the next worker, taking 20 seconds between two successive workers. When a worker

finishes processing an item, the item leaves the system. If an item passes by the last worker, it is recirculated on a loop conveyor and will return to the first worker after 5 minutes.

Management is interested in having a balanced workload—that is, management would like worker utilizations to be equal. Let ρ_i be the long-run utilization of worker i, and let ρ be the average utilization of all workers. Thus, $\rho = (\rho_1 + \rho_2 + \rho_3 + \rho_4)/4$. According to queueing theory, ρ can be estimated by $\rho = \lambda/c\mu$, where $\lambda = 2$ arrivals per minute, $c = 4$ servers, and $1/\mu = 1.6$ minutes is the mean service time. Thus, $\rho = \lambda/c\mu = (2/4)1.6 = 0.8$; so, on the average, a worker will be busy 80% of the time.

(a) Make 10 independent replications, each of run length 40 hours preceded by a one-hour initialization period. Compute 95% confidence intervals for ρ_1 and ρ_4. Draw conclusions concerning workload balance.

(b) Based on the same 10 replications, test the hypothesis $H_0{:}\rho_1 = 0.8$ at a level of significance $\alpha = 0.05$. If a difference of $\pm.05$ is important to detect, determine the probability that such a deviation is detected. In addition, if it is desired to detect such a deviation with probability at least 0.9, figure out the sample size needed to do so. [*Hint*: See any basic statistics textbook for guidance on hypothesis testing.]

(c) Repeat (b) for $H_0{:}\rho_4 = 0.8$.

(d) From the results of (a)–(c), draw conclusions for management about the balancing of workloads.

18. At a small rock quarry, a single power shovel dumps a scoop full of rocks at the loading area approximately every 10 minutes, with the actual time between scoops modeled well as being exponentially distributed, with mean 10 minutes. Three scoops of rocks make a pile; whenever one pile of rocks is completed, the shovel starts a new pile.

The quarry has a single truck that can carry one pile (3 scoops) at a time. It takes approximately 27 minutes for a pile of rocks to be loaded into the truck and for the truck to drive to the processing plant, unload, and return to the loading area. The actual time to do these things altogether is modeled well as being normally distributed, with mean 27 minutes and standard deviation 12 minutes.

When the truck returns to the loading area, it will load and transport another pile if one is waiting to be loaded; otherwise, it stays idle until another pile is ready. For safety reasons, no loading of the truck occurs until a complete pile is waiting.

The quarry operates in this manner for an 8-hour day. We are interested in estimating the utilization of the trucks and the expected number of piles waiting to be transported if an additional truck is purchased.

19. Big Bruin, Inc. plans to open a small grocery store in Juneberry, NC. They expect to have two checkout lanes, with one lane being reserved for customers paying with cash. The question they want to answer is: How many grocery carts do they need?

During business hours (6 A.M.–8 P.M.), cash-paying customers are expected to arrive at 8 per hour. All other customers are expected to arrive at 9 per hour. The time between arrivals of each type can be modeled as exponentially distributed random variables.

The time spent shopping is modeled as normally distributed, with mean 40 minutes and standard deviation 10 minutes. The time required to check out after shopping can be modeled as lognormally distributed, with (a) mean 4 minutes and standard deviation 1 minute for cash-paying customers; (b) mean 6 minutes and standard deviation 1 minute for all other customers.

We will assume that every customer uses a shopping cart and that a customer who finishes shopping leaves the cart in the store so that it is available immediately for another customer. We will also assume that any customer who cannot obtain a cart immediately leaves the store, disgusted.

The primary performance measures of interest to Big Bruin are the expected number of shopping carts in use and the expected number of customers lost per day. Recommend a number of carts for the store, remembering that carts are expensive, but so are lost customers.

20. Develop a simulation model of the total time in the system for an $M/M/1$ queue with service rate $\mu = 1$; therefore, the traffic intensity is $\rho = \lambda/\mu = \lambda$, the arrival rate. Use the simulation, in conjunction with the technique of plotting ensemble averages, to study the effect of traffic intensity on initialization bias when the queue starts empty. Specifically, see how the initialization phase T_0 changes for $\rho = 0.5, 0.7, 0.8, 0.9, 0.95$.

21. Many simulation-software tools come with built-in support for output analysis. Use the web to research the features of one of these products.

12

Estimation of Relative Performance

Chapter 11 dealt with the precise estimation of measures of performance. This chapter discusses a few of the many statistical methods that can be used to compare the relative performance of two or more system designs. This is one of the most important uses of simulation. Because the observations of the response variables contain random variation, statistical analysis is needed to discover whether any observed differences are due to differences in design or merely to the random fluctuation inherent in the models.

The comparison of two system designs is easier than the simultaneous comparison of multiple system designs. Section 12.1 discusses the case of two system designs, using two possible statistical techniques: *independent sampling* and *correlated sampling*. Correlated sampling is also known as the *common random numbers* (CRN) technique; simply put, the same random numbers are used to simulate both alternative system designs. If implemented correctly, CRN usually reduces the variance of the estimated difference of the performance measures and thus can provide, for a given sample size, more precise estimates of the mean difference than can independent sampling. Section 12.2 extends the statistical techniques of Section 12.1 to the comparison of multiple (more than two) system designs using simultaneous confidence intervals, screening and selection of the best. These approaches are somewhat limited with respect to the number of system designs that can be considered, so Section 12.3 describes how a large number of complex system designs can sometimes be represented by a simpler metamodel. Finally, for comparison and evaluation of a very large number of system designs that are related in a less structured way, Section 12.4 presents optimization via simulation.

12.1   Comparison of Two System Designs

Suppose that a simulation analyst desires to compare two possible configurations of a system. In a queueing system, perhaps two possible queue disciplines, or two possible sets of servers, are to be compared. In a supply-chain inventory system, perhaps two possible ordering policies will be compared. A job shop could have two possible scheduling rules; a production system could have in-process inventory buffers of various capacities. Many other examples of alternative system designs can be provided.

The method of replications will be used to analyze the output data. The mean performance measure for system i will be denoted by θ_i, $i = 1, 2$. If it is a steady-state simulation, it will be assumed that deletion of data, or other appropriate techniques, have been used to ensure that the point estimators are approximately unbiased estimators of the mean performance measures θ_i. The goal of the simulation experiments is to obtain point and interval estimates of the difference in mean performance, namely $\theta_1 - \theta_2$. Two methods of computing a confidence interval for $\theta_1 - \theta_2$ will be discussed, but first an example and a general framework will be given.

Example 12.1
Recall the Software Made Personal (SMP) call center problem, Example 11.7: SMP has a customer support call center where 7 operators handle questions from owners of their software from 8 A.M. to 4 P.M. Eastern Time. When a customer calls they use an automated system to select between the two product lines, finance and contact management. Currently each product line has its own operators (4 for financial, 3 for contact management) and hold queue. SMP wonders if they can reduce the total number of operators needed by cross-training them so that they can answer calls for any product line. This is expected to increase the time to process a call by about 10%. The current system is illustrated in Figure 12.1(a). and the alternative system design is shown in Figure 12.1(b).

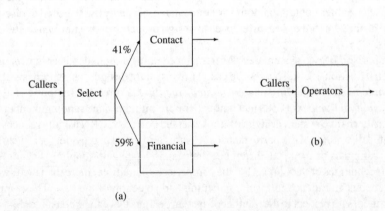

(a)

Figure 12.1   Current SMP call center (a) and an alternative design (b).

Table 12.1 Simulation Output Data and Summary Measures for Comparing Two Systems

System	Replication 1	2	$\cdots$	R_i	Sample Mean	Sample Variance
1	Y_{11}	Y_{21}	$\cdots$	$Y_{R_1 1}$	$\bar{Y}_{\cdot 1}$	S_1^2
2	Y_{12}	Y_{22}	$\cdots$	$Y_{R_2 2}$	$\bar{Y}_{\cdot 2}$	S_2^2

Before considering reducing operators, SMP wants to know what their quality of service would be if they simply cross-train the 7 operators they currently have. Incoming calls have been modeled as a nonstationary Poisson arrival process and distributions have been fit to operator time to answer questions. Comparisons will be based on the average time from call initiation until the customer's questions are answered.

When comparing two systems, such as those in Example 12.1, the simulation analyst must decide on a run length $T_E^{(i)}$ for each model $i = 1, 2$, a number of replications R_i to be made of each model, or both. From replication r of system i, the simulation analyst obtains an estimate Y_{ri} of the mean performance measure θ_i. In Example 12.1, Y_{ri} would be the average call response time observed during replication r for system i, where $r = 1, 2, \ldots, R_i$; and $i = 1, 2$. The data, together with the two summary measures, the sample means $\bar{Y}_{\cdot i}$, and the sample variances S_i^2, are exhibited in Table 12.1. Assuming that the estimators Y_{ri} are, at least approximately, unbiased, it follows that

$$\theta_1 = E(Y_{r1}), \quad r = 1, \ldots, R_1; \quad \theta_2 = E(Y_{r2}), r = 1, \ldots, R_2$$

In Example 12.1, SMP is initially interested in a comparison of two system designs (same number of operators, but specialized vs. cross trained), so the simulation analyst decides to compute a confidence interval for $\theta_1 - \theta_2$, the difference between the two mean performance measures. This will lead to one of three possible conclusions:

1. If the confidence interval (c.i.) for $\theta_1 - \theta_2$ is totally to the left of zero, as shown in Figure 12.2(a), then there is strong evidence for the hypothesis that $\theta_1 - \theta_2 < 0$, or equivalently $\theta_1 < \theta_2$.

In Example 12.1, $\theta_1 < \theta_2$ implies that the mean response time for system 1, the original system, is smaller than for system 2, the alternative system.

2. If the c.i. for $\theta_1 - \theta_2$ is totally to the right of zero, as shown in Figure 12.2(b), then there is strong evidence that $\theta_1 - \theta_2 > 0$, or equivalently, $\theta_1 > \theta_2$.

In Example 12.1, $\theta_1 > \theta_2$ can be interpreted as system 2 being better than system 1, in the sense that system 2 has a smaller mean response time.

3. If the c.i. for $\theta_1 - \theta_2$ contains zero, then, with the data at hand, there is no strong statistical evidence that one system design is better than the other.

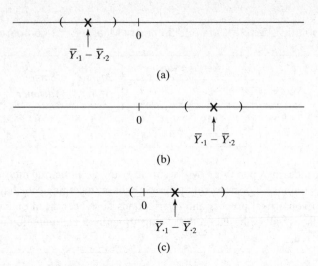

Figure 12.2   Three confidence intervals that can occur in the comparing of two systems.

Some statistics textbooks say that the weak conclusion $\theta_1 = \theta_2$ can be drawn, but such statements can be misleading. A "weak" conclusion is often no conclusion at all. Most likely, if enough additional data were collected (i.e., R_i increased), the c.i. would shift, and definitely shrink in length, until conclusion 1 or 2 would be drawn. Since the confidence interval provides a measure of the precision of the estimator of $\theta_1 - \theta_2$, we want to shrink it until we can either conclude that there is a difference that matters, or the difference is so small that we do not need to separate them.

In this chapter, a two-sided $100(1 - \alpha)\%$ c.i. for $\theta_1 - \theta_2$ will always be of the form

$$\bar{Y}_{\cdot 1} - \bar{Y}_{\cdot 2} \pm t_{\alpha/2,\nu}\,\text{s.e.}(\bar{Y}_{\cdot 1} - \bar{Y}_{\cdot 2}) \tag{12.1}$$

where $\bar{Y}_{\cdot i}$ is the sample mean performance measure for system i over all replications

$$\bar{Y}_{\cdot i} = \frac{1}{R_i} \sum_{r=1}^{R_i} Y_{ri} \tag{12.2}$$

ν is the degrees of freedom associated with the variance estimator, $t_{\alpha/2,\nu}$ is the $100(1-\alpha/2)$ percentage point of a t distribution with ν degrees of freedom, and s.e.$(\cdot)$ represents the standard error of the specified point estimator. To obtain the standard error and the degrees of freedom, the analyst uses one of two statistical techniques. Both techniques assume that the basic data, Y_{ri} of Table 12.1, are approximately normally distributed. This assumption is reasonable provided that each Y_{ri} is itself a sample mean of observations from replication r, which is indeed the situation in Example 12.1.

By design of the simulation experiment, $Y_{r1}(r = 1, \ldots, R_1)$ are independently and identically distributed (i.i.d.) with mean θ_1 and variance σ_1^2, say. Similarly, $Y_{r2}(r = 1, \ldots, R_2)$ are i.i.d. with mean θ_2 and variance σ_2^2, say. The two techniques for computing the confidence interval in Equation (12.1), which are based on different sets of assumptions, are discussed in the following subsections.

There is an important distinction between *statistically significant* differences and *practically significant* differences in systems performance. Statistical significance answers the following question: Is the observed difference $\bar{Y}_{.1} - \bar{Y}_{.2}$ larger than the variability in $\bar{Y}_{.1} - \bar{Y}_{.2}$? This question can be restated as: Have we collected enough data to be confident that the difference we observed is real, or just chance? Conclusions 1 and 2 above imply a statistically significant difference, while conclusion 3 implies that the observed difference is not statistically significant, even though the systems may indeed be different. Statistical significance is a function of the simulation experiment and the output data.

Practical significance answers the following question: Is the true difference $\theta_1 - \theta_2$ large enough to matter for the decision we need to make? In Example 12.1, we may reach the conclusion that $\theta_1 > \theta_2$ and decide that system 2 is better (smaller expected response time). However, if the actual difference $\theta_1 - \theta_2$ is very small—say, small enough that a customer would not notice the improvement—then it might not be worth the cost to replace system 1 with system 2. Practical significance is a function of the actual difference between the systems and is independent of the simulation experiment.

Confidence intervals do not answer the question of practical significance directly. Instead, they bound, with probability $1 - \alpha$, the true difference $\theta_1 - \theta_2$ within the range

$$\bar{Y}_{.1} - \bar{Y}_{.2} - t_{\alpha/2,\nu}\text{s.e.}(\bar{Y}_{.1} - \bar{Y}_{.2}) \leq \theta_1 - \theta_2 \leq \bar{Y}_{.1} - \bar{Y}_{.2} + t_{\alpha/2,\nu}\text{s.e.}(\bar{Y}_{.2})$$

Whether a difference within these bounds is practically significant depends on the particular problem.

12.1.1   Independent Sampling

Independent sampling means that different and independent random number streams will be used to simulate the two systems. This implies that all the observations of simulated system 1, namely $\{Y_{r1}, r = 1, \ldots, R_1\}$, are statistically independent of all the observations of simulated system 2, namely $\{Y_{r2}, r = 1, \ldots, R_2\}$. By Equation (12.2) and the independence of the replications, the variance of the sample mean $\bar{Y}_{.i}$ is given by

$$\text{V}(\bar{Y}_{.i}) = \frac{\text{V}(Y_{ri})}{R_i} = \frac{\sigma_i^2}{R_i}, \qquad i = 1, 2$$

For independent sampling, $\bar{Y}_{.1}$ and $\bar{Y}_{.2}$ are statistically independent; hence,

$$\text{V}(\bar{Y}_{.1} - \bar{Y}_{.2}) = \text{V}(\bar{Y}_{.1}) + \text{V}(\bar{Y}_{.2})$$

$$= \frac{\sigma_1^2}{R_1} + \frac{\sigma_2^2}{R_2} \tag{12.3}$$

There are confidence-interval procedures that are appropriate when the two variances are equal but unknown in value; that is, $\sigma_1^2 = \sigma_2^2$. However, since we would rarely know this to be true in practice, and the advantages of exploiting equal variances are slight provided the number of replications is not too small, we refer the reader to any statistics textbook for this case (for instance Hines *et al.* [2002]).

An approximate $100(1 - \alpha)\%$ c.i. for $\theta_1 - \theta_2$ can be computed as follows. The point estimate is computed by

$$\bar{Y}_{\cdot 1} - \bar{Y}_{\cdot 2} \tag{12.4}$$

with $\bar{Y}_{\cdot i}$ given by Equation (12.2), while the sample variance for system i is

$$S_i^2 = \frac{1}{R_i - 1} \sum_{r=1}^{R_i} (Y_{ri} - \bar{Y}_{\cdot i})^2$$

$$= \frac{1}{R_i - 1} \left(\sum_{r=1}^{R_i} Y_{ri}^2 - R_i \bar{Y}_{\cdot i}^2 \right) \tag{12.5}$$

Thus, the standard error of the point estimate is given by

$$\text{s.e.}(\bar{Y}_{\cdot 1} - \bar{Y}_{\cdot 2}) = \sqrt{\frac{S_1^2}{R_1} + \frac{S_2^2}{R_2}} \tag{12.6}$$

with degrees of freedom ν approximated by the expression

$$\nu = \frac{(S_1^2/R_1 + S_2^2/R_2)^2}{[(S_1^2/R_1)^2/(R_1 - 1)] + [(S_2^2/R_2)^2/(R_2 - 1)]} \tag{12.7}$$

rounded to an integer. The confidence interval is then given by Expression (12.1), using the standard error of Equation (12.6). A minimum number of replications $R_1 \geq 6$ and $R_2 \geq 6$ is recommended for this procedure.

12.1.2   Common Random Numbers (CRN)

CRN (also known as *correlated sampling*) means that, for each replication, the same random numbers are used to simulate both systems. Therefore, R_1 and R_2 must be equal, say, $R_1 = R_2 = R$. Thus, for each replication r, the two estimates Y_{r1} and Y_{r2} are no longer independent, but rather are correlated. However, independent random numbers are used on different replications, so the pairs (Y_{r1}, Y_{s2}) are mutually independent when $r \neq s$. For example, in Table 12.1, the observation Y_{11} is correlated with Y_{12}, but Y_{11} is independent of all other observations. The purpose of using CRN is to induce a positive correlation between Y_{r1} and Y_{r2} for each r and thus to achieve a variance reduction in the point estimator of mean difference, $\bar{Y}_{\cdot 1} - \bar{Y}_{\cdot 2}$. In general, this variance is given by

$$V(\bar{Y}_{\cdot 1} - \bar{Y}_{\cdot 2}) = V(\bar{Y}_{\cdot 1}) + V(\bar{Y}_{\cdot 2}) - 2\text{cov}(\bar{Y}_{\cdot 1}, \bar{Y}_{\cdot 2})$$

$$= \frac{\sigma_1^2}{R} + \frac{\sigma_2^2}{R} - \frac{2\rho_{12}\sigma_1\sigma_2}{R} \tag{12.8}$$

where ρ_{12} is the correlation between Y_{r1} and Y_{r2}. [By definition, $\rho_{12} = \text{cov}(Y_{r1}, Y_{r2})/\sigma_1\sigma_2$, which does not depend on r.]

Now, compare the variance of $\bar{Y}_{.1} - \bar{Y}_{.2}$ arising from the use of CRN [Equation (12.8), call it V_{CRN}] to the variance arising from the use of independent sampling with equal sample sizes [Equation (12.3) with $R_1 = R_2 = R$, call it V_{IND}]. Notice that

$$V_{CRN} = V_{IND} - \frac{2\rho_{12}\sigma_1\sigma_2}{R} \tag{12.9}$$

If CRN works as intended, the correlation ρ_{12} will be positive; hence, the second term on the right side of Equation (12.9) will be positive, and, therefore,

$$V_{CRN} < V_{IND}$$

That is, the variance of the point estimator will be smaller with CRN than with independent sampling. A smaller variance, for the same sample size R, implies that the estimator based on CRN is more precise, leading to a shorter confidence interval on the difference, which implies that smaller differences in performance can be detected.

To compute a $100(1 - \alpha)\%$ c.i. with correlated data, first compute the differences

$$D_r = Y_{r1} - Y_{r2} \tag{12.10}$$

which, by the definition of CRN, are i.i.d.; then compute the sample mean difference as

$$\bar{D} = \frac{1}{R}\sum_{r=1}^{R} D_r \tag{12.11}$$

Thus, $\bar{D} = \bar{Y}_{.1} - \bar{Y}_{.2}$. The sample variance of the differences $\{D_r\}$ is computed as

$$S_D^2 = \frac{1}{R-1}\sum_{r=1}^{R}(D_r - \bar{D})^2$$

$$= \frac{1}{R-1}\left(\sum_{r=1}^{R} D_r^2 - R\bar{D}^2\right) \tag{12.12}$$

which has degrees of freedom $\nu = R-1$. The $100(1-\alpha)\%$ c.i. for $\theta_1 - \theta_2$ is given by Expression (12.1), with the standard error of $\bar{Y}_{.1} - \bar{Y}_{.2} = \bar{D}$, estimated by

$$\text{s.e.}(\bar{D}) = \text{s.e.}(\bar{Y}_{.1} - \bar{Y}_{.2}) = \frac{S_D}{\sqrt{R}} \tag{12.13}$$

Because $S_D/\sqrt{R}$ of Equation (12.13) is an estimate of $\sqrt{V_{CRN}}$ and Expression (12.6) is an estimate of $\sqrt{V_{IND}}$, CRN typically will produce a c.i. that is shorter for a given sample size than the c.i. produced by independent sampling if $\rho_{12} > 0$. In fact, the expected length of the c.i. will be shorter with the use of CRN if $\rho_{12} > 0.1$, provided $R > 10$. The larger R is, the smaller ρ_{12} can be and still yield a shorter expected length [Nelson 1987].

For any problem, there are many ways of implementing common random numbers. It is never enough to simply use the same seed on the random-number generator(s). Each random number used in one model for some purpose should be used for the same purpose in the second model—that is, the use of the random numbers must be synchronized. For example, if the ith random number is used to generate the call service time of an operator for the 5th caller in model 1, then the ith random number should be used for the very same purpose in model 2. For queueing systems or service facilities, synchronization of the common random numbers guarantees that the two systems face identical work loads: both systems face arrivals at the same instants of time, and these arrivals demand equal amounts of service. (The actual service times of a given arrival in the two models may not be equal; they could be proportional if the server in one model were faster than the server in the other model.) For an inventory system, in comparing different ordering policies, synchronization guarantees that the two systems face identical demand for a given product. For production or reliability systems, synchronization guarantees that downtimes for a given machine will occur at exactly the same times and will have identical durations, in the two models. On the other hand, if some aspect of one of the systems is totally different from the other system, synchronization could be inappropriate—or even impossible to achieve. In summary, those aspects of the two system designs that are sufficiently similar should be simulated with common random numbers in such a way that the two models behave similarly; but those aspects that are totally different should be simulated with independent random numbers.

Implementation of common random numbers is model-dependent, but certain guidelines can be given that will make CRN more likely to yield a positive correlation. The purpose of the guidelines is to ensure that synchronization occurs:

1. Dedicate a random-number stream to a specific purpose, and use as many different streams as needed. (Different random-number generators, or widely spaced seeds on the same generator, can be used to get two different, nonoverlapping streams.) In addition, assign independently chosen seeds to each stream at the beginning of each replication. It is not sufficient to assign seeds at the beginning of the first replication and then let the random-number generator merely continue for the second and subsequent replications. If simulation is conducted in this manner, the first replication will be synchronized, but subsequent replications might not be.
2. For systems (or subsystems) with external arrivals: As each entity enters the system, the next interarrival time is generated, and then immediately all random variables (such as service times, order sizes, etc.) needed by the arriving entity and identical in both models are generated in a fixed order and stored as attributes of the entity, to be used later as needed. Apply guideline 1: Dedicate one random-number stream to these external arrivals and all their attributes.
3. For systems having an entity performing given activities in a cyclic or repeating fashion, assign a random-number stream to this entity. (Example: a machine that cycles between two states: up–down–up–down–.... Use a dedicated random-number stream to generate the uptimes and downtimes.)
4. If synchronization is not possible, or if it is inappropriate for some part of the two models, use independent streams of random numbers for this subset of random variates.

Unfortunately, there is no guarantee that CRN will always induce a positive correlation between comparable runs of the two models. It is known that if, for each input random variate X, the estimators Y_{r1} and Y_{r2} are increasing functions of the random variate X (or both are decreasing functions of X), then ρ_{12} will be positive. The intuitive idea is that both models, that is, both Y_{r1} and Y_{r2}, respond in the same direction to each input random variate, and this results in positive correlation. This increasing or decreasing nature of the response variables, called *monotonicity*, with respect to the input random variables, is known to hold for certain queueing systems, such as the $GI/G/c$ queues, when the response variable is customer delay, so some evidence exists that common random numbers is a worthwhile technique for queueing simulations. (For simple queues, customer delay is an increasing function of service times and a decreasing function of interarrival times.) Wright and Ramsay [1979] reported a negative correlation for certain inventory simulations, however. In summary, the guidelines recently described should be followed, and some reasonable notion that the response variable of interest is a monotonic function of the random input variables should be evident.

Example 12.1:   (cont.) _____

SMP's call center designs in Figure 12.1 will be compared by simulating them with CRN. To illustrate several issues we will look at two comparisons: the current system with 7 operators vs. cross-trained system with 7 operators; and a cross-trained system with 7 operators vs. cross-trained system with 6 operators. We will also assess how much benefit was obtained by using CRN relative to independent sampling

There are four driving input processes in the SMP problem: The caller interarrival times, the caller type, the call times for financial software callers, and the call times for the contact management software callers. By "call time" we mean the time that an operator spends with a caller, excluding the time the caller spends on hold.

To implement CRN, we dedicate a distinct random number stream to each input process. This guarantees that precisely the same sequence of calls and caller types will be seen by each alternative on each replication. The call times could be generated in one of two ways, however: Either generate them as needed (that is, when a caller reaches an operator, generate the call time from the appropriate distribution), or generate the call time at the moment that the caller arrives, save it as an attribute of the caller, and use it when the caller reaches an operator. The latter approach (generate the call time at arrival and save it) provides better synchronization in this case; in fact, if we generate call times for the cross-trained alternative by multiplying call times from the current system by 1.1, then this approach guarantees that each caller in the cross-trained alternatives has call time that is precisely 10% greater than they would have experienced in the current design.

Table 12.2 shows the first 10 replications from a run of 300 replications of the three system designs (the complete data are available from www.bcnn.net). Figure 12.3 plots the average response times for the current system and the cross-trained system with 7 operators for these same 10 replications. The desired impact of CRN is apparent in the figure: The response averages from the two system designs tend to go up and down together, making their differences easier to observe. The sample correlation between them, based on the entire 300 replications, is 0.24, showing a modest impact of CRN. The sample correlation, again based on all 300 replications, between the proposed systems with 7 and 6 operators is a substantial 0.94, however. As shown in Figure 12.4, the responses of these two system designs track distinctly up and down together, revealing a strong positive correlation.

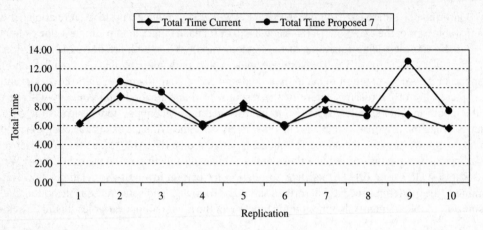

Figure 12.3   Average response times from the first 10 replications of the current and proposed (7 operators) SMP call center

A more quantitative way to see the impact of CRN is to look at the standard errors, because they directly impact the length of the confidence interval. For instance, the standard error for the comparison of the two proposed systems is 0.88 minutes. We can approximate what the standard error would have been with independent sampling by using the sample variances and Equation (12.6):

$$\text{s.e.} = \sqrt{\frac{4.86}{10} + \frac{23.99}{10}} = 1.70$$

The reduction is almost 50%, making it far easier to detect differences.

Table 12.2   Comparison of System Designs for the SMP Call Center

	Average Response Time			Differences	
Replication	Current	Proposed 7	Proposed 6	C - P7	P7 - P6
1	6.24	6.19	8.35	0.05	−2.16
2	9.06	10.64	18.03	−1.59	−7.39
3	8.02	9.53	16.17	−1.51	−6.64
4	5.93	6.15	7.40	−0.22	−1.25
5	8.31	7.83	12.70	0.48	−4.87
6	5.91	6.09	8.26	−0.17	−2.17
7	8.74	7.62	12.32	1.12	−4.70
8	7.78	7.03	11.40	0.75	−4.37
9	7.15	12.79	23.04	−5.64	−10.24
10	5.72	7.57	14.30	−1.85	−6.73
Sample mean	7.29	8.14	13.20	−0.86	−5.05
Sample variance	1.60	4.86	23.99	3.87	7.72
Standard error				0.62	0.88

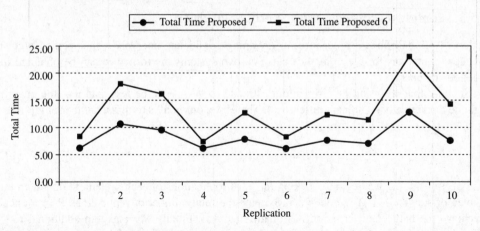

Figure 12.4   Average response times from the first 10 replications of the proposed designs with 7 and 6 operators for the SMP call center

Understanding why the strength of the induced correlation due to CRN is different in these two cases helps in understanding when CRN will be most effective. The two proposed designs (6 and 7 operators) are structurally the same system, differing only in the number of operators. Therefore, when the random inputs are such that the system with 7 operators is congested, the system with 6 operators is certain to be at least as congested, and probably more so. The current system is structurally different from the alternatives, having a distinct queue for each type of caller. When congestion occurs in the financial software queue it may not occur in the contact management queue, so long response times for the financial callers could be averaged with shorter response times for the contact management callers. Thus, the relationship with the proposed system that has a single queue is not as strong.

Let the subscript C denote the current call center, $P7$ the proposal with 7 operators, and $P6$ the proposal with 6 operators. Then a 95% confidence interval for the difference $\theta_C - \theta_{P7}$ is

$$-0.86 \pm 2.26(0.62)$$

or

$$-2.26 \le \theta_C - \theta_{P7} \le 0.54 \tag{12.14}$$

Therefore, from these 10 replications we cannot say for certain which design has a lower response time, although we can say with high confidence that the difference is within ± 1.4 minutes.

On the other hand, and as we might expect, the 7 and 6 operator systems are distinctly different, with 95% confidence interval

$$-7.02 \le \theta_{P7} - \theta_{P6} \le -3.06$$

showing that having a 7th operator is at least 3 minutes faster, on average, than only having 6.

12.1.3   Confidence Intervals with Specified Precision

Section 11.4.2 described a procedure for obtaining confidence intervals with specified precision. Confidence intervals for the *difference* between two systems' performance can be obtained in an analogous manner.

Suppose that we want the error in our estimate of $\theta_1 - \theta_2$ to be less than $\pm\epsilon$ (the quantity ϵ might be a practically significant difference). Therefore, our goal is to find a number of replications R such that

$$H = t_{\alpha/2,\nu}\text{s.e.}(\bar{Y}_{\cdot 1} - \bar{Y}_{\cdot 2}) \leq \epsilon \qquad (12.15)$$

As in Section 11.4.2, we begin by making $R_0 \geq 10$ replications of each system, to obtain an initial estimate of s.e.$(\bar{Y}_{\cdot 1} - \bar{Y}_{\cdot 2})$. We then approximate the total number of replications $R \geq R_0$ needed to achieve the half-length criterion in Equation (12.15). Finally, we make an additional $R - R_0$ replications (or a fresh R replications) of each system, compute the confidence interval, and check that the half-length criterion has been attained.

Example 12.1:   (cont.)

Recall that $R_0 = 10$ replications and CRN yielded the 95% confidence interval for the difference in expected response time of the current and 7-operator cross-trained system in Equation (12.14); this interval can be rewritten as -0.86 ± 1.40 minutes. Although the current call center design appears to have the smaller expected response time, the difference is not statistically significant, since the confidence interval contains 0. Suppose that a difference larger than ± 0.25 minutes is considered to be practically significant. We therefore want to make enough replications to obtain $H \leq \epsilon = 0.25$.

The confidence interval used in Example 12.1 was of the form $\bar{D} \pm t_{\alpha/2,R_0-1}S_D/\sqrt{R_0}$, with the specific values $\bar{D} = -0.86$, $R_0 = 10$, $t_{0.025,9} = 2.26$, and $S_D^2 = 3.87$. To obtain the desired precision, we need to find R such that

$$\frac{t_{\alpha/2,R-1}S_D}{\sqrt{R}} \leq \epsilon$$

We can approximate this by finding the smallest integer R satisfying $R \geq R_0$ and

$$R \geq \left(\frac{z_{\alpha/2}S_D}{\epsilon}\right)^2$$

Substituting $z_{0.025} = 1.96$ and $S_D^2 = 3.87$, we obtain

$$R \geq \frac{(1.96)^2(3.87)}{(0.25)^2} = 238$$

implying that 238 replications are needed, 228 more than in the initial experiment. There are 300 replications available at www.bcnn.net and the reader is encouraged to check the result.

12.2   Comparison of Several System Designs

Suppose that a simulation analyst desires to compare K alternative system designs. The comparison will be made on the basis of some specified performance measure θ_i of system i, for $i = 1, 2, \ldots, K$. Many different statistical procedures have been developed that can be used to analyze simulation data and draw statistically sound inferences concerning the parameters θ_i. These procedures can be classified as being either *fixed-sample-size procedures* or *sequential-sampling* (or *multistage*) procedures.

In the first type, a predetermined sample size (i.e., run length and number of replications) is used to draw inferences via hypothesis tests or confidence intervals. Examples of fixed-sample-size procedures include the interval estimation of a mean performance measure (Section 11.3.2) and the interval estimation of the difference between mean performance measures of two systems [as by Expression (12.1) in Section 12.1]. Advantages of fixed-sample-size procedures include a known or easily estimated cost in terms of computer time before running the experiments. When computer time is limited, or when a pilot study is being conducted, a fixed-sample-size procedure might be appropriate. In some cases, clearly inferior system designs may be ruled out at this early stage. A major disadvantage is that a strong conclusion could be impossible. For example, the confidence interval could be too wide for practical use, since the width is an indication of the precision of the point estimator. A hypothesis test may lead to a failure to reject the null hypothesis, a weak conclusion in general, meaning that there is no strong evidence one way or the other about the truth or falsity of the null hypothesis.

A sequential sampling scheme is one in which more and more data are collected until an estimator with a prespecified precision is achieved or until one of several alternatives is selected, with the probability of correct selection being larger than a prespecified value. A two-stage (or multistage) procedure is one in which an initial sample is used to estimate how many additional observations are needed to draw conclusions with a specified precision. An example of a two-stage procedure for estimating the performance measure of a single system was given in Sections 11.4.2 and for two systems in Section 12.1.3.

The proper procedure to use depends on the goal of the simulation analyst. Some possible goals are the following:

1. estimation of each parameter, θ_i;
2. comparison of each performance measure θ_i to a control θ_1 (where θ_1 could represent the mean performance of an existing system);
3. all pairwise comparisons, $\theta_i - \theta_j$, for $i \neq j$;
4. selection of the best θ_i (largest or smallest).

The first three goals will be achieved by the construction of confidence intervals. The number of such confidence intervals is $C = K$, $C = K-1$, and $C = K(K-1)/2$, respectively. Hochberg and Tamhane [1987] and Hsu [1996] are comprehensive references for such multiple-comparison procedures. The fourth goal requires the use of a type of statistical procedure known as a multiple ranking and selection procedure. Procedures to achieve these and other goals are discussed by Kleijnen [1975, Chapters II and V], who also discusses their relative merit and disadvantages. Goldsman and Nelson [1998] and Law [2007] discuss those selection procedures most relevant to simulation. A comprehensive reference is Bechhofer, Santner, and Goldsman [1995]. The next subsection presents a fixed-sample-size procedure that can be used to meet goals 1, 2, and 3, and is applicable in a wide range of circumstances. Subsection 12.2.2 presents a procedure to achieve goal 4.

12.2.1   Bonferroni Approach to Multiple Comparisons

Suppose that C confidence intervals are computed and that the ith interval has confidence coefficient $1 - \alpha_i$. Let S_i be the statement that the ith confidence interval contains the parameter (or difference of two parameters) being estimated. This statement might be true or false for a given set of data, but the procedure leading to the interval is designed so that statement S_i will be true with probability $1 - \alpha_i$. When it is desired to make statements about several parameters simultaneously, as in goals 1, 2 and 3, the analyst would like to have high confidence that *all* statements are true simultaneously. The Bonferroni inequality states that

$$P(\text{all statements } S_i \text{ are true}, i = 1, \dots, C) \geq 1 - \sum_{j=1}^{C} \alpha_j = 1 - \alpha_E \qquad (12.16)$$

where $\alpha_E = \sum_{j=1}^{C} \alpha_j$ is called the *overall error probability*. Expression (12.16) can be restated as

$$P(\text{one or more statements } S_i \text{ is false}, i = 1, \dots, C) \leq \alpha_E$$

or equivalently,

$$P(\text{one or more of the } C \text{ confidence intervals does not} \\ \text{contain the parameter being estimated}) \leq \alpha_E$$

Thus, α_E provides an upper bound on the probability of a false conclusion. To conduct an experiment that involves making C comparisons, first select the overall error probability, say, $\alpha_E = 0.05$ or 0.10. The individual α_j may be chosen to be equal ($\alpha_j = \alpha_E/C$), or unequal, as desired. The smaller the value of α_j, the wider the jth confidence interval will be. For example, if two 95% c.i.s ($\alpha_1 = \alpha_2 = 0.05$) are constructed, the overall confidence level will be 90% or greater ($\alpha_E = \alpha_1 + \alpha_2 = 0.10$). If ten 95% c.i.s are constructed ($\alpha_i = 0.05, i = 1, \dots, 10$), the resulting overall confidence level could be as low as 50% ($\alpha_E = \sum_{i=1}^{10} \alpha_i = 0.50$), which is far too low for practical use. To guarantee an overall confidence level of 95%, when 10 comparisons are being made, one approach is to construct ten 99.5% confidence intervals for the parameters (or differences) of interest.

   The Bonferroni approach to multiple confidence intervals is based on Expression (12.16). A major advantage is that it holds whether the models for the alternative designs are run with independent sampling or with common random numbers.

   The major disadvantage of the Bonferroni approach in making a large number of comparisons is the increased width of each individual interval. For example, for a given set of data and a large sample size, a 99.5% c.i. will be $z_{0.0025}/z_{0.025} = 2.807/1.96 = 1.43$ times longer than a 95% c.i. For small sample sizes, say, for a sample of size 5, a 99.5% c.i. will be $t_{0.0025,4}/t_{0.025,4} = 5.598/2.776 = 1.99$ times longer than an individual 95% c.i. The width of a c.i. is a measure of the precision of the estimate. For these reasons, it is recommended that the Bonferroni approach be used only when a small number of comparisons are being made. Twenty or so comparisons appear to be the practical upper limit.

Corresponding to goals 1, 2, and 3, there are at least three possible ways of using the Bonferroni Inequality (12.16) when comparing K alternative system designs:

1. *Individual c.i.s*: Construct a $100(1 - \alpha_i)\%$ c.i. for parameter θ_i by using Expression (11.10), in which case the number of intervals is $C = K$. If independent sampling were used, the K c.i.s would be mutually independent, and thus the overall confidence level would be $(1 - \alpha_1) \times (1 - \alpha_2) \times \cdots \times (1 - \alpha_C)$, which is larger, but not by much, than the right side of Expression (12.16). This type of procedure is most often used to estimate multiple parameters of a single system, rather than to compare systems; because multiple parameter estimates from the same system are likely to be dependent, the Bonferroni inequality typically is needed.

2. *Comparison to an existing system*: Compare all designs to one specific design, usually to an existing system. That is, construct a $100(1 - \alpha_i)\%$ c.i. for $\theta_i - \theta_1 (i = 2, 3, \ldots, K)$, using Expression (12.1). (System 1 with performance measure θ_1 is assumed to be the existing system.) In this case, the number of intervals is $C = K - 1$. This type of procedure is most often used to compare several competitors to the present system in order to learn which are better.

3. *All pairwise comparisons*: Compare all designs to each other. That is, for any two system designs $i \neq j$, construct a $100(1 - \alpha_{ij})\%$ c.i. for $\theta_i - \theta_j$. With K designs, the number of confidence intervals computed is $C = K(K - 1)/2$. The overall confidence coefficient would be bounded below by $1 - \alpha_E = 1 - \sum\sum_{i \neq j} \alpha_{ij}$ [which follows by Expression (12.16)]. It is generally believed that CRN will make the true overall confidence level larger than the right side of Expression (12.16), and usually larger than will independent sampling. The right side of Expression (12.16) can be thought of as giving the worst case, that is, the lowest possible overall confidence level.

Example 12.2

Reconsider the call center design problem of Example 12.1. The alternative system designs are the following:

a. C: existing system (distinct operator queues);
b. P7: cross-trained system with 7 operators (single queue);
c. P6: cross-trained system with 6 operators (single queue).

Using the data in Table 12.2, confidence intervals for $\theta_C - \theta_{P7}, \theta_C - \theta_{P6}$, and $\theta_{P7} - \theta_{P6}$ will be constructed, having an overall confidence level of 95%. Recall that CRN was used in all models, but this does not affect the overall confidence level, because, as mentioned, the Bonferroni Inequality (12.16) holds regardless of the statistical independence or dependence of the data.

Since the overall error probability is $\alpha_E = 0.05$ and $C = 3$ confidence intervals are to be constructed, let $\alpha_i = 0.05/3 = 0.0167$. Then use Expression (12.1), with proper modifications, to construct $C = 3$ confidence intervals with $\alpha = \alpha_i = 0.0167$ and degrees of freedom $\nu = 10 - 1 = 9$.

The value of $t_{\alpha_i/2, R-1} = t_{0.0083, 9} = 2.97$ is obtained from Table A.5 by interpolation; the point estimates and standard errors are obtained from Table 12.2

The three confidence intervals, with overall confidence coefficient at least 95%, are given by

$$\theta_C - \theta_{P7} : -0.86 \pm 2.97(0.62)$$

$$\theta_C - \theta_{P6} : -5.91 \pm 2.97(1.42)$$

$$\theta_{P7} - \theta_{P6} : -5.05 \pm 2.97(0.88)$$

or

$$\theta_C - \theta_{P7} : -0.86 \pm 1.84$$

$$\theta_C - \theta_{P6} : -5.91 \pm 4.22$$

$$\theta_{P7} - \theta_{P6} : -5.05 \pm 2.61$$

The simulation analyst has high confidence (at least 95%) that all three confidence statements are correct. Notice that the c.i. for $\theta_C - \theta_{P7}$ again contains zero; thus, there is no statistically significant difference between the current design and the alternative with 7 cross-trained operators, a conclusion that supports the previous results in Example 12.1. The other confidence intervals lie completely to the left of 0, indicating that C and P7 both dominate P6.

Some of the exercises at the end of this chapter provide an opportunity to compare CRN and independent sampling and to compute simultaneous confidence intervals under the Bonferroni approach.

12.2.2   Selection of the Best

Suppose that there are K system designs, and the unknown expected value of the ith system's performance is θ_i. We are interested in which system design is best, where "best" means having the maximum or minimum θ_i, depending on the problem. For instance, in Example 12.3 we will compare $K = 8$ designs for the semiconductor fabrication facility described in Example 11.8 of Chapter 11, in which θ_i is the steady-state mean cycle time of one product family for the ith design, and smaller θ_i is better. We want a procedure that is capable of handling large K, say, on the order of 100

Let B denote the (unknown) index of the best system design. The smaller the true differences $|\theta_B - \theta_i|, i \neq B$ are, and the more certain we want to be that we find the best system, the more replications are required to achieve our goal. Therefore, instead of demanding that we find the best design B no matter what, we compromise and ask to find B with high probability whenever the difference between system B and the others is significantly large. More precisely, we want the probability that we select the best system to be at least $1 - \alpha$ whenever $|\theta_B - \theta_i| \geq \epsilon$ for all $i \neq B$, where ϵ depends on the problem. If there are systems that are within ϵ of the best, then we will be satisfied to select either the best or any one of the near-best designs. Both the probability of correct selection $1 - \alpha$ and the practically significant difference ϵ will be under our control.

The following procedure accomplishes this goal by undertaking two stages of simulation: In the first stage, R_0 replications are obtained from each of the system designs. Those systems that are statistically significantly inferior to the others are screened out, meaning that they are eliminated from further consideration, with confidence $1 - \alpha/2$. If more than one system survives screening, then the survivors receive enough additional replications to select the best, or a system design within ϵ of the best, with confidence $1 - \alpha/2$. Then, using reasoning related to the Bonferroni inequality, the selected system is the best or a near-best system with confidence at least $1 - \alpha$ for the combination of screening and selection.

The first-stage screening can be very important because, as Example 12.3 will illustrate, the number of replications required in the second stage to select the best can be quite large, and there is no reason to waste time and effort on system designs that are obviously not competitive. This is especially important when K is large.

The procedure below is implemented in `SimulationTools.xls`, which is available at `www.bcnn.net`. The `SimulationTools.xls` implementation guides the user as to what data need to be obtained for each stage.

Select-the-Best Procedure

1. Specify the desired probability of correct selection $1/k < 1 - \alpha < 1$, the practically significant difference $\epsilon > 0$, an initial number of replications $R_0 \geq 10$, and the number of competing systems K. Set

$$t = t_{1-(1-\alpha/2)^{\frac{1}{k-1}}, R_0 - 1}$$

and obtain Rinott's constant $h = h(R_0, K, 1 - \alpha/2)$. [*Hint:* Both critical values are calculated automatically by `SimulationTools.xls`; small tables of t and h are given in the Appendix.]

2. Make R_0 replications of each system. Calculate the first-stage sample means and sample variances

$$\bar{Y}_{\cdot i} = \frac{1}{R_0} \sum_{r=1}^{R_0} Y_{ri}$$

$$S_i^2 = \frac{1}{n_0 - 1} \sum_{r=1}^{n_0} \left(Y_{ri} - \bar{Y}_{\cdot i}\right)^2$$

for $i = 1, 2, \ldots, K$.

3. Calculate the screening thresholds

$$W_{ij} = t \left(\frac{S_i^2 + S_j^2}{R_0}\right)^{1/2}$$

for all $i \neq j$.

(a) If bigger is better, then form the survivor subset S containing every system design i such that

$$\bar{Y}_{\cdot i} \geq \bar{Y}_{\cdot j} - \max\{0, W_{ij} - \epsilon\} \quad \text{for all } j \neq i$$

That is, retain every system i whose sample mean is no more than $\max\{0, W_{ij} - \epsilon\}$ smaller than any other system's sample mean.

(b) If smaller is better, then form the survivor subset S containing every system design i such that

$$\bar{Y}_{.i} \le \bar{Y}_{.j} + \max\{0, W_{ij} - \epsilon\} \quad \text{for all } j \ne i$$

That is, retain every system i whose sample mean is no more than $\max\{0, W_{ij} - \epsilon\}$ larger than any other system's sample mean.

4. If the set S contains only one system, then stop and return that system as the best. Otherwise, for all designs i in S, compute the second-stage sample sizes

$$R_i = \max\left\{R_0, \lceil(hS_i/\epsilon)^2\rceil\right\}$$

where $\lceil\cdot\rceil$ means to round up.

5. Take $R_i - R_0$ additional replications from all systems i in S, or, if it is more convenient, obtain a total of R_i replications from system i by starting over.

6. Compute the overall sample means

$$\bar{\bar{Y}}_{.i} = \frac{1}{R_i} \sum_{r=1}^{R_i} Y_{ri}$$

for all i in S. If bigger is better, then select the system with the largest $\bar{\bar{Y}}_{.i}$ as best. Otherwise select the system with the smallest $\bar{\bar{Y}}_{.i}$ as best.

The critical values t and h cannot usually be obtained directly from tables so interpolation is often required. As an alternative, algorithms for computing quantiles from the t distribution can be found in most numerical analysis and spreadsheet software; unfortunately, this is not the case for Rinott's constant h. Therefore, use of `SimulationTools.xls` is highly recommended for this procedure.

Nelson et al. [2001] prove that the Select-the-Best Procedure finds the system with the largest or smallest θ_i, or one within ϵ of the best θ, with confidence level at least $1 - \alpha$, provided the data are normally distributed and the systems are simulated independently. The procedure will also work if CRN is applied.

Example 12.3

Recall the FastChip, Inc. semiconductor manufacturing example from Chapter 11, in which FastChip, Inc. wants to use simulation to design a new "fab" by evaluating the steady-state mean cycle time (time from wafer release to completion) that they can expect under a particular loading of two distinct products, called C-chip and D-chip. The system design we evaluated in Chapter 11 is actually a base case, and there are seven additional designs that arise by making different choices about how to implement certain steps in the manufacturing process; these steps are called Clean, Load Quartz, Oxidize, Unload Quartz, Coat, and Develop. The question is, which design leads to the smallest steady-state mean cycle time? As in Chapter 11 we will focus on the cycle time for C-chip, and we will use the warmup period and run length obtained there.

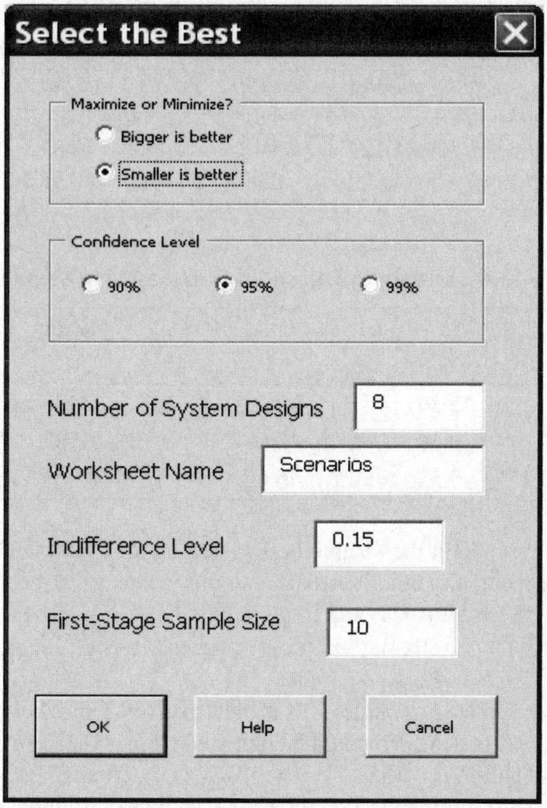

Figure 12.5   Input form for Select-the-Best Procedure in SimulationTools.xls

Including the base case there are $K = 8$ system designs, and the performance measures $\theta_1, \theta_2, \ldots, \theta_8$ are the steady-state mean cycle times for C-chip under each design. FastChip decides that even very small improvements in cycle time are important, so they set $\epsilon = 0.15$ hours (or 9 minutes). This is a very tight requirement when cycle times average around 45 hours, and we will see the consequences of this choice below. FastChip would like 95% confidence they have selected the best design, or one within 9 minutes of the best. Figure 12.5 shows the input screen to set up this problem in SimulationTools.xls, which uses the term "indifference level" for the practically significant difference ϵ.

Table 12.3 contains the first-stage results obtained from $R_0 = 10$ replications of each system design. The alternative that adjusts the Oxidize step seems to provide the smallest average cycle time, but all of the alternatives appear close. The complete data from this example are available at www.bcnn.net.

The t value we need for screening is

$$t = t_{1-(1-\alpha/2)^{\frac{1}{k-1}}, R_0-1} = t_{1-(0.975)^{\frac{1}{7}}, 9} = t_{0.0036, 9} = 3.455$$

Table 12.3   First-Stage Data from FastChip Simulation

Scenario	Base	Clean	Load	Oxidize	Unload	Coat	Step	Develop
i	1	2	3	4	5	6	7	8
$\bar{Y}_{.i}$	46.86	45.70	47.23	45.13	46.80	47.81	47.41	46.94
S_i^2	0.21	0.37	0.29	0.09	0.28	0.98	0.12	0.28

Table 12.4   Screening Threshold from FastChip Simulation

Scenario	Base	Clean	Load	Oxidize	Unload	Coat	Step	Develop
i	1	2	3	4	5	6	7	8
$\bar{Y}_{.i}$	46.86	45.70	47.23	45.13	46.80	47.81	47.41	46.94
W_{i4}	0.60	0.74	0.68		0.67	1.13	0.50	0.67
$\leq \bar{Y}_{.4} + \max\{W_{i4} - 0.15\}$	45.58	45.72	45.66		45.65	46.11	45.48	45.65

The subset S will always contain the sample best system, which is system $i = 4$ (Oxidize) here. To avoid being screened out, the other systems' sample means must be small enough to satisfy $\bar{Y}_{.i} \leq \bar{Y}_{.j} + \max\{W_{ij} - \epsilon\}$ for all $j \neq i$, and in particular $\bar{Y}_{.i} \leq \bar{Y}_{.4} + \max\{0, W_{i4} - \epsilon\}$. Table 12.4 contains the thresholds $\bar{Y}_{.4} + \max\{0, W_{i4} - \epsilon\}$; only alternative 2 with sample mean 45.70 hours is small enough to survive, so $S = \{2, 4\}$.

To determine the second-stage sample sizes for alternatives 2 and 4, we need $h = h(R_0, K, 1 - \alpha/2) = h(10, 8, 0.975) = 4.635$, which could be obtained from `SimulationTools.xls`. Then $R_2 = \lceil (hS_2/\epsilon)^2 \rceil = 349$ and $R_4 = \lceil (hS_4/\epsilon)^2 \rceil = 90$. Thus, alternative 2 requires $349 - 10 = 339$ additional replications, while alternative 4 requires $90 - 10 = 80$ additional replications. Notice that the number of replications for alternative 2 is quite large. In general, the second-stage sample size is large if the sample variance is large, making it is difficult to detect differences, or if ϵ is small, even small differences matter, as in this example.

After obtaining the additional replications, $\bar{\bar{Y}}_{.2} = 45.88$ and $\bar{\bar{Y}}_{.4} = 45.00$; therefore, alternative 4, Oxidize, is selected. The procedure guarantees, with 95% confidence, that θ_4 is the smallest steady-state mean cycle time of the 8 alternatives, or, if it is not, then it is within $\epsilon = 0.15$ hours of being the best.

The Select-the-Best Procedure, as presented here, does both screening and selection. This makes it particularly useful for finding the best system design from among a large number of alternatives since many of them may be screened out in the first stage. The procedure can be used for screening alone, without selection, if the objective is to eliminate designs that are not competitive. This might be worthwhile when we want to choose from a collection of competitive alternatives based on criteria beyond mean performance alone. For screening only, the procedure stops at Step 3 and can use the critical value

$$t = t_{1-(1-\alpha)^{\frac{1}{k-1}},R_0-1}$$

which provides somewhat tighter screening.

Similarly, if the number of system designs K is small, and the goal is to select the best, then screening can be skipped and all systems receive second-stage replications. For selection only, skip Step 3 and use the somewhat smaller critical value $h = h(R_0, K, 1 - \alpha)$.

12.3 Metamodeling

Suppose that there is a simulation output response variable Y that is related to k independent variables, say, $x_1, x_2, \ldots, x_k$. The dependent variable Y is a *random variable*, while the independent variables $x_1, x_2, \ldots, x_k$ are called *design variables* and are usually subject to control. The true relationship between the variables Y and x is represented by the often complex simulation model. Our goal is to approximate this relationship by a simpler mathematical function called a *metamodel*. In some cases, the analyst will know the exact form of the functional relationship between Y and $x_1, x_2, \ldots, x_k$, say, $Y = f(x_1, x_2, \ldots, x_k)$. However, in most cases, the functional relationship is unknown, and the analyst must select an appropriate f containing unknown parameters, and then estimate those parameters from a set of data $\{Y, x\}$. Regression analysis is one method for estimating the parameters.

Example 12.4

An insurance company promises to process all claims it receives each day by the end of the next day. It has developed a simulation model of its proposed claims-processing system to evaluate how hard it will be to meet this promise. The actual number and types of claims that will need to be processed each day will vary, and the number may grow over time. Therefore, the company would like to have a model that predicts the total processing time as a function of the number of claims received.

The primary value of a metamodel is to make it easy to answer "what-if" questions, such as what the processing time will be if there are x claims. Evaluating a function f, or perhaps its derivatives, at a number of values of x is typically much easier than running a simulation experiment for each value.

12.3.1 Simple Linear Regression

Suppose that it is desired to estimate the relationship between a single independent variable x and a dependent variable Y, and suppose that the true relationship between Y and x is suspected to be linear. Mathematically, the expected value of Y for a given value of x is assumed to be

$$E(Y|x) = \beta_0 + \beta_1 x \tag{12.17}$$

where β_0 is the intercept on the Y axis, an unknown constant, and β_1 is the slope, or change in Y for a unit change in x, also an unknown constant. It is further assumed that each observation of Y can be described by the model

$$Y = \beta_0 + \beta_1 x + \varepsilon \tag{12.18}$$

where ε is a random error with mean zero and constant variance σ^2. The regression model given by Equation (12.18) involves a single variable x and is commonly called a *simple linear regression model*.

Suppose that there are n pairs of observations $(Y_1, x_1), (Y_2, x_2), \ldots, (Y_n, x_n)$. These observations can be used to estimate β_0 and β_1 in Equation (12.18). The method of least squares is commonly used to form the estimates. In the method of *least squares*, β_0 and β_1 are estimated in such a way that the sum of the squares of the deviations between the observations and the regression line is minimized. The individual observations in Equation (12.18) can be written as

$$Y_i = \beta_0 + \beta_1 x_i + \varepsilon_i, \qquad i = 1, 2, \ldots, n \tag{12.19}$$

where $\varepsilon_1, \varepsilon_2, \ldots$ are assumed to be uncorrelated random variables.

Each ε_i in Equation (12.19) is given by

$$\varepsilon_i = Y_i - \beta_0 - \beta_1 x_i \tag{12.20}$$

and represents the difference between the observed response Y_i and the expected response $\beta_0 + \beta_1 x_i$, predicted by the model in Equation (12.17). Figure 12.6 shows how ε_i is related to x_i, Y_i, and $E(Y_i|x_i)$.

The sum of squares of the deviations given in Equation (12.20) is given by

$$L = \sum_{i=1}^{n} \varepsilon_i^2 = \sum_{i=1}^{n} (Y_i - \beta_0 - \beta_1 x_i)^2 \tag{12.21}$$

and L is called the *least-squares function*. It is convenient to rewrite Y_i as follows:

$$Y_i = \beta_0' + \beta_1 (x_i - \bar{x}) + \varepsilon_i \tag{12.22}$$

where $\beta_0' = \beta_0 + \beta_1 \bar{x}$ and $\bar{x} = \sum_{i=1}^{n} x_i / n$. Equation (12.22) is often called the *transformed linear regression model*. Using Equation (12.22), Equation (12.21) becomes

$$L = \sum_{i=1}^{n} [Y_i - \beta_0' - \beta_1 (x_i - \bar{x})]^2$$

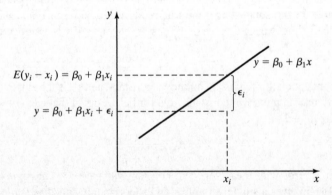

Figure 12.6   Relationship of ε_i to x_i, Y_i, and $E(Y_i|x_i)$.

To minimize L, find $\partial L/\partial \widehat{\beta}'_0$ and $\partial L/\partial \beta_1$, set each to zero, and solve for $\widehat{\beta}'_0$ and $\widehat{\beta}_1$. Taking the partial derivatives and setting each to zero yields

$$n\widehat{\beta}'_0 = \sum_{i=1}^{n} Y_i$$

$$\widehat{\beta}_1 \sum_{i=1}^{n} (x_i - \bar{x})^2 = \sum_{i=1}^{n} Y_i(x_i - \bar{x}) \tag{12.23}$$

Equations (12.23) are often called the *normal equations*, which have the solutions

$$\widehat{\beta}'_0 = \bar{Y} = \sum_{i=1}^{n} \frac{Y_i}{n} \tag{12.24}$$

and

$$\widehat{\beta}_1 = \frac{\sum_{i=1}^{n} Y_i(x_i - \bar{x})}{\sum_{i=1}^{n} (x_i - \bar{x})^2} \tag{12.25}$$

The numerator in Equation (12.25) is rewritten for computational purposes as

$$S_{xy} = \sum_{i=1}^{n} Y_i(x_i - \bar{x}) = \sum_{i=1}^{n} x_i Y_i - \frac{\left(\sum_{i=1}^{n} x_i\right)\left(\sum_{i=1}^{n} Y_i\right)}{n} \tag{12.26}$$

where S_{xy} denotes the corrected sum of cross products of x and Y. The denominator of Equation (12.25) is rewritten for computational purposes as

$$S_{xx} = \sum_{i=1}^{n} (x_i - \bar{x})^2 = \sum_{i=1}^{n} x_i^2 - \frac{\left(\sum_{i=1}^{n} x_i\right)^2}{n} \tag{12.27}$$

where S_{xx} denotes the corrected sum of squares of x. The value of $\widehat{\beta}_0$ can be retrieved easily as

$$\widehat{\beta}_0 = \widehat{\beta}'_0 - \widehat{\beta}_1 \bar{x} \tag{12.28}$$

Example 12.5: Calculating $\widehat{\beta}_0$ and $\widehat{\beta}_1$ _____

The simulation model of the claims-processing system in Example 12.4 was executed with initial conditions $x = 100, 150, 200, 250,$ and 300 claims received the previous day. Three replications were obtained at each setting. The response Y is the number of hours required to process x claims. The results are shown in Table 12.5. The graphical relationship between the number of claims received and total processing time is shown in Figure 12.7. Examination of this scatter diagram indicates that there is a strong relationship between number of claims and processing time. The tentative assumption of the linear model given by Equation (12.18) appears to be reasonable.

Table 12.5   Simulation Results for Processing Time Given x Claims

Number of Claims x	Hours of Processing Time Y
100	8.1
100	7.8
100	7.0
150	9.6
150	8.5
150	9.0
200	10.9
200	13.3
200	11.6
250	12.7
250	14.5
250	14.7
300	16.5
300	17.5
300	16.3

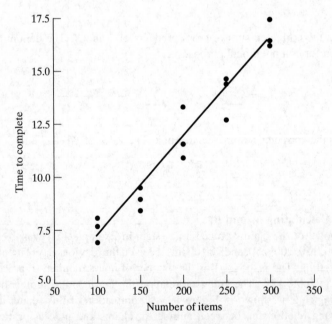

Figure 12.7   Relationship between number of claims and hours of processing time.

With the processing times as the Y_i values (the dependent variables) and the number of claims as the x_i values (the independent variables), $\widehat{\beta}_0$ and $\widehat{\beta}_1$ can be found by the following computations:

$$n = 15$$

$$\sum_{i=1}^{15} x_i = 3000$$

$$\sum_{i=1}^{15} Y_i = 178$$

$$\sum_{i=1}^{15} x_i^2 = 675,000$$

$$\sum_{i=1}^{15} x_i Y_i = 39080$$

and $\bar{x} = 3000/15 = 200$

From Equation (12.26) S_{xy} is calculated as

$$S_{xy} = 39,080 - \frac{(3000)(178)}{15} = 3480$$

From Equation (12.27), S_{xx} is calculated as

$$S_{xx} = 675,000 - \frac{(3000)^2}{15} = 75,000$$

Then, $\widehat{\beta}_1$ is calculated from Equation (12.25) as

$$\widehat{\beta}_1 = \frac{S_{xy}}{S_{xx}} = \frac{3480}{75,000} = 0.0464$$

As shown in Equation (12.24), $\widehat{\beta}_0'$ is just $\bar{Y}$, or

$$\widehat{\beta}_0' = \frac{178}{15} \approx 11.8667$$

To express the model in the original terms, compute $\widehat{\beta}_0$ from Equation (12.28) as

$$\widehat{\beta}_0 = 11.8667 - 0.0464(200) = 2.5867$$

Then an estimate of the mean of Y given x, $E(Y|x)$, is given by

$$\hat{y} = \widehat{\beta}_0 + \widehat{\beta}_1 x = 2.5867 + 0.0464x \tag{12.29}$$

For a given number of claims x this model can be used to predict the number of hours required to process them. The coefficient $\widehat{\beta}_1$ has the interpretation that each additional claim received adds an expected 0.0464 hours, or 2.8 minutes, to the expected total processing time.

Regression analysis is widely used and frequently misused. Several of the common abuses are briefly mentioned here. Relationships derived in the manner of Equation (12.29) are valid for values of the independent variable within the range of the original data. The linear relationship that has been tentatively assumed may not be valid outside the original range. In fact, we know from queueing theory that mean processing time may increase rapidly as the number of claims approaches the capacity of the system. Therefore, Equation (12.29) can be considered valid only for $100 \le x \le 300$. Regression models are not advised for extrapolation purposes.

Care should be taken in selecting variables that have a plausible causal relationship with one another. It is quite possible to develop statistical relationships that are unrelated in a practical sense. For example, an attempt might be made to relate monthly output of a steel mill to the weight of reports appearing on a manager's desk during the month. A straight line may appear to provide a good model for the data, but the relationship between the two variables is tenuous. A strong observed relationship does not imply that a causal relationship exists between the variables. Causality can be inferred only when analysis uncovers some plausible reasons for its existence. In Example 12.4 it is reasonable that starting with more claims implies that more time is needed to process them. Therefore, a relationship of the form of Equation (12.29) is at least plausible.

Testing for Significance of Regression

In Section 12.3.1, it was assumed that a linear relationship existed between Y and x. In Example 12.4, a scatter diagram, shown in Figure 12.7, relating number of claims and processing time was prepared, to evaluate whether a linear model was a reasonable tentative assumption prior to the calculation of $\widehat{\beta}_0$ and $\widehat{\beta}_1$. However, the adequacy of the simple linear relationship should be tested prior to using the model for predicting the response Y_i given an independent variable, x_i. There are several tests which may be conducted to aid in determining model adequacy. Testing whether the order of the model tentatively assumed is correct is commonly called *lack-of-fit testing*.

Testing for the significance of regression provides another means for assessing the adequacy of the model. In particular, tests to establish that the estimated regression coefficients $\widehat{\beta}_0$ and $\widehat{\beta}_1$ are statistically different from 0 are essential, particularly for multiple linear regression, described below, when a large number of design variables x are included in the metamodel but it is not certain that all of them have a substantial explanatory effect.

Tests for significance and lack of fit are included in all regression software. See Box and Draper [1987], Hines *et al.* [2002], and Montgomery [2000] for details about these tests.

Multiple Linear Regression

If the simple linear regression model of Section 12.3.1 is inadequate, several other possibilities exist. There could be several independent variables, so that the relationship is of the form

$$Y = \beta_0 + \beta_1 x_1 + \beta_2 x_2 + \cdots + \beta_m x_m + \varepsilon \tag{12.30}$$

Notice that this model is still linear, but has more than one independent variable. Regression models having the form shown in Equation (12.30) are called *multiple linear regression models*. Another possibility is that the model is of a quadratic form such as

$$Y = \beta_0 + \beta_1 x + \beta_2 x^2 + \varepsilon \qquad (12.31)$$

Equation (12.31) is also a linear model which may be transformed to the form of Equation (12.30) by letting $x_1 = x$ and $x_2 = x^2$.

Yet another possibility is a model of the form such as

$$Y = \beta_0 + \beta_1 x_1 + \beta_2 x_2 + \beta_3 x_1 x_2 + \varepsilon$$

which is also a linear model. The analysis of these three models with the forms just shown, and related models, can be found in Box and Draper [1987], Hines *et al.* [2002], Montgomery [2000], and other applied statistics texts; and also in Kleijnen [1987, 1998], which is concerned primarily with the application of these models in simulation.

12.3.2 Metamodeling and Computer Simulation

The material in Section 12.3.1 applies to regression metamodeling in general, not just computer simulation. However, there are special issues and opportunities that arise when goal is to provide a metamodel for a computer simulation; we describe these below.

Random-Number Assignment for Regression

The assignment of random-number seeds or streams is part of the design of a simulation experiment. (This section is based on Nelson [1992].) Assigning a different seed or stream to different design points (settings for $x_1, x_2, \ldots, x_m$ in a multiple linear regression) guarantees that the responses Y from different design points will be statistically independent. Similarly, assigning the same seed or stream to different design points induces dependence among the corresponding responses, by virtue of their all having the same source of randomness.

Many textbook experiment designs assume independent responses across design points. To conform to this assumption, we must assign different seeds or streams to each design point. However, it is often useful to assign the same random number seeds or streams to all of the design points, in other words, to use common random numbers.

The intuition behind common random numbers for metamodels is that a fairer comparison among design points is achieved if the design points are subjected to the same experimental conditions, specifically the same source of randomness. The mathematical justification is as follows: Suppose we fit the simple linear regression $Y_i = \beta_0 + \beta_1 x_i + \varepsilon_i$ and obtain least squares estimates $\widehat{\beta}_0$ and $\widehat{\beta}_1$. Then an estimator of the expected difference in performance between design points i and j is

$$\widehat{\beta}_0 + \widehat{\beta}_1 x_i - (\widehat{\beta}_0 + \widehat{\beta}_1 x_j) = \widehat{\beta}_1 (x_i - x_j)$$

when x_i and x_j are fixed design points, $\widehat{\beta}_1$ determines the estimated difference between design points i and j, or for that matter between any other two values of x. Therefore, common random numbers can be expected to reduce the variance of $\widehat{\beta}_1$ and, more generally, reduce the variance of all of the slope terms in a multiple linear regression. Common random numbers typically do not reduce the variance of the intercept term $\widehat{\beta}_0$.

The least-squares estimators $\widehat{\beta}_0$ and $\widehat{\beta}_1$ are appropriate regardless of whether we use common random numbers, but the associated statistical analysis is affected by that choice. For statistical analysis of a metamodel under common random numbers, see Kleijnen [1988] and Nelson [1992].

Queueing Simulation

Many computer simulation models are of systems that involve queues or networks of queues, and metamodels of these simulations may include design variables related to capacity (numbers of servers and service rates) and customer load (arrival rates). Such models violate standard regression assumptions as described in Section 12.3.1. A small example illustrates the difficulties and possible solutions.

Recall the $M/M/1$ queue described in Chapter 6: A single-server, first-come-first-served queue with a Poisson arrival process having rate λ customers/time and exponentially distributed service time having rate μ customers/time. Let $x = \lambda/\mu$, which was called the utilization ρ in Chapter 6. Then as shown in Table 6.4, the steady-state expected number of customers in the queue for $0 < x < 1$ is

$$L = \frac{x}{1-x} = x + x^2 + x^3 + \cdots \tag{12.32}$$

Now, suppose that we did not know the relationship (12.32) and were trying to build a metamodel for L as a function of $0 < x < 1$ by running simulations of an $M/M/1$ queue. Clearly the usual low-order polynomial models, including only the terms x and x^2 described above, will not fit very well, particularly for x close to 1, where the higher-order terms matter.

There are at least two solutions to this problem. One is to break up the design space ($0 < x < 1$ in this example) into smaller pieces for which low-order polynomials are adequate. The second approach is to fit a nonlinear model motivated by queueing theory. Yang, Ankenman, and Nelson [2007] showed that

$$Y = \frac{\sum_{i=0}^{m} \beta_i x^i}{(1-x)^p} + \varepsilon \tag{12.33}$$

with $m \leq 3$, works well for many queueing outputs Y, including average number in the queue and average time in the queue. Notice that the unknown parameters are m, β_i, and p, and therefore nonlinear least squares methods are required to fit them (Bates and Watts [1988]). When Y is the average number in queue, an estimator of L, then $m = 1$, $\beta_0 = 0$, $\beta_1 = 1$, and $p = 1$.

Unfortunately, the problems go even deeper. If we estimate L by making a simulation run of length T time units and record the average number of customers in queue (call it Y), then it can be shown (Whitt [1989]) that

$$\mathrm{var}(Y) \approx \frac{2x(1+x)}{(1-x)^4 \, T} \tag{12.34}$$

Thus, instead of the variance being constant across the design space, as assumed in regression, it increases, and increases explosively, as x approaches 1. As a consequence, a run of, say, $T = 100$ time units may yield a pretty good estimate of L at $x = 0.1$ but a very poor estimate at $x = 0.9$.

The standard statistical inference produced by regression software will be invalid when response variance depends strongly on x, meaning we may not recognize lack of fit or poor fit.

A remedy for unequal response variance in regression is to apply a variance-stabilizing transformation to the data to make the variance more nearly constant across the design space. In simulation we can attack the problem directly by using different run lengths T or different numbers of replications, depending on x.

We suggest using the $M/M/1$ result as a rough guide for more general networks of queues. If we let x denote the network utilization, that is, the most heavily utilized queue, then Approximation (12.34) implies that the run lengths T_1 and T_2 at two design points x_1 and x_2 should satisfy

$$\frac{T_2}{T_1} = \frac{x_2}{x_1}\left(\frac{1+x_2}{1+x_1}\right)\left(\frac{1-x_1}{1-x_2}\right)^4$$

For instance, if we were to simulate $T_1 = 1$ time unit at utilization $x_1 = 0.1$, then we would need about $T_2 = 10$ time units at $x_2 = 0.3$, or $T_2 = 72$ time units at $x_2 = 0.5$, or $T_2 = 876$ time units at $x_2 = 0.7$, or $T_2 = 101,944$ time units at $x_2 = 0.9$ to stabilize variances.

The central message is this: In queueing simulation, if the metamodel includes decision variables that affect customer load—by changing the arrival rates, service rates, or numbers of servers—then a low-order polynomial may not be a good choice for the metamodel, and equal response variance across the design space cannot be assumed without exercising run-length control.

12.4   Optimization via Simulation

Consider the following examples. (Some of these descriptions are based on Boesel, Nelson, and Ishii [2003].

Example 12.6:   Materials Handling System (MHS) _____
Engineers need to design an MHS consisting of a large automated storage and retrieval device, automated guided vehicles (AGVs), AGV stations, lifters, and conveyors. Among the design variables they can control are the number of AGVs, the load per AGV, and the routing algorithm used to dispatch the AGVs. Alternative designs will be evaluated according to AGV utilization, transportation delay for material that needs to be moved, and overall investment and operation costs.

Example 12.7:   Liquefied Natural Gas (LNG) Transportation _____
An LNG transportation system will consist of LNG tankers and of loading, unloading, and storage facilities. In order to minimize cost, designers can control tanker size, number of tankers in use, number of jetties at the loading and unloading facilities, and capacity of the storage tanks.

Example 12.8:   Automobile Engine Assembly —————————————————————————

In an assembly line, a large buffer (queue) between workstations could increase station utilization—because there will tend to be something waiting to be processed—but drive up space requirements and work-in-process inventory. An allocation of buffer capacity that minimizes the sum of these competing costs is desired.

Example 12.9:   Traffic Signal Sequencing —————————————————————————

Civil engineers want to sequence the traffic signals along a busy section of road to reduce driver delay and the congestion occurring along narrow cross streets. For each traffic signal, the length of the red, green, and green-turn-arrow cycles can be set individually.

Example 12.10:   Online Services —————————————————————————

A company offering online information services over the Internet is changing its computer architecture from central mainframe computers to distributed workstation computing. The numbers and types of CPUs, the network structure, and the allocation of processing tasks all need to be chosen. Response time to customer queries is the key performance measure.

What do these design problems have in common? Clearly, a simulation model could be useful in each, and all have an implied goal of finding the best design relative to some performance measures (cost, delay, etc.). In each example, there are potentially a very large number of alternative designs, ranging from tens to thousands, and certainly more than the 2 to 100 we considered in Section 12.2.2. Some of the examples contain a diverse collection of decision variables: discrete (number of AGVs, number of CPUs), continuous (tanker size, red-cycle length), and qualitative (routing strategy, algorithm for allocating processing tasks). This makes developing a metamodel, as described in Section 12.3, difficult.

All of these problems fall under the general topic of "optimization via simulation," where the goal is to minimize or maximize some measures of system performance, and system performance can be evaluated only by running a computer simulation. Optimization via simulation is a relatively new but already vast topic, and commercial software has become widely available. In this section, we describe the key issues that should be considered in undertaking optimization via simulation, provide some pointers to the available literature, and give one example algorithm.

12.4.1   What Does "Optimization via Simulation" Mean?

Optimization is a key tool used by operations researchers and management scientists, and there are well-developed algorithms for many classes of problems, the most famous being linear programming. Much of the work on optimization deals with problems in which all aspects of the system are treated as being known with certainty; most critically, the performance of any design (cost, profit, makespan, etc.) can be evaluated exactly.

In stochastic, discrete-event simulation, the result of any simulation run is a random variable. For notation, let $x_1, x_2, \ldots, x_m$ be the m controllable design variables and let $Y(x_1, x_2, \ldots, x_m)$ be the observed simulation output performance on one run. To be concrete, x_1, x_2, x_3 might denote the number of AGVs, the load per AGV, and the routing algorithm used to dispatch the AGVs, respectively, in Example 12.6, while $Y(x_1, x_2, x_3)$ could be total MHS acquisition and operation cost.

What does it mean to "optimize" $Y(x_1, x_2, \ldots, x_m)$ with respect to $x_1, x_2, \ldots, x_m$? Since Y is a random variable we cannot optimize the *actual* value of Y. The most common definition of optimization is

$$\text{maximize or minimize } E\left(Y(x_1, x_2, \ldots, x_m)\right) \qquad (12.35)$$

In other words, the mathematical expectation, or long-run average, of performance is maximized or minimized. This is the default definition of optimization used in all commercial packages of which we are aware. In our example, $E\left(Y(x_1, x_2, x_3)\right)$ is the expected, or long-run average cost of operating the MHS with x_1 AGVs, x_2 load per AGV, and routing algorithm x_3.

It is important to note that Formulation (12.35) is not the only possible definition, however. For instance, we might want to select the MHS design that has the best chance of costing less than $\$D$ to purchase and operate, changing the objective to

$$\text{maximize } \Pr\left(Y(x_1, x_2, x_3) \leq D\right)$$

We can fit this objective into formulation (12.35) by defining a new performance measure

$$Y'(x_1, x_2, x_3) = \begin{cases} 1, & \text{if } Y(x_1, x_2, x_3) \leq D \\ 0, & \text{otherwise} \end{cases}$$

and maximizing $E\left(Y'(x_1, x_2, x_3)\right)$ instead.

A more complex optimization problem occurs when we want to select the system design that is most likely to be the best. Such an objective is relevant when one-shot, rather than long-run average, performance matters. Examples include a Space Shuttle launch, or the delivery of a unique, large order of products. Bechhofer, Santner, and Goldsman [1995] address this problem under the topic "multinomial selection".

We have been assuming that a system design $x_1, x_2, \ldots, x_m$ can be evaluated in terms of a single performance measure Y, such as cost. Obviously, this may not always be the case. In the MHS example, we might also be interested in some measure of system productivity, such as throughput or cycle time. At present, multiple objective optimization via simulation is not well developed. Therefore, one of three strategies is typically employed:

1. Combine all of the performance measures into a single measure, the most common being cost. For instance, the revenue generated by each completed product in the MHS could represent productivity and be included as a negative cost.
2. Optimize with respect to one key performance measure, but then evaluate the top solutions with respect to secondary performance measures. For instance, the MHS could be optimized with respect to expected cost, and then the cycle time could be compared for the top 5 designs. This approach requires that information on more than just the best solution be maintained.
3. Optimize with respect to one key performance measure, but consider only those alternatives that meet certain constraints on the other performance measures. For instance, the MHS could be optimized with respect to expected cost for those alternatives whose expected cycle time is less than a given threshold.

Finally, there are typically constraints on the system design variables $x_1, x_2, \ldots, x_m$. In the online services in Example 12.10 there will be a budget constraint that limits the number of workstations that can be purchased. Complex, and especially nonlinear, constraints make the feasible space more difficult to search. In some cases the constraints, like the objective function, cannot be evaluated with certainty and must be estimated from the simulation. For instance, while the online services company may want to minimize overall customer response time, they may have to do so subject to a constraint that the average response time for a particular, say, government, customer be below a certain level.

12.4.2   Why is Optimization via Simulation Difficult?

Even when there is no uncertainty, optimization can be very difficult if the number of design variables is large, if the problem contains a diverse collection of design variable types, or if little is known about the structure of the performance function. Optimization via simulation adds an additional complication: The performance of a particular design cannot be evaluated exactly, but instead must be estimated. Because we have estimates, it is not possible to conclude with assurance that one design is better than another, and this uncertainty frustrates optimization algorithms that try to move in improving directions. In principle, one can eliminate this complication by making so many replications, or such long runs, at each design point that the performance estimate has essentially no variance. In practice, this could mean that very few alternative designs will be explored because of the time required to simulate each one.

The existence of sampling variability forces optimization via simulation to make compromises. The following are the standard ones:

1. *Guarantee a prespecified probability of correct selection.* The Select-the-Best Procedure in Section 12.2.2 is an example of this approach, which allows the analyst to specify the desired chance of being right. Such algorithms typically require either that every possible design be simulated or that a strong functional relationship among the designs (such as a metamodel) apply. Other algorithms can be found in Goldsman and Nelson [1998] and Kim and Nelson [2006].

2. *Guarantee asymptotic convergence.* There are many algorithms that guarantee convergence to the global optimal solution as the simulation effort (number of replications, length of replications) becomes infinite. These guarantees are useful because they indicate that the algorithm tends to get to where the analyst wants it to go. However, convergence can be slow, and there is often no guarantee as to how good the reported solution is when the algorithm is terminated in finite time (as it must be, in practice). See Andradóttir [1998] for specific algorithms that apply to discrete- or continuous-variable problems.

3. *Optimal for deterministic counterpart.* The idea here is to use an algorithm that would find the optimal solution if the performance of each design could be evaluated with certainty. An example might be applying a standard nonlinear programming algorithm to the simulation optimization problem. It is typically up to the analyst to make sure that enough simulation effort is expended (replications or run length) to ensure that such an algorithm is not misled by sampling variability. Direct application of an algorithm that assumes deterministic evaluation to a stochastic simulation is not recommended.

4. *Robust heuristics.* Many heuristics have been developed for deterministic optimization problems that do not guarantee finding the optimal solution, but nevertheless have been shown to be very effective on difficult, practical problems. Some of these heuristics use randomness as part of their search strategy, so one might argue that they are less sensitive to sampling variability than other types of algorithms. Nevertheless, it is still important to make sure that enough simulation effort is expended (replications or run length) to ensure that such an algorithm is not misled by sampling variability.

Robust heuristics are the most common algorithms found in commercial optimization via simulation software. We provide some guidance on their use in the next section. See Fu [2002] for a comprehensive discussion of optimization theory versus practice and Ólafsson [2006] for more about robust heuristics.

12.4.3   Using Robust Heuristics

By a "robust heuristic" we mean a procedure that does not depend on strong problem structure—such as continuity or convexity of $E(Y(x_1, \ldots, x_m))$—to be effective, can be applied to problems with mixed types of decision variables, and, ideally, is tolerant of some sampling variability. Genetic algorithms (GA) and tabu search (TS) are two prominent examples, but there are many others and many variations of them. Such heuristics form the core of most commercial implementations. To give a sense of these heuristics, we describe GA and TS next. We caution the reader that only a high-level description of the simplest version of each procedure is provided. The commercial implementations are much more sophisticated.

Suppose that there are k possible solutions to the optimization via simulation problem. Let $\mathbf{X} = \{\mathbf{x}_1, \mathbf{x}_2, \ldots, \mathbf{x}_k\}$ denote the solutions, where the ith solution $\mathbf{x}_i = (x_{i1}, x_{i2}, \ldots, x_{im})$ provides specific settings for the m decision variables. The simulation output at solution $\mathbf{x}_i$ is denoted $Y(\mathbf{x}_i)$; this could be the output of a single replication, or the average of several replications. Our goal is to find the solution $\mathbf{x}^*$ that minimizes $E(Y(\mathbf{x}))$.

On each iteration, known as a *generation*, a GA operates on a "population" of p solutions. Denote the population of solutions on the jth iteration as $\mathbf{P}(j) = \{\mathbf{x}_1(j), \mathbf{x}_2(j), \ldots, \mathbf{x}_p(j)\}$. There may be multiple copies of the same solution in $\mathbf{P}(j)$, and $\mathbf{P}(j)$ may contain solutions that were discovered on previous iterations. From iteration to iteration, this population evolves in such a way that good solutions tend to survive and give birth to new, and hopefully better, solutions, while inferior solutions tend to be removed from the population. The basic GA is given here.

Basic Genetic Algorithms (GA)

Step 1. Set the iteration counter $j = 0$, and select, perhaps randomly, an initial population of p solutions $\mathbf{P}(0) = \{\mathbf{x}_1(0), \ldots, \mathbf{x}_p(0)\}$.

Step 2. Run simulation experiments to obtain performance estimates $Y(\mathbf{x})$ for all p solutions $\mathbf{x}(j)$ in $\mathbf{P}(j)$.

Step 3. Select a population of p solutions from those in $\mathbf{P}(j)$ in such a way that those with smaller $Y(\mathbf{x})$ values are more likely, but not certain, to be selected. Denote this population of solutions as $\mathbf{P}(j+1)$.

Step 4.   Recombine the solutions in $\mathbf{P}(j+1)$ via crossover (which joins parts of two solutions $\mathbf{x}_i(j+1)$ and $\mathbf{x}_\ell(j+1)$ to form a new solution) and mutation (which randomly changes a part of a solution $\mathbf{x}_i(j+1)$).

Step 5.   Set $j = j+1$ and go to Step 2.

The GA can be terminated after a specified number of iterations, when little or no improvement is noted in the population, or when the population contains p copies of the same solution. At termination, the solution $\mathbf{x}^*$ that has the smallest $Y(\mathbf{x})$ value in the last population is chosen as best, or alternatively, the solution with the smallest $Y(\mathbf{x})$ over all iterations could be chosen.

GAs are applicable to almost any optimization problem, because the operations of selection, crossover, and mutation can be defined in a very generic way that does not depend on specifics of the problem. However, when these operations are not tuned to the specific problem, a GA's progress can be very slow. Commercial versions are often self-tuning, meaning that they update selection, crossover, and mutation parameters during the course of the search. There is some evidence that GAs are tolerant of sampling variability in $Y(\mathbf{x})$ because they maintain a population of solutions rather than focusing on improving a current-best solution. In other words, it is not critical that the GA rank the solutions in a population of solutions perfectly, because the next iteration depends on the entire population, not on a single solution.

Tabu search (TS), on the other hand, identifies a current best solution on each iteration and then tries to improve it. Improvements occur by changing the solution via "moves." For example, the solution (x_1, x_2, x_3) could be changed to the solution $(x_1 + 1, x_2, x_3)$ by the move of adding 1 to the first decision variable (perhaps x_1 represents the number of AGVs in Example 12.6, so the move would add one more AGV). The "neighbors" of solution $\mathbf{x}$ are all of those solutions that can be reached by legal moves. TS finds the best neighbor solution and moves to it. However, to avoid making moves that return the search to a previously visited solution, moves may become "tabu" (not usable) for some number of iterations. Conceptually, think about how you would find your way through a maze: If you took a path that lead to a dead end, then you would avoid taking that path again—it would be tabu.

The basic TS algorithm is given next. The description is based on Glover [1989].

Basic Tabu Search (TS)

Step 1.   Set the iteration counter $j = 0$ and the list of tabu moves to empty. Select an initial solution $\mathbf{x}^*$ in $\mathbf{X}$, perhaps randomly.

Step 2.   Find the solution $\mathbf{x}'$ that minimizes $Y(\mathbf{x})$ over all of the neighbors of $\mathbf{x}^*$ that are not reached by tabu moves, running whatever simulations are needed to do the optimization.

Step 3.   If $Y(\mathbf{x}') < Y(\mathbf{x}^*)$, then $x^* = x'$; move the current best solution to x'.

Step 4.   Update the list of tabu moves and go to Step 2.

The TS can be terminated when a specified number of iterations have been completed, when some number of iterations has passed without changing $\mathbf{x}^*$, or when there are no more feasible moves. At termination, the solution $\mathbf{x}^*$ is chosen as best.

TS is fundamentally a discrete-decision-variable optimizer, but continuous decision variables can be discretized, as described in Section 12.4.4. TS aggressively pursues improving solutions, and therefore tends to make rapid progress. However, it is more sensitive to random variability in $Y(\mathbf{x})$, because $\mathbf{x}^*$ is taken to be the *true* best solution so far and attempts are made to improve it. There are probabilistic versions of TS that should be less sensitive, however. An important feature of commercial implementations of TS, which is not present in the Basic TS, is a mechanism for overriding the tabu list when doing so is advantageous.

Next, we offer three suggestions for using commercial products that employ a GA, TS, or other robust heuristic: controlling sampling variability, restarting, and clean up.

Control Sampling Variability

In many cases, it will up to the user to determine how much sampling (replications or run length) will be undertaken at each potential solution. This is a difficult problem in general. Ideally, sampling should increase as the heuristic closes in on the better solutions, simply because it is much more difficult to distinguish solutions that are close in expected performance from those that differ widely. Early in the search, it may be easy for the heuristic to identify good solutions and search directions, because clearly inferior solutions are being compared to much better ones; however, late in the search this might not be the case.

If the analyst must specify a fixed number of replications per solution that will be used through the search, then a preliminary experiment should be conducted. Simulate several designs, some at the extremes of the solution space and some nearer the center. Compare the apparent best and apparent worst of these designs, using the approaches in Section 12.1. Using the technique described in Section 12.1.3, find the minimum for the number of replications required to declare these designs to be statistically significantly different. This is the minimum number of replications that should be used.

Restarting

Because robust heuristics provide no guarantees that they converge to the optimal solution for optimization via simulation, it makes sense to run the optimization two or more times to see which run yields the best solution. Each optimization run should use different random number seeds or streams and, ideally, should start from different initial solutions. Try starting the optimization at solutions on the extremes of the solution space, in the center of the space, and at randomly generated solutions. If people familiar with the system suspect that certain designs will be good, be sure to include them as possible starting solutions for the heuristic.

Clean Up

After the optimization run or runs have completed, it is critical to perform a second set of experiments on the top designs identified by the heuristic, to avoid two types of errors: Failing to recognize the best system design that was simulated during the search, and poorly estimating the performance of the design that is selected at the end. These errors occur because no optimization algorithm can hope to make any progress while at the same time maintaining statistical error control every step of the way. Therefore, we recommend cleaning up after the search, by which we mean performing a rigorous statistical analysis employing the comparison techniques in Section 12.2, and specifically the Select-the-Best Procedure in Section 12.2.2, to evaluate which are the best or near-best of the designs discovered during the search. The top 5 to 10% of the system designs simulated during the search should be subjected to this controlled experiment. The "clean up" concept was introduced in Boesel, Nelson and Kim [2003].

12.4.4   An Illustration: Random Search

In this section, we present an algorithm for optimization via simulation known as random search. The specific implementation is based on Algorithm 2 in Andradóttir [1998], which provides guaranteed asymptotic convergence under certain conditions. Thus, it will find the true optimal solution if permitted to run long enough. However, in practice, convergence can be slow, and the memory requirements of this particular version of random search can be quite large. Even though random search is not a robust heuristic, we will also use it to demonstrate some strategies we would employ in conjunction with such heuristics and to demonstrate why optimization via simulation is tricky even with what appears to be an uncomplicated algorithm.

The random-search algorithm that we present requires that there be a finite number of possible system designs, although that number may be quite large. This might seem to rule out problems with continuous decision variables, such as conveyor speed. In practice, however, apparently continuous decision variables can often be discretized in a reasonable way. For instance, if conveyor speed can be anything from 60 to 120 feet per minute, little may be lost by treating the possible conveyor speeds as $60, 61, 62, \ldots, 120$ feet per minute (61 possible values). Note, however, that there are algorithms designed specifically for continuous-variable problems (Andradóttir [1998]).

Again, let the k possible solutions to the optimization via simulation problem be denoted $\{\mathbf{x}_1, \mathbf{x}_2, \ldots, \mathbf{x}_k\}$, where the ith solution $\mathbf{x}_i = (x_{i1}, x_{i2}, \ldots, x_{im})$ provides specific settings for the m decision variables. The simulation output at solution $\mathbf{x}_i$ is denoted $Y(\mathbf{x}_i)$; this could be the output of a single replication or the average of several replications. Our goal is to find the solution $\mathbf{x}^*$ that minimizes $E(Y(\mathbf{x}))$.

On each iteration of the random-search algorithm, we compare a current good solution to a randomly chosen competitor. If the competitor is better, then it becomes the current good solution. When we terminate the search, the solution we choose is the one that has been visited most often— which means that we expect to revisit solutions many times.

Random-Search Algorithm

Step 1.   Initialize counter variables $C(i) = 0$ for $i = 1, 2, \ldots, k$. Select an initial solution i^0, and set $C(i^0) = 1$. [*Hint*: $C(i)$ counts the number of times we visit solution i.]

Step 2.   Choose another solution i' from the set of all solutions *except* i^0 in such a way that each solution has an equal chance of being selected.

Step 3.   Run simulation experiments at the two solutions i^0 and i' to obtain outputs $Y(i^0)$ and $Y(i')$. If $Y(i') < Y(i^0)$, then set $i^0 = i'$. (See note following Step 4.)

Step 4.   Set $C(i^0) = C(i^0) + 1$. If not done, then go to Step 2. If done, then select as the estimated optimal solution x_{i^*} such that $C(i^*)$ is the largest count.

Note that, if the problem is a maximization problem, then replace Step 3 with

Step 3′.   Run simulation experiments at the two solutions i^0 and i' to obtain outputs $Y(i^0)$ and $Y(i')$. If $Y(i') > Y(i^0)$, then set $i^0 = i'$.

One of the difficult problems with many optimization-via-simulation algorithms is knowing when to stop. (Exceptions include algorithms that guarantee a probability of correct selection.) Typical rules might be to stop after a certain number of iterations, stop when the best solution has not changed much in several iterations, or stop when all time available to solve the problem has been exhausted. Whatever rule is used, we recommend applying a statistical selection procedure, such as the Select-the-Best Procedure in Section 12.2.2, to the top 5 to 10% of the apparently best solutions. This is done to evaluate which among them is the true best with guaranteed confidence, and to precisely estimate their performance. If the raw data from the search have been saved, then these data can be used as the first-stage sample for a two-stage selection procedure.

Example 12.11:   Implementing Random Search _____
Suppose that a manufacturing system consists of 4 stations in series. The zeroth station always has raw material available. When the zeroth station completes work on a part, it passes the part along to the first station, then the first passes the part to the second, and so on. Buffer space between stations 0 and 1, 1 and 2, and 2 and 3 is limited to 50 parts in total. If, say, station 2 finishes a part but there is no buffer space available in front of station 3, then station 2 is blocked, meaning that it cannot do any further work. The question is how to allocate these 50 spaces to minimize the expected cycle time per part over one shift.

Let x_i be the number of buffer spaces in front of station i. Then the decision variables are x_1, x_2, x_3 with the constraint that $x_1 + x_2 + x_3 = 50$ (it makes no sense to allocate fewer buffer spaces than we have available). This implies a total of 1326 possible designs (can you figure out how this number is computed?).

To simplify the presentation of the random-search algorithm, let the counter for solution (x_1, x_2, x_3) be denoted as $C(x_1, x_2, x_3)$.

Random-Search Algorithm

Step 1.   Initialize 1326 counter variables $C(x_1, x_2, x_3) = 0$, one for each of the possible solutions (x_1, x_2, x_3). Select an initial solution, say, $(x_1 = 20, x_2 = 15, x_3 = 15)$ and set $C(20, 15, 15) = 1$.

Step 2.   Choose another solution from the set of all solutions *except* $(20, 15, 15)$ in such a way that each solution has an equal chance of being selected. Suppose $(11, 35, 4)$ is chosen.

Step 3.   Run simulation experiments at the two solutions to obtain estimates of the expected cycle time $Y(20, 15, 15)$ and $Y(11, 35, 4)$. Suppose that $Y(20, 15, 15) < Y(11, 35, 4)$. Then $(20, 15, 15)$ remains as the current good solution.

Step 4.   Set $C(20, 15, 15) = C(20, 15, 15) + 1$.

Step 2.   Choose another solution from the set of all solutions *except* (20, 15, 15) in such a way that each solution has an equal chance of being selected. Suppose (28, 12, 10) is chosen.

Step 3.   Run simulation experiments at the two solutions to obtain estimates of the expected cycle time $Y(20, 15, 15)$ and $Y(28, 12, 10)$. Suppose that $Y(28, 12, 10) < Y(20, 15, 15)$. Then (28, 12, 10) becomes the current good solution.

Step 4.   Set $C(28, 12, 10) = C(28, 12, 10) + 1$.

Step 2.   Choose another solution from the set of all solutions *except* (28, 12, 10) in such a way that each solution has an equal chance of being selected. Suppose (0, 14, 36) is chosen.

Step 3.   Continue...

When the search is terminated, we select the solution (x_1, x_2, x_3) that gives the largest $C(x_1, x_2, x_3)$ count. As we discussed earlier, the top 5 to 10 solutions should then be subjected to a separate statistical analysis to determine which among them is the true best, with high confidence. In this case, the solutions with the largest counts would receive the second analysis.

Despite the apparent simplicity of the Random-Search Algorithm, we have glossed over a subtle issue that often arises in algorithms with provable performance. In Step 2, the algorithm must randomly choose a solution such that all are equally likely to be selected (except the current one). How can this be accomplished in Example 12.11? The constraint that $x_1 + x_2 + x_3 = 50$ means that x_1, x_2 and x_3 cannot be sampled independently. One might be tempted to sample x_1 as a discrete uniform random variable on 0 to 50, then sample x_2 as a discrete uniform on 0 to $50 - x_1$, and finally set $x_3 = 50 - x_1 - x_2$. But this method does not make all solutions equally likely, as the following illustration shows: Suppose that x_1 is randomly sampled to be 50. Then the trial solution must be (50, 0, 0); there is only one choice. But if $x_1 = 49$, then both (49, 1, 0) and (49, 0, 1) are possible. Thus, $x_1 = 49$ should be more likely than $x_1 = 50$ if all solutions with $x_1 + x_2 + x_3 = 50$ are to be equally likely.

12.5   Summary

This chapter provided a basic introduction to the comparative evaluation of alternative system designs based on data collected from simulation runs. It was assumed that a fixed set of alternative system designs had been selected for consideration. Comparisons based on confidence intervals and the use of common random numbers were emphasized. A brief introduction to metamodels—whose purpose is to describe the relationship between design variables and the output response—and to optimization via simulation—whose purpose is to select the best from among a large and diverse collection of system designs—was also provided. There are many additional topics of potential interest, beyond the scope of this text, in the realm of statistical analysis techniques relevant to simulation. Some of these topics are:

a. experimental design models, whose purpose is to discover which factors have a significant impact on the performance of system alternatives;

b. output-analysis methods other than the methods of replication and batch means;

c. variance-reduction techniques, which are methods to improve the statistical efficiency of simulation experiments (common random numbers being an important example).

The reader is referred to Banks [1998] and Law [2007] for discussions of these topics and of others relevant to simulation.

The most important idea in Chapters 11 and 12 is that simulation output data require a statistical analysis in order to be interpreted correctly. In particular, a statistical analysis can provide a measure of the precision of the results produced by a simulation and can provide techniques for achieving a specified precision.

REFERENCES

ANDRADÓTTIR, S. [1998], "Simulation Optimization," Chapter 9 in *Handbook of Simulation*, J. Banks, ed., Wiley, New York.

ANDRADÓTTIR, S. [1999], "Accelerating the Convergence of Random Search Methods for Discrete Stochastic Optimization," *ACM TOMACS*, Vol. 9, pp. 349–380.

BANKS, J., ed. [1998], *Handbook of Simulation*, Wiley, New York.

BATES, D. M., AND D. G. WATTS [1988], *Nonlinear Regression Analysis and its Applications*, Wiley, New York.

BECHHOFER, R. E., T. J. SANTNER, AND D. GOLDSMAN [1995], *Design and Analysis for Statistical Selection, Screening and Multiple Comparisons*, Wiley, New York.

BOESEL, J., B. L. NELSON, AND N. ISHII [2003], "A Framework for Simulation-Optimization Software," *IIE Transactions*, Vol. 35, pp. 221–229.

BOESEL, J., B. L. NELSON, AND S. KIM [2003], "Using Ranking and Selection to 'Clean Up' After Simulation Optimization," *Operations Research*, Vol. 51, pp. 814–825.

BOX, G. E. P., AND N. R. DRAPER [1987], *Empirical Model-Building and Response Surfaces*, Wiley, New York.

FU, M. C. [2002], "Optimization for Simulation: Theory vs. Practice," *INFORMS Journal on Computing*, Vol. 14, pp. 192–215.

GLOVER, F. [1989], "Tabu Search—Part I," *ORSA Journal on Computing*, Vol. 1, pp. 190–206.

GOLDSMAN, D., AND B. L. NELSON [1998], "Comparing Systems via Simulation," Chapter 8 in *Handbook of Simulation*, J. Banks, ed., Wiley, New York.

HINES, W. W., D. C. MONTGOMERY, D. M. GOLDSMAN, AND C. M. BORROR [2002], *Probability and Statistics in Engineering*, 4th ed., Wiley, New York.

HOCHBERG, Y., AND A. C. TAMHANE [1987], *Multiple Comparison Procedures*, Wiley, New York.

HSU, J. C. [1996], *Multiple Comparisons: Theory and Methods*, Chapman & Hall, New York.

KIM, S., AND B. L. NELSON [2006], "Selecting the Best System," Chapter 17 in *Handbooks in Operations Research and Management Science: Simulation*, S. G. Henderson and B. L. Nelson, eds., North-Holland, New York.

KLEIJNEN, J. P. C. [1975], *Statistical Techniques in Simulation, Parts I and II*, Dekker, New York.

KLEIJNEN, J. P. C. [1987], *Statistical Tools for Simulation Practitioners*, Dekker, New York.

KLEIJNEN, J. P. C. [1988], "Analyzing Simulation Experiments with Common Random Numbers," *Management Science*, Vol. 34, pp. 65–74.

KLEIJNEN, J. P. C. [1998], "Experimental Design for Sensitivity Analysis, Optimization, and Validation of Simulation Models," Chapter 6 in *Handbook of Simulation*, J. Banks, ed., Wiley, New York.

LAW, A. M. [2007], *Simulation Modeling and Analysis*, 4th ed., McGraw-Hill, New York.

MONTGOMERY, D. C. [2000], *Design and Analysis of Experiments*, 5th ed., Wiley, New York.

NELSON, B. L., J. SWANN, D. GOLDSMAN, AND W.-M. T. SONG [2001], "Simple Procedures for Selecting the Best System when the Number of Alternatives is Large," *Operations Research*, Vol. 49, pp. 950–963.

NELSON, B. L. [1987], "Some Properties of Simulation Interval Estimators Under Dependence Induction," *Operations Research Letters*, Vol. 6, pp. 169–176.

NELSON, B. L. [1992], "Designing Efficient Simulation Experiments," *Proceedings of the 1992 Winter Simulation Conference*, J. J. Swain, D. Goldsman, R. C. Crain, and J. R. Wilson, eds., Arlington, VA, Dec. 13–16, pp. 126–132.

ÓLAFSSON, S. [2006], "Metaheuristics," Chapter 21 in *Handbooks in Operations Research and Management Science: Simulation*, S. G. Henderson and B. L. Nelson, eds., North-Holland, New York.

WHITT, W. [1989], "Planning Queueing Simulations," *Management Science*, Vol. 35, pp. 1341–1366.

WRIGHT, R. D., AND T. E. RAMSAY, JR. [1979], "On the Effectiveness of Common Random Numbers," *Management Science*, Vol. 25, pp. 649–656.

YANG, F., B. ANKENMAN, AND B. L. NELSON [2007], "Efficient Generation of Cycle Time-Throughput Curves through Simulation and Metamodeling," *Naval Research Logistics*, Vol. 54, pp. 78–93.

EXERCISES

1. A company has dump trucks that repeatedly go through three activities: loading, weighing, and traveling. Assume that there are eight trucks and that, at time 0, all eight are at the loaders. Weighing time per truck on the single scale is uniformly distributed between 1 and 9 minutes, and travel time per truck is exponentially distributed with mean 85 minutes. An unlimited queue is allowed before the loader(s) and before the scale. All trucks can be traveling at the same time. Management desires to compare one fast loader against the two slower loaders currently being used. Each of the slow loaders can fill a truck in from 1 to 27 minutes, uniformly distributed. The new fast loader can fill a truck in from 1 to 19 minutes, uniformly distributed. The basis for comparison is mean system response time over a 40-hour time horizon, where a response time is defined as the duration of time from a truck arrival at the loader queue to that truck's departure from the scale. Perform a statistically valid comparison of the two options simulated using common random numbers.

2. In Exercise 11.6, consider the following alternative (M, L) policies:

			L	
			Low 30	High 40
M	Low	50	(50,30)	(50, 40)
	High	100	(100,30)	(100, 40)

Investigate the relative costs of these policies, using suitable modifications of the simulation model developed in Exercise 11.6. Compare the four system designs on the basis of long-run mean monthly cost. First, make four replications of each (M, L) policy, using common random numbers to the greatest extent possible. Each replication should have a 12-month initialization phase followed by a 100-month data-collection phase. Compute confidence intervals having an overall confidence level of 90% for mean monthly cost for each policy. Then estimate the additional replications needed to achieve confidence intervals that do not overlap. Draw conclusions as to which is the best policy.

3. Reconsider Exercise 11.7. Compare the four inventory policies studied in Exercise 2, taking the cost of rush orders into account when computing monthly cost.

4. Reconsider Exercise 11.8. Compare the four monthly inventory policies studied in Exercise 2, taking into account the selling price of the perishable items.

5. In Exercise 11.9, investigate the effect of the order quantity on long-run mean daily cost. Each order arrives on a pallet on a delivery truck, so the permissible order quantities Q are multiples of 10; that is, Q may equal 10, or 20, or 30, ... In Exercise 11.9, the policy $Q = 20$ was investigated.

   (a) First, investigate the two policies $Q = 10$ and $Q = 50$. Use the run lengths, and so on, suggested in Exercise 11.9. On the basis of these runs, decide whether the optimal Q, say, Q^*, is between 10 and 50 or is greater than 50. (The cost curve as a function of Q should have what kind of shape?)
   (b) Using the results in part (a), suggest two additional values for Q and simulate the two policies. Draw conclusions. Include an analysis of the strength of your conclusions.

6. In Exercise 11.10, find the number of cards Q that the card-shop owner should purchase to maximize the profit within an error (for total profit) of approximately $5.00 at most. First, make runs for the policy of ordering $Q = 250$ and $Q = 350$ cards. With these results, plus the results of Exercise 11.10 for ordering $Q = 300$ cards, decide on a range of Q worth considering further: $200 \le Q \le 250$, or $250 \le Q \le 300$, etc. Then investigate this restricted range for two additional policies: $Q = 200$ and $Q = 225$, or $Q = 265$ and $Q = 285$, etc.

7. In Exercise 11.11, investigate the effect of target level M and review period N on mean monthly cost. Consider two target levels M, determined by ± 10 from the target level used in Exercise 11.11, and consider review periods N of 1 month and 3 months. Which (N, M) pair is best, according to these simulations?

8. Reconsider Exercises 11.13 and 11.14, which involved the scheduling rules (or queue disciplines) first-in–first-out (FIFO) and priority-by-type (PR) in a job shop. In addition to these two rules, consider a shortest imminent operation (SIO) scheduling rule: For a given station, all jobs of the type with the smallest mean processing time are given highest priority. For example, when using an SIO rule at station 1, jobs are processed in the following order: type II first, then type I, and type III last. Two jobs of the same type are processed on a FIFO basis. Develop a simulation experiment to compare the FIFO, PR, and SIO rules on the basis of mean total response time over all jobs.

9. In Exercise 11.13 (the job shop with FIFO rule), find the minimum number of workers needed at each station to avoid bottlenecks. A bottleneck occurs when average queue lengths at a station increase steadily over time. [*Hint*: Do not confuse increasing average queue length due to an inadequate number of servers with increasing average queue length due to initialization bias. In the former case, average queue length continues to increase indefinitely and server utilization is 1.0. In the latter case, average queue length eventually levels off and server utilization is less than 1.] Report on utilization of workers and total time it takes for a job to get through the job shop, by type, and over all types.

   [*Hint*: If server utilization at a work station is 1.0, and if average queue length tends to increase linearly as simulation run length increases, it is a good possibility that the work station is unstable and therefore is a bottleneck. In this case, at least one additional worker is needed at the work station. Use queueing theory, namely $\lambda/c_i\mu < 1$, to suggest the minimum number of workers needed at station 1. Recall that λ is the arrival rate, $1/\mu$ is the overall mean service time for one job with one worker, and c_i is the number of workers at station i. Attempt to use the same basic condition, $\lambda/c_i\mu < 1$, to suggest an initial number of servers at station i for $i = 2, 3, 4$.]

10. (a) Repeat Exercise 9 for the PR scheduling rule (see Exercise 11.14).
    (b) Repeat Exercise 9 for the SIO scheduling rule (see Exercise 12.8).
    (c) Compare the minimum required number of workers for each scheduling rule: FIFO vs. PR vs. SIO.

11. With the minimum number of workers found in Exercises 9 and 10 for the job shop of Exercise 11.13, consider adding one worker to the entire shop. This worker can be trained to handle the processing at only one station. At which station should this worker be placed? How does this additional worker affect mean total response time over all jobs? Over type I jobs? Investigate the job shop with and without the additional worker for each scheduling rule: FIFO, PR, SIO.

12. In Exercise 11.17, suppose that a buffer of capacity one item is constructed in front of each worker. Design an experiment to investigate whether this change in system design has a significant impact upon individual worker utilizations ρ_1, ρ_2, ρ_3, and ρ_4. At the very least, compute confidence intervals for $\rho_1^0 - \rho_1^1$ and $\rho_4^0 - \rho_4^1$, where ρ_i^s is utilization for worker i when the buffer has capacity s.

13. A clerk in the admissions office at Small State University processes requests for admissions materials. The time to process requests depends on the program of interest (e.g., industrial engineering, management science, computer science, etc.) and on the level of the program (Bachelor's, Master's, Ph.D.). Suppose that the processing time is modeled well as normally distributed, with mean 7 minutes and standard deviation 2 minutes. At the beginning of the day it takes the clerk some time to get set to begin working on requests; suppose that this time is modeled well as exponentially distributed, with mean 20 minutes. The admissions office typically receives between 40 and 60 requests per day.

Let x be the number of applications received in a day, and let Y be the time required to process them (including the set-up time). Fit a metamodel for $E(Y|x)$ by making n replications at the design points $x = 40, 50, 60$. Notice that, in this case, we know that the correct model is

$$E(Y|x) = \beta_0 + \beta_1 x = 20 + 7x$$

(Why?) Begin with $n = 2$ replications at each design point and estimate β_0 and β_1. Gradually increase the number of replications and observe how many are required for the estimates to be close to the true values.

14. Repeat the previous exercise using CRN. How do the results change?

15. Consider the simple linear regression model $Y = \beta_0 + \beta_1 x + \varepsilon$. Find a statistics textbook that covers regression modeling. There you will find a test for the hypothesis $\beta_1 = 0$ (rejecting this hypothesis is evidence that x and Y are linearly related). The usual assumptions behind this test do not hold if we use CRN. Where do they break down?

16. Riches and Associates retains its cash reserves primarily in the form of certificates of deposit (CDs), which earn interest at an annual rate of 8%. Periodically, however, withdrawals must be made from these CDs in order to pay suppliers, etc. These cash outflows are made through a checking account that earns no interest. The need for cash cannot be predicted with certainty. Transfers from CDs to checking can be made instantaneously, but there is a "substantial penalty" for early withdrawal from CDs. Therefore, it might make sense for R&A to make use of the overdraft protection on their checking account, which charges interest at a rate of $0.00033 per dollar per day (i.e., 12% per year) for overdrafts.

R&A likes simple policies in which it transfers a fixed amount, a fixed number of times, per year. Currently, it makes 6 transfers per year, of $18,250 each time. Your job is to find a policy that reduces its long-run cost per day.

Judging from historical patterns, demands for cash arrive a rate of about 1 per day, with the arrivals being modeled well as a Poisson process. The amount of cash needed to satisfy each demand is reasonably represented by a lognormally distributed random variable with mean $300 and standard deviation $150.

The penalty for early withdrawal is different for different CDs. It averages $150 for each withdrawal, regardless of size, but the actual penalty can be modeled as a uniformly distributed random variable with range $100 to $200.

Use cash level in checking to determine the length of the initialization phase. Make enough replications that your confidence interval for the difference in long-run cost per day does not contain zero. Be sure to use CRN in your experiment design.

17. If you have access to commercial optimization-via-simulation software, test how well it works as the variability of the simulation outputs increases. Use a simple model, such as $Y = x^2 + \varepsilon$, where ε is a random variable with a $N(0, \sigma^2)$ distribution, and for which the optimal solution is known ($x = 0$ for minimization, in this case). See how quickly, or whether, the software can

find the true optimal solution as σ^2 increases. Next, try more complex models with more than one design variable.

18. For Example 12.11, show why there are 1326 solutions. Then derive a way to sample x_1, x_2, and x_3 such that $x_1 + x_2 + x_3 = 50$ and all outcomes are equally likely.

19. A critical electronic component with mean time to failure of x years can be purchased for $2x$ thousand dollars (thus, the more reliable the component, the more expensive it is). The value of x is restricted to being between 1 to 10 years, and the actual time to failure is modeled as exponentially distributed. The mission for which the component is to be used lasts one year; if the component fails in less than one year, then there is a cost of $20,000 for early failure. What value of x should be chosen to minimize the expected total cost (purchase plus early failure)?

To solve this problem, develop a simulation that generates a total cost for a component with mean time to failure of x years. This requires sampling an exponentially distributed random variable with mean x, and then computing the total cost as $2000x$ plus 20,000 if the failure time is less than 1. Fit a quadratic metamodel in x and use it to find the value of x that minimizes the fitted model. [*Hints*: Select several values of x between 1 and 10 as design points. At each value of x, let the response variable $Y(x)$ be the average of at least 30 observations of total cost.]

20. Use optimization-via-simulation software to solve Exercise 19. If you do not have access to such software, use the random-search algorithm, and let the possible values of x be {1.00, 1.25, 1.50, ..., 10.00}.

21. Suppose that demand for a certain product has a Poisson distribution with mean 10 units. Use optimization-via-simulation software (or the random-search algorithm, if you do not have access to such software) to find the order size x that maximizes the probability that demand equals x units, assuming that $0 \le x \le 100$, and x is an integer. (This problem is based on Example 5.1 in Andradóttir [1999].) [*Hint*: On each replication of the simulation for a trial value of x, a random demand is generated from the Poisson distribution with mean 10. If this random demand equals x, then the response is $Y(x) = 1$; otherwise it is $Y(x) = 0$.]

22. In Exercise 16, consider the following 5 policies: 6 transfers of $18,250; 5 transfers of $21,900; 4 transfers of $27,375; 3 transfers of $36,500; and 2 transfers of $54,750. Make 10 replications of each policy and construct simultaneous confidence intervals for the differences in expected cost of all pairs of policies.

23. In Exercise 16, consider the following 5 policies: 6 transfers of $18,250; 5 transfers of $21,900; 4 transfers of $27,375; 3 transfers of $36,500; and 2 transfers of $54,750. Make 10 replications of each policy and then apply the Select-the-Best Procedure to select the policy with the least expected cost per day. Use $\epsilon = \$2$ per day.

Part V

Applications

13

Simulation of Manufacturing and Material-Handling Systems

Manufacturing and material-handling systems provide one of the most important applications of simulation. Simulation has been used successfully as an aid in the design of new production facilities, warehouses, and distribution centers. It has also been used to evaluate suggested improvements to existing systems. Engineers and analysts using simulation have found it valuable for evaluating the impact of capital investments in equipment and physical facility and of proposed changes to material handling and layout. They have also found it useful to evaluate staffing and operating rules and proposed rules and algorithms to be incorporated into production control systems, warehouse-management control software, and material-handling controls. Managers have found simulation useful in providing a "test drive" before making capital investments, without disrupting the existing system with untried changes.

Section 13.1 provides an introduction and discusses some of the features of simulation models of manufacturing and material-handling systems. Section 13.2 discussed the goals of manufacturing simulation and the most common measures of system performance. Section 13.3 discusses a number of the issues common to many manufacturing and material-handling simulations, including the treatment of downtimes and failure, and trace-driven simulations using actual historical data or historical order files. Section 13.4 provides brief abstracts of a number of reported simulation projects, with references for additional reading. Section 13.5 gives an extended example of a simulation of a small production line, emphasizing the experimentation and analysis of system performance to achieve a desired throughput. For an overview of simulation software for manufacturing and material-handling applications, see Section 4.7.

13.1   Manufacturing and Material-Handling Simulations

As do all modeling projects, manufacturing and material-handling simulation projects need to address the issues of scope and level of detail. Consider scope as analogous to breadth and level of detail as analogous to depth. Scope describes the boundaries of the project: what's in the model, and what's not. For a subsystem, process, machine, or other component, the project scope determines whether the object is in the model. Then, once a component or subsystem is treated as part of a model, often it can be simulated at many different levels of detail.

The proper scope and level of detail should be determined by the objectives of the study and the questions being asked. On the other hand, level of detail could be constrained by the availability of input data and the knowledge of how system components work. For new, nonexistent systems, data availability might be limited, and system knowledge might be based on assumptions.

Some general guidelines can be provided, but the judgment of experienced simulation analysts, working with the customer to define, early in the project, the questions the model is being designed to address, provides the most effective basis for selecting a proper scope and a proper level of detail.

Should the model simulate each conveyor section or vehicle movement, or can some be replaced by a simple time delay? Should the model simulate auxiliary parts, or the handling of purchased parts, or can the model assume that such parts are always available at the right location when needed for assembly?

At what level of detail does the control system need to be simulated? Many modern manufacturing facilities, distribution centers, baggage-handling systems, and other material-handling systems are computer controlled by a management-control software system. The algorithms built into such control software play a key role in system performance. Simulation is often used to evaluate and compare the effectiveness of competing control schemes and to evaluate suggested improvements. It can be used to debug and fine-tune the logic of a control system before it is installed.

These questions are representative of the issues that need to be addressed in choosing the proper scope and level of detail for a modeling project. In turn, the scope and level of model detail limit the type of questions that can be addressed by the model. In addition, models can be developed in an iterative fashion, adding detail for peripheral operations at later stages if such operations are later judged to affect the main operation significantly. It is good advice to start as simply as possible and add detail only as needed.

13.1.1   Models of Manufacturing Systems

Models of manufacturing systems might have to take into account a number of characteristics of such systems, including some of the following:

1. Physical layout
2. Labor
    a. Shift schedules
    b. Job duties and certification
3. Equipment
    a. Rates and capacities
    b. Breakdowns
        i. Time to failure
        ii. Time to repair
        iii. Resources needed for repair
4. Maintenance
    a. Preventative maintenance (PM) schedule
    b. Time and resources required
    c. Tooling and fixtures
5. Workcenters
    a. Processing
    b. Assembly
    c. Disassembly
6. Product
    a. Product flow, routing, and resources needed
    b. Bill of materials
7. Production schedules
    a. Made-to-stock
    b. Made-to-order
        i. Customer orders
        ii. Line items and quantities
8. Production control
    a. Assignment of jobs to work areas
    b. Task selection at workcenters
    c. Routing decisions
9. Supplies
    a. Ordering
    b. Receipt and storage
    c. Delivery to workcenters
10. Storage
    a. Supplies
    b. Spare parts
    c. Work-in-process (WIP)
    d. Finished goods
11. Packing and shipping
    a. Order consolidation
    b. Paperwork
    c. Loading of trailers

13.1.2 Models of Material Handling Systems

In manufacturing systems, it is not unusual for 80 to 85% of an item's total time in system to be expended on material handling or on waiting for material handling to occur. This work-in-process (WIP) represents a vast investment, and reductions in WIP and associated delays can result in large cost savings. Therefore, for some studies, detailed material-handling simulations are cost effective.

In some production lines, the material-handling system is an essential component. For example, automotive paint shops typically consist of a power-and-free conveyor system that transports automobile bodies or body parts through the paint booths.

In warehouses, distribution centers, and flow-through and cross-docking operations, material handling is clearly a key component of any material-flow model. Manual warehouses typically use manual fork trucks to move pallets from receiving dock to storage and from storage to shipping dock. More automated distribution centers might use extensive conveyor systems to support putaway, order picking, order sortation, and consolidation.

Models of material-handling systems often have to contain some of the following types of subsystems:

1. Conveyors
   a. Accumulating
   b. Nonaccumulating
   c. Indexing and other special purpose
   d. Fixed window or random spacing
   e. Power and free
2. Transporters
   a. Unconstrained vehicles (e.g., manually guided fork trucks)
   b. Guided vehicles (automated or operator controlled, wire guided chemical paths, rail guided)
   c. Bridge cranes and other overhead lifts
3. Storage systems
   a. Pallet storage
   b. Case storage
   c. Small-part storage (totes)
   d. Oversize items
   e. Rack storage or block stacked
   f. Automated storage and retrieval systems (AS/RS) with storage–retrieval machines (SRM)

13.1.3 Some Common Material-Handling Equipment

There are numerous types of material-handling devices common to manufacturing, warehousing, and distribution operations. They include unconstrained transporters, such as carts, manually driven fork-lift trucks, and pallet jacks; guided path transporters, such as AGVs (automated guided vehicles); and fixed-path devices, such as various types of conveyor.

The class of unconstrained transporters, sometimes called free-path transporters, includes carts, fork-lift trucks, pallet jacks, and other manually driven vehicles that are free to travel throughout a

facility unconstrained by a guide path of any kind. Unconstrained transporters are not confined to a network of paths and may choose an alternate path or move around an obstruction. In contrast, the guided-path transporters move along a fixed path, such as chemical trails on the floor, wires embedded in the floor, or infrared lights placed strategically, or by self-guidance, using radio communications, laser guidance and dead reckoning, and rail. Guided-path transporters sometimes contend with one another for space along their paths and usually have limited options upon meeting obstacles and congestion. Examples of guided-path transporters include the automated guided vehicle (AGV), a rail-guided turret truck for storage and retrievals of pallets in rack storage, and a crane in an AS/RS (automated storage and retrieval system).

The conveyor is a fixed-path device for moving entities from point to point, following a fixed path with specific load, stopping or processing points, and unload points. A conveyor system can consist of numerous connected sections with merges and diverts. Each section can be of one of a number of different types. Examples of conveyor types include belt, powered and gravity roller, bucket, chain, tilt tray, and power-and-free, each with its own characteristics that must be modeled accurately.

Most conveyor sections can be classified as either accumulating or nonaccumulating. An accumulating conveyor section runs continuously. If the forward progress of an item is halted while on the accumulating conveyor, slippage occurs, allowing the item to remain stationary and items behind it to continue moving until they reach the stationary item. Some belt and most roller conveyors operate in this manner. Only items that will not be damaged by bumping into each other can be placed on an accumulating conveyor.

In contrast, after an item is on a nonaccumulating conveyor section, its spacing relative to other items does not change. If one item stops moving, the entire section stops moving, and hence all items on the section stop. For example, nonaccumulating conveyor is used for moving televisions not yet in cartons, for they must be kept at a safe distance from one another while moving from one assembly or testing station to the next. Bucket conveyors, tilt-tray conveyors, some belt conveyors, and conveyors designed to carry heavy loads (usually pallets) are nonaccumulating conveyors.

Conveyors can also be classified as fixed-window or random spacing. In fixed-window spacing, items on the conveyor must always be within zones of equal length, which can be pictured as lines drawn on a belt conveyor or trays pulled by a chain. For example, in a tilt-tray conveyor, continuously moving trays of fixed size are used to move items. The control system is designed to induct items in such a way that each item is in a separate tray; thus it is a nonaccumulating fixed-window conveyor. In contrast, with random spacing, items can be anywhere on the conveyor section relative to other items. To be inducted, they simply require sufficient space.

Besides these basic types, there are innumerable types of specialized conveyors for special purposes. For example, a specialized indexing conveyor may move forward in increments, always maintaining a fixed distance between the trailing edge of the load ahead and the leading edge of the load behind. Its purpose is to form a "slug" of items, equally spaced apart, to be inducted all together onto a transport conveyor. For the local behavior of some systems—that is, the performance at a particular workstation or induction point—a detailed understanding and accurate model of the physical workings and the control logic are essential for accurate results.

13.2   Goals and Performance Measures

The purpose of simulation is insight, not numbers. Those who purchase and use simulation software and services want to gain insight and understanding into how a new or modified system will work. Will it meet throughput expectations? What happens to response time at peak periods? Is the system resilient to short-term surges? What is the recovery time when short-term surges cause congestion and queueing? What are the staffing requirements? What problems occur? If problems occur, what is their cause and how do they arise? What is the system capacity? What conditions and loads cause a system to reach its capacity?

Simulations are expected to provide numeric measures of performance, such as throughput under a given set of conditions, but the major benefit of simulation comes from the insight and understanding gained regarding system operations. Visualization through animation and graphics provides major assistance in the communication of model assumptions, system operations, and model results. Often, visualization is the major contributor to a model's credibility, which in turn leads to acceptance of the model's numeric outputs. Of course, a proper experimental design that includes the right range of experimental conditions plus a rigorous analysis and, for stochastic simulation models, a proper statistical analysis, is of utmost importance for the simulation analyst to draw correct conclusions from simulation outputs.

The major goals of manufacturing-simulation models are to identify problem areas and quantify system performance. Common measures of system performance include the following:

- throughput under average and peak loads;
- system cycle time (how long it takes to produce one part);
- utilization of resources, labor, and machines;
- bottlenecks and choke points;
- queueing at work locations;
- queueing and delays caused by material-handling devices and systems;
- WIP storage needs;
- staffing requirements;
- effectiveness of scheduling systems;
- effectiveness of control systems.

Often, material handling is an important part of a manufacturing system and its performance. Non-manufacturing material-handling systems include warehouses, distribution centers, cross-docking operations, baggage-handling systems at airports and container terminals. The major goals of these nonmanufacturing material-handling systems are similar to those identified for manufacturing systems. Some additional considerations are the following:

- how long it takes to process one day of customer orders;
- effect of changes in order profiles (for distribution centers);
- truck/trailer queueing and delays at receiving and shipping docks;
- effectiveness of material-handling systems at peak loads;
- recovery time from short-term surges (for example, with baggage-handling).

13.3   Issues in Manufacturing and Material-Handling Simulations

There are a number of modeling issues especially important for the achievement of accurate and valid simulation models of manufacturing and material-handling systems. Two of these issues are the proper modeling of downtimes and whether, for some inputs, to use actual system data or a statistical model of those inputs.

13.3.1   Modeling Downtimes and Failures

Unscheduled random downtimes can have a major effect on the performance of manufacturing systems. Many authors have discussed the proper modeling of downtime data (Williams [1994]; Clark [1994]; Law and Kelton [2000]). This section discusses the problems that can arise when downtime is modeled incorrectly and suggests a number of ways to model machine and system downtimes correctly.

Scheduled downtime, such as for preventive maintenance, or periodic downtime, such as for tool replacement, also can have a major effect on system performance. But these downtimes are usually (or should be) predictable and can be scheduled to minimize disruptions. In addition, engineering efforts or new technology might be able to reduce their duration.

There are a number of alternatives for modeling random unscheduled downtime, some better than others:

1. Ignore it.
2. Do not model it explicitly, but increase processing times in appropriate proportion.
3. Use constant values for time to failure and time to repair.
4. Use statistical distributions for time to failure and time to repair.

Of course, alternative (1) generally is not the suggested approach. This is certainly an irresponsible modeling technique if downtimes have an impact on the results, as they do in almost all situations. One situation in which ignoring downtimes could be appropriate, with the full knowledge of the customer, is to leave out those catastrophic downtimes that occur rarely and leave a production line or plant down for a long period of time. In other words, the model would incorporate normal downtimes but ignore those catastrophic downtimes, such as general power failures, snow storms, cyclones, and hurricanes, that occur rarely but stop all production when they do occur. The documented scope of the project should clearly state the assumed operating conditions and those conditions that are not included in the model. If it is generally known that a plant will be closed for some number of snow days per year, then the simulation need not take these downtimes into account, for the effect of any given number of days can easily be factored into the simulation results when making annual projections.

The second possibility, to factor into the model the effect of downtimes by adjusting processing times applied to each job or part, might be an acceptable approximation under limited circumstances. If each job or part is subject to a large number of small delays associated with downtime of equipment or tools, then the total of such delays may be added to the pure processing time to arrive at an adjusted processing time. If total delay time and pure processing time are random in nature, then an appropriate statistical distribution should be used for the total adjusted processing time. If the pure processing time is constant while the total delay time in one cycle is random and variable, it is almost never accurate to adjust the processing time by a constant factor. For example, if processing time is usually 10 minutes but the equipment is subject to downtimes that cause about a 10% loss in capacity, it is not appropriate to merely change the processing time to a constant 11 minutes. Such a deterministic adjustment might provide reasonably accurate estimates of overall system throughput, but will not provide accurate estimates of such local behavior as queue and buffer space needed at peak times. Queueing and short-term congestion are strongly influenced by randomness and variability.

The third possibility, using constant durations for time to failure and time to repair, might be appropriate when, for example, the downtime is actually due to preventive maintenance that is on a fixed schedule. In almost all other circumstances, the fourth possibility, modeling time to failure and time to repair by appropriate statistical distributions, is the appropriate technique. This requires either actual data for choosing a statistical distribution based on the techniques in Chapter 11, or, when data is lacking, a reasonable assumption based on the physical nature of the causes of downtimes.

The nature of time to failure is also important. Are times to failure completely random in nature, a situation due typically to a large number of possible causes of failure? In this case, exponential distribution might provide a good statistical model. Or are times to failure, rather, more regular— typically, due to some major component, say, a tool, wearing out? In this case, a uniform or (truncated) normal distribution could be more nearly appropriate. In the latter case, the mean of the distribution represents the average time to failure, and the distribution places a plus or minus around the mean.

Time to failure can be measured in a number of different ways:

1. by wall-clock time;
2. by machine or equipment busy time;
3. by number of cycle times;
4. by number of items produced.

Breakdowns or failures can be based on clock time, actual usage, or cycles. Note that the word *breakdown* or *failure* is used, even though preventive maintenance could be the reason for a downtime. As mentioned, breakdowns or failures can be probabilistic or deterministic in duration.

Actual usage breakdowns are based on the time during which the resource is used. For example, wear on a machine tool occurs only when the machine is in use. Time to failure is measured against machine-busy time and not against wall-clock time. If the time to failure is 90 hours, then the model keeps track of total busy time since the last downtime ended, and, when 90 hours is reached, processing is interrupted and a downtime occurs.

Clock-time breakdowns might be associated with scheduled maintenance; for example, changes of fluids every three months when a complete lubrication is required. Downtimes based on wall-clock time may also be used for equipment that is always busy or equipment that runs even when it is not processing parts.

Cycle breakdowns or failures are based on the number of times the resource is used. For example, after every 50 uses of a tool, it needs to be sharpened. Downtimes based on number of cycle times or number of items produced are implemented by generating the number of times or items and, in the model, simply counting until this number is reached. Typical uses of downtimes based on busy time or cycle times may be for maintenance or tool replacement.

Another issue is what happens to a part at a machine when the breakdown or failure occurs. Possibilities include scrapping the part, rework, or simply continuing processing after repair. In some cases—for example, when preventive maintenance is due—the part in the machine may complete processing before the repair (or maintenance activity) begins.

Time to repair can also be modeled in two fundamentally different ways:

1. as a pure time delay (no resources required);
2. as a wait time for a resource (e.g., maintenance person) plus a time delay for actual repair.

Of course, there are many variations on these methods in actual modeling situations. When a repair or maintenance person is a limited resource, the second approach will be a more accurate model and provide more information.

The next example illustrates the importance of using the proper approach for modeling downtimes and the consequences and inaccurate results that sometimes result from inappropriate assumptions.

Example 13.1:   Effect of Downtime on Queueing

Consider a single machine that processes a wide variety of parts that arrive in random mixes at random times. Data analysis has shown that an exponentially distributed processing time with a mean of 7.5 minutes provides a fairly accurate representation. Parts arrive at random, time between arrivals being exponentially distributed with mean 10 minutes. The machine fails at random times. Downtime studies have shown that time-to-failure can be reasonably approximated by an exponential distribution with mean time 1000 minutes. The time to repair the resource is also exponentially distributed, with mean time 50 minutes. When a failure occurs, the current part in the machine is removed from the machine; when the repair has been completed, the part resumes its processing.

When a part arrives, it queues and waits its turn at the machine. It is desired to estimate the size of this queue. An experiment was designed to estimate the average number of parts in the queue. To illustrate the effect of an accurate treatment of downtimes, the model was run under a number of different assumptions. For each case and replication, the simulation run length was 100,000 minutes.

Table 13.1 shows the average number of parts in the queue for six different treatments of the time between breakdowns. For each treatment that involves randomness, five replications of those treatments and the average for those five replications are shown.

Case A ignores the breakdowns. The average number in the queue is 2.31 parts. Across the 5 independent replications, the averages range from 2.05 to 2.70 parts. This treatment of breakdowns is not recommended.

Case B increases the average service time from 7.5 minutes to 8.0 minutes in an attempt to approximate the effect of downtimes. On average, each downtime and repair cycle is 1050 minutes, with the machine down for 50 minutes. Thus the machine is down, on the average in the long run, $50/1050 = 4.8\%$ of total time. Thus, some have argued that downtime has approximately the same effect as increasing the processing time of each part by 4.8%, which is about 7.86 minutes. Therefore, an assumed constant 8 minutes per part should be (it is argued) a conservative approach. For this treatment of downtimes, the average number of parts in the queue, over the five replications, is about 3.26 parts. Across the 5 replications, the range is from 2.81 to 4.03 parts. (Note that the variability as shown in the range of values is very small compared to the other cases.) The treatment in Case B

Table 13.1 Average Number of Parts in Queue for Machines with Breakdowns

Case	1st Rep	2nd Rep	3rd Rep	4th Rep	5th Rep	Avg Rep
A. Ignore the breakdowns	2.36	2.05	2.38	2.05	2.70	2.31
B. Increase service time to 8.0	3.32	2.82	3.32	2.81	4.03	3.26
C. All random	4.05	3.77	4.36	3.95	4.43	4.11
D. Random processing, deterministic breakdowns	3.24	2.85	3.28	3.05	3.79	3.24
E. All deterministic						0.52
F. Deterministic processing, random breakdowns	1.06	1.04	1.10	1.32	1.16	1.13

might be appropriate under some limited circumstances, but, as was discussed in a previous section, it is not appropriate under the assumptions of this example.

The proper treatment, shown as Case C, treats the randomness in processing and breakdowns properly, with the assumed correct exponential distributions. The average value is about 4.11 parts waiting for the machine. Across the 5 replications, the average queue length ranges from 3.77 to 4.43 parts. The average number waiting differs from that of Case B by almost one part.

Case D is a simplification that treats the processing randomly, but treats the breakdowns as deterministic. The results average about 3.24 parts in the queue. The range of averages is from 2.85 to 3.79 parts, quite a reduction in variability from Case C.

Case E treats all of the times as deterministic. Only one replication is needed, because additional replications (using the same seed) will reproduce the result. The average value in the queue is 0.52 parts, well below the value in Case C, or any other case for that matter. The conclusion: Ignoring randomness is dangerous and leads to totally unrealistic results.

Case F treats arrivals and processing as deterministic, but breakdowns are random. The average number of parts in the queue at the machine is about 1.13. The range is from 1.04 to 1.32 parts. For some machines and processing in manufacturing environments, Case F is the realistic situation: Processing times are constant, and arrivals are regulated—that is, are also constant. The reader is left to consider the inaccuracies that would result from making faulty assumptions regarding the nature of time to failure and time to repair.

In conclusion, there can be significant differences between the estimated average numbers in a queue, based on the treatment of randomness. The results using the correct treatment of randomness can be far different from those using alternatives. Often, one is tempted by the unavailability of detailed data and the availability of averages to want to use average time to failure as if it were a constant. Example 13.1 illustrates the dangers of inappropriate assumptions. Both the appropriate technique to use and the appropriate statistical distribution depend on the available data and on the situation at hand.

As discussed by Williams [1994], the accurate treatment of downtimes is essential for achieving valid models of manufacturing systems. Some of the essential ingredients are the following:

1. avoidance of oversimplified and inappropriate assumptions;
2. careful collection of downtime data;
3. accurate representation of time to failure and time to repair by statistical distributions;
4. accurate modeling of system logic when a downtime occurs, in terms both of the repair-time logic and of what happens to the part currently processing.

13.3.2 Trace-Driven Models

A model driven by actual historical data is called a trace-driven model. In this section, we provide an example and a description of some modeling situations in which a trace-driven approach for some input data items is beneficial compared to other approaches.

Consider a model of a distribution center that receives customer orders that must be processed and shipped in one day. One modeling question is how to represent the day's set of orders. A typical order will contain one or more line items, and each line item can have a quantity of one or more pieces. For example, when you buy a new stereo, you might purchase an amplifier, a tuner, and a CD player (all separate line items, each having a quantity of one piece), and 4 identical speakers (another line item with a quantity of 4 pieces). The overall order profile can have a major impact on the performance of a particular system design. A system designed to handle large orders going to a small number of customers might not perform well if order profiles shift toward a larger number of customers (or larger number of separate shipments) with one or two items per order.

One approach is to characterize the order profile by using a discrete statistical distribution for each variable in an order:

1. the number of line items;
2. for each line item, the number of pieces.

If these two variables are statistically independent, then this approach might provide a valid model of the order profile. For many applications, however, these two variables may be highly correlated in ways that could be difficult to characterize statistically. For example, an apparel and shoe company has six large customers (the large department stores and discount chains), representing 50% of sales volume, which typically order dozens or hundreds of line items and large quantities of many of the items. At the opposite pole, on any given day approximately 50% of the orders are for one or two pairs of shoes (just-in-time with a vengeance!). For this company, the number of line items in an order is highly positively correlated with the quantity ordered; that is, large orders with a large number of line items also usually have large quantities of many of the line items. And small orders with only a few line items typically order small quantities of each item.

What would happen if the two variables, number of line items and quantity per line item, were modeled by independent statistical distributions? When an order began processing, the model would make two random uncorrelated draws, which could result in order profiles quite different from those found in practice. Such an erroneous assumption could result, for example, in far too large a proportion of orders having one or two line items with large unrealistic quantities.

Another common but more serious error is to assume that there is an average order and to simulate only the number of orders in a day with each being the typical order. In the author's experience, analyses of many order profiles have shown (1) that there is no such thing as a typical order and (2) that there is no such thing as a typical order profile.

An alternative approach, and one that has proven successful in many studies, is for the company to provide the actual orders for a sample of days over the previous year. Usually, it is desirable to simulate peak days. A model driven in this manner is called a *trace-driven model*.

A trace-driven model eliminates all possibility of error due to ignoring or misestimating correlations in the data. One apparent limitation could be a customer's desire, at times, to be able to simulate hypothesized changes to the order profile, such as a higher proportion of smaller orders in terms of both line items and quantities. In practice, this limitation can be removed by adding "dials" to the order-profile portion of the model, so that a simulation analyst can dial up more or less of certain characteristics, as desired. One approach is to treat the day's orders as a statistical population from which the model draws samples in a random fashion. This approach makes it easy to change overall order volume without modifying the profile. A second related approach would be to subdivide a day's orders into subgroups based on number of line items, quantities or other numeric parameters, and then sample in a specified proportion from each subgroup. By changing the proportion of each subgroup, different order profiles can be "dialed up" and fed into the model. A third approach is to use factors to adjust the number of daily orders, the number of line items, and/or the quantities. In practice one of these approaches might be as accurate as can be expected for hypothesized future order profiles and might provide a cost effective and reasonably accurate model, especially for testing the robustness of a system design for assumed changes in order characteristics.

Other examples of input variables that may be used in trace-driven models include the following:

- orders to a custom job shop, using actual historical orders;
- product mix and quantities, and production sequencing, for an assembly line making 100 styles and sizes of hot-water heaters;
- time to failure and downtime, using actual maintenance records;
- truck arrival times to a warehouse, using gate records.

Whether to make an input variable trace-driven or to characterize it as a statistical distribution depends on a number of issues, including the nature of the variable itself, whether it is correlated with or independent of other variables, the availability of accurate data, and the questions being addressed.

13.4 Case Studies of the Simulation of Manufacturing and Material Handling Systems

The *Winter Simulation Conference Proceedings*, *IIE Magazine*, *Modern Material Handling* and other periodicals are excellent sources of information for short cases in the simulation of manufacturing and material-handling systems.

An abstract of some of the papers from past *Winter Simulation Conference Proceedings* will provide some insight into the types of problems that can be addressed by simulation. These abstracts have been paraphrased and shortened where appropriate; our goal is to provide an indication of the breadth of real-world applications of simulation.

Session: Semiconductor Wafer Manufacturing
Paper: Modeling and Simulation of Material Handling for Semiconductor
             Wafer Fabrication
Authors: Neal G. Pierce and Richard Stafford [1994]

Abstract: This paper presents the results of a design study to analyze the interbay
             material-handling systems for semiconductor wafer manufacturing.
             The authors developed discrete-event simulation models of the perfor-
             mance of conventional cleanroom material handling including manual
             and automated systems. The components of a conventional clean-
             room material-handling system include an overhead monorail system
             for interbay (bay-to-bay) transport, work-in-process stockers for lot
             storage, and manual systems for intrabay movement. The authors
             constructed models and experiments that assisted with analyzing
             cleanroom material-handling issues such as designing conventional
             automated material-handling systems and specifying requirements for
             transport vehicles.

Session: Simulation in Aerospace Manufacturing
Paper: Modeling Aircraft Assembly Operations
Authors: Harold A. Scott [1994]

Abstract: A simulation model is used to aid in the understanding of complex
             interactions of aircraft assembly operations. Simulation helps to iden-
             tify the effects of resource constraints on dynamic process capacity
             and cycle time. To analyze these effects, the model must capture job
             and crew interactions at the control code level. This paper explores five
             aspects of developing simulation models to analyze crew operations
             on aircraft assembly lines:

             • Representing job precedence relationships;
             • Simulating crew members with different skill and job
               proficiency levels;
             • Reallocating crew members to assist ongoing jobs;
             • Depicting shifts and overtime;
             • Modeling spatial constraints and crew movements in the
               production area.

Session: Control of Manufacturing Systems
Paper: Discrete Event Simulation for Shop Floor Control
Authors: J. S. Smith, R. A. Wysk, D. T. Sturrock, S. E. Ramaswamy,
             G. D. Smith, S. B. Joshi [1994]

Abstract: This paper describes an application of simulation to shop floor control
             of a flexible manufacturing system. The simulation is used not only
             as an analysis and evaluation tool, but also as a task generator for
             the specification of shop floor control tasks. Using this approach, the
             effort applied to the development of control logic in the simulation is

not duplicated in the development of the control system. Instead, the same control logic is used for the control system as was used for the simulation. Additionally, since the simulation implements the control, it provides very high fidelity performance predictions. The paper describes implementation experience in two flexible manufacturing laboratories.

Session: Flexible Manufacturing
Paper: Developing and Analyzing Flexible Cell Systems Using Simulation
Authors: Edward F. Watson and Randall P. Sadowski [1994]
Abstract: This paper develops and evaluates flexible cell alternatives to support an agile production environment at a mid-sized manufacturer of industrial equipment. Three work-cell alternatives were developed, based on traditional flow-analysis studies, past experience, and common sense. The simulation model allowed the analyst to evaluate each cell alternative under current conditions as well as anticipated future conditions that included changes to product demand, product mix, and process technology.

Session: Modeling of Production Systems
Paper: Inventory Cost Model for "Just-in-Time" Production
Authors: Mahesh Mathur [1994]
Abstract: This paper presents a simulation model used to compare setup and inventory carrying costs with varying lot sizes. While reduction of lot sizes is a necessary step towards implementation of Just-in-Time (JIT) production in a job shop environment, a careful cost study is required to determine the optimum lot size under the present set up conditions. The simulation model graphically displays the fluctuation of carrying costs and accumulation of set up costs on a time scale in a dynamic manner. The decision of the optimum lot size can then be based on realistic cost figures.

Session: Analysis of Manufacturing Systems
Paper: Modelling Strain of Manual Work in Manufacturing Systems
Authors: I. Ehrhardt, H. Herper, and H. Gebhardt [1994]
Abstract: This paper describes a simulation model that considers manual operations for increasing the effectiveness of planning logistic systems. Even though there is ever-increasing automation, there are vital tasks in production and logistics that are still assigned to humans. Present simulation modeling efforts rarely concentrate on the manual activities assigned to humans.

Session: Manufacturing Case Studies
Paper: Simulation Modeling for Quality and Productivity in Steel Cord Manufacturing

Authors: C. H. Turkseven and G. Ertek [2003]
Abstract: The paper describes the application of simulation modeling to esti-
            mate and improve quality and productivity performance of a steel cord
            manufacturing system. It focuses on wire fractures, which can be an
            important source of system disruption.

Session: Manufacturing Analysis and Control
Paper: Shared Resource Capacity Analysis in Biotech Manufacturing
Author: P. V. Saraph [2003]
Abstract: This paper discusses an application of simulation in analyzing the
            capacity needs of a shared resource, the Blast Freezer, at one of the
            Bayer Corporation's manufacturing facilities. The simulation model
            was used to analyze the workload patterns, run different workload
            scenarios (taking into consideration uncertainty and variability), and
            provide recommendations on a capacity increase plan. This analysis
            also demonstrated the benefits of certain operational scheduling poli-
            cies. The analysis outcome was used to determine capital investments
            for 2002.

Session: Manufacturing Analysis and Control
Paper: Behavior of an Order Release Mechanism in a Make-to-Order
            Manufacturing System with Selected Order Acceptance
Authors: A. Nandi and P. Rogers [2003]
Abstract: The authors used a simulation model to evaluate a controversial policy,
            namely, holding orders in a preshop pool prior to their release to the
            factory floor. In a make-to-order manufacturing system, if capacity
            is fixed and exogenous due dates are inflexible, having orders wait
            in a preshop pool may cause the overall due-date performance of the
            system to deteriorate. The model was used to evaluate an alternative
            approach, the selective rejection of orders for dealing with surges in
            demand while maintaining acceptable due-date performance.

13.5   Manufacturing Example: An Assembly-line Simulation

This section describes a model of a production line for the final assembly of "gizmos". It then focuses on how simulation can be used to analyze system performance.

13.5.1   System Description and Model Assumptions

At a manufacturing facility, an engineering team has designed a new production line for the final assembly of gizmos. Before making the investment to install the new system, some team members propose using simulation to analyze the system's performance, specifically to predict system through-put (gizmos per 8-hour shift, on the average). In addition, the engineers desire to evaluate potential improvements to the designed system. One such potential improvement is adding buffer space for holding work-in-process (WIP) between adjacent workstations.

The team decides to develop a simulation model and conduct an analysis. Their primary objective is to predict throughput (completed gizmos per shift, on the average) for the given system design and to evaluate whether it meets the desired throughput. In addition, should throughput be less than expected, the team wants to use the model to help in identifying bottlenecks, gaining insight into the system's dynamic behavior, and evaluating potential design improvements.

The proposed production line has six workstations and a special rack for WIP storage between adjacent stations. There are four manual stations, each having its own operator, and two automated stations, which share a single operator. The six stations perform production tasks in the following sequence:

Station 1: initial manual station begins final assembly of a new gizmo
Station 2: manual assembly station
Station 3: manual assembly station
Station 4: automated assembly station
Station 5: automated testing station
Station 6: manual packing station

At each manual station, an operator loads a gizmo onto a workbench, performs some tasks, and on completion unloads the gizmo and places it into the WIP storage for the next workstation. The operator takes 10 seconds and 5 seconds, respectively, for the loading and unloading tasks.

The WIP storage racks between each pair of adjacent stations have limited capacity. If a station completes its tasks on a gizmo but the downstream rack is full, the gizmo must remain in the station, blocking any further work. In the initial design, the WIP storage racks have the capacities shown in Table 13.2. (By assumption, the WIP storage preceding Station 1 is always kept full at 4 units; since it is assumed to always be full, its specific capacity plays no role.) The system design with capacities given in Table 13.2 is called the *Baseline configuration*.

From time to time, a tool will fail, causing unscheduled downtime or unexpected extra work at a manual or automated station. In addition, all operators are scheduled to take a 30-minute lunch break at the same time. Work is interrupted and resumes where it left off after lunch. This interrupt/resume rule applies to operator tasks including assembly work, parts resupply, and repairs during a downtime.

At the automatic stations, a machine performs an assembly or testing task. The automatic stations might have unscheduled (random) downtimes, but they continue to operate during the operator's lunch break. One operator services both machines to load and unload gizmos (10 seconds and 5 seconds, respectively). After being loaded, a machine processes the gizmo without further operator intervention unless a downtime occurs. At all stations, the operator performs repairs as needed whenever the station experiences a downtime.

Table 13.2   Capacity of WIP Storage Buffers for Baseline Configuration

Rack Before Station	1	2	3	4	5	6
Buffer Capacity	4	2	2	2	1	2

Table 13.3   Assembly and Parts Resupply Times

Station	Assembly per Gizmo (seconds)	Part Number	Parts Resupply Time (seconds per batch)	No. of Parts per batch
1	40	A	10	15
		B	15	10
2	38	C	20	8
		D	15	14
3	38	E	30	25
4	35			
5	35			
6	40	F[a]	30	32

[a] At Station 6, the part number (F) represents the shipping containers.

Table 13.3 gives the total assembly time and parts resupply times for each station, plus the number of parts in a batch. The assembly time for the manual stations is assumed to vary by plus/minus 2 seconds (uniformly distributed) from the times given in Table 13.3. Parts resupply time does not occur for each gizmo, but rather after a batch of parts has been assembled onto the gizmo. The machines at stations 4 and 5 do not consume parts.

Each station is subject to unscheduled (random) downtime. Manual Stations 1, 2, and 3 have tool failures or other unexpected problems. The automatic stations occasionally jam or have some other problem that requires the assigned operator to fix it. Station 6 (packing) is not subject to these downtimes. Table 13.4 shows time-to-failure (TTF) and time-to-repair (TTR) distributional assumptions and the assumed mean-time-to-failure (MTTF), mean-time-to-repair (MTTR) and spread ($+/-$) of repair times. For example, at Station 1, repair time is uniformly distributed with mean 4.0 minutes plus or minus 1.0 minutes; that is, uniformly distributed between 3.0 and 5.0 minutes. Failure can only occur when an operator or machine is working; hence, TTF is modeled by measuring only busy or processing time until a failure occurs.

The primary model output or response is average throughput during the assumed 7.5 working hours per 8-hour shift. The model also measures detailed station utilization, including busy or

Table 13.4   Assumptions and Data for Unscheduled Downtimes

Station	TTF	MTTF (Minutes)	TTR	MTTR (Minutes)	$+/-$	Expected Availability
1	Exponential	36.0	Uniform	4.0	1.0	90%
2	Exponential	4.5	Uniform	0.5	0.1	90%
3	Exponential	27.0	Uniform	3.0	1.0	90%
4	Exponential	9.0	Uniform	1.0	0.5	90%
5	Exponential	18.0	Uniform	2.0	1.0	90%

processing time, idle or starved time (no parts ready for processing), blocked time (part cannot leave station, because downstream WIP buffer is full), unscheduled downtime, and time waiting for an operator.

Station starvation occurs when the operator and station are ready to work on the next gizmo, the just-completed gizmo leaves the station, but upstream conditions cause no gizmo to be ready for this production step. In short, the upstream WIP buffer is empty.

Station blockage occurs when a station completes all tasks on a gizmo, but cannot release the part because the downstream WIP buffer is full. For both starvation and blockage, production time is lost at the given station and cannot be made up.

When an operator services more than one station, as does the operator servicing Stations 4 and 5, it is possible for both stations to need the operator at the same time. This could cause additional delay at the station and is measured by a "wait-for-operator" state. Blockage, starvation, and wait-for-operator at each station will be measured in order to help explain any throughput shortfall, should it occur, and to assist in identifying potential system improvements.

13.5.2   Presimulation Analysis

A presimulation analysis, taking into account the average station cycle time as well as expected station availability (90%), indicates that each station, if unhindered, can achieve the desired throughput. This initial analysis is carried out as described in this section.

From the assumed downtime data, the team was able to estimate expected station availability, under the (ideal) assumption of no interaction between stations. The expected availability shows each station's individual availability during working (nonlunch, nonbreak) hours, assuming that the operator can always place a completed gizmo into the downstream rack storage and the next gizmo is ready to begin work at the station. Expected availability is computed by MTTF/(MTTF + MTTR), or expected busy time during a downtime cycle divided by the length of a downtime cycle (a busy cycle plus a repair cycle), and is given in Table 13.4. This calculation ignores certain aspects of the problem, including the parts resupply times and any delay caused by having only one operator to service both Stations 4 and 5.

The design goal for the modeled system is 390 finished gizmos per 8-hour shift. After taking lunch into account, each shift has up to 7.5 hours of available work time. With unscheduled (random) downtime expected to be 10% of available time, this further reduces working time to 0.90×7.5 hours $= 6.75$ hours. This implies that the station with the slowest total cycle time must be able to produce 390 gizmos in the available 6.75 hours. Therefore, the total cycle time per gizmo at each station must not exceed 6.75 hours/390 = 62.3 seconds.

Now, total cycle time consists of assembly, testing or packing time, and parts resupply time (as given in Table 13.3), plus gizmo loading time of 10 seconds and unload time of 5 seconds. Parts resupply is not taken on every gizmo, but rather after a given number of gizmos corresponding to using all parts in a given batch of parts. For example, using the values in Table 13.3 for Station 1, parts resupply will take 10 seconds every 15 gizmos for Part A, plus 15 seconds every 10 gizmos for Part B, for a total time, on the average, of $10/15 + 15/10$ seconds per gizmo.

Using this information, the (minimum) total cycle time for each station is estimated in Table 13.5. These presimulation estimates indicate, first, that each theoretical cycle time is well below the

Table 13.5   Estimated Total Cycle Time at Each Station

Station	Formula to Estimate Cycle Time (seconds)	Estimate (seconds)
1	$10 + 40 + 5 + 10/15 + 15/10$	57.2
2	$10 + 38 + 5 + 20/8 + 15/14$	56.6
3	$10 + 38 + 5 + 30/25$	54.2
4	$10 + 35 + 5$	50.0
5	$10 + 35 + 5$	50.0
6	$10 + 40 + 5 + 30/32$	55.9

requirement of 62.3 seconds. Secondly, they indicate that Stations 1 and 2 are potential bottlenecks, if there are any.

As the simulation analysis will later show, Station 1 experiences blockage due to Station 2 downtime, and Station 2 occasionally experiences starvation due to downtime at Station 1 and blockage due to downtime at Station 3. These blockage and starvation conditions reduce the available work time below the calculated 90%; hence, for the Baseline Configuration, they reduce the design throughput well below the desired value: 390 gizmos per shift. In summary, a presimulation analysis, although valuable, at best can provide a rough estimate of system performance. As the simulation will show, ignoring blockage and starvation gives an overly optimistic estimate of system throughput.

13.5.3   Simulation Model and Analysis of the Designed System

Using the simulation model, the first experiment was conducted to estimate system performance of the system as designed. The simulation analyst on the team made 10 replications of the model, each having a 2-hour warm-up or initialization followed by a 5-day simulation (each day being 24 hours). A 95% confidence interval (CI) was computed for mean throughput per shift:

$$95\% \text{ CI for mean throughput: } (364.5, 366.8), \text{ or } 365.7 \pm 1.14.$$

With 95% confidence, the model predicts that mean (or long-run average) throughput will be between 364.5 and 366.8 gizmos per 8-hour shift with the system as designed. This is well below the design throughput of 390 gizmos per shift.

The team decided to conduct further analyses to identify possible bottlenecks and potential areas of improvement.

13.5.4   Analysis of Station Utilization

At this point, the team desired to have some explanation of the shortfall in throughput. They suspected that perhaps it had to do with the small WIP buffer capacity and the resulting blockage and starvation. The same model was used to estimate detailed workstation utilization in hopes that it would provide an explanation of throughput shortfall. Table 13.6 contains 95% confidence-interval estimates for the first five workstations for percent of time down, blocked, starved, and waiting for an operator. (Waiting for operator affects only Stations 4 and 5, as these two stations share one operator. The other

Table 13.6   Confidence Intervals for Station Utilizations for the Baseline Configuration

Station	% Down	% Blocked	% Starved	% Wait for Operator
1	(8.8,9.6)	(11.4,12.5)	(0.0,0.0)	(0.0,0.0)
2	(8.2,8.4)	(8.0,8.8)	(4.9,5.6)	(0.0,0.0)
3	(7.9,8.6)	(9.9,10.4)	(6.1,6.9)	(0.0,0.0)
4	(8.9,9.6)	(2.0,2.8)	(7.5,8.2)	(13.1,14.4)
5	(8.3,9.0)	(0.0,0.2)	(19.4,20.4)	(3.9,4.7)

stations have a dedicated operator. In addition to the utilization statistics in Table 13.6, the operators have a 30-minute lunch per 8-hour shift, representing 6.25% of available time.)

From the results in Table 13.6, it appears that blockage and starvation explain some portion of the shortfall in throughput. In addition, another possible explanation surfaces: Station 4 experiences a significant time waiting for the single operator that services Stations 4 and 5. This delay at Station 4 could result in a full WIP buffer, which in turn would help explain the blockage at Station 3 preceding it. Percent of time blocked is higher than percent starved for Stations 1 to 3, so it appears that downstream delays could be a significant bottleneck.

The team proposed some possible system improvements:

1. having two operators to service Stations 4 and 5 (instead of the currently proposed one operator);
2. increasing the capacity of some of the WIP buffers;
3. a combination of both.

The expense of additional WIP storage space prompted the team to want to keep total buffer space as small as possible, and to require an additional operator only if absolutely necessary, while achieving the design goal of 390 gizmos per shift.

13.5.5   Analysis of Potential System Improvements

To evaluate the addition of an operator and larger WIP buffers, the model was revised appropriately to allow these changes, and a new analysis was conducted. In this analysis, the capacity of each WIP buffer for Stations 2–6 was allowed to increase by one unit above the Baseline value given in Table 13.2. In addition, the effect of a second operator at Stations 4 and 5 is considered. These possibilities result in a total of 64 scenarios or model configurations. (Why?) Making 10 replications per scenario results in a total of 640 simulation runs.

To facilitate the analysis, the team decided to use the Common Random Number technique discussed in Section 12.1.2. To implement it with proper synchronization, each source of random variability was identified and assigned a dedicated random-number stream. In this model, processing time, TTF, and TTR are modeled by statistical distributions at each of the six workstations. Therefore, a total of 18 random-number streams were defined, with 3 used at each workstation. In this way, in each set of runs, each workstation experienced the same workload and random downtimes no matter

Table 13.7   Improvement in System Throughput for Alternative Configurations

Number of Operators at Stations 4 & 5	Buffer Capacities						Increase in Mean Throughput per Shift (compared to baseline)		
	Buffer 2	Buffer 3	Buffer 4	Buffer 5	Buffer 6	Total	Ave. Diff.	CI Low	CI High
2	3	3	3	2	2	13	31.7	30.3	33.1
2	3	3	3	2	3	14	31.7	30.4	33.0
2	3	3	2	2	3	13	30.0	28.6	31.3
2	3	3	3	1	3	13	29.8	28.6	31.0
2	3	3	2	2	2	12	29.7	28.1	31.3
2	3	3	3	1	2	12	29.5	28.1	31.0
2	3	3	2	1	3	12	26.6	25.4	27.9
2	2	3	3	2	2	12	26.6	25.1	28.1
2	2	3	3	2	3	13	26.6	25.0	28.1
2	3	2	3	2	3	13	26.5	25.0	28.0
2	3	2	3	2	2	12	26.4	25.3	27.5
2	3	3	2	1	2	11	26.3	25.1	27.5

which configuration was being simulated. For a given number of replications the CRN technique, also known as correlated sampling, is expected to give shorter confidence intervals for differences in system performance.

The model configurations with the most improvement in system throughput, compared with the Baseline Configuration, are shown in Table 13.7. These configurations were chosen for further evaluation because each shows a potential improvement in throughput of approximately 25 units or more; that is, the lower end of the 95% confidence interval is 25 or higher. The values shown for "Ave Diff." represent the increase in throughput compared to the Baseline Configuration. Recall that the Baseline throughput was previously estimated, with 95% confidence, to be in the interval (364.5, 366.8). Being conservative, the engineering team would like to see an improvement of $390 - 364.5 = 25.5$ gizmos per shift. The top six configurations in Table 13.7 have a lower confidence interval larger than 25.5 and hence are likely candidates for achieving the desired throughput. Interpreted statistically: The lower end of the confidence interval is larger than 25.5, so the results yield a 95% confidence that mean throughput will increase by 25.5 or more in the top six configurations listed in Table 13.7.

Note that all the most improved configurations include two operators at Stations 4 and 5. The simulation results for configurations with one operator (not shown here) indicate that a 390 throughput cannot be achieved with one operator, at least not with the buffer sizes considered.

Some configurations can be ruled out because a less expensive option achieves a similar throughput. Consider, for example, the first two configurations in Table 13.7. They are identical except for Buffer 6 capacity. Since WIP buffer capacity is expensive, the smaller total buffer capacity will be the less expensive option. Clearly, there is no need to expand from 2 to 3 units at Buffer 6. The

"Total" column can assist in quickly ruling out configurations that do no better than a similar one with smaller total buffer capacity.

The model configuration that increases throughput by 25.5 or better and has the smallest total buffer capacity is the fifth one in Table 13.7, with capacities of (3,3,2,2,2) for Buffers 2 to 6, respectively. On these considerations, this system design becomes the team's top candidate for further evaluation. The next step (not included here) would be to conduct a financial analysis of each alternative configuration.

13.5.6   Concluding Words: The Gizmo Assembly-Line Simulation

Real-life examples similar to this example model include assembly lines for automotive parts and automobile bodies, automotive pollution-control assemblies, consumer items such as washing machines, ranges, and dishwashers, and any number of other assembly operations with a straight flow and limited buffer space between workstations. Similar models and analyses may also apply to a job shop with multiple products, variable routing, and limited work-in-process storage.

13.6   Summary

This chapter introduced some of the ideas and concepts most relevant to manufacturing and material handling simulation. Some of the key points are the importance of modeling downtimes accurately, the advantages of trace-driven simulations with respect to some of the inputs, and the need in some models for accurate modeling of material-handling equipment and the control software.

REFERENCES

BANKS, J. [1994], "Software for Simulation," in *Proceedings of the 1994 Winter Simulation Conference*, J. D. Tew, S. Manivannan, D. A. Sadowski, and A. F. Seila, eds., Lake Buena Vista, Fl, Dec. 11–14, pp. 26–33.

CLARK, G. M. [1994], "Introduction to Manufacturing Applications," in *Proceedings of the 1994 Winter Simulation Conference*, J. D. Tew, S. Manivannan, D. A. Sadowski, and A. F. Seila, eds., Lake Buena Vista, Fl, Dec. 11–14, pp. 15–21.

EHRHARDT, I., H. HERPER, AND H. GEBHARDT [1994], "Modelling Strain of Manual Work in Manufacturing Systems," in *Proceedings of the 1994 Winter Simulation Conference*, J. D. Tew, S. Manivannan, D. A. Sadowski, and A. F. Seila, eds., Lake Buena Vista, Fl, Dec. 11–14, pp. 1044–1049.

LAW, A. M. AND W. D. KELTON [2000], *Simulation Modeling and Analysis*, 3d ed., McGraw–Hill, New York.

MATHUR, M. [1994], "Inventory Cost Model for 'Just-in-time" Production," in *Proceedings of the 1994 Winter Simulation Conference*, J. D. Tew, S. Manivannan, D. A. Sadowski, and A. F. Seila, eds., Lake Buena Vista, Fl, Dec. 11–14, pp. 1020–1026.

NANDI, A., AND P. ROGERS [2003], "Behavior of an Order Release Mechanism in a Make-to-Order Manufacturing System with Selected Order Acceptance," in *Proceedings of the 2003 Winter Simulation Conference*, S. E. Chick, P. J. Sánchez, D. Ferrin, and D. J. Morrice, eds., New Orleans, La, Dec. 7–10, pp. 1251–1259.

PIERCE, N. G., AND R. STAFFORD [1994], "Modeling and Simulation of Material Handling for Semiconductor Wafer Fabrication," in *Proceedings of the 1994 Winter Simulation Conference*, J. D. Tew, S. Manivannan, D. A. Sadowski, and A. F. Seila, eds., Lake Buena Vista, Fl, Dec. 11–14, pp. 900–906.

SARAPH, P. V. [2003], "Shared Resource Capacity Analysis in Biotech Manufacturing," in *Proceedings of the 2003 Winter Simulation Conference*, S. E. Chick, P. J. Sánchez, D. Ferrin, and D. J. Morrice, eds, New Orleans, La, Dec. 7–10, pp. 1247–1250.

SCOTT, H. A. [1994], "Modeling Aircraft Assembly Operations," in *Proceedings of the 1994 Winter Simulation Conference*, J. D. Tew, S. Manivannan, D. A. Sadowski, and A. F. Seila, eds., Lake Buena Vista, Fl, Dec. 11–14, pp. 920–927.

SMITH, J. S., *et al.* [1994], "Discrete Event Simulation for Shop Floor Control," in *Proceedings of the 1994 Winter Simulation Conference*, J. D. Tew, S. Manivannan, D. A. Sadowski, and A. F. Seila, eds, Lake Buena Vista, Fl, Dec. 11–14, pp. 962–969.

TURKSEVEN, C. H., AND G. ERTEK [2003], "Simulation Modeling for Quality and Productivity in Steel Cord Manufacturing," in *Proceedings of the 2003 Winter Simulation Conference*, S. E. Chick, P. J. Sánchez, D. Ferrin, and D. J. Morrice, eds., New Orleans, La, Dec. 7–10, pp. 1225–1229.

WATSON, E. F., AND R. P. SADOWSKI [1994], "Developing and Analyzing Flexible Cell Systems Using Simulation," in *Proceedings of the 1994 Winter Simulation Conference*, J. D. Tew, S. Manivannan, D. A. Sadowski, and A. F. Seila, eds., Lake Buena Vista, Fl, Dec. 11–14, pp. 978–985.

WILLIAMS, E. J. [1994], "Downtime Data— Its Collection, Analysis, and Importance," in *Proceedings of the 1994 Winter Simulation Conference*, J. D. Tew, S. Manivannan, D. A. Sadowski, and A. F. Seila, eds., Lake Buena Vista, Fl, Dec. 11–14, pp. 1040–1043.

EXERCISES

Instructions to the student: Many of the following exercises contain material-handling equipment such as conveyors and vehicles. The student is expected to use any simulation language or simulator that supports modeling conveyors and vehicles at a high level.

Some of the following exercises use the uniform, exponential, normal, or triangular distributions. Virtually all simulation languages and simulators support these, plus other distributions. The use of the first three distributions was explained in the note to the exercises in Chapter 4; the use of the triangular is explained in the exercise that requires it. For reference, the properties of these distributions, plus others used in simulation, are given in Chapter 5, and random-variate generation is covered in Chapter 8.

1. A case sortation system consists of one infeed conveyor and 12 sortation lanes, as shown in the following schematic (not to scale):

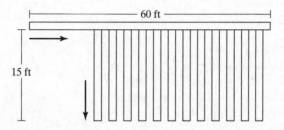

Cases enter the system from the left at a rate of 50 per minute at random times. All cases are 18 by 12 inches and travel along the 18-inch dimension. The incoming mainline conveyor is 20 inches wide and 60 feet in length (as shown). The sortation lanes are numbered 1 to 12 from left to right, and are 18 inches wide and 15 feet in length, with 2 feet of spacing between adjacent lanes. (Estimate any other dimensions that are needed.) The infeed conveyor runs at 180 feet/minute, the sortation lanes at 90 feet/minute. All conveyor sections are accumulating, but, upon entrance at the left, incoming cases are at least 2 feet apart from leading edge to leading edge. On the sortation lanes, the cases accumulate with no gap between them.

Incoming cases are distributed to the 12 lanes in the following proportions:

1	6%	7	11%
2	6%	8	6%
3	5%	9	5%
4	24%	10	5%
5	15%	11	3%
6	14%	12	0%

The 12th lane is an overflow lane; it is used only if one of the other lanes fill and a divert is not possible.

At the end of the sortation lanes, there is a group of operators who scan each case with a bar-code scanner, apply a label, and then place it on a pallet. Operators move from lane to lane as necessary to avoid allowing a lane to fill. There is one pallet per lane, each holding 40 cases. When a pallet is full, assume a new empty one is immediately available. If a lane fills to 10 cases and another case arrives at the divert point, this last case continues to move down the 60-foot mainline conveyor and is diverted into lane 12, the overflow lane.

Assume that one operator can handle 8.5 cases per minute, on the average. Ignore walking time and assignment of an operator to a particular lane; in other words, assume the operators work as a group uniformly spread over all 12 lanes.

(a) Set up an experiment that varies the number of operators and addresses the question: How many operators are needed? The objective is to have the minimum number of operators but also to avoid overflow.

(b) For each experiment in part (a), report the following output statistics:

- operator utilization;
- total number of cases palletized;
- number of cases palletized by lane;
- number of cases to the overflow lane.

(c) For each experiment in part (a), verify that all cases are being palletized. In other words, verify that the system can handle 50 cases per minute, or explain why it cannot.

2. Redo Exercise 1 to a greater level of detail by modeling operator walking time and operator assignment to lanes. Assume that operators walk at 200 feet per minute and that the walking distance from one lane to the next is 5 feet. Handling time per case is now assumed to be 7.5 cases per minute. Devise a set of rules that can be used by operators for lane changing. (*Hint*: For example, change lanes to that lane with the greatest number of cases only when the current lane is empty or the other lane reaches a certain level.) Assume that each operator is assigned to a certain number of adjacent lanes and handles only those lanes. However, if necessary, two operators (but no more) may be assigned to one lane; that is, operator assignments may overlap.

   (a) If your lane-changing rule has any numeric parameters, experiment to find the best settings. Under these circumstances, how many operators are needed? What is the average operator utilization?
   (b) Does a model that has more detail, as does Exercise 2(a) when compared to Exercise 1, always have greater accuracy? How about this particular model? Compare the results of Exercise 2(a) to the results for Exercise 1. Are the same or different conclusions drawn?
   (c) Devise a second lane-changing rule. Compare results between the two rules. Compare total walking time or percent of time spent walking between the two rules.
   Suggestion: A lane-changing rule could have one or two "triggers." A one-trigger rule might state that, if a lane reached a certain level, the operator moved to that lane. (*Hint*: Without modification, such a rule could lead to excessive operator movement, if two lanes had about the same number of cases near the trigger level.) A two-trigger rule might state that, if a lane reached a certain level and the operator's current lane became empty, then change to the new lane; but if a lane reaches a specified higher critical level, then the operator immediately changes lanes.
   (d) Compare your results with those of other students who may have used a different lane-changing rule.

3. Redo Exercise 2 with a different operator-assignment rule. Basically, operators can be assigned to any lane as conditions warrant, but, as before, no more than two operators can be assigned to a lane at the same time. Address the same questions (a)–(d) as in Exercise 2.

4. A package-sortation system consists of one in-feed conveyor, 12 sortation lanes (or chutes), and a take-away conveyor, as shown in the following schematic (not to scale):

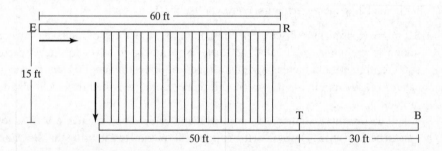

Packages enter the system from the left (at point E) at a rate of 100 per minute at random times. A package is diverted to a sortation lane on the basis of its shipping destination; in the model, a given percentage of the packages chosen randomly, according to the proportions in the next table, are assigned to each lane. The percentages for each lane/destination is as follows:

1	6%	7	10%
2	6%	8	6%
3	5%	9	4%
4	24%	10	4%
5	15%	11	3%
6	14%	12	3%

After a slug of 8 packages has accumulated, they are all released together onto a 50-foot section of takeaway conveyor and head toward the point labeled T; they transfer there (at conveyor speed) onto a second section of takeaway conveyor at the end of which is a bar-code scanner (at the point labeled B). A slug cannot be released until the previous slug has cleared the point T. In this model, packages disappear at point B.

All packages are 18 by 12 inches and travel along their 18-inch dimension. The incoming mainline conveyor is 20 inches wide and 60 feet in length (as shown). The 12 sortation lanes are 18 inches wide and 15 feet in length with 2 feet of spacing between adjacent lanes. (Estimate any other dimensions that are needed.) The infeed conveyor runs at 240 feet/minute, the sortation lanes at 90 feet/minute. The infeed and sortation conveyor sections are accumulating, but, upon entrance at the left, incoming packages are at least 2 feet apart from leading edge to leading edge. On the sortation lanes, the packages accumulate with no gap between them. The 50-foot section of takeaway conveyor is a nonaccumulating belt-type conveyor; the 30-foot section is accumulating.

When a lane contains 8 packages, no more diverts are allowed until the lane is released and the slug completely clears the bottom of the chute. When a lane is full or before the slug has cleared, if another package arrives to divert, the affected package does not divert, but rather moves to the end of the infeed conveyor and then recirculates to the start of the infeed conveyor on a conveyor that is not shown in the schematic. Simulate recirculation as a time delay of 45 seconds without explicitly modeling the recirculation conveyor itself.

(a) Set up an experiment to discover the minimum speed required of the takeaway conveyor to minimize or eliminate recirculation. Conveyors available from vendors have speeds starting at 60 feet/minute and increasing in increments of 30 feet/minute. (Faster conveyor is more expensive, so minimizing the speed will reduce costs, provided that recirculation is not a problem.)

(b) Could destination assignment to available lane be modified in such a way as to improve system performance? Devise a better assignment rule and test it, using the simulation model.

5. A fleet of AGVs (Automated Guided Vehicles) services 5 workstations in series by transporting parts on fixtures from one workstation to the next, remaining with a part from the pickup point through all workstations to the offload point. The incoming staging conveyor, the AGV guidepath, the five workstations (A–E), and the offload conveyor are shown in the following schematic (not to scale):

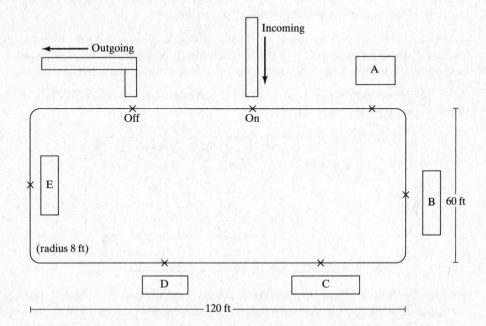

AGVs travel in a clockwise direction. Initially, all are empty and queued at the incoming conveyor at the "On" point. The workstations are labeled A–E. Parts on fixtures are staged on the incoming conveyor and picked up by an AGV at the point labeled "On". The AGV then travels to machine A and stops at the point marked on the guidepath. The part and fixture remain on the AGV during processing, then travel together to the next workstation. After processing is completed at workstation E, the AGV travels to the point labeled "Off" the part with fixture is automatically unloaded onto the offload conveyor, and the empty AGV queues at the point labeled "On" for the next part.

Assume that the incoming conveyor remains full. It is 15 feet long and can hold 3 parts with fixtures, each taking 5 feet of conveyor space. Therefore, when one part with fixture is picked up by an AGV, another part and fixture are placed on the end of the conveyor. Pickup time is 30 seconds. The incoming conveyor runs at 30 feet per minute.

When a part has completed processing at all workstations and has been transported to the offload point labeled "Off" the offload operation takes 45 seconds. The offload conveyor itself runs at 30 feet per minute. The parts and fixtures, although taking up a space of 3 feet by 3 feet, require 5 feet on the conveyor.

Initial workstation placements are shown in approximate fashion, with no exact locations given. Constraints are that each workstation must be placed along the straight section of guidepath on which they are shown on the schematic, with the AGV stopping point no closer than 5 feet from the entrance to or exit from a curve. The incoming and offload conveyor must be placed as shown, with the offload point 20 feet from the end of the preceding curve and the pickup point 40 additional feet to the right. Otherwise, AGV stopping points at workstations can be placed at any desired location on the straight section of guidepath.

AGVs travel at 15 feet per second and accelerate and decelerate at 3 ft/sec/sec. They are 7 feet in length and queue 8 feet apart (1 foot gap between two AGVs) when waiting for the processing position at a workstation and when in the pickup/offload position at a conveyor.

At each of the workstations, processing-time-per-part, mean-time-to-failure (MTTF), and mean-time-to-repair (MTTR) are as follows:

Workstation	Processing Time (sec.)	MTTF (min.)	MTTR (min.)
A	120	90	3.1
B	100	82	4.0
C	140	110	4.5
D	45	20	5.0
E	100	240	9.0

Detailed data is not available, so certain assumptions regarding time to failure and time to repair are made: Assume that actual time to failure is exponentially distributed, with the specified mean (MTTF). Assume that repair time requires a maintenance person and that there is only one available. Actual repair times are uniformly distributed, with the mean as shown and a half-width of one-half the mean. Time to failure is measured against wall-clock time, not against station-busy time. After a repair, any remaining processing time is taken, and the part continues.

There are three basic questions:

(1) What is maximum system throughput?
(2) How many AGVs are needed?
(3) What is the best workstation placement (within the stated constraints)?

For each set of experiments in parts (a) and (b), consider the following issues:

(1) Should there be a warmup period for loading the system during which no statistics are collected?
(2) How long should the simulation run?

(One suggestion: The warmup period should last at least until several parts have completed processing. The run length should allow for at least 10 downtimes on each machine. Simulate with as many AGVs as will reasonably fit on the guidepath and then try to reduce the number without reducing throughput.)

Your experiments should vary the number of AGVs and the workstation placement. Before simulating, make your best estimate regarding workstation placement, and justify that choice. Also, make a reasonable estimate regarding the likely range for number of AGVs.

(a) In the first set of experiments, use a rough-cut model that does not model the AGVs explicitly. Compute the approximate travel time between adjacent workstations and use this value as a time delay in the model. Assume there are sufficient AGVs. (*Hint*: Use this model to address the question of maximum throughput, but not the number of AGVs needed.) What are the advantages and limitations of this approach?

(b) In the second set of experiments, model the AGVs and guidepath explicitly. Design an experiment to find out the maximum system throughput (in parts per hour) with the smallest number of AGVs.

6. More accurate information has been obtained on failure times for the workstations in Exercise 5. Redo Exercise 5 with the new information. Failures appear to be related more closely to the number of parts processed than to the amount of time passed. The new data is as follows:

Workstation	MTTF (parts)	MTTR (min.)
A	45	3.1
B	49	4.0
C	47	4.5
D	27	5.0
E	144	9.0

Note that MTTF (mean-time-to-failure) is measured in terms of number of parts, but assumptions about time to repair are unchanged from Exercise 5. For example, for Workstation A, if time to failure happened to be 45 parts, then after a failure, the 45th part would experience the next failure. In reality, failure occurs during processing, but for modeling purposes, assume the failure occurs just as processing begins. (Does this make any difference?)

For the distribution of time to failure, assume a discrete uniform distribution with mean (MTTF) over the range from 1 to 2*MTTF − 1. For example, for Workstation A, the actual times to failure are equally likely to be any value from 1 to 89 parts.

Address the same questions as in Exercise 5. Do these new assumptions make any difference? How many failures are there per 8-hour shift for the first set of assumptions as compared with those for the revised assumptions?

7. For the problem in Exercises 5 and 6, a study was done to evaluate the repair times more accurately for each workstation. It was found that a triangular distribution provided a closer fit

to the actual repair times. Assume the following parameters for the triangular distribution:

Workstation	Repair Time (Triangular Distribution)			
	Minimum (min.)	Most Likely (min.)	Maximum (min.)	Average (min.)
A	1.0	2.3	6.0	3.1
B	2.0	3.0	7.0	4.0
C	2.5	3.5	7.5	4.5
D	2.0	4.0	9.0	5.0
E	4.0	7.0	16.0	9.0

Note that the average repair times are identical to the assumed averages (MTTR) in Exercises 5 and 6, but the distribution is triangular instead of uniform. (Most simulation languages or simulators have a built-in triangular distribution that requires specification of the minimum value, most likely value or mode, and maximum value. The average value is given for informational purposes only; it is the sum of the other three parameters divided by 3. The triangular distribution is discussed in Section 5.4.7.)

(a) Redo Exercise 5, but with the new assumptions for repair times.
(b) Redo Exercise 6, but with the new assumptions for repair times.

8. One approach to modeling downtimes discussed in this chapter, but not recommended, was to replace the processing time with a larger adjusted processing time and to leave out the downtimes. The adjusted processing time is meant to account for downtimes, at least in the long run and on the average. How would this method have affected the results for Exercise 5?

(a) For Exercise 5, compute the adjusted processing times for each workstation.
(b) For Exercise 5, use the adjusted processing times and leave out all explicit downtimes due to failure. How does this change the results? Is this a good approach for this model?

9. Suppose that improved maintenance and better machines could totally eliminate downtime due to failure for the system in Exercise 5. By how much would throughput improve?

10. Develop a model for Example 13.1 and attempt to reproduce qualitatively the results found in the text regarding different assumptions for simulating downtimes. Do not attempt to get exactly the same numerical results, but rather to show the same qualitative results.

(a) Do your models support the conclusions discussed in the text? Provide a discussion and conclusions.
(b) Make a plot of the number of entities in the queue versus time. Can you tell when failures occurred? After a repair, about how long does it take for the queue to get back to normal?

11. In Example 13.1, the failures occurred at low frequency compared with the processing time of an entity. Time to failure was 1000 minutes, and interarrival time was 10 minutes,

implying that few entities would experience a failure. But, when an entity did experience a failure (of 50 minutes, on average), it was several times longer than the processing time of 7.5 minutes.

Redo the model for Example 13.1, assuming high-frequency failures. Specifically, assume that the time to failure is exponentially distributed, with mean 2 minutes, and the time to repair is exponentially distributed, with mean 0.1 minute or 6 seconds. As compared with the low-frequency case, entities will tend to experience a number of short downtimes.

For low-frequency versus high-frequency downtimes, compare the average number of down-times experienced per entity, the average duration of downtime experienced, the average time to complete service (including downtime, if any), and the percent of time down.

Note that the percentage of time the machine is down for repair should be the same in both cases:

$$50/(1000 + 50) = 4.76\%$$
$$6\sec /(2\min + 6\sec) = 4.76\%$$

Verify percentage downtime from the simulation results. Are the results identical? ... close? Should they be identical, or just close? As the simulation run length increases, what should happen to the percentage of time down?

With high-frequency failures, do you come to the same conclusions as were drawn in the text regarding the different ways to simulate downtimes? Make recommendations regarding how to model low-frequency versus high-frequency failures.

12. Redo Exercise 11 (based on Example 13.1), but with one change: When an entity experiences a downtime, it must be reprocessed from the beginning. If service time is random, take a new draw from the assumed distribution. If service time is constant, it starts over again. How does this assumption affect the results?

13. Redo Exercise 11 (based on Example 13.1), but with one change: When an entity experiences a downtime, it is scrapped. How does scrapping entities on failure affect the results in both the low-frequency and in the high-frequency situations? What are your recommendations regarding the handling of low- versus high-frequency downtimes when parts are scrapped?

14. Sheets of metal pass sequentially through 4 presses: shear, punch, form, and bend. Each machine is subject to downtime and die change. The parameters for each machine are as follows:

Press	Process Rate (per min.)	Time to Failure (min.)	Time to Repair (min.)	No. of Sheets to a Die Change (no. sheets)	Time to Change Die (min.)
Shear	4.5	100	8	500	25
Punch	5.5	90	10	400	25
Form	3.8	180	9	750	25
Bend	3.2	240	20	600	25

Note that processing time is given as a rate; for example, the shear press works at a rate of 4.5 sheets per minute. Assume that processing time is constant. The automated equipment makes the time to change a die fairly constant, so it is assumed to be always 25 minutes. Die changes occur between stamping of two sheets after the number shown in the table have gone through a machine. Time to failure is assumed to be exponentially distributed, with the mean given in the table. Time to repair is assumed to be uniformly distributed, with the mean taken from the table and a half-width of 5 minutes. When a failure occurs, 20% of the sheets are scrapped. The remaining 80% are reprocessed at the failed machine after the repair.

Assume that an unlimited supply of material is available in front of the shear press, which processes one sheet after the next as long as there is space available between itself and the next machine, the punch press. In general, one machine processes one sheet after another continuously, stopping only for a downtime, for a die change, or because the available buffer space between itself and the next machine becomes full. Assume that sheets are taken away after bending at the bend press. Buffer space is divided into 3 separate areas, one between the shear and the punch presses, the second between the punch and form presses, and the last between form and bend.

(a) Assume that there is an unlimited amount of space between machines. Run the simulation for 480 hours (about 1 month with 24-hour days, 5 days per week). Where do backups occur? If the total buffer space for all three buffers is limited to 15 sheets (not counting before shear or after bend), how would you recommend dividing this space among the three adjacent pairs of machines? Does this simulation provide enough information to make a reasonable decision?

(b) Modify the model so that there is a finite buffer between adjacent machines. When the buffer becomes full and the machine feeding the buffer completes a sheet, the sheet is not able to exit the machine. It remains in the machine, blocking additional work. Assume that total buffer space is 15 sheets for the 3 buffers.

Use the recommendation from part (a) as a starting point for each buffer size. Attempt to minimize the number of runs. You are allowed to experiment with a maximum of 3 buffer sizes for each buffer. (How many runs does this make?) Run a set of experiments to determine the allocation of buffer space that maximizes production. Simulate each alternative for at least 1000 hours.

Report total production per hour on the average, press utilization (broken down by percentage of time busy, down, changing dies, and idle), and average number of sheets in each buffer.

14

Simulation of Networked Computer Systems

It is only natural that simulation be extensively used to simulate networked computer systems, because of their great importance to the everyday operations of business, industry, government, and universities. In this chapter we look at the motivations for simulating networked computer systems, the different types of approaches used, and the interplay between characteristics of the model and implementation strategies. We begin the discussion by looking at general characteristics of simulations in this domain. Next, we lay the groundwork in section 14.2 by looking at various types of simulation tools used to perform those simulations. In section 14.3, we describe different ways that input is presented or generated to drive them. We examine mobility models often used in ad-hoc mobile network simulations, in section 14.4. Next, in section 14.5 we talk about the OSI layer approach to networking; the OSI framework guides the discussion into topics in the physical layer in section 14.6, media access control in section 14.7, the data link layer in section 14.8, and TCP (in the transport layer) in section 14.9. We discuss how to describe models using SSFNet in section 14.10, and finally summarize the chapter.

14.1   Introduction

Computer systems are incredibly complex. A computer system exhibits complicated behavior at time scales from the time to flip a transistor's state (on the order of 10^{-11} seconds) to the time it takes a packet of information to travel from computer to computer (on the order of 10^{-4} seconds), to the time it takes a human to interact with it (on the order of seconds or minutes). These systems are designed hierarchically, in an effort to manage this complexity. Figure 14.1 illustrates the point. At a high level of abstraction (the system level), one might view computational activity in terms of tasks circulating among servers, queueing for service when a server is busy. At a lower level in the hierarchy one can view the activity in terms of the packets streaming between processors carrying those requests, and the responses produced by their execution. At a lower level still, one views the activity of functional units that, together, make up a central processing unit, and, at an even lower level, one can view the logical circuitry that makes it all happen.

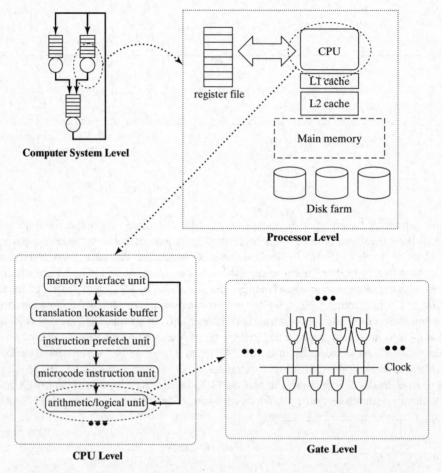

Figure 14.1 Different levels of abstraction in networked computer systems.

Simulation is used extensively at every level of this hierarchy, with some results from one level being used at another. For instance, engineers working on designing a new chip will begin by partitioning the chip functionally (e.g., the subsystem that does arithmetic, the subsystem that interacts with memory, and so on), establish interfaces between the subsystems, then design and test the subsystems individually. Given a subsystem design, the electrical properties of the circuit are first studied by using a circuit simulator that solves differential equations describing electrical behavior. At this level, engineers work to ensure the correctness of signal timing throughout the circuit and to ensure that the electrical properties fall within the parameters intended by the design. Once this level of validation has been achieved, the electrical behavior is abstracted into logical behavior (e.g., signals formerly thought of as electrical waveforms are now thought of as logical 1s and 0s). A different type of simulator is next used to test the correctness of the circuit's logical behavior. A common testing technique is to present the design with many different sets of logical inputs or *test vectors* for which the desired logical outputs are known. Discrete-event simulation is used to evaluate the logical response of the circuit to each test vector and is also used to evaluate timing (e.g., the time required to load a register with a datum from the main memory). Once a chip's subsystems are designed and tested, the designs are integrated, and then the whole system is subjected to testing, again by simulation.

At a higher level, one simulates by using functional abstractions. For instance, a memory chip could be modeled simply as an array of numbers, and a reference to memory as just an indexing operation. A special type of description language exists for this level, called "register-transfer-language." This is like a programming language, with reassigned names for registers and other hardware specific entities, and with assignment statements used to indicate data transfer between hardware entities. For example, the following sequence loads into register $r3$ the data whose memory address is in register $r6$, subtracts one from it, and writes the result into the memory location that is word adjacent (a word in this example is 4 bytes in size) to the location first read:

```
r3 = M[r6];
r3 = r3-1;
r6 = r6+4;
M[r6] = r3;
```

A simulator of such a language might ascribe deterministic time constants to the execution of each of these statements. This is a useful level of abstraction to use when one needs to express sequencing of data transfers at a low level, but not so low as the gates themselves. The abstraction makes sense when one is content to assume that the memory works and that the time to put a datum in or out is a *known constant*. The "known constant" is a value resulting from analysis at a lower level of abstraction. Functional abstraction is also commonly used to simulate subsystems of a CPU (central processing unit), in the study of how an executing program exercises special architectural features of the CPU.

At a higher level still, one might study how a communication system behaves in response to workload demands placed on it by web services. The communication traffic may be abstracted to the point of being *modeled*, but with some detailed description of the service demands (e.g., with a Markov chain that, with some specificity, describes the sizes of data segments returned in response to queries). The behavior of the web server may be abstracted to the point that all that is considered is how long it takes to complete a specified query, and the volume of data returned. Because of these abstractions, one can simulate larger systems, and simulate them more quickly.

Different levels of abstraction serve to answer different sorts of questions about a computer system, and different simulation tools exist for each level. Highly abstract models rely on stochastically modeled behavior to estimate high-level system performance, such as throughput (average number of "jobs" processed per unit time) and round-trip-delay (for a packet crossing a network). Such models can also incorporate system failure and repair and can estimate metrics such as mean time to failure and availability. Less abstract models are used to evaluate specific systems components. A study of an advanced CPU design might be aimed at estimating the throughput (instructions executed per unit time); a study of a network switch might seek to estimate the fraction of packets that are lost due to traffic congestion.

14.2   Simulation Tools

Hand in hand with different abstraction levels, one finds different tools used to perform and evaluate simulations. We next examine different types of tools and identify important characteristics about their function and their use.

An important characteristic of a tool is how it supports model building. In many tools, one constructs networks of components whose local behavior is already known and already programmed into the tool. This is a powerful paradigm for complex model construction. At the low end of the abstraction hierarchy, electrical circuit simulators and gate-level simulators are driven by network descriptions. Likewise, at the high end of the abstraction hierarchy, tools that simulate queueing networks and Petri nets are driven by network descriptions, as are sophisticated commercial communication-system simulators that have extensive libraries of preprogrammed protocol behaviors. Some of these tools allow one to incorporate user-programmed behavior, but it appears this is not the norm as a usage pattern.

A very significant player in computer-systems design at lower levels of abstraction is VHDL (very high-level design language; e.g., see Ashenden [2001]). VHDL is the result of a U.S. effort in the 1980s to standardize the languages used to build electronic systems for the government. It has since undergone the IEEE standardization process and is widely used throughout the industry. As a language for describing digital electronic systems, VHDL serves as both a design specification and a simulation specification. VHDL is a rich language, full both of constructs specific to digital systems as well as the constructs one expects to find in a procedural programming language. It achieves its dual role by imposing a clear separation between system topology and system behavior. *Design* specification is a matter of topology; *simulation* specification is a matter of behavior. Libraries of predefined subsystems and behaviors are widely available, but the language itself very much promotes user-defined programmed behavior. VHDL is also innovative in its use of abstract interfaces (e.g., to a functional unit) to which different architectures at different levels of abstraction may be attached. For instance, the interface to the ALU (arithmetic logical unit) would be VHDL signals that identify the input operands, the operation to be applied to them, and the output. One could attach to this interface an architecture that, in a few lines of code, just performs the operation—if an addition is specified, just one VHDL statement assigns the output signal to be the sum (using the VHDL addition operator) of the two input signals. An alternative architecture could completely specify the gate-level logical design of the ALU. Models that interact with the ALU interface cannot tell how the semantics of the interface are implemented. This separation of interface from architecture supports modular construction of models and allows one to validate a new submodel architecture by comparing the results it returns to the interface with those returned by a different architecture given the same inputs. A substantive treatment of VHDL is well beyond the scope of this book. VHDL is widely used in the electrical and computer engineering community, but is hardly used outside of it.

One drawback to VHDL is that it is a big language, requires a substantial VHDL compiler, and vendors typically target the commercial market at prices that exclude academic research. Of course, other simulation languages exist, and this text describes several in Chapter 4. Such languages are good for modeling certain types of computer systems at a high level, but are not designed or suited for expression of computer-systems modeling at lower levels of the abstraction hierarchy. As a result, when computer scientists need to simulate specialized model behavior, they will often write a simulation (or a simulator) from scratch. For example, if a new policy for moving data between memories in a hierarchy is to be considered, an existing language will not have that policy preprogrammed; when a new architectural feature in a CPU is designed, the modeler will have to describe that feature and its interaction with the rest of the CPU, using a general programming language. A class of tools exists that use a general programming language to express simulation-model behavior, among them SimPack (Fishwick [1992]), C++SIM (Little and McCue [1994]), CSIM (Schwetman [1986]), Awesime (Grunwald [1995]), and SSF (Cowie, Ogielski, and Nicol [1999]). In the networking domain the ns-2 simulator is widely used (Issarlykul and Hossian [2008]). This type of tool defines objects and libraries for use with such languages as C, C++, and Java. Model behavior is expressed as computer code that manipulates these predefined objects. The technique is especially powerful when used with object-oriented languages, because the tool can define base-class objects whose behavior is extended by the modeler.

Some commercial simulation languages do support interaction with general programming languages; however, simulation languages are not frequently used in the academic computer science world. Cost is a partial explanation. Commercial packages are developed with commercial needs and commercial budgets in mind, while computer scientists can usually develop what they need relatively quickly, themselves. Another explanation is a matter of emphasis: Simulation languages tend to include a rich number of predefined simulation objects and actions and allow access to a programming language to express object behavior; a simulation model is expressed primarily in the constructs of the simulation language, and the model is evaluated either by compiling the model (using a simulation-language-specific compiler) and running it or by using a simulation-language-specific interpreter.

One of the many advantages to such an approach is that the relative rigidity of the programming model makes possible graphical model building, thereby raising the whole model-building endeavor to a higher level of abstraction. Some tools have so much preprogrammed functionality that it is possible to design and run a model without writing a single line of computer code.

By contrast, programming languages with simulation constructs tend to define a few elemental simulation objects; a simulation model is expressed principally via the notions and control flow of the general programming language, with references to simulation objects interspersed. To evaluate the model, one compiles or interprets the program, using a compiler or interpreter associated with the general programming language, as opposed to one associated with the simulation language. The former approach supports more rapid model development in contexts where the language is tuned to the application; the latter approach supports much greater generality in the sorts of models that can be expressed.

Among tools supporting user-programmed behavior, a fundamental characteristic is the world-view that is supported. In the following two subsections, we look closely at process orientation as it is expressed in SSF, then at an event-oriented approach using a Java-based framework.

14.2.1   Process Orientation

A process-oriented view (see Chapter 3) implies that the tool must support separately schedulable threads of control. *Threading* is a fundamental concept in programming, and a discussion of its capabilities and implementation serves to highlight important issues in simulation modeling. Fundamentally, a *thread* is a separately schedulable unit of execution control, implemented as part of a single executing process (as seen by the operating system; see Nutt [2004]). An operating system has the notion of separate processes (which might interact), which typically have their own separate and independent memory spaces. A group of threads operate in the same process memory space, with each thread having allocated to it a relatively small portion of that space for its own use. That space is used to contain the thread's *state*, which is the full set of all information needed to restart the thread after it is suspended. State would include register values and the thread's runtime stack, which holds variables that are local to the procedures called by the thread. Once a thread is given control, it runs until it yields up control, either via an explicit statement that serves simply to relinquish control or by *blocking* until signaled by another thread to continue.

These ideas are made more concrete by discussing them in the context of a Java implementation of SSF. Java defines the `Thread` class; a subclass of `Thread` defines the `execute` method, which is defined in the thread body. Threads coordinate with each other through "locks," which provide mutually exclusive access to code segments. Every instance of a Java object has an associated lock (and almost every variable in Java is an object). A thread tries to execute a code fragment protected by the lock for object `obj` via the Java statement

```
synchronized(obj) { /* code fragment */ }
```

A thread must acquire the lock before executing the code fragment, and only one thread has the lock at a time. A thread that executes a `synchronized` statement at an instant at which another thread holds the lock blocks—which could mean suspension, depending on the thread scheduler. Java threads can also coordinate through `wait` and `notify` method calls, also associated with an object's lock. A thread that executes `obj.wait()` suspends. Actually, multiple threads can execute `obj.wait()`, and each will suspend. Eventually, some thread executes `obj.notify()`, and the thread scheduler releases one of the suspended threads to continue.

These notions can be used to implement process orientation in a Java simulator. Each simulation process derives from the Java `Thread` class. One additional thread will maintain an event list; processing for that thread involves removing the least-time event from the event list, reanimating the simulation process thread (or threads) associated with that event, and blocking until those threads have completed. While a process thread is executing, it may cause additional events to be inserted into the scheduler thread's event list. When a process thread completes, it needs to block and to signal the scheduler thread that it is finished. We accomplish all this by using two locks per simulation process. One of these locks is the one Java provides automatically for every object (and a simulation process thread is an object). The other lock is a variable each simulation object defines, which we'll call `lock`. A suspended process thread blocks on a call `lock.wait()`; it remains blocked there

```
public void waitFor(long timeinterval){
  time = owner.owner.clock + timeinterval;
  owner.owner.insertProcess(this);
  if (!isSimple()){
    synchronized(lock){
     synchronized(this){
        notify();
     }
     try{lock.wait();}
     catch(InterruptedException e){}
    }
  }
}
```

Figure 14.2   SSF implementation of `waitFor` statement.

until the scheduling thread executes `notify()` on that same object variable. After the scheduler does this, it blocks by calling `wait()` on the simulation process object's own built-in lock. So the simulation process thread notifies the scheduler that it is finished by calling `notify()` on its own built-in lock.

SSF code we discussed earlier in Chapter 4 (Figures 4.14 and 4.15) illustrates some of these points. Recall that this code models a single server with exponentially distributed interarrival times and positive normal service times. A cursory glance shows the model to be legitimate Java code that uses SSF base classes.

SSF defines five base classes around which simulation frameworks are built. The key one for discussing process orientation is the *process* class; derived classes *Arrivals* in Figure 4.14 and *Server* in Figure 4.15 are examples of it. The base class specifies that method *action* be the thread body; each derived class overrides the base-class definition to specify its own thread's behavior. Every object of a given class derived from *process* defines a separate thread of control, but all execute the same thread code body.

The `waitFor` statement used in `Arrival`'s thread body suspends the thread; its argument specifies how long in simulation time the thread suspends. The Java thread-based scheduling mechanism we described earlier enables implementation of `waitFor` to cause a "wake up" event to be inserted into the scheduling thread's event list, time-stamped with the current time plus the `waitFor` argument. Here variable `time` is the future-event time; method `insertProcess` puts the process into the event queue. A non-Simple process (e.g., one implemented with a Java thread) goes through a sequence of synchronization steps to reach the `notify()` method. (We will say more about Simple processes in 14.2.2.) The scheduler thread has blocked on the process's native lock; this `notify()` releases it. The process then immediately calls `wait()` on its `lock` variable, which suspends the thread until the scheduler executes `notify()` on that same variable. From the point of view of the code fragment executing `waitFor`, the statement following the `waitFor` call executes precisely at the time implied by the `waitFor` argument. The code in Figure 14.2 (taken from an SSF implementation) illustrates this.

The call to `waitOn` in the **Server**'s `action` has a slightly different implementation. The code implementing `waitOn` first attaches the process to the `inChannel`'s list of processes that are

blocked on it, then engages in the same lock synchronization sequence as `waitFor` to block itself
and release the scheduler thread. The semantics of releasing a blocked process are defined in terms
of SSF `Events`. An `outChannel` object to which an `Event` object is written has almost always
been mapped to an `inChannel` object. When an `Event` is written to an `outChannel` at time t,
the `outChannel`'s `write` method computes the time $t + d$ at which the `Event` is available on
the associated `inChannel` (d is a function of delays declared when the `outChannel` is created,
the `mapTo` method is called, and the `write` method is called), and an internal event is put on the
scheduler's event list, with time stamp $t + d$. The scheduler executes this event (no SSF process does)
and releases all processes blocked on the `inChannel` to which the `Event` arrives. Each of these is
able to get a copy of the `Event` so delivered, by calling the `inChannel`'s `activeEvents` method.

These descriptions show us that, normally, each event has a thread overhead cost: 2 thread
reanimations and 2 thread suspensions. Depending on how thread context switching is implemented,
this cost ranges from heavy to very heavy, as compared with a purely event-oriented view. These
costs can be avoided in SSF by designing processes to be *simple*, as described below.

14.2.2   Event Orientation

From a methodological point of view, the process-oriented view is distinguished from the event-
oriented view in terms of the focus of the model description. Process orientation allows for a continuous description, with pauses or suspensions. Event orientation does not. From an *implementation*
point of view, the key distinguishing feature of process-oriented simulation is the need to support
suspension and reanimation, which leads us to threads, as we have seen. In SSF, though, we see that
the difference between process and event orientation is not very large: The SSF world encompasses
both. The only difference is that, for SSF to be event-oriented, its processes need to be *simple*, a
technical term for the case when every statement in `action` that might suspend the process would
be the last statement executed under normal execution semantics.

The implementation of `waitFor` in Figure 14.2 computes the time when the suspension is lifted
and puts a reanimation event in the event list. Synchronization by threads through locks is used only
if the process is not *simple*. An implementation of `waitOn` would be entirely similar. If every SSF
process in a model is *simple*, there is no true code suspension, and the model is essentially event-
oriented. The `action` body for a *simple* process is just executed from its normal entry point when
the condition that releases that process from "suspension" is satisfied. The only way an **Event** that is
written into an `outChannel` is delivered is if the recipient had called `waitOn` for the corresponding
`inChannel` at a time prior to that at which the `Event` was written. Thus, we see that some of the
"events" implicit in an SSF model with event orientation are *kernel events*, which decide whether
model events ought to be executed as a result. Writing to an `outChannel` schedules a kernel event
at the `Event`'s receive time, but the kernel's processing of that event determines whether an `action`
body is called. Nevertheless, execution of `action` bodies constitutes the essential "event processing"
when SSF is used in a purely event-oriented view. It is interesting that, from a conceptual point of
view, there is very little difference between process-oriented and event-oriented SSF.

To conclude this discussion on tools, we remark that *flexibility* is the key requirement in computer-
systems simulation. Flexibility in most contexts means the ability to use the full power of a general
programming language. This requires a level of programming expertise that is not needed by users
of commercial graphically oriented modeling packages. The implementation requirements of an

object-oriented event-oriented approach are much less delicate than those of a threaded simulator, and the amount of simulator overhead involved in delivering an event to an object is considerably less than the cost of a context switch in a threaded system. For these reasons, most of the simulators written from scratch take the event-oriented view. However, the underlying simulation framework necessarily provides a lower level of abstraction and so forces a modeler to design and implement more model-management logic. The choice between using a process-oriented or an event-oriented simulator—or writing one's own—is a function of the level of modeling ease versus execution speed.

14.3   Model Input

Just as there are different levels of abstraction in systems simulation, there are different means of providing input to a model of a computer and/or a network. The input might represent service demands from users for CPU resources, it might represent volumes of traffic offered by computers to a network to be carried. The input model might be driven by stochastically generated input, or it might be given trace input, measured from actual systems. Simulations at the high end of the abstraction hierarchy most typically use stochastic input; simulations at lower levels of abstraction commonly employ trace input. Stochastic input models are particularly useful when one wishes to study system behavior over a range of scenarios; it could be that all that is required is to adjust an input model parameter and rerun the simulation. Of course, using randomly generated input raises the question of how real or representative the input is; that doubt frequently leads systems people to prefer trace data on lower level simulations. Using a trace means one cannot explore different input scenarios, but traces are useful when directly comparing two different implementations of some policy or mechanism on the same input. The realism of the input gives the simulation added authority.

In all cases, the data used to drive the simulation is intended to exercise whatever facet of the networked system is of interest. High-level systems simulations accept a stream of job descriptions; CPU simulations accept a stream of instruction descriptions; network simulations accept a stream of communication requests; network device simulations accept a stream of packets.

Computer systems modeled as queueing networks typically interpret "customers" as computer programs; servers typically represent services such as attention by the CPU or an I/O (input–output) system. Random sampling generates customer interarrival times; it may also be used to govern routing and time-in service. However, it is common in computer-systems contexts to have routing and service times be state-dependent (e.g., the next server visited is already specified in the customer's description, or could be the attached server with least queue length).

Interarrival processes have historically been modeled as Poisson processes (where times between successive arrivals have an exponential distribution). However, this assumption has fallen from favor as a result of empirical observations that significantly contradict Poisson assumptions in current computer and communication systems. The real value of Poisson assumptions lies in tractability for mathematical analysis, so, as simulationists, we can discard them with little loss.

In the subsections to follow, we look at the mathematical formulation of common input models for networked computer systems.

14.3.1   Modulated Poisson Process

Stochastic input models ought to reflect the real-life phenomenon called *burstiness*—that is, brief periods when demand intensity is much higher than normal. An input model sometimes used to support this, retaining a useful level of mathematical tractability, is a Modulated Poisson Process, or MPP. (See Fischer and Meier-Hellstern, [1993].) Burstiness arises in many different applications. The MPP is a general model that is suitable for a variety of input processes encountered within a networked computer system. However, following our description of MPPs, we will develop models that derive from observed characteristics of networks and are used to generate input demands for networks.

An MPP is built up from a CTMC (continuous-time Markov chain), whose details we sketch so as to employ the concept later. A CTMC is always in some *state*; for descriptive purposes, states are named by the integers: 1, 2, ... The CTMC remains in a state for a random period of time, transitions randomly to another state, stays there for a random period of time, transitions again, and so on. The CTMC behavior is completely described by its *generator matrix*, $\mathcal{Q} = \{q_{i,j}\}$. For states $i \neq j$, entry $q_{i,j}$ describes the rate at which the chain transitions from state i into state j (this is the total transition rate out of state i, times the probability that it transitions then into state j). The rate describes how quickly the transition is made; its units are transitions per unit simulation time. Diagonal element $q_{i,i}$ is the negated sum of all rates out of state $i : q_{i,i} = -\sum_{j \neq i} q_{i,j}$. An operational view of the CTMC is that, upon entering a state i, it remains in that state for an exponentially distributed period of time, the exponential having rate $-q_{i,i}$. When making the transition, it chooses state j with probability $-q_{i,j}/q_{i,i}$. Many CTMCs are *ergodic*, meaning that, if it is left to run forever, every state is visited infinitely often. In an ergodic chain, π_i denotes state i's *stationary probability*, which we can interpret as the long-term average fraction of time the CTMC is in state i. A critical relationship exists between stationary probabilities and transition rates: For every state i,

$$\pi_i \sum_{j \neq i} q_{i,j} = \sum_{j \neq i} \pi_j q_{j,i}$$

If we think of $q_{i,j}$ as describing a probability flow that is enabled when the CTMC is in state i, then these equations say that, in the long term, the sum of all flows out of state i is the same as the sum of all flows into the state. We will see in the example that follows that we can use the balance equations to build a stochastic input with desired characteristics. To complete the definition of an MPP, it remains only to associate a customer arrival rate λ_i with state i. When the CTMC is in state i, customers are generated as a Poisson process with rate λ_i.

To illustrate the use of balanced equations, let us consider an input process that is either OFF, ON, or BURSTY (the output rate is much higher in the BURSTY state than in the ON state). We wish for the process to be OFF half of the time—on average, for 1 second—and, when it is not OFF, we wish for it to be BURSTY for 10% of the time. We will assume that the CTMC transitions into BURSTY only from the ON state and transitions out of BURSTY only into the ON state. We will say that state 0 corresponds to OFF, 1 to ON, and 2 to BURSTY. Our problem statement implies that $\pi_0 = 0.5$, $\pi_1 = 0.45$, and $\pi_2 = 0.05$. The only transition from OFF is to ON, and the mean OFF time is 1, so we infer that $q_{0,1} = 1$. The balance equation for state 0 can be rewritten as

$$0.5 = 0.45q_{1,0}$$

and hence $q_{1,0} = (0.5/0.45)$. The balance equation for state 1 can be rewritten as

$$0.45((0.5/0.45) + q_{1,2}) = 0.5 + 0.05q_{2,1}$$

and the balance equation for state 2 is

$$0.05q_{2,1} = 0.45q_{1,2}$$

The equations for states 1 and 2 are identical; mathematically, we don't have enough conditions to force a unique solution. If we add the constraint that a BURSTY period lasts, on average, $1/10$ of a second, we thereby define that $q_{2,1} = 10$ and, hence, that $q_{1,2} = (0.5/0.45)$. Operationally, the simulation of this CTMC is straightforward. In state 0, one samples an exponential with mean 1 to determine the state's holding time. Following this period, the CTMC transitions into state 1 and samples a holding time from an exponential with mean 0.45, after which it transitions to OFF or BURSTY with equal probability. In the BURSTY state, it samples an exponential holding time with mean 0.1. Now all that is left is for us to define the state-dependent customer arrival rates. Obviously, $\lambda_0 = 0$; for illustration, we choose $\lambda_1 = 10$ and $\lambda_2 = 500$.

Figure 14.3 presents a snippet of code used to generate times of arrivals in this process. Transitions between states are sampled by using the inverse-transform technique. (The variable `acc` computes the cumulative probability function in the distribution described by the row vector `P[state]`.) Figure 14.4 plots total customers generated as a function of time—for a short period of a sample run, and for a longer period. In the shorter run, we see regions where the graph increases sharply; they correspond to periods in the BURSTY state. While the CTMC is not in this state, a mixture of OFF and ON periods moves the accumulated packet count up at a much more gradual rate. The MPP model can describe burstiness, but the burstiness is limited in the time scale. The longer run views the data at a time scale that is two orders of magnitude larger, and we see that the irregularities are largely smoothed.

14.3.2   Poisson-Pareto Process

The Internet is increasingly supporting telephony—VoIP (voice over IP, Black [2001]), and so attendant models are being developed. A sampling of the current literature suggests that a VoIP source be modeled as an on–off process, where both phases have distributions with tails somewhat heavier than exponential (e.g., an appropriate Weibull). Increasingly, the Internet will be used to stream video content. Models for video are more complex, because they must capture a number of facets of video compression, at different time scales.

The MPP model is typically used to describe traffic workload offered to a network by an individual user or application. There are contexts in which a modeler needs, instead, to consider the impact of aggregated application flows on a network device. One could create the aggregate stream by piecing together many individual application streams—or one could start with an aggregated model in the first place. We next consider direct models of aggregated offered load.

Classic models of telephone traffic assume that aggregated call arrivals to the telephony network follow a Poisson distribution and that call completions likewise are Poisson. The early days of modeling and engineering *data* networks made the same assumption. However, with time, it became clear that this assumption did not match reality well. In telephony, the increased use of faxes, and then

```
class mpp {

 public static double Finish;          // sim termination
 public static double time = 0.0;      // current clock
 public static double htime, etime;    // transition times
 public static int state = 0;          // current state id
 public static int total = 0;          // total pkts emitted
 public static Random stream;
 ...

public static void main(String argv[]) {
 ...
while( time < Finish ) {

 // generate exponential holding time, state-dependent mean
 htime = time+exponential( stream, hold[state] );

 // emit packets until state transition time. State dependent
 // rate. Note assignment made to etime in while condition test
 while( (etime = time+exponential( stream, 1.0/rate[state]))
                 < min( htime, Finish) )   {
        System.out.println( etime + `` '' + total);
    total++;
    time = etime;    // advance to packet issue time
 }
 time = htime;

 // select next state
 double trans = stream.nextDouble();
 double acc = P[state][0];
 int i = 0;

 while( acc < trans ) acc += P[state][++i];
 state = i;
  }
 }
 ...
}
```

Figure 14.3   Java code generating MPP trace.

Internet connections, radically transformed the statistical behavior of traffic. Two things emerged as being particularly different: First, data traffic exhibits a burstiness that flies in the face of the exponential's memoryless property. MPP processes described earlier can be used to introduce bursti-ness explicitly into the arrival pattern of packets to a data network. However, studies indicate that the durations of burstiness aren't Markovian, as in the MPP model. Instead, traffic seems to exhibit

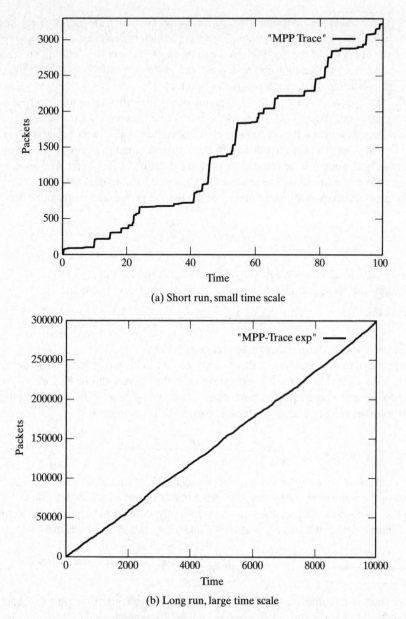

(a) Short run, small time scale

(b) Long run, large time scale

Figure 14.4   Sample runs from MPP model.

long-term temporal dependence—correlations in the number of active sessions that extend past what, statistically, can be expected from MPP models.

Researchers noticed that there was tremendous variance in the size of files transferred within a session. It seemed that a heavy-tailed distribution such as the Pareto did a good job of capturing this spread. Heavy-tailed distributions have the characteristic that, infrequently, very large samples emerge. These large samples are large enough, relative to their probability, to exert a very significant influence on the moments of the distribution; in some cases, the integral defining variance diverges. It was hypothesized, then, that long-range dependence in session counts was due to the correlations induced by the concurrency of very long-lived sessions.

A model that appears to capture these explanations is the PPBP (Poisson Pareto Burst Process, Zukeman, Neame, and Addie, [2003]), in which bursts (e.g., sessions) of traffic arrive as a Poisson process. Each session length has duration sampled from a Pareto distribution. Bursts may be concurrent. More formally, let t_i be the arrival time of the ith burst, equal to $t_{i-1} + e_i$, where e_i is sampled from an exponential, and let b_i be the Pareto-sampled duration of that burst, and let $d_i = t_i + b_i$ be the finishing time of the ith burst. The state at t, $X(t)$, is the number of bursts t_j with $t_j \leq t < d_j$.

The Pareto distribution with parameters a and b has the probability distribution function

$$D(x) = 1 - \left(\frac{b}{x}\right)^a$$

for $x \geq b$. The distribution has mean $(ab)/(a-1)$ and variance $ab^2/((a-1)^2(a-2))$. One can sample a Pareto with these parameters, using the inverse transform technique:

$$x = b \times (1.0 - U)^{-1.0/a}$$

In this equation U is a uniformly distributed random variable.

It is instructive to consider how traffic is analyzed for evidence of long-range dependence and whether the style of synthetic traffic generation described here exhibits it. Let $X_1, X_2, \ldots$, be a stationary time series, whose samples have mean μ and variance σ^2. The autocorrelation function $\rho(k)$ describes how well correlated samples k apart in the time series are:

$$\rho(k) = \frac{E[(X_t - \mu)(X_{t+k} - \mu)]}{\sigma^2}$$

The sample autocorrelation function can be constructed from an actual sample by estimating the expectation in the numerator. Long-range dependence is observed when $\rho(k)$ decays slowly as a function of k. Long-range dependence is more formally defined in terms of the autocorrelation function, if there exists a real number $\alpha \in (0, 1)$, and a constant $\beta > 0$ such that

$$\lim_{k \to \infty} \frac{\rho(k)}{\beta k^{-\alpha}} = 1$$

The denominator of this limit describes how slowly $\rho(k)$ needs to go to zero as k increases. The smaller α is, the slower the degradation. $H = 1 - \alpha/2$ is known as the *Hurst parameter* for the sequence. Values of H with $0.5 < H < 1.0$ define long-range dependence; the larger H is, the more significant is the long-range dependence.

To see evidence that PPBP does yield long-range dependence, we ran an experiment where the mean burst interarrival time was 1 second and the Pareto parameters were $a = 1.1$ and $b = 10$. We computed the sample autocorrelation function, shown in Figure 14.5. Here we see directly that the autocorrelation decays very slowly. We also used the SELFIS tool (Karagiannis, Faloutsos, and Molle

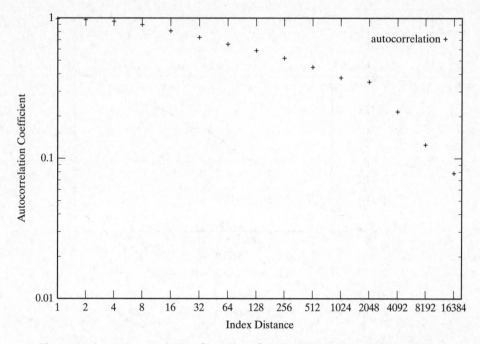

Figure 14.5   Autocorrelation function of aggregated stream of 50 sources.

[2003]) to estimate the Hurst parameter—all of its estimators indicate strong long-range dependence in the sampled series.

Burstiness is not the only consideration in traffic modeling. Traffic intensity exhibits a strong diurnal characteristic—that is, source intensity varies with the source's time of day; furthermore, weekends and holidays behave differently still. To accommodate time-of-day considerations, one can allow the exponential burst interarrival distribution of the PPBP to have a parameter that is dependent on the time of day.

The PPBP describes the number of active sessions $X(t)$ as a function of time. $X(t)$ may be transformed into packet arrival rates, and hence into packets, by including a packet-rate parameter λ. The process $\lambda X(t)$ thus gives an arrival rate of packets from an aggregated set of sources to a network device that handles such.

14.3.3   Pareto-length Phase Time

One of the easiest models of traffic-load generation is that of moving files across the network. Our interest here is not in the mechanics of the protocols that accomplish the movement so much as it is in the model of the traffic load that is offered to the network. Simulation studies that model file transfers typically are focused on the impact that the traffic has on servers holding the files. A given transfer can be characterized by the size of the file and by the rate at which its bytes are presented to the network. We usually also characterize how often a user initiates an ftp transfer. A simple model of a file-transfer request process is as an on–off source, whose *off* period is randomly

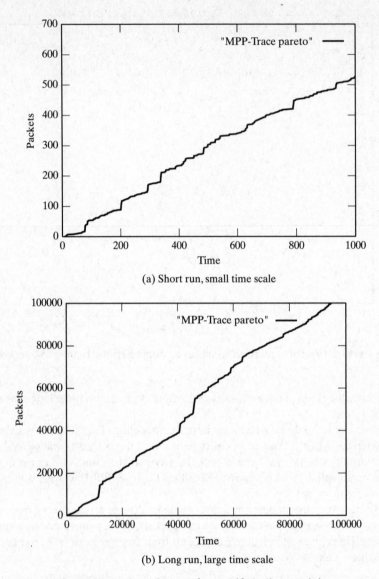

(a) Short run, small time scale

(b) Long run, large time scale

Figure 14.6   Sample runs from self-similar model.

distributed (e.g., an exponential think time) and whose *on* period is driven by the arrival of a file. The *on* period lasts as long as needed to push or pull a file of the referenced length. File size is sampled from another probability distribution. Measurements suggest that a heavy-tailed distribution like the Pareto is appropriate. This is especially appropriate given the level of music sharing activity on the Internet.

For example, consider a traffic source that remains OFF for an exponentially distributed period of time with mean 1.0, but, when it comes ON, remains on for a period of time sampled from a Pareto distribution. While it is ON, packets are generated as a Poisson process. This is different from the Poisson-Pareto burst process, but has related self-similar behavior.

Figure 14.6 parallels the MPP data, displaying accumulated packet counts as a function of time; it presents behavior for the first 1000 units of time and for the first 100,000 units of time. Here, despite two orders of magnitude of difference in run length, the visual impression of behavior is much the same between the two traces. This sort of behavior is frequently seen in computer and communication systems; the long lengths reflect burstiness of packets, file lengths, and demand on a server. Use of the Pareto distribution yields self-similar behavior at different time scales.

Another model couples the behavior of the traffic source with the network. The source has a random OFF time, and the ON time is the time required by the network to transfer a block of data whose size has a Pareto distribution. In the simplest model of a network, that time is just the block size divided by some bandwidth number. Feedback to the source generation occurs if the network model is more sophisticated, and induces some variability—possibly dependent on the network state—into the transfer times.

14.3.4   WWW Traffic

Accesses to the World Wide Web are another significant source of application traffic. Models of accessing web pages are more complex than individual file transfers and so bear separate treatment. We describe a model expounded upon in (Barford and Crovella [1998]), called Surge. Here we model the delay between successive sessions with an intersession delay distribution. Within a session, a number of different URLs will be accessed, with another delay time between each such access; this is illustrated in Figure 14.7.

The Surge model incorporates a number of important characteristics of files, the more important of which includes

- the distribution of file sizes, among all files on a web server;
- the distribution of the file sizes of those files that are actually *requested*;
- temporal locality of file-referenced file.

The first and third characteristics, coupled with a model of referencing pattern, essentially define the second characteristic. Suppose that we have selected the first k files already—call them $f_1, f_2, \ldots, f_k$— and suppose that this set of references is organized in a LRU (least recently used) stack. We select the $(k + 1)^{\text{st}}$ file by sampling an integer from a stack-distance distribution. If that sample has value j, the next file selected has position j in the LRU stack (position 1 being the last file referenced). Empirical studies of reference strings of files suggest that a lognormal distribution is appropriate.

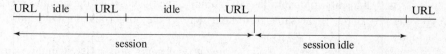

Figure 14.7   Nested on–off periods in Surge WWW traffic generation.

This distribution places significant weight on small values; hence, it induces temporal locality of reference. When the stack-distance sample is larger than the number of files in the LRU stack, a new file is sampled from the set of files not yet in the reference stream.

This description gives a general, but simplistic idea of the structure of Surge. Its authors pay much attention to issues of identifying distributional parameters that are internally consistent and that produce traffic that can be validated against real traffic. Our goal here is to introduce the fundamental notions behind a model of web traffic.

The main point to be understood about traffic-modeling is that models of aggregated traffic ought to exhibit characteristics of aggregation, whereas application traffic ought to focus on what makes the applications distinct.

14.4   Mobility Models in Wireless Systems

Another important area in modeling and simulation of networked systems is mobility in simulations of wireless networks. Cellphone, wireless hot-spots, ad-hoc networks—much of the communication we do with personal electronic devices uses radio. Modeling wireless communication differs from modeling wireline communication in a number of ways, including the possibility that communication endpoints are in motion. This affects signal strength, and so affects quality of communication. Models of mobile wireless networks need to include a representation of mobility; we now consider modeling options for mobility.

One selects a mobility model based on the physical domain in which users move, and the granularity needed for movement. We have done simulation studies of mobile phone users in a shopping mall, where there are main throughfares, with stores of various sizes attached. In this particular application we studied how malicious software might spread from phone to phone using the Bluetooth medium. For this study the most important thing to capture was proximity—Bluetooth radio does not carry far, so for one infected phone to infect another it is necessary that they be physically close. Our model of the domain and mobility within the domain is based on a grid of cells that tesselate the domain. The state of each cell is a pair (n_i, n_v), where n_i is the number of infected phones in the cell and n_v is the number of vulnerable phones. Mobility is modeled with a time-stepped mechanism where we assume that in one unit of time, a cell user could traverse the cell. Then at each time-step, each mobile stays in its current cell or moves to an adjacent cell; cell and mobile type-specific probabilities are given for every possible movement. We see immediately that this construction forms a discrete-time Markov chain. Cell-to-cell movement probabilities can be assigned to constrain movement, e.g., to not leave a store through an interior wall, or to have faster movement in the mall corridor than within a store. In our application, an infected mobile can try to infect another if they are close enough, i.e., in cells that are close enough.

Figure 14.8 illustrates the cell-based mobility structure. The dark lines delimit walls through which mobile users do not pass; the black block is a region where users cannot go. The figure highlights one small region of the grid, showing the states of two cells, (2, 4) and (1, 0); it also labels cell boundaries with transition probabilities. To minimize the busyness of the graphic we assume that the crossing probability is the same from both sides of the border and for both infected and noninfected mobiles.

Another common mobility model is call the *random way point* model. Here a mobile's position is created with an algorithm that moves the mobile in a straight line from way point to way point. The

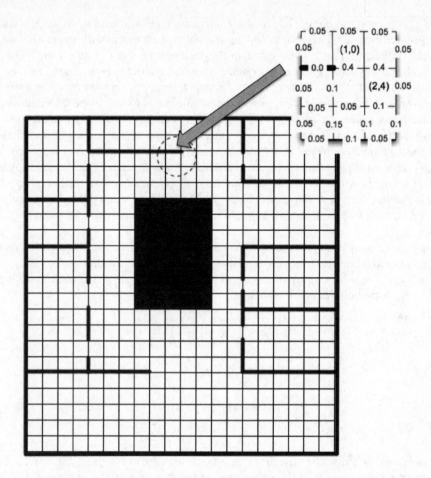

Figure 14.8   Cell-based model of mobility in a shopping mall.

mobile's speed may be randomly chosen on a per-segment basis. On reaching a waypoint, the mobile remains stationary for a random period of time (possibly zero), then randomly chooses a different waypoint and a speed towards it, and begins to move again. The random waypoint model is general enough so that random waypoint selection may be constrained, or not. For example, a mobile's path might be constrained to follow the edges of a graph, where the graph describes a transportation network. Waypoints are then intersections in that network. Alternatively, a mobile at waypoint (x, y) may choose the next waypoint (x', y') any way it likes, e.g., sampling uniformly on the range allowed for x, and on the range allowed for y. In a variation on this method the mobile randomly chooses a direction, a speed, and a duration for travel; the target waypoint (x', y') can be computed from this information. This version of the random waypoint model raises issues on how to deal with boundary constraints and obstacles.

It is possible for the trajectory to cross a domain boundary or to encounter an obstacle in the domain. In the former case it is sometimes reasonable to assume that there are no real boundary constraints, that the top and bottom edges of a rectangular domain touch, as do the left and right edges. Topologically the domain is a *torus*, and the destination waypoint coordinates can be computed using modular arithmetic. That is, if the extent of the domain is D_x and D_y in the x and y directions, respectively, then a waypoint $(\hat{x}, \hat{y})$ can be computed without respect to boundaries, and then transformed into a waypoint within the domain $x' = \hat{x} \bmod D_x$ and $y' = \hat{y} \bmod D_y$. Of course, in the case of wrap-around, the assumed trajectory of the mobile needs to follow the direction randomly chosen that actually causes the mobile to hit the domain boundary before wrapping around to the opposite side of the domain. If, however, a trajectory hitting a boundary means that the trajectory must stop, then the mobility model defines the next waypoint to be the point of intersection with the boundary, after which another waypoint is chosen. Likewise, if the domain contains obstacles, then the next waypoint will be either the original (x', y') chosen, or an intersection point with the first obstacle that interferes with the chosen trajectory. We will have more to say on the problem of detecting intersections with obstacles when we discuss ray-tracing algorithms for radio propagation (Section 4.6.1).

The random waypoint model has a hidden danger that must be appreciated. If a mobile's *speed S* is chosen randomly and is intended to be non-zero, then the speed distribution must satisfy $E[1/S] < \infty$ for the mobility system to have an equilibrium state. A notable example of a distribution that fails to meet this criteria is the uniform distribution on $(0, a)$. For

$$E[1/S] = \int_0^a \left(\frac{1}{a}\right) \left(\frac{1}{x}\right) \, dx$$

$$= \frac{1}{a} \Big|_0^a ln \, x$$

$$= \frac{1}{a}(ln \, a - ln \, 0)$$

$$= \infty$$

Intuitively, this condition arises because the expected time required to traverse a distance L is $E[L/S]$. If $E[1/S]$ is not bounded then $E[L/S]$ is not bounded, and mobiles ultimately get trapped in arbitrarily long traversals of the domain. On the other hand, a sufficient condition for $E[1/S] < \infty$ is that $S > \epsilon$ for some $\epsilon > 0$. In this case the traversal of a distance L is never longer than L/ϵ, and so mobiles are not trapped in arbitrarily long traversal times.

14.5 The OSI Stack Model

Networked computer systems exhibit complexity at multiple layers. To manage that complexity they are usually designed (with varying degrees of fidelity) in accordance with the so-called OSI (open system interconnection) stack model (Zimmerman [1980]). The fundamental idea is that each layer provides certain services and guarantees to the layer above it. An application or protocol at a particular layer communicates only with protocols directly above and below it in the stack, implementing communication with a corresponding application or protocol at the same stack layer in a different device. Simulation is used to study behavior at all these layers, although not generally all in the same model. Different layers encapsulate different levels of communication abstraction.

The *physical layer* is concerned with the communication of a raw bit stream over a physical medium. The specification of a physical layer has to address all the physical aspects of the communication: voltage or radio signal strength, standards for connecting a physical device to the medium, and so on. Models of this layer describe physics.

The *data link layer* implements the communication of so-called *data frames*, which contain a limited chunk of data and some addressing information. Protocols at the Data Link Layer interact with the physical layer to send and receive frames, and also provide the service of error-free communication to the layer above it. Protocols at the Data Link Layer must therefore implement error-detection and retransmission when needed. A critical component of avoiding errors is access control, which ensures that, at most, one device is transmitting at a time on a shared medium. Techniques for access control have significant impact on how long it takes to deliver data and on the overall capacity of the network to move data. Simulation plays an important role in understanding tradeoffs between access-control techniques; in Section 14.7, we will look at some protocols and the characteristics that simulation reveals.

The *network layer* is responsible for all aspects of delivering data frames across subnetworks. A given frame may cross multiple physical mediums en route to its final destination; the Network Layer is responsible for logical addresses across subnets, for routing across subnets, for flow control, and so on. The success of the Internet is due in no small part to widespread adaption of the Internet Protocol, more commonly known as the IP (Comer [2000]). In IP specifies a global addressing scheme that allows communication between devices across the globe. The specification of IP packets includes fields that describe the type of data being carried in the packet, the size of the packet, the protocol suitable for interpreting the packet, source/destination addressing information, and more. The Network Layer provides error-free end-to-end delivery of packets to the layer above it. Simulation is frequently used to study algorithms that manage devices (routers) that implement the Network Layer.

The *transport layer* accepts a message from the layer above, segments it into packets that are passed to the Network Layer for transmission, and provides the assurance that received packets are delivered to the layer above in the order in which they appear in the original message, error-free, without loss, and without duplication. Thus, the Transport Layer protocol in the sending device coordinates with the Transport Layer protocol in the receiving device in such a way that the receiving device can infer packet-order information. Variants of the TCP (transmission control protocol) are most commonly used at this layer of the stack (Comer [2000]). Dealing with packet loss is the responsibility of the transport layer. Packet loss is distinct from error-free transmission—a packet could be transmitted to a routing device without error, only to find that device does not have the buffering capacity to store it; the packet is received without error, but is deliberately dropped. Transport layer protocols need to detect and react to packet loss, because they are responsible for replacing the packets that are dropped. One of the ways they do this is to apply flow-control algorithms that simultaneously try to utilize the available bandwidth fully, yet avoid the loss of packets. Simulation has historically played a critical role in studying the behavior of different transport protocols, and in this chapter we will examine simulation of TCP.

The first four OSI layers are well defined and separated in actual implementation. The remaining three have not emerged so strongly in practice. Officially, the *session layer* is responsible for the creation, maintenance, and termination of a "session" abstraction, a session being a prolonged period of interaction between two entities. Above this one finds the *presentation layer*, whose specification includes conversion between data formats. An increasingly important conversion function is encryption/decryption. Finally, the *application layer* serves as the interface between users and network services. Services typically associated with the application layer include email, network management tools, remote printer access, and sharing of other computational resources.

Wireline networks transmit data over wires or cable; wireless networks transmit using radio frequencies. The last few years have seen an explosion of research problems explored in the domain of wireless networks, and it is appropriate for us to look at issues related to simulation of wireless traffic as an example of simulation being used to model the physical layer.

Devices with traffic they wish to transmit must somehow gain access to the networks that carry traffic. We next consider the problem of how devices coordinate with one another to use the network medium, sometimes called MAC (media access control) protocols. Historically, simulation has played an important role in helping engineers to understand the performance of different MAC protocols.

Moving up the OSI stack we next describe the TCP (transport control protocol) and discuss how simulation plays an important role in its study. This leads us to a description of network sockets, and how applications higher still in the OSI stack use them. In particular we consider applications that provide web services, and modeling issues related to their simulation.

14.6   Physical Layer in Wireless Systems

Wireless networks differ from wireline networks also in the modeling of the physical layer. Packet transmission within a wireline network is usually so reliable that, for many purposes, the only aspects of the wireline physical layer that need to be included are network bandwidth, and packet latency. The physical layer in a wireless system is not so easily treated. The quality of communication degrades markedly as a function of distance and clutter in the radio domain, and so simulations of wireless networks need to explicitly account for these degradations. There is a wide spectrum of models available for the physical layer, ranging from the ab initio Maxwell equations to very simple models that degrade received signal strength as a simple function of distance. Detailed models require sophisticated computer programs and tremendous amounts of computation to give answers based on minute details that affect propagation, while simple models are easier to program and execute quickly by glossing over details, with the corresponding risk of inaccuracy.

The choice of model for a simulation depends largely on the objectives of the simulation study. Companies that build hand-held radios or those that deploy wireless networks are very interested in details that might possibly affect their designs, so they use detailed models of the physical domain, but typically use simple models of traffic as drivers, to create the detailed radio behavior. Researchers interested in higher level software protocols (e.g., routing or services that manage quality of service) typically focus their modeling efforts on the protocols, and use simple models of the wireless physical layer.

14.6.1   Propagation Models

Wireless models of all kinds are treated extensively in Rappaport [2002], and we use it as a basis for our discussion on the physical layer. We will discuss the simpler models here, and refer those more interested in detailed electrical engineering to Rappaport.

The basic functionality of the physical layer is to carry a bit transmitted at some power P_t to the intended receiver. The basic question a model of the physical layer must answer is what the power level of that signal is at the receiver. The problem is that the electromagnetic waves emitted by the transmitter are subject to refraction, reflection, diffraction, and scattering as they encounter obstacles in the domain. The received signal is a combination of waves that have taken different paths and undergone different transformations as a function of the objects on those paths, that are skewed in time, and that arrive at different strengths. Normal use of the received power P_r at a receiver is to determine whether a transmitted packet is recognized. This determination has subtleties and complexities we discuss later.

The simplest propagation model in use is the *free-space propagation* model. This assumes that the transmitting and receiving antennas are in line-of-sight of each other, and that the degradation of signal strength from transmitter to receiver is caused entirely by dissipation due to distance. Contexts that adhere most closely to free-space assumptions include satellite communication systems and line-of-sight microwave systems. In its simplest form the power received P_r is related to the power transmitted P_t by

$$P_r = cP_t/d^2$$

where d is the distance between transmitter and receiver, and c is a constant. This is the inverse-square law frequently encountered in physics.

A slightly more general model allows for faster degradation of strength with distance

$$P_r = cP_t/d^\alpha$$

with $\alpha \geq 2$. In this form, by simply increasing α, the model accounts for other phenomena that degrade the signal. An example of this is a *two-wave* model, designed to include effects of destructive interference from the wave that reflects off the ground from the transmitter. A distance threshold D_t is defined as a function of the heights of the antenna and the wavelength. For receivers closer than D_t to the transmitter, the destructive interference effect is not important. For these, the received power is obtained using $\alpha = 2$. For receivers farther than D_t from the transmitter, we use $\alpha = 4$. While stated at this level it seems a curious hack, the choice is based on a detailed geometric model of the electric fields involved with a wave bouncing off the ground, interfering with the line-of-sight path, and the cumulative effect on the received power. Note, however, that there is a discontinuity in received power as a function of distance, at D_t.

Both the free-space and two-wave models are deterministic. Research has shown that obstacles near a receiver often impact received power by creating shadows where the received signal is significantly degraded. One way we can account for these degradations—but avoid having to explicitly compute the geometric effects of the obstacles on radio waves—is to model the degradation as a random variable. The *lognormal shadowing* model (also known as the shadow-fading model) specifies that

$$P_r = kP_t 10^{-X/10}/d^\alpha \tag{14.1}$$

where X is a Gaussian random variable with mean 0, and standard deviation σ, and k is constant. The phenomenon that 10^X models is deterministic in time, but random in space. That is, if we were to leave the transmitter and receiver in the exact same position, the received power for a transmission of strength P_t is always the same. However, the shadow fading a receiver suffers (with respect to a given transmitter) at (x, y) is correlated with the fading it suffers at a nearby position (x', y')—if a wall shadows (x, y) from the transmitter it is likely to shadow (x', y') as well. One of the problems of using lognormal shadowing is correlation—the random shadowing suffered by one receiver is correlated with the shadowing suffered by one that is nearby.

Correlation in lognormal shadow random variable problem can be addressed, in principle, by specifying the correlation structure on a spatial structure. Imagine tesselating the domain into an $N \times N$ grid of cells. For any transmitter in cell c_A and any receiver in cell c_B we assume the shadow fading is the same as with any other transmitter in c_A and receiver in c_B. So we have on the order of $(N^2)^2 = N^4$ cell pairs that define a particular sample from the lognormal shadow distribution. In the general case we have to explicitly specify the covariance of any pair of shadow random variables. But, this is unwieldy—there are on the order of N^8 such covariances to specify! However, as shadow-fading is based on local effects, it is natural to assume that the correlation between a pair of shadow variables is zero if they are defined by different transmitters, decreases rapidly as a function of distance between variables defined by the same transmitter, and the distance is measured in cells. Further, it is not unreasonable to assume that the correlation is only a function of the distance. Consider: given an $n \times n$ covariance matrix Q, a vector U of n independent normal 0–1 random variables, and an n-vector $\bar{\mu}$, the equation $X = \bar{\mu} + QU$ produces a sample vector of n normal variables whose means are described by $\bar{\mu}$ and whose covariance structure is Q. To implement correlated shadow fading we can at the start of the simulation sample U—a vector of N^4 0–1 normal variables. When the simulator needs to compute Equation (14.1) for a receiver, it must identify which row in Q corresponds to the transmitter–receiver cell-pair, compute the indices in U of the neighboring cells for which there is a nonzero correlation, and compute the (sparse) dot-product of the row in Q with the vector U. Because of the sparsity (i.e., many values of the row in Q are zero), the number of terms to compute in the dot-product can be quite small. Finally, upon computing the normal variable this way, the natural logarithm of it is taken to determine L in Equation (14.1).

A more computationally intensive technique called *ray-tracing* is also used to model propagation (McKnown and Hamilton [1991]). The idea is to compute the various paths that radio waves may travel from source to destination, by shooting rays out from the source at different angles, and then compute what happens to the signal strength of the rays as they reflect off and penetrate through obstacles.

Consider the example shown in Figure 14.9. The black circle in the lower right represents the transmitter. Two receivers are represented as white circles. During transmission, radio waves emanate from the source in all directions, changing direction and intensity as they spread throughout the domain. In a ray-tracing algorithm we computationally track discrete representations of these rays, determine where they encounter obstacles on their path, and model the impact on the ray of encountering each obstacle.

Signal strength degrades with distance and with interactions with obstacles (as a function of material properties of the obstacle). Furthermore, interaction with an obstacle may produce not only a new direction for a ray, but an additional set of rays altogether, heading in different directions. Some of the many rays traced will pass close enough to a receiver to be considered as having hit the receiver, and contributing to the strength of the received signal. For any given receiver there may be multiple paths that reach it; in the simplest model the received signal strength is the sum of the signal strengths of the rays that reach it. Reality is more complex, however, and models may be as well. Waves modulate, and as they interact with other waves they may re-enforce wave peaks and deepen wave valleys, they may cancel one another out, or some combination in the middle. A detailed ray-tracing algorithm can account for these phenomena, but it is beyond the scope of the present discussion.

It is reasonable to assume that signal strength degrades from point to point on a ray, following the free-space propagation model. When the ray encounters a flat surface at an angle θ to a surface, we compute that it may (depending on θ and the electrostatic property of the surface, and strength of the signal)

- reflect symmetrically and not pass through the surface;
- reflect symmetrically and also pass through the surface;
- simply pass through the surface;
- not proceed further.

Any post-reflection rays have a signal strength that depends on the pre-reflection strength and the property of the material. If the model accounts for refraction, a ray passing through may have a different angle as well, which again is a function of the property of the material it interacts with.

Figure 14.9 illustrates some of these points. Both receivers are shown to have direct line-of-sight rays with the transmitter, and these rays will contribute most to the overall received signal strength. At the surface labelled (a) we see a ray that reflects and passes through the surface. The reflected

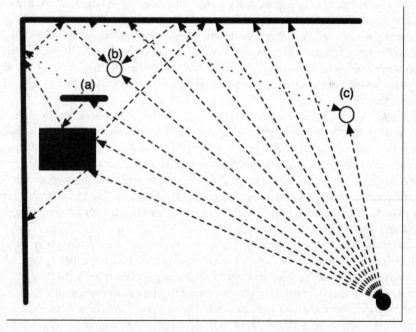

Figure 14.9   Ray-tracing for wireless propagation.

ray encounters another obstacle and reflects again, and continues to bounce twice more off walls, to reach the receiver (b). That ray has a smaller intensity than the line-of-sight ray, owing to the fact that it has traveled further and lost intensity with each reflection. The ray that passes through the surface at (a) has its signal strength degraded by the encounter, and then it reflects twice more before hitting the receiver marked (c).

The ray-tracing algorithm maintains a list of active rays. The list is initialized by a set of all rays separated by an angle α that emanate from the transmitter. The algorithm then enters a loop where, with each iteration, a ray from the list is selected and removed. The next intersection of that ray with an obstacle or receiver is computed. Depending on the angle of intersection, strength of signal (after being degraded by distance from the ray's source point), and material properties, a variable number of new rays are computed and added to the active list, with source endpoints being the point of intersection. That variable number may be zero, which happens when the signal strength drops below some threshold (possibly dependent also on the angle of intersection and material property). If the ray is actually intersecting a region of space close to a receiver, the signal strength at that intersection is added to the receiver's received signal strength. After all new rays have been added to the list, the algorithm returns to the top of the loop to select another ray. The algorithm terminates when the active list is empty.

The computation of a ray's new direction and signal strength is usually elementary trigonometry and arithmetic. The heart of the challenge in ray-tracing is determining where a ray's "next" inter-section lies. This is usually accomplished with the aid of an auxiliary tree-like data structure. The domain is recursively subdivided in a binary fashion, as illustrated in Figure 14.10(a). First, a vertical cut is found such that the number of obstacles or radios on each side of the cut is approximately the same. In the diagram that cut is a thick dashed line, with label "0" on the left side and label "1" on the right side. Descending recursively, each of the pieces is partitioned in a way that balances the obstacles and radios on each side; this time the cut is horizontal, and the labels "00," "01," "10," and "11" refer to the four regions created in this way. This process continues, alternating cuts vertically or horizontally, until every region has 1 or 0 obstacles or radios. In the course of the algorithm, a region with one or zero objects within is not subdivided. Note also that we permit the partitioning to dissect an obstacle.

From the partitioning we can construct a tree similar to that shown in Figure 14.10(b). Associated with each node in the tree is the physical extent of the region with the label on the node, and the location/extent of the obstacle or radio in that region, if extant. A tree like this has the property that, given an (x, y) coordinate, we can descend the tree to find the leaf node that contains it. Starting at the root node, we determine whether (x, y) falls to the left or right of the partition, between region 0 and 1. Having identified that, we compare (x, y) to the partition line in the region just selected, determine which subtree contains (x, y), and descend: We will find the region containing (x, y) when the process ends in a leaf node.

Now, consider how the tree structure can be used to find the next intersection for a ray. Given a ray and its starting point, we can descend the tree to find the leaf node that contains that starting point. The algorithm now is to push the ray through the regions described by the leaf nodes, until the next obstacle (or radio) is intersected. At each step in this process we start with the ray's entry point into the region, and, given the direction, compute the exit point. As there is, at most, one obstacle in the region, we determine whether there is an intersection or not. If so, yes, we can compute the point of intersection. If not, we determine the exit point, and do a little more computation to

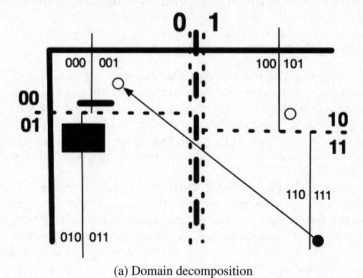

(a) Domain decomposition

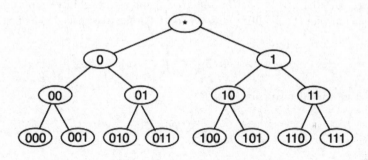

(b) Tree used to describe decomposition

Figure 14.10   Tracing a ray through the domain.

determine the identity of the region it enters. Consider a ray that travels straight from the transmitter in Figure 14.10(a) to the receiver in the upper left. That ray exits region 111 on the line that partitions the parent region 11 into 110 and 111, from which we can infer that the next region is 110. This region has no obstacles, and so we compute where the ray exits the region. We can tell it enters the "0" super-region, and can use the tree to find that region 011 contains the entry point. The ray does not intersect the obstacle in 011, but intersects the line that partitions super-regions 00 and 01, and so enters region 001. Finally, we find that the ray hits the radio.

In summary, we see that there are a variety of ways of modeling radio propagation. They differ in the level of detail, and hence accuracy. In choosing the method to use, a modeler needs to understand whether the inaccuracies of a fast simple model are likely to significantly affect the results of the experiment.

14.6.2 Determining the Receivers

Radio propagation models tell us how to compute the power P_r at a receiver as a function of the transmission power P_t and the distance between them. Typically the decision to accept a packet is based on computing a SNR (signal-to-noise ratio) ρ_t at the receiver, and accept the packet if ρ_t is sufficiently high. The value of P_r gives the ratio's numerator. To compute the denominator we add in noise n_r from the radio itself—a value that is independent of any transmission anywhere, to the sum of signal powers from other concurrent transmissions. These transmissions interfere with reception of the signal of interest.

When the simulation has a transmitter begin to broadcast, our first problem is to identify the set of receivers for whom the signal might possibly be strong enough to have the packet delivered. A necessary condition is that the received power be large enough so that the SNR computed, assuming that the only noise is n_r, is at least as large as ρ_t. We can compute the minimal power P_m, and then transform the question into one that asks *how close* the receiver needs to be to the transmitter in order to receive power P_m, in the absence of interference.

For deterministic propagation models it is straightforward to compute that distance, simply by solving for d in the propagation equation, using $P_r = P_m$. For the lognormal shadow-fading model, we can choose U_t such that $10^{-X/10} < U_t$ (with overwhelming probability), replace $10^{-X/10}$ in the propagation equation with U_t, and likewise solve for d. For the free-space model that solution is:

$$d_t = \left(\frac{cP_t}{t} \right)^{1/\alpha}.$$

For lognormal shadowing that solution is:

$$d_t = \left(\frac{kU_tP_t}{t} \right)^{1/\alpha}.$$

For the two-wave model with cut-off D_t, the solution is:

$$d_t = \begin{cases} \left(\frac{cP_t}{t} \right)^{1/2} & \text{if } P_m \geq cP_t/(D_t)^2 \\ D_t & \text{if } cP_t/(D_t)^4 \leq P_m < cP_t/(D_t)^2 \\ \left(\frac{cP_t}{t} \right)^{1/4} & \text{if } P_m \geq cP_t/(D_t)^2 \end{cases}.$$

This encodes the three possibilities of the minimum power for reception: (i) being less than the two-wave transition point D_t, (ii) being so small that it avoids the discontinuity in received power that is inherent in the two-wave model, and (iii) having a value that the two-wave model never predicts because of the discontinuity.

When the simulation involves many mobiles it is inefficient to compute the distance between the transmitter and every other mobile in order to determine those that are within distance d_t of the transmitter. Auxiliary data structures can be used to narrow the search. Recall that the mobility model is based on cells, and imagine that the domain is tesselated by cells, but that precise mobile positions are maintained. A cell will have a list of mobiles in that cell, with their exact positions. Based on mobile trajectories, the simulator will schedule events to cause a mobile to transition from one cell to another. When a transmission occurs at time t from a mobile at position (x, y), the simulator need only update the positions and check the distances of mobiles in cells that have any area within distance d_t of position (x, y)—because of the events that move mobiles from one cell to another, we are assured that if a mobile is within range of the transmitter, it is found in a cell that the simulator checks at the time of transmission.

When the simulation identifies a mobile m as potentially being within range, it must also compute the signal-to-noise ratio at m to determine whether the signal is actually strong enough for reception. If the simulator keeps lists of current transmissions (one per unique channel), it can access the list for the channel used by the transmission of interest, and compute, for every transmitter on that list, the power of its signal received at m. Accumulating these results we compute the actual SNR at m for the transmission of interest. One downside of this scheme is seen when the number of concurrent transmissions is large. The computational workload associated with determining which receivers recognize a packet is proportional to the number of receivers close enough to recognize the packet, times the total number of concurrent transmissions on that same channel. One approach that is sometimes taken is to assume that a transmission that is farther away than some threshold does not contribute to the SNR of m. There are concerns, however, about the impact this assumption may have on the accuracy of the simulation.

A moment's reflection shows that there is a tradeoff between the granularity of the cells (larger cells implying fewer cross over events on any particular trajectory) and the number of mobiles whose positions are updated and checked at the point of a transmission. The sweet spot in that tradeoff is a function of the application itself. A good simulator will allow the user some flexibility in choosing that granularity, so that a good selection might be obtained for a particular application based on some experiments that determine performance as a function of granularity.

14.7   Media Access Control

Computers in an office or university environment are usually integrated into a LAN (local area network). Computers access the network through cables (a.k.a. wireline), although an increasing fraction access it through radio (wireless). In either case, when a computer wishes to use the network to transmit some information, it engages in a MAC protocol (see Section 14.5).

Different MAC protocols give traffic different characteristics. Simulation is an extremely important tool for assessing the behavior of a given protocol. A MAC protocol gives traffic-specific qualities of latency (average and maximum are usually interesting) and throughput. The behavior of these qualities as a function of "offered load" (traffic intensity) is of critical interest, because some protocols allow throughput to actually *decrease* as the demands on the network go up—a lose–lose situation.

14.7.1   Token-Passing Protocols

One class of MAC protocols is based on the notion of a *token*, or permission to transmit. These types of protocols were developed and first used on local area networks, but have been replaced in those contexts by protocols such as Ethernet. In token-passing every device knows if it has permission to transmit, and so the one holding the token may do so without consideration of interference. The permission-to-transmit token is passed to another device in a way that ensures the token is not lost in the exchange. We will speak more of Ethernet in Section 14.7.2. The key point to appreciate here is that under Ethernet a device wishing to transmit senses the medium to determine if there is already an active transmission occurring. Within small timing windows the medium may appear to be idle to the device, but a transmission has already started and will be seen shortly. There is no such ambiguity in a token-passing protocol, and it remains useful in contexts where the medium is very noisy, making a sense-based access control problematic.

Token-passing protocols vary in how the token is managed. In the *polling protocol* variation, a master controller governs which device on the shared medium may transmit (Kurose and Ross [2002]). The controller selects a network device and sends it the token. If the recipient has frames (the basic unit of transmission) buffered up, it sends them, up to a maximum number of frames. The controller listens to the network and detects when the token holder either has selected not to transmit or has finished transmission. The controller then selects another network device and sends it the token. Devices are visited in round-robin fashion.

One drawback of the polling protocol is that the controller is a device with its functionality separate from the others. A more homogeneous approach is achieved by using a *token bus* protocol. In this approach a device is programmed to transmit frames (again, up to a maximum number) when it receives a token, but is programmed to pass the token directly to a different specified network device after it is finished. There is no controller; the network devices pass the token among themselves, effectively creating a decentralized round-robin polling scheme.

One drawback to both types of token-passing protocols is that a single failure can stop the network in its tracks—in the case of the polling protocol, the network stops if the controller dies; in the case of the token bus, a token passed to a dead device effectively gets lost. In the latter case, one can detect that a device failed to pass the token on and so amend the protocol to deal with similar failures.

Token-passing networks are "fair," in the sense that each device is assured its turn within each round. The overhead of access control is the time that the network spends on transmitting the token (rather than data) and the time that the network is idle long enough for a device to ascertain that a transmission has ended or is not going to occur. An important characteristic of token-passing protocols is that the throughput (bits per second of useful traffic) is monotone nondecreasing as a function of the "offered load" (traffic that the network is requested to carry).

To illustrate this point, Figure 14.11 plots data from a set of experiments on a modeled 10 Mbits (10 million bits per second) network, with 10 devices, evenly spaced, having a latency delay of 25.6 μsec between the most distant pair. (We use this figure in order to compare this network with one managed by using Ethernet, later.) Five different experiments are displayed on the graph; right now, we are interested only in the one labeled "token bus, Poisson." The experiments assume that the data frame is 1500 bytes long and that the token is 10 bytes long. They assume that once a device gains the token, it may send at most one frame and then must release the token. This set of data uses a Poisson process to generate frame arrivals. The x-axis gives the offered load, measured here as the total sum of bits presented to the network before the simulation end time, divided by the length of the simulation run. The y-axis plots the measured throughput. For each off-time rate, we run 10 independent experiments. For each experiment, we plot the observed pair (offered load, throughput). For the experiment of interest, the throughput increases linearly with the offered load, right up to network saturation. It is interesting to note, however, the impact of a change in the traffic-arrival pattern.

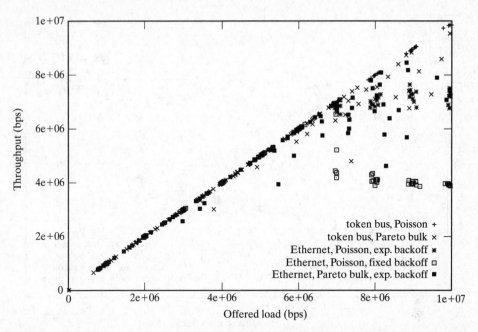

Figure 14.11 Throughput versus offered load, for token bus and Ethernet MAC protocols, Poisson and bulk Pareto arrival processes, exponential and fixed backoff (for Ethernet)

We replaced the Poisson arrival process with the arrival process that defines PPBP (Section 14.3.2), a Poisson bulk arrival process, where the number of frames in each bulk arrival is a truncated Pareto. We use the same Pareto parameters as before ($a = 1.1$ and $b = 10$) and reduced the arrival by a factor of the inverse Pareto mean ($(a - 1)/(ab)$) to obtain the same average bit-arrival rate. The set of data points associated with the label "token bus, Pareto bulk" reflect the impact of this change. Throughput grows linearly with offered load until the bus is roughly 60% utilized. For larger loads, we begin to see some deviations from linear. For a point (x, y) off the diagonal, the difference between x and y reflects the volume of unserved frames at the end of the simulation—the frames in queue. This is no surprise; queueing theory tells us that we should expect significant queue lengths when the arrival pattern is highly variant.

Another important aspect is the average time a frame awaits transmission after arrival. Knowledge of queueing theory and the protocol's operation identifies two factors that ought to contribute to growth in the queueing length. One factor is the time required by a token to reach a new frame arrival. As the offered load increases, the amount of work that the token encounters and must serve prior to reaching the new arrival increases linearly. A second factor is from queueing theory; the view from a station is of an $M/G/1$ queue. In this view, the service time incorporates the time spent waiting for a token to arrive, a mean that increases with the offered load. A job's average time in an $M/G/1$ queue grows with $1/(1 - \rho)$, where $\rho = \lambda/\mu$ is the ratio of arrival rate to service-completion rate. As the offered load grows, ρ increases; this fact explains the second factor of waiting-time growth. As ρ approaches unity, the asymptotic waiting time increases rapidly.

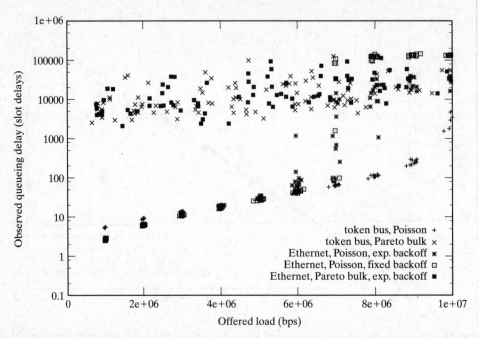

Figure 14.12 Average queue delay versus offered load: for token bus and Ethernet MAC protocols, for Poisson and bulk Pareto arrival processes, and for exponential and fixed backoff (for Ethernet).

Figure 14.12 confirms this intuition, plotting the average time a frame waits in queue between the time of its arrival and the time at which it begins transmission. Again, we execute 10 independent experiments for a given offered load and plot the raw pair (x, q), where x is the average number of bits presented to the network per second in that run and q is the average time a job is enqueued. Units of queueing delay are *slot times*, the length of time required for a bit to traverse a cable at the limit of what is permissible for Ethernet (25.6 μsec). The extreme range of queueing delays observed for the five experiment types encourages our use of a log scale on the y-axis. Tracking data from the experiments by using Poisson arrivals, we see stability in the growth pattern, up to the point where the bus is fully saturated. We know to expect extremes there. What is very interesting, though, are the extremely high average queueing delays experienced under the "bulk Pareto" assumptions. If nothing else, these kinds of experiments point out the importance of the traffic model when analyzing of network behavior.

A straight forward implementation of a token-bus protocol models devices, the bus, and the explicit and continuous passing of the token among stations. However, such an implementation has an undesirable characteristic: under low traffic load, the model creates a discrete event approximately every 10.84 μsec, the time it takes to transmit a token between adjacent stations. Under low-traffic load, the token could completely cycle through the network many times before reaching a point in simulation time when there is a frame available for transmission. Unless the simulation has some particular reason for pushing the token around an otherwise idle network (e.g., if, at each hop, there is a nonzero probability of the token's being lost or corrupted, forcing the protocol to detect and react), there are more efficient ways of executing the simulation, at the cost of incorporating extra logic. We may suppose that each device samples the next future time at which a batch of frames arrives. Before that time, if the device has no frames to transmit, it will make no further demands on the network. When the simulation has reached a time at which no frame is being transmitted and no device has a frame waiting for transmission, we perform a calculation to advance simulation time past during which the only activity is token passing. Because the time required to circulate the token around the ring is computable, and the next time at which a frame is available at any station is known, we can advance simulation time to the cycle in which the next frame is transmitted and save ourselves the computational effort of getting to that place by pushing a token around.

14.7.2   Ethernet

Token-based access protocols have been popular, but they have drawbacks when it comes to network management. In particular, every time a device is added to or removed from the network, configuration actions must be taken to ensure that a new device gets the token and that a removed device is never again sent the token. The Ethernet access protocol is a solution to this problem (Spurgeon [2000]). A device attached to an Ethernet cable has no specific idea of other devices on that cable; however, when it wants to use the cable, it must coordinate with such other devices. Consider the problem—a device has a frame to send; when can it send it? Ethernet is a decentralized protocol, meaning that there is no controller granting access. A device can "listen" to the Ethernet cable to see whether it is currently in use. If the cable is already in use, the device holds off until the cable is free. However, two or more devices could independently and more or less simultaneously decide to transmit, shortly after which the transmission on the cable is garbled. Both devices can detect this "collision"—by comparing what they are transmitting on the cable with what they are receiving from the cable. Collision detection and reaction to it is one of the key components of the Ethernet protocol; it is a so-called CSMA/CD (carrier sense multiple access/collision detection) protocol.

The format of an Ethernet frame is illustrated in Figure 14.13. The 8-byte preamble is a special sequence of bits (alternating 1s and 0s, except for the last bit, which is also a 1) that listeners on the cable recognize and use to prepare to examine the next frame field, a 6-byte destination address that may specify one device, a group of devices, or a broadcast to all listening devices. After scanning the full Destination address, a device listening to the cable knows whether it is an intended recipient. The next 6 bytes identify the sending device; then comes a 2-byte field describing the number of data bytes. The data follow, and the frame is terminated with a 4-byte code used for error detection.

When a device decides to transmit, it begins in the knowledge that it is possible for another device to begin also, not yet having heard the new transmission. Ethernet specifications on network design ensure that any transmission will be heard by another device within $\delta = 25.6\ \mu$sec. This is called a slot time. The worst case is that the device begins to transmit at time t, yet before time $t + \delta$,

size (bytes)	8	6	6	2		4
field	Preamble	Destination MAC adrs	Source MAC adrs	Length	Data	Cyclic Redundancy Check

Figure 14.13   Format of an Ethernet frame

a device at the other end of the cable decides to transmit and does so just before time $t + \delta$, and another δ time is needed by the first device to detect the collision.

The length of the data portion of an Ethernet frame is not specified by the protocol. However, there is a lower bound on the allowable length of the data portion. The frame must be large enough so that it takes longer than 2 slot times to transmit it. This bound ensures that, if a collision does occur, the sending device will be transmitting when the effects of the collision reach it, and hence it can detect the collision. This minimum is 46 bytes of data; furthermore, a frame is not permitted to carry more than 1500 bytes of data.

Some of the complexities of Ethernet exist because of physics. An accurate simulation of Ethernet must therefore pay attention to the delicacies of signal latency. The model used to generate Ethernet performance figures specifically accounts for signal latency. It assumes that the devices are evenly spaced along a cable that requires a full slot time (25.6 μsec) for a signal to traverse. When a device listens to the cable to see whether it is free, the model really answers the question of whether the device can, at that instant, hear any transmission that might have already started. This is a matter of measuring the distance between a sending device and a listening device, computing the signal latency time between them, and working out whether the sender started longer ago than that latency. Likewise, when a device has a frame to send and is listening to the cable to find out when it is idle, its view of the cable state is one that accounts for a certain delay between when a transmission ends and when that end is seen by an observer.

A device with a frame to send listens to the cable and, if it hears nothing, begins to transmit. If it successfully transmits the frame without collision and has another frame to send, it waits 2 slot times before making the attempt. If a device wanting to send a frame hears that the cable is in use, it simply waits until the cable is quiet and then begins to transmit. The most interesting part of Ethernet is its approach to collisions. If a device transmitting a frame F detects a collision, it continues to transmit—but jumbled—long enough to ensure that it transmits a full minimum frame's worth of bits. This "jamming" ensures that all devices on the cable detect the collision. Next, it backs off and waits a while before trying to send F again.

The backoff period following a collision has been a topic of some study, one in which simulation has played an important role. If the backoff time is short, there is a chance of not overly increasing the delay time of a frame, but there is also a significant chance of incurring another collision. On the other hand, if the backoff time is large, one reduces the risk of a subsequent collision, but ensures that the delay of the frame in the system will be large. Over time, the following strategy, called "exponential backoff", has become the Ethernet standard. Following the mth collision while attempting to transmit frame F, the device randomly samples an integer k from $[0, 2^m - 1]$, and waits $2k$ slot times before making another attempt. If 10 attempts are made without success, the frame is simply dropped. The term *exponential backoff* describes the doubling in length of the mean backoff time on each successive collision. Successive collisions are measures, of a sort, of the level of congestion in the network. A device strives to reduce its contribution to the congestion, and thereby enable other frames to get through and relieve the congestion.

Simulation is a useful tool to investigate both backoff schemes and other variants of Ethernet that one might consider. We did experiments (assuming Poisson arrivals) on exponential backoff and on "fixed" backoff—where, after a collision occurs, the sender chooses $k \in [0, 4]$ slot times to wait, uniformly at random. Figure 14.11 illustrates the effects on throughput. Under exponential backoff, throughput increases linearly with offered load until after about 60% utilization. For greater load, throughput hovers in the 70% of available bandwidth, without significant degradation. The story is quite different under fixed backoff. When offered load is 70% of the network bandwidth, the throughput plummets from 60+% and settles down at around 40% of bandwidth—under higher load, the network delivers poorer service. Queueing delays are affected too, as one would expect. Under high load, the delays under fixed backoff are an order of magnitude larger than those under exponential backoff.

A final set of experiments used the same Poisson bulk arrival process, with Pareto-based bulk arrivals, assuming exponential backoff. The results are similar to those for the token bus: large and highly variable queueing delays, and some deviation of throughput from linear at high load. This set of experiments suggests that Ethernet may be more sensitive to the Pareto's high variance than is the token-bus protocol.

14.8   Data Link Layer

A network is far more complicated than the single channel seen by a MAC protocol. A frame might be sent and received many times, by many devices, before it reaches its ultimate destination. Consequently, data traveling at the physical layer contains at least two addresses. One address is a hardware address of the intended endpoint of the current hop. This address (like an Ethernet address) is recognizable by a device's network-interface hardware. The second address is the ultimate destination's network address, typically an IP address. Different types of devices make up the network. A *hub* is a device that simply copies every bit received on one interface to all its other interfaces. Hubs are useful for connecting separated networks, but have the disadvantage that the connection brings those networks into the same Ethernet collision domain.

A *bridge* makes the same sort of connection, but keeps component subnetworks in different collision domains. For every frame heard on one interface, the bridge takes the destination address and looks up in a table the interface through which that destination can be reached. The bridge has nothing to do if one reaches the destination through the same interface as that through which the frame was observed—the destination will recognize the frame for itself. However, if the destination is reached through a different interface, the bridge takes the responsibility of injecting the frame through that interface, moving it closer to its ultimate goal. In injecting the frame, the bridge acts like a source on that subnetwork, engaging in that subnetwork's MAC protocol. The bridge in effect moves a frame from one collision domain and puts it into another. It can also bridge different subdomain technologies (e.g., different types of Ethernet). Contexts where one would consider simulation study of MAC protocols on one subdomain include those where one would use simulation and involve models of bridges.

A bridge involves only the physical layer and the data link layer. There is a practical limit on devices retaining the physical addresses of other devices, particularly devices that are in different administrative domains.

A *router* is a device that can connect more widely dispersed networks, by making its connections at the Network Layer. A frame coming in to a router on one interface is pushed up to the IP layer, where the IP destination address is extracted; the IP address determines which interface should be used to forward the packet. The *forwarding tables* used to direct traffic flow are the result of complex *routing algorithms*, such as OSPF (Moy [1998]) and BGP (van Beijum [2002]). Simulation is frequently used to study variants and optimizations of these protocols.

We will see that commonly used network services provide users with delivery of data error-free and in the order it was sent. These attributes are provided in spite of the real possibility that data will be corrupted or lost in transmission. A router is one place where a frame might be lost, because, if the router experiences a temporary burst of traffic, all to be routed through a particular interface, buffers holding frames waiting to be forwarded could become exhausted. We think of the traffic flowing through a router as being a set of flows, each flow being defined by the source–destination pair involved. When the arrivals become bursty, and the router's buffer becomes saturated, arrivals that cannot be buffered are deliberately dropped. Most flows actively involved in the burst will lose frames. Under TCP, data loss is the signal that congestion exists, and TCP reacts by significantly decreasing the rate at which it injects traffic into the network. But it takes time to detect this loss—a lot more time than it takes to route frames through the router. One idea that has been studied extensively (by using simulation) is RED (random early detection, Floyd and Jacobson [1993]) queue management. The idea behind RED is to have a router continuously monitor the number of frames enqueued for transmission and, when the average length exceeds a threshold, proactively attempt to throttle back arrival rates before the arrivals overwhelm the buffer and cause all of the flows to suffer. RED visits each frame and, with some probability, either preemptively discards it, or marks a "congestion bit" that is available in the TCP header, but is not much used by most TCP implementations. RED chooses a few flows to suffer for the hoped-for sake of the network as a whole. Complexity abounds in finding effective RED parameters (e.g., threshold queue length, probability of dropping a visited frame) and in assessing tradeoffs and impacts that the use of RED could have. Simulation, of course, has played and will continue to play a key role in making these assessments.

14.9   TCP

The TCP (transport control protocol, Comer [2000]) establishes a connection between two devices, both of which view the communication as a stream of bytes. TCP ensures error-free, in-order delivery of that stream. As we have seen, data frames might be discarded (in response to congestion) somewhere between the sender and the receiver; TCP is responsible for recognizing when data loss occurs and for retransmitting data that have gone missing. TCP mechanics are focused on avoiding loss, detecting it, and rapidly responding to it. A number of TCP variants have been proposed and studied; all of these studies use simulation extensively to determine the protocol's behavior under different operating conditions.

Our discussion of TCP serves to illustrate further how different components of networking layers come together. Figure 14.14 illustrates data flow from a server to a client. Two applications intending to communicate establish *sockets* at each side. Sockets are viewed by the applications as buffers into which data could be written and out of which data might be read. Calls to sockets are sometimes blocking calls, in the sense that, if a socket buffer cannot accept more data on a write, or has no data to provide on a read, the calling processes block. On the server side, the TCP implementation is responsible for removing data from the socket's buffer and sending it down through the protocol stack to the network. Once on the network, the data pass through different devices. In this figure, we illustrate a bridge (which involves remapping of hardware addresses and does not look at the IP address) and a router (which must decode the IP address to find out the interface through which the data is passed). The client host's IP recognizes that the data ought to go up the stack to TCP,

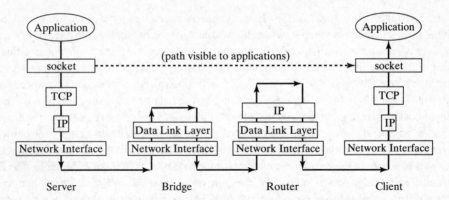

Figure 14.14 Data flow from TCP sender to TCP receiver, passing through network devices.

and the client side TCP is responsible for releasing the data to the socket—but only a contiguous stream of data. If the router drops a frame of this flow, the client-side TCP must somehow detect and communicate this absence to the server-side TCP.

TCP partitions the data flow into *segments*. Figure 14.15 illustrates the header (in 32-bit words) that is placed around the data. First, note that the only addressing information is "port number" at the source and destination machines—IP is responsible for knowing (and remembering) the identity of the machines involved. From TCP's point of view, there is just a source and a destination. SeqN and

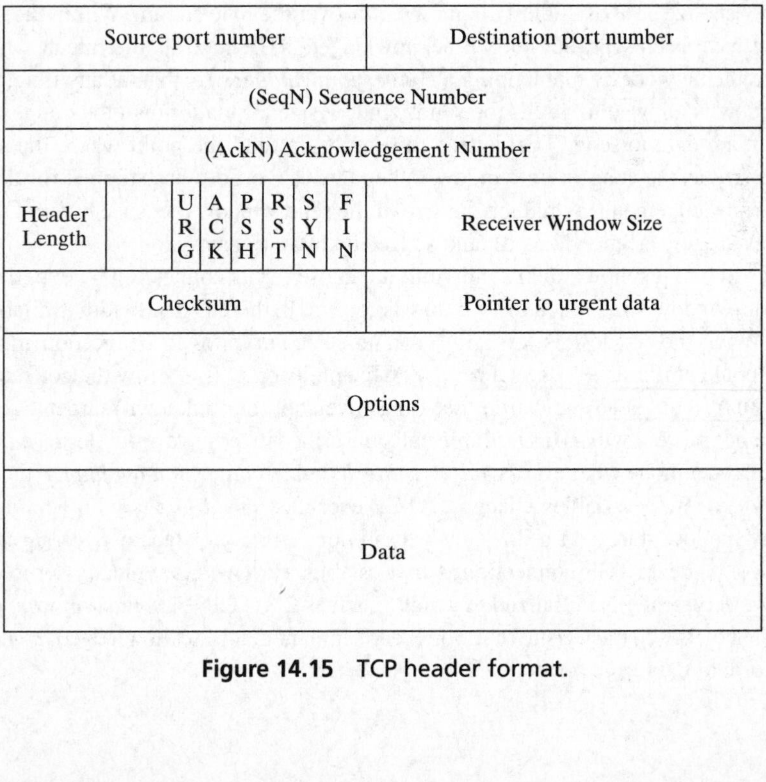

Source port number		Destination port number
(SeqN) Sequence Number		
(AckN) Acknowledgement Number		
Header Length	U R G / A C K / P S H / R S T / S Y N / F I N	Receiver Window Size
Checksum		Pointer to urgent data
Options		
Data		

Figure 14.15 TCP header format.

AckN are descriptors of points in the data flow, viewed as a stream of bytes, each numbered. SeqN is then the *sequence number* of the first byte in the segment. At the beginning of a connection, a sender and receiver agree upon an initial sequence number (usually random); the SeqN value is this initial number plus the byte index within the stream of the first byte carried in the segment. Because the segment size is fixed, the receiver can infer the precise subsequence of the byte stream contained in the segment. The AckN field is critical for detecting lost segments. Every time a TCP receiver sends a header about the flow (e.g., in accordance with acknowledgement rules), it puts into the AckN field the sequence number of the next byte it needs to receive in order to maintain a contiguous flow. Since TCP provides a contiguous data stream to the layer above, the value in AckN is the initial sequence number plus the index of the next byte it would provide to that layer, if it were available. The linkage of this value with packet loss is subtle. TCP requires a receiver to send an acknowledge for every segment it receives and requires a sender to detect within a certain time limit whether a segment it has sent has been acknowledged. Now imagine the effect if 3 segments are sent, and the second one is lost en route. Assume the initial sequence number is 0. The first segment is received, and the receiver sends back an acknowledgement with AckN equal to, say, 961 (and the ACK flag set to 1 to indicate that the AckN field is valid). The third segment is received, but the receiver notices that the value of SeqN is a segment larger than expected—it notices the hole. So it sends back an acknowledgement, but AckN in that header is again 961. The second segment sent is not acknowledged, of course, but interestingly, neither is the third. Eventually, the TCP sender times out while waiting for these acknowledgements and resends the unacknowledged packets. The only other field in this header, that is critical to our discussion, is the Receiver window size, which is included in an acknowledgement to report how many bytes of buffer are currently available to receive data from the sender.

One can visualize TCP as sliding a *send window* over the byte stream. Within the send window are bytes that have been sent, but not yet acknowledged. TCP controls the rate at which it injects segments into the network by maintaining a *congestion window size*, that, at any time, is the largest the send window is allowed to get. If the send-window size is smaller than the congestion-window size and there are data to send, TCP is free to send it, up until the point where the send window has the same size as the congestion window. When the TCP sender has stopped for this reason, an incoming acknowledgement can reduce the size of the send window (because bytes at the lower end of the window are now acknowledged), and so free up more transmission.

TCP tries to find just how much bandwidth it can use for its connection by experimenting with the congestion-window size. When the window is too small, there is bandwidth available but it isn't being used. When the window is too large, the sender contributes to congestion in the network, and the flow could suffer data loss as a result. TCP's philosophy is to grow the congestion window aggressively until there is an indication that it has overshot the (unknown) target size, then to fall back and advance more slowly. This is all formally described in terms of variables $cwnd$ and $ssthresh$. TCP is in *slow start* mode whenever $cwnd < ssthresh$, but in *congestion avoidance* mode whenever $cwnd > ssthresh$. Both variables change as TCP executes; $cwnd$ grows with acknowledgements a certain way in slow start, and a different way in congestion avoidance; $ssthresh$ changes when packets are lost. When a TCP connection is first established, $cwnd$ is typically set to one segment size and $ssthresh$ typically is initialized to a value such as 2^{16}. TCP starts in *slow-start* mode, which is distinguished by the characteristic that, for every segment that is acknowledged, $cwnd$ grows by a segment's worth of data.

Consider how *cwnd* behaves during slow start by thinking about TCP sending out segments in rounds. In the first round, it sends out one segment, then immediately stalls, because the send-window and congestion-window sizes are equal. When the acknowledgement eventually returns, the sender issues *two* segments as the second round—it replaces the segment that was acknowledged and sends another, because *cwnd* increased by 1. The sender stalls until acknowledgements come in. The *two* acknowledgements for the second round allow the sender to issue *four* segments: half of these due to replacing the ones acknowledged, the other half due to the one-per-acknowledgement increase of the *cwnd* rule during slow start. The number of segments issued thus doubles in successive rounds.

Any one of a number of things can halt the doubling of the number of segments sent each round. One is detection of packet loss, the effects of which are to set *ssthresh* to half the size of the send window, set the send-window size to zero, and set *cwnd* to allow retransmission of one segment (the one in the lost packet). Another way TCP ceases to double the number of segments sent each round is due to the rule that the congestion window may not be increased to exceed certain limits—an internally imposed buffer size at the sender side, or the size of the "receiver window"—the field in ACKs that reports how much space is available for new data. Finally, the doubling effect changes hence if *cwnd* grows to exceed *ssthresh*, and so puts TCP into congestion-avoidance mode. Within congestion-avoidance mode, *cwnd* increases, but much more slowly. Intuitively, *cwnd* increases by one segment for every full round that is sent and acknowledged (as opposed to increasing by one segment with every segment that is acknowledged). This is sometimes described as increasing *cwnd* by *1/cwnd* with every acknowledgement.

Simulation is an excellent tool for understanding how TCP works and many of the subtleties of its behavior; we now examine simple examples of that behavior. The first topology is that of a server, a client, and a 800 kbps link between them. The server is to send a 300000 byte file to the client. We attach a monitor that emits a tcpdump formatted trace (see www.tcpdump.org) of every TCP packet that passes (in either direction) through the server's network interface. Postprocessing of this trace yields information about how TCP variables of interest behave. In the first situation, we plot the values of SeqN in packets sent by the server and the values of AckN in packets sent by the receiver in response for the first six rounds, assuming an initial sequence number of 0. This is illustrated in Figure 14.16, where the Y-axis is logarithmic in order to illustrate interesting behavior at different scales. The TCP connection is requested by the client at time 192, the first step in TCP's three-way handshake that results in the server sending the first segment at time 192.3 (not actually shown in the graph, to allow higher resolution to later rounds). The SeqN in the header of that segment is 1, the index of the first byte in the segment. It takes approximately 100 ms for the segment to reach the client, and another 100 ms for the client's acknowledgment to reach the monitoring point, at time 192.5. (The exact figures are a little different, as they account for the transmission delay caused by the link bandwidth.) The ACK bit of that segment is set, and the AckN value in the header is 961—the index of the next byte the receiver expects to see. The server's send window now being empty, and *cwnd* having advanced from 1 to 2 by virtue of the received acknowledgement, the server immediately sends *two* segments, one with SeqN equal to 961, the next with SeqN equal to $961 + 960 = 1921$. The graph shows overlapping marks for byte index 961, one from the acknowledgement header, and one from the next segment the server sends. The delay between the server's sending of a segment and the ultimate acknowledgement of that segment is known as the RTT (round-trip time). In this example, the network is as simple as it can be, and the RTT is just the sum of the time to send a segment across the link plus the time to send an acknowledgement back—here, a value very close

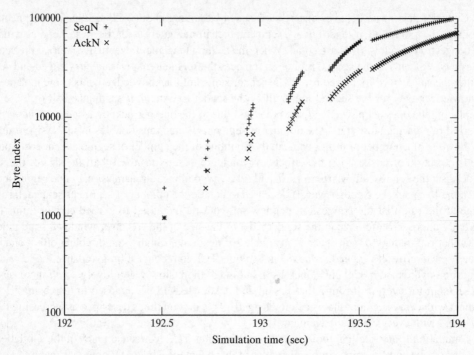

Figure 14.16   Early rounds of TCP connection on 800 kbps/100 ms link, with a tcpdump probe at the server's network interface.

to 200 ms. At times 192.3 and 192.5, the server stopped sending segments just as soon as its send window and the congestion window were the same size. After one RTT, acknowledgments from the previous round come in; they allow the server to double the number of segments sent from one round to the next. For rounds three and four (at times 192.7 and 192.9, approximately), the graph shows the slight staggering of times associated with acknowledgments coming in and new segments going out.

Figure 14.16 shows how, in slow-start mode, upon receiving a burst of acknowledgments, the server generates a burst of new segments. A moment's reflection shows that, if the acknowledgement for the first segment in that burst is received while the burst is continuing, then the burst will continue ad infinitum. This is because, at the instant that the critical acknowledgement is received, the send window must be smaller than the congestion window, and the send window will not grow after this point, while the congestion window will. We can compute the size of the congestion window when this phenomenon occurs—it is when the congestion window is large enough that the time needed to transmit that many bytes is precisely the RTT. Back-of-the-envelope calculations indicate that this is 20,000 bytes, or just under 21 segments. In these experiments, the receiver window is limited to 32 segments, so this saturation happens before the flow is limited by that buffer. SSFNet initializes *ssthresh* to 65,396 bytes, so this saturation point is reached in slow-start, before *cwnd* reaches *ssthresh* and can trigger congestion-avoidance mode. Since *cwnd* starts with value 1 and doubles with every round, the server saturates its sends in the middle of the 6th round. This is observed in Figure 14.16, in the round that starts just after time 193.5.

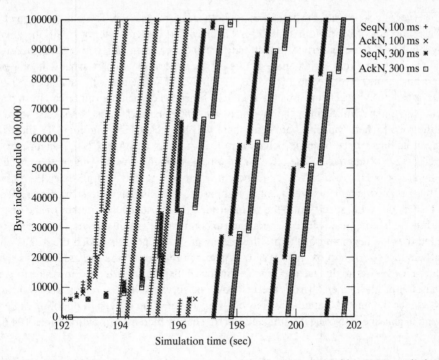

Figure 14.17 TCP connections between server and client: an 800 kbps/100-ms link and an 800 kbps/300-ms link.

In Figure 14.17, we illustrate this same experiment, along with another that is identical—save that the link latency is 300 ms. A larger span of simulation time is illustrated. There is an interesting kink in the SeqN data set for the 100 ms network, in the vicinity of $SeqN = 65K$. The "slope" of the data set decreases perceptibly. Up to this point, for every acknowledgement received two new segments are transmitted, and they are marked in the tcpdump trace as occurring at the same instant (SSFNet does not ascribe time advance to protocol actions, only to network transmission). At the point of the kink, the value of *cwnd* becomes equal to the receiver window, 32 segments. The sender window becomes limited by the size of the receiver window, rather than by *cwnd*, so, after the kink, there is a one-to-one correspondence between receipt of an acknowledgement and transmission of another segment. Now, consider the experiments using a 300-ms latency. As we would expect, rounds happen approximately every 600 ms. To saturate the link, the sender window has to become three times as large as in the first experiment—almost 64 segments. However, this will never happen, because the send window will be limited by the receiver window, at 32 segments. Indeed, we see that the change in slope of the SeqN trace happens at the same byte index as it did with the first experiment. Likewise, we see visually that there is a gap in transmission time between each successive round.

As a final example of how simulation illustrates the behavior of TCP, we consider an experiment designed to induce packet loss. The topology is that of a server, a router, and a client. Again, the server is to send 300,000 bytes to the client. Both server and client connect with the router. The link between server and router has 8 Mbps of bandwidth and 5-ms latency. The link between client and router has 800 kbps of bandwidth and 100-ms latency. The router's interface with the client has a 6000-byte buffer. If a packet arrives to that interface and there is insufficient buffer space available, the packet is dropped. From earlier analysis of TCP, we can foresee, in part, what will happen. In the slow-start phase, the server begins to double the number of segments with each successive round. However, it can push packets towards the router 10 times faster than the router can push packets to the client, so a queue will form at the interface. The buffer holds at most 6 packets, so we expect that, in the round where 8 packets are sent, there will be packet loss. Figure 14.18 illustrates this experiment, adding a trace of *cwnd* behavior to that of SeqN and AckN (once again measured at the server's network interface). The effects of the packet loss are visually distinctive. Around time 193.5, the server begins to receive a sequence of acknowledgments that all carry the same AckN value. These acknowledgments were sent in response to packets that were sent *after* a loss. Recall that TCP rules on AckN specify that the receiver identify the sequence number of the next byte it needs to receive in order to advance the sequence of contiguously received bytes; hence, the repeated AckN identifies the beginning of the first lost segment. At the point at which the loss is observed, the send-window size is approximately 25,000 bytes; in reaction to the loss, *ssthresh* is set to half this value, *cwnd* is set to 1, and the sender window collapses to size zero in order to cause the retransmission of all segments (from the first lost one forward). In the region between times 193 and 194, we see the impact that this loss has on *cwnd* and how the slow-start doubling of *cwnd* with each round begins anew. (Notice the small periods of sharply increased growth at times 194.6, 194.8,

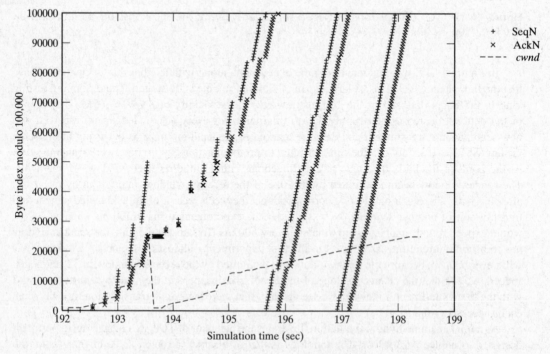

Figure 14.18   TCP connection suffering loss.

and 195.) However, this time, congestion-avoidance mode is entered when *cwnd* reaches *ssthresh*, shortly after time 195; thereafter, it grows more or less linearly with time. This particular transfer ends just before *cwnd* reaches a size that will allow loss once again; had the transfer advanced that far, TCP's treatment of *cwnd* would look very much like the period from 193.8 on.

As these simple examples show, TCP's relatively simple rules create complex behavior. Simulation is an indispensable tool for predicting how TCP will behave in any given context and for understanding that behavior.

14.10   Model Construction

Used in the previous section to look at how TCP behaves, SSFNet is a versatile tool for building and analyzing network simulations. Exercises at the end of this chapter encourage use of SSFNet, and so we describe the general process SSFNet uses in constructing a simulation from an input model. We then illustrate this process, in part, by describing the contents of one input file used in the last subsection. This is not a users' manual for SSFNet; very complete documentation exists at www.ssfnet.org. Our aim here is to give a sense of the approach and to encourage readers to investigate further.

14.10.1   Construction

Input to SSFNet is in the form of so-called DML (domain modeling language) files. At the simplest level, a DML file contains just a recursively defined list of attribute–value pairs, where an attribute is a string and a value may be either a string or a list of attribute–value pairs. This structure naturally induces a tree, where interior nodes are attributes (labeled with the attribute string name) and leaves are values of type *string* (rather than of type *list*). To illustrate, consider this DML list:

```
Net [
    frequency 1000000
    host [ id 0
          interface [ id 0 bitrate 800000]
          nhi_route [ dest 1(0) interface 0 ]
    ]
    host [ id 1
          interface [ id 0 bitrate 800000]
          nhi_route [ dest default interface 0 ]
    ]
    link [ attach 0(0) attach 1(0) latency 0.1 ]
  ]
```

This has some elements of SSFNet DML structure worth noting. Description of a network, elements within the network, and connections between them use a hierarchical naming convention known as the Network-Host-Interface convention, or just NHI. The network is defined in terms of links between interfaces, and each interface has an id number that is unique among all interfaces owned by a common host. That host has an id number that is unique among all hosts in a common net. Each net has an id, unique among all nets contained in the same parent net, and so on. The NHI address

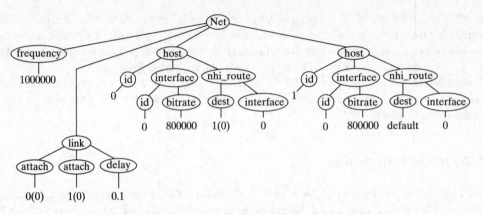

Figure 14.19   Parse tree of simple DML example.

0.1.2(4) refers to an interface named 4, within a host named 2, within a net named 1, within a net named 0. Within a net, a reference such as 2(4) is understood to mean interface 4 associated with the uniquely named host 2 within that understood net. The NHI address of an interface is derived from the nesting described within a DML file. The first interface to appear in the preceding example has NHI address 0(0); the second interface to appear has address 1(0). The link attribute in this example specifies two endpoints of the link, in NHI addressing (using the attach attribute), and a link latency of 100 ms.

The recursive structure of DML allows it be parsed easily and allows one to construct a parse tree whose interior nodes are attributes and whose leaves are string-valued values. The parse tree associated with the previous example is illustrated in Figure 14.19. This data structure gives a handy way of methodically building a model from a DML description. The SSFNet engine recursively traverses the tree and configures core SSFNet objects (such as host). Attributes or values within the tree can be referenced globally by the sequence of attribute labels on nodes from the root to the target. This proves to be useful: one can embed in a DML file a "library" of attribute–value pairs and reference elements of that library.

SSFNet recognizes a variety of attributes, many of which are described in Table 14.1.

14.10.2   DML Example

Finally, we illustrate some of these ideas by looking at the DML input file for one of our TCP examples. The file is presented in Figure 14.20 (annotated with line numbers for easier reference).

In this particular file, lines 1-8 are comments describing the architecture. Line 10 tells the SSFNet model parser where to find format descriptions of certain constructs; when the parser encounters these constructs in the DML file, it will check against the schema to ensure format correctness. Line 12 starts the overarching list: Net followed by a list. Line 14 specifies a clock resolution of 1 microsecond. Lines 15–20 describe the network's traffic, a single pattern that includes host 0 as client. The servers attribute gives a list of servers, in this case a single one at NHI address 1(0) (meaning host 1, interface), using port 10.

Table 14.1   Common Attributes in SSFNet DML Models

Attribute	Value
net	list describing a network
frequency	number of discrete ticks per simulation second
traffic	list of traffic patterns
pattern	description of traffic pattern, in terms of receiver (client) and server (sender).
servers	list describing a set of servers to which a client might connect—including their NHI identities and port numbers
link	list describing interfaces to be connected, and associated latency
host	list describing a host, and diverse attributes it may have
graph	list of protocols in a host's protocol stack
ProtocolSession	list specifying a protocol
interface	a list describing a connection to the network; attributes include connection bandwidth and target file for storing monitoring information
route	description of a forwarding table entry for IP. The *dest* attribute identifies the destination being described; the interface attribute describes which interface packets for that destination should be routed
dictionary	a list of constants that can be referenced elsewhere within the DML file

The link attribute at line 24 describes two interfaces to be connected: the one at NHI address 0(0), and the one at NHI address 1(0). The latency across this link is specified to be 0.1 seconds.

A host contains protocols and interfaces to the network. The host beginning at line 28 is given NHI id 1 and contains a graph of protocol sessions. Each model of a software component is described as such a session. The order of appearance in the graph is important, descending from higher to lower in the stack. Each protocol session describes its type (e.g., server, client, TCP, IP), and the Java class that describes its behavior. By using certain methologies, these classes are constructed to be composable; builders of simulation models (in contrast to developers of modeling components these builders use) need not develop new classes, but the methodology specifies how one does. A protocol session of a given type may include attributes specific to that type. For example, the tcpServer protocol beginning at line 32 specifies the port through which it is accessible (10). Line 37 begins the declaration of the tcpSessionMaster, a component that manages all TCP sessions. Characteristics of its version of TCP are described by including a list of attributes defined in a list held elsewhere in the DML file. The statement _find .dictionary.tcpinit causes the contents of the named list essentially to be inserted at the point of the statement. The string .dictionary.tcpinit names the list in terms of how to find it in the file: . is the highest level list, dictionary is the name of an attribute in that list, tcpinit is the attribute associated with the sought list, an attribute of the value-list of dictionary. This list starts at line 82.

We quickly describe the meaning of each attribute not obvious from the comments, in order to illustrate the diversity of parameters in SSFNet's implementation of TCP.

RcvWndSize, SendWndSize, and SendBufferSize describe units of *MSS* and limit buffer usage (which affects TCP behavior, as we have already seen). A missing segment will be retransmitted up to MaxRexmitTimes times before the TCP session is aborted. TCP_SLOW_INTERVAL and TCP_FAST_INTERVAL give timer values used to determine that a transmitted segment has

```
     # basic3a.dml
     #
     # topology
     # client              ...              server
5    #              800kb  100ms
     #  0(0) ··································· 1(0)
     #                                          |
     #                                        tcpdump

10   _schema [ _find .schemas.Net ]

     Net [

         frequency 1000000
15       traffic [
             pattern [
                 client 0
                 servers [ port 10 nhi 1(0) ]
             ]
20       ]

     # ·············································· LINKS

         link[attach 0(0) attach 1(0) delay 0.1]
25   # ·············································· SERVER

         host [
             id 1
30           graph [
                 ProtocolSession [name server use SSF.OS.TCP.test.tcpServer port 10
                     request_size 4
                     show_report  true
                     debug        false
35               ]
                 ProtocolSession [name socket use SSF.OS.Socket.socketMaster]
                 ProtocolSession [name tcp use SSF.OS.TCP.tcpSessionMaster
                     _find .dictionary.tcpinit
                     debug    true
40                  rttdump  "basic3a.rtt"
                     rexdump  "basic3a.rto"
                     cwnddump "basic3a.wnd"
                     con_count "basic3a.con_count"
                 ]
45               ProtocolSession [name ip use SSF.OS.IP]
             ]
             interface [id 0 bitrate 800000 tcpdump "basic3a.tcpdump"]
             route [dest default interface 0 ]
         ]
50   # ·············································· CLIENT

         host [
             id 0
55           graph [
                 ProtocolSession [
                     name client use SSF.OS.TCP.test.tcpClient
                     start_time 192      # earliest time to send request to server
                     start_window 0      # start time jitter window size
60                  file_size 300000    # file size
                     request_size  4     # client request datagram size (bytes)
                     show_report true    # print client-server session summary report
                     debug false         # print verbose client/server diagnostics
                 ]
65               ProtocolSession [name socket use SSF.OS.Socket.socketMaster]
                 ProtocolSession [name tcp use SSF.OS.TCP.tcpSessionMaster
                         _find .dictionary.tcpinit
                 ]
                 ProtocolSession [name ip use SSF.OS.IP]
70           ]
             interface [id 0 bitrate 800000]
             route [dest default interface 0 ]
         ]
     ]
75



80   dictionary [
             tcpinit[
                 ISS 0                  # initial sequence number
                 MSS 960                # maximum segment size
85               RcvWndSize  32         # receive buffer size
                 SendWndSize 32         # maximum send window size
                 SendBufferSize 32      # send buffer size
                 MaxRexmitTimes 12      # maximum retransmission times before drop
                 TCP_SLOW_INTERVAL 0.5  # granularity of TCP slow timer
90               TCP_FAST_INTERVAL 0.2  # granularity of TCP fast(delayack) timer
                 MSL 60.0               # maximum segment lifetime
                 MaxIdleTime 600.0      # maximum idle time for drop a connection
                 delayed_ack false      # delayed ack option
                 fast_recovery true     # implement fast recovery algorithm
95               show_report true       # print terse TCP connection diagnostics
             ]
     ]
```

Figure 14.20 Domain Modeling Language specification of a simple network experiment.

not yet been acknowledged within the required interval of time. If a TCP session is inactive for `MaxIdleTime` seconds, it is terminated. `delayed_ack` and `fast_recovery` are Boolean flags that describe whether to use particular optimizations known for TCP.

Back within the specification of the host (at lines 40–43), we find attributes whose values are files into which the system saves descriptions of how TCP variables behaved during the simulation. Following this (at line 40) is the inclusion of the IP protocol. In turn, this is followed by declaration of the server's single interface, given id 0 for NHI coordinates and specified to have a bandwidth of 800 Kbits per second. The last attribute for the server is an `nhi_route`, an element in IP's forwarding table, described in NHI coordinates. The server is not a router and so only needs to direct traffic from IP to one interface. The attribute–value pair `dest default` (line 48) says to route *everything* through the interface to follow, which is 0.

Specification of the second host is similar. In this case, the uppermost ProtocolSession is that of a client that requests data, through a socket. Attributes for the client include the simulation time at which it initiates the request. (It actually specifies a window of simulation time in which this occurs, to provide some jitter when multiple clients are to start more or less simultaneously). The length of the transfer being requested is an attribute (line 60). The rest of this host's ProtocolSessions are similar to the server's, although we do not save as much information about TCP's behavior at this host.

These simple examples illustrate how a compact DML format can be used to assemble networks with complex behaviors. Another advantage of the DML format is that one can easily write programs that assemble descriptions of fairly large networks, a capability that has proven to be very useful to us in simulation experiments that examine how certain network protocols scale up with larger sized networks.

14.11   Summary

This chapter outlined fundamental implementation issues behind networked computer-system simulators—principally, how process orientation is implemented and how object-oriented concepts such as inheritance are fruitfully employed. Input modeling—at different levels of abstraction—is a crucial element of simulating and modeling networks. We emphasized the importance of non-Poisson arrival models; in some cases to better match characteristics of specific applications, in others to be sure to explain and capture long-range dependence.

We discussed mobility models commonly used in simulations of ad-hoc mobile networks. We turned next to the OSI stack and discussed simulation models, and the use of simulation at the lower layers. We described commonly used models of radio propagation to model the Physical Layer in wireless networks. We then focused on the data link layer and on the media access control algorithms. We examined the token-bus and Ethernet protocols, discussed subleties of their simulation, and showed by example how significant an impact traffic-model assumptions can have on network performance. Following this, we mentioned issues at the data link layer for which simulation has been a critical tool for investigation.

Much of the traffic on the Internet is carried by using TCP. We described TCP's basic rules and used simulation to illustrate some of the consequences of these rules. Finally, we sketched how one builds network models in the SSFNet simulator.

This chapter has barely scratched the surface of how one uses simulation to study networked computer systems. Our hope is that what we discuss leads a student to explore more deeply any one of a number of fascinating areas of networking that can be examined only with simulation.

REFERENCES

ASHENDEN, P.J. [2001], *The Designer's Guide to VHDL*, 2d ed., Morgan Kaufmann, San Fransisco.

BARFORD, P., AND M. CROVELLA [1998], "An Architecture for a WWW Workload Generator," *Proceedings of the 1998 SIGMETRICS Conference*, Madison, WI, pp. 151-160.

BLACK, U. [2001], *Voice Over IP*, Prentice Hall, Upper Saddle River, NJ.

COMER, D. [2000], *Networking with TCP/IP Vol. 1: Principles, Protocols, and Architecture*, 4th ed., Prentice Hall, Upper Saddle River, NJ.

COWIE, J., A. OGIELSKI, AND D. NICOL [1999], "Modeling the Global Internet," *Computing in Science and Engineering*, Vol. 1, No. 1, pp. 42–50.

FISCHER, W., AND K. MEIER-HELLSTERN [1993], "The Markov-Modulated Poisson Process (MMPP) Cookbook," *Performance Evaluation*, Vol. 18, No. 2, pp. 149–171.

FISHWICK, P. [1992], "SIMPACK: Getting Started with Simulation Programming in C and C++," *Proceedings of the 1992 Winter Simulation Conference*, Arlington, VA, Dec. 13–16, pp. 154–162.

FLOYD, S., AND V. JACOBSON [1993], "Random Early Detection Gateways for Congestion Avoidance," *IEEE/ACM Transactions on Networking*, Vol. 1, No. 4, pp. 397-413.

GRUNWALD, D. [1995], *User's Guide to Awesime-II*, Department of Computer Science, Univ. of Colorado, Boulder, CO.

ISSARLYKUL AND HOSSIAN [2008], *Introduction to Network Simulator NS2*, Springer, New York.

KARAGIANNIS T., M. FALOUTSOS, AND M. MOLLE [2003], "A User-Friendly Self-Similarity Analysis Tool," *ACM SIGCOMM Computer Communication Review*, Vol. 33, No. 3, pp. 81-93.

KUROSE, J., AND K. ROSS [2002], *Computer Networking: A Top-Down Approach Featuring the Internet*, 2d ed., Addison-Wesley, Reading, MA.

LITTLE, M.C., AND D.L. MCCUE [1994], "Construction and Use of a Simulation Package in C++," *C User's Journal*, Vol. 3, No. 12.

McKNOWN, J.W. AND R.L. HAMILTON [1991], "Ray-tracing as a Design Tool for Radio Networks," *IEEE Networks Magazine*, Vol 5, pp 27–30.

MOY, J. [1998],
  OSPF: Anatomy of an Internet Routing Protocol, Addison-Wesley, Reading, MA.

NUTT, G. [2004], *Operating Systems, A Modern Prespective*, 3d ed., Addison-Wesley, Reading, MA.

RAPPAPORT, T.S. [2002], *Wireless Communications*, Prentice-Hall, Upper Saddle River, NJ.

SPURGEON, C. [2000], *Ethernet: The Definitive Guide*, O'Reilly, Cambridge, MA.

SCHWETMAN, H. [1986], "CSIM: AC-Based, Process-Oriented Simulation Language," *Proceedings of the 1986 Winter Simulation Conference*, Washington DC, Dec 8–10, pp. 387–396.

SCHWETMAN, H.D. [2001], "CSIM 19: A Powerful Tool for Building Systems Models," *Proceedings of the 2001 Winter Simulation Conference*, Arlington VA, Dec. 9–12, pp. 250–255.

VAN BEIJNUM, I. [2002], *BGP: Building Reliable Networks with the Border Gateway Protocol*, O'Reilly, Cambridge, MA.

ZIMMERMAN, H. [1980], "OSI reference model—The ISO model of architecture for open system interconnection." *IEEE Transactions on Communications*, Vol. COM-28, No. 4, pp. 425–432.

ZUKEMAN, M., D. NEAME, AND R. ADDIE [2003], "Internet Traffic Modeling and Future Technology Implications," *Proceedings of the 2003 InfoCom Conference*, San Franciso, CA.

EXERCISES

1. Sketch the logic of an event-oriented model of an $M/M/1$ queue. Estimate the number of events executed when processing the arrival of 5000 jobs. On average, how many context switches does a process-oriented implementation of this queue incur if patterned after the SSF implementation of the single-server queue in Chapter 4?

2. For each of the systems listed, sketch the logic of a process-oriented model and of an event-oriented model. For both approaches, develop and simulate the model in any language:

   (a) a central-server queueing model: when a job leaves the CPU queue, it joins the I/O queue with shortest length.
   (b) a queueing model of a database system, that implements fork join: a job receives service in two parts. When it first enters the server, it spends a small amount of simulation time generating a random number of requests to disks. It then suspends (freeing the server) until such time as *all* the requests it made have finished, and then enqueues for its second phase of service, where it spends a larger amount of simulation time, before finally exiting. Disks may serve requests from various jobs concurrently, but serve them using FCFS ordering. Your model should report on the statistics of a job in service—how long (on average) it waits for phase 1, how long it waits on average for its I/O requests to complete, and how long it waits on average for service after its I/O requests complete.

3. Consider a Markov-Modulated Poisson Process with three states (OFF, ON, BURSTY) and having the following characteristics:

   (a) The MMP is in the OFF state for 25% of the time, on average;
   (b) The MMP is in the BURSTY state for 5% of the time, on average;
   (c) The MMP transitions from OFF to BURSTY with probability 0.05, and from OFF to ON with probability 0.95;
   (d) The MMP transitions from ON to OFF with probability 0.9, and from ON to BURSTY with probability 0.1;
   (e) The MMP transitions from BURSTY to ON with probability 0.5, and from BURSTY to OFF with probability 0.5.

   If the time spent in the OFF state is an exponential with mean 0.25, what are the means of the exponential times spent in states ON and BURSTY? Simulate and plot the results.

4. Survey the literature in models of Voice-over-IP traffic, and build a simulator that creates traffic load corresponding to one model of particular interest.

5. Create a Markov-Modulated Poisson Process and a Poisson-Pareto Burst Process that yield the same average bit-rate traffic demand. Acquire the SELFIS tool for analyzing long-range dependence (available for free), and compare traces from the MPP and PPBP models.

6. Develop simulations of movements of:

   (a) cell-phone users within a train system;
   (b) cell-phone users within a city highway system;
   (c) cell-phone users wandering around a mall;
   (d) university students using laptops on campus, connecting to wifi points.

A

Appendix

Table A.1   Random Digits and Random Numbers

94737	08225	35614	24826	88319	05595	58701	57365	74759
87259	85982	13296	89326	74863	99986	68558	06391	50248
63856	14016	18527	11634	96908	52146	53496	51730	03500
66612	54714	46783	61934	30258	61674	07471	67566	31635
30712	58582	05704	23172	86689	94834	99057	55832	21012
69607	24145	43886	86477	05317	30445	33456	34029	09603
37792	27282	94107	41967	21425	04743	42822	28111	09757
01488	56680	73847	64930	11108	44834	45390	86043	23973
66248	97697	38244	50918	55441	51217	54786	04940	50807
51453	03462	61157	65366	61130	26204	15016	85665	97714
92168	82530	19271	86999	96499	12765	20926	25282	39119
36463	07331	54590	00546	03337	41583	46439	40173	46455
47097	78780	04210	87084	44484	75377	57753	41415	09890
80400	45972	44111	99708	45935	03694	81421	60170	58457
94554	13863	88239	91624	00022	40471	78462	96265	55360
31567	53597	08490	73544	72573	30961	12282	97033	13676
07821	24759	47266	21747	72496	77755	50391	59554	31177
09056	10709	69314	11449	40531	02917	95878	74587	60906
19922	37025	80731	26179	16039	01518	82697	73227	13160
29923	02570	80164	36108	73689	26342	35712	49137	13482
29602	29464	99219	20308	82109	03898	82072	85199	13103
94135	94661	87724	88187	62191	70607	63099	40494	49069
87926	34092	34334	55064	43152	01610	03126	47312	59578
85039	19212	59160	83537	54414	19856	90527	21756	64783
66070	38480	74636	45095	86576	79337	39578	40851	53503
78166	82521	79261	12570	10930	47564	77869	16480	43972
94672	07912	26153	10531	12715	63142	88937	94466	31388
56406	70023	27734	22254	27685	67518	63966	33203	70803
67726	57805	94264	77009	08682	18784	47554	59869	66320
07516	45979	76735	46509	17696	67177	92600	55572	17245
43070	22671	00152	81326	89428	16368	57659	79424	57604
36917	60370	80812	87225	02850	47118	23790	55043	75117
03919	82922	02312	31106	44335	05573	17470	25900	91080
46724	22558	64303	78804	05762	70650	56117	06707	90035
16108	61281	86823	20286	14025	24909	38391	12183	89393
74541	75808	89669	87680	72758	60851	55292	95663	88326
82919	31285	01850	72550	42986	57518	01159	01786	98145
31388	26809	77258	99360	92362	21979	41319	75739	98082
17190	75522	15687	07161	99745	48767	03121	20046	28013
00466	88068	68631	98745	97810	35886	14497	90230	69264

To get a random number between 0 and 1, take a grouping of digits, for example 5 at a time, and place a decimal point in front.

Table A.2 Random Normal Numbers

0.23	−0.17	0.43	2.18	2.13	0.49	2.72	−0.18	0.42
0.24	−1.17	0.02	0.67	−0.59	−0.13	−0.15	−0.46	1.64
−1.16	−0.17	0.36	−1.26	0.91	0.71	−1.00	−1.09	−0.02
−0.02	−0.19	−0.04	1.92	0.71	−0.90	−0.21	−1.40	−0.38
0.39	0.55	0.13	2.55	−0.33	−0.05	−0.34	−1.95	−0.44
0.64	−0.36	0.98	−0.21	−0.52	−0.02	−0.15	−0.43	0.62
−1.90	0.48	−0.54	0.60	−0.35	−1.29	−0.57	0.23	1.41
−1.04	−0.70	−1.69	1.76	0.47	−0.52	−0.73	0.94	−1.63
−.78	0.11	−0.91	−1.13	0.07	0.45	−0.94	1.42	0.75
0.68	1.77	−0.82	−1.68	−2.60	1.59	−0.72	−0.80	0.61
−0.02	0.92	1.76	−0.66	0.18	−1.32	1.26	0.61	0.83
−0.47	1.04	0.83	−2.05	1.00	−0.70	1.12	0.82	0.08
−0.40	1.40	1.20	0.00	0.21	−2.13	−0.22	1.79	0.87
−0.75	0.09	−1.50	0.14	−2.99	−0.41	−0.99	−0.70	0.51
−0.66	−1.97	0.15	−1.16	−0.60	0.50	1.36	1.94	0.11
−0.44	−0.09	−0.59	1.37	0.18	1.44	−0.80	2.11	−1.37
1.41	−2.71	−0.67	1.83	0.97	0.06	−0.28	0.04	−0.21
1.21	−0.52	−0.20	−0.88	−0.78	0.84	−1.08	−0.25	0.17
0.07	0.66	−0.51	−0.04	−0.84	0.04	1.60	−0.92	1.14
−0.08	0.79	−0.09	−1.12	−1.13	0.77	0.40	0.69	−0.12
0.53	−0.36	−2.64	0.22	−0.78	1.92	−0.26	1.04	−1.61
−1.56	1.82	−1.03	1.14	−0.12	−0.78	−0.12	1.42	−0.52
0.03	−1.29	−0.33	2.60	−0.64	1.19	−0.13	0.91	0.78
1.49	1.55	−0.79	1.37	0.97	0.17	0.58	1.43	−1.29
−1.19	1.35	0.16	1.06	−0.17	0.32	−0.28	0.68	0.54
−1.19	−1.03	−0.12	1.07	0.87	−1.40	−0.24	−0.81	0.31
0.11	−1.95	−0.44	−0.39	−0.15	−1.20	−1.98	0.32	2.91
−1.86	0.06	0.19	−1.29	0.33	1.51	−0.36	−0.80	−0.99
0.16	0.28	0.60	−0.78	0.67	0.13	−0.47	−0.18	−0.89
1.21	−1.19	−0.60	−1.22	0.07	−1.13	1.45	0.94	0.54
−0.82	0.54	−0.98	−0.13	1.52	0.77	0.95	−0.84	2.40
0.75	−0.80	−0.28	1.77	−0.16	−0.33	2.43	−1.11	1.63
0.42	0.31	1.56	0.56	0.64	−0.78	0.04	1.34	−0.01
−1.50	−1.78	−0.59	0.16	0.36	1.89	−1.19	0.53	−0.97
−0.89	0.08	0.95	−0.73	1.25	−1.04	−0.47	−0.68	−0.87
0.19	0.85	1.68	−0.57	0.37	−0.48	−0.17	2.36	−0.53
0.49	0.32	−2.08	−1.02	2.59	−0.53	0.15	0.11	0.05
−1.44	0.07	−0.22	−0.93	−1.40	0.54	−1.28	−0.15	0.67
−0.21	−0.48	1.21	0.67	−1.10	−0.75	−0.37	0.68	−0.02
−0.65	−0.12	0.94	−0.44	−1.21	−0.06	−1.28	−1.51	1.39
0.24	−0.83	1.55	0.33	−0.59	−1.24	0.70	0.01	0.15
−0.73	1.24	0.40	−0.61	0.68	0.69	0.07	−0.23	−0.66
−1.93	0.75	−0.32	0.95	1.35	1.51	−0.88	0.10	−1.19
0.08	0.16	0.38	−0.96	1.99	−0.20	0.98	0.16	0.26
−0.47	−1.25	0.32	0.51	−1.04	0.97	2.60	−0.08	1.19

Table A.3   Cumulative Normal Distribution

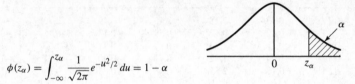

$$\phi(z_\alpha) = \int_{-\infty}^{z_\alpha} \frac{1}{\sqrt{2\pi}} e^{-u^2/2} \, du = 1 - \alpha$$

z_α	0.00	0.01	0.02	0.03	0.04	z_α
0.0	0.500 00	0.503 99	0.507 98	0.511 97	0.515 95	**0.0**
0.1	0.539 83	0.543 79	0.547 76	0.551 72	0.555 67	**0.1**
0.2	0.579 26	0.583 17	0.587 06	0.590 95	0.594 83	**0.2**
0.3	0.617 91	0.621 72	0.625 51	0.629 30	0.633 07	**0.3**
0.4	0.655 42	0.659 10	0.662 76	0.666 40	0.670 03	**0.4**
0.5	0.691 46	0.694 97	0.698 47	0.701 94	0.705 40	**0.5**
0.6	0.725 75	0.729 07	0.732 37	0.735 65	0.738 91	**0.6**
0.7	0.758 03	0.761 15	0.764 24	0.767 30	0.770 35	**0.7**
0.8	0.788 14	0.791 03	0.793 89	0.796 73	0.799 54	**0.8**
0.9	0.815 94	0.818 59	0.821 21	0.823 81	0.826 39	**0.9**
1.0	0.841 34	0.843 75	0.846 13	0.848 49	0.850 83	**1.0**
1.1	0.864 33	0.866 50	0.868 64	0.870 76	0.872 85	**1.1**
1.2	0.884 93	0.886 86	0.888 77	0.890 65	0.892 51	**1.2**
1.3	0.903 20	0.904 90	0.906 58	0.908 24	0.909 88	**1.3**
1.4	0.919 24	0.920 73	0.922 19	0.923 64	0.925 06	**1.4**
1.5	0.933 19	0.934 48	0.935 74	0.936 99	0.938 22	**1.5**
1.6	0.945 20	0.946 30	0.947 38	0.948 45	0.949 50	**1.6**
1.7	0.955 43	0.956 37	0.957 28	0.958 18	0.959 07	**1.7**
1.8	0.964 07	0.964 85	0.965 62	0.966 37	0.967 11	**1.8**
1.9	0.971 28	0.971 93	0.972 57	0.973 20	0.973 81	**1.9**
2.0	0.977 25	0.977 78	0.978 31	0.978 82	0.979 32	**2.0**
2.1	0.982 14	0.982 57	0.983 00	0.983 41	0.983 82	**2.1**
2.2	0.986 10	0.986 45	0.986 79	0.987 13	0.987 45	**2.2**
2.3	0.989 28	0.989 56	0.989 83	0.990 10	0.990 36	**2.3**
2.4	0.991 80	0.992 02	0.992 24	0.992 45	0.992 66	**2.4**
2.5	0.993 79	0.993 96	0.994 13	0.994 30	0.994 46	**2.5**
2.6	0.995 34	0.995 47	0.995 60	0.995 73	0.995 85	**2.6**
2.7	0.996 53	0.996 64	0.996 74	0.996 83	0.996 93	**2.7**
2.8	0.997 44	0.997 52	0.997 60	0.997 67	0.997 74	**2.8**
2.9	0.998 13	0.998 19	0.998 25	0.998 31	0.998 36	**2.9**
3.0	0.998 65	0.998 69	0.998 74	0.998 78	0.998 82	**3.0**
3.1	0.999 03	0.999 06	0.999 10	0.999 13	0.999 16	**3.1**
3.2	0.999 31	0.999 34	0.999 36	0.999 38	0.999 40	**3.2**
3.3	0.999 52	0.999 53	0.999 55	0.999 57	0.999 58	**3.3**
3.4	0.999 66	0.999 68	0.999 69	0.999 70	0.999 71	**3.4**
3.5	0.999 77	0.999 78	0.999 78	0.999 79	0.999 80	**3.5**
3.6	0.999 84	0.999 85	0.999 85	0.999 86	0.999 86	**3.6**
3.7	0.999 89	0.999 90	0.999 90	0.999 90	0.999 91	**3.7**
3.8	0.999 93	0.999 93	0.999 93	0.999 94	0.999 94	**3.8**
3.9	0.999 95	0.999 95	0.999 96	0.999 96	0.999 96	**3.9**

continues...

Table A.3   Continued

z_α	0.05	0.06	0.07	0.08	0.09	z_α
0.0	0.519 94	0.523 92	0.527 90	0.531 88	0.535 86	**0.0**
0.1	0.559 62	0.563 56	0.567 49	0.571 42	0.575 34	**0.1**
0.2	0.598 71	0.602 57	0.606 42	0.610 26	0.614 09	**0.2**
0.3	0.636 83	0.640 58	0.644 31	0.648 03	0.651 73	**0.3**
0.4	0.673 64	0.677 24	0.680 82	0.684 38	0.687 93	**0.4**
0.5	0.708 84	0.712 26	0.715 66	0.719 04	0.722 40	**0.5**
0.6	0.742 15	0.745 37	0.748 57	0.751 75	0.754 90	**0.6**
0.7	0.773 37	0.776 37	0.779 35	0.782 30	0.785 23	**0.7**
0.8	0.802 34	0.805 10	0.807 85	0.810 57	0.813 27	**0.8**
0.9	0.824 94	0.831 47	0.833 97	0.836 46	0.838 91	**0.9**
1.0	0.853 14	0.855 43	0.857 69	0.859 93	0.862 14	**1.0**
1.1	0.874 93	0.876 97	0.879 00	0.881 00	0.882 97	**1.1**
1.2	0.894 35	0.896 16	0.897 96	0.899 73	0.901 47	**1.2**
1.3	0.911 49	0.913 08	0.914 65	0.916 21	0.917 73	**1.3**
1.4	0.926 47	0.927 85	0.929 22	0.930 56	0.931 89	**1.4**
1.5	0.939 43	0.940 62	0.941 79	0.942 95	0.944 08	**1.5**
1.6	0.950 53	0.951 54	0.952 54	0.953 52	0.954 48	**1.6**
1.7	0.959 94	0.960 80	0.961 64	0.962 46	0.963 27	**1.7**
1.8	0.967 84	0.968 56	0.969 26	0.969 95	0.970 62	**1.8**
1.9	0.974 41	0.975 00	0.975 58	0.976 15	0.976 70	**1.9**
2.0	0.979 82	0.980 30	0.980 77	0.981 24	0.981 69	**2.0**
2.1	0.984 22	0.984 61	0.985 00	0.985 37	0.985 74	**2.1**
2.2	0.987 78	0.988 09	0.988 40	0.988 70	0.988 99	**2.2**
2.3	0.990 61	0.990 86	0.991 11	0.991 34	0.991 58	**2.3**
2.4	0.992 86	0.993 05	0.993 24	0.993 43	0.993 61	**2.4**
2.5	0.994 61	0.994 77	0.994 92	0.995 06	0.995 20	**2.5**
2.6	0.995 98	0.996 09	0.996 21	0.996 32	0.996 43	**2.6**
2.7	0.997 02	0.997 11	0.997 20	0.997 28	0.997 36	**2.7**
2.8	0.997 81	0.997 88	0.997 95	0.998 01	0.998 07	**2.8**
2.9	0.998 41	0.998 46	0.998 51	0.998 56	0.998 61	**2.9**
3.0	0.998 86	0.998 89	0.998 93	0.998 97	0.999 00	**3.0**
3.1	0.999 18	0.999 21	0.999 24	0.999 26	0.999 29	**3.1**
3.2	0.999 42	0.999 44	0.999 46	0.999 48	0.999 50	**3.2**
3.3	0.999 60	0.999 61	0.999 62	0.999 64	0.999 65	**3.3**
3.4	0.999 72	0.999 73	0.999 74	0.999 75	0.999 76	**3.4**
3.5	0.999 81	0.999 81	0.999 82	0.999 83	0.999 83	**3.5**
3.6	0.999 87	0.999 87	0.999 88	0.999 88	0.999 89	**3.6**
3.7	0.999 91	0.999 92	0.999 92	0.999 92	0.999 92	**3.7**
3.8	0.999 94	0.999 94	0.999 95	0.999 95	0.999 95	**3.8**
3.9	0.999 96	0.999 96	0.999 96	0.999 97	0.999 97	**3.9**

Source: W. W. Hines and D. C. Montgomery, *Probability and Statistics in Engineering and Management Science*, 2d ed., © 1980, pp. 592–3. Reprinted by permission of John Wiley & Sons, Inc., New York.

Table A.4   Cumulative Poisson Distribution

x	.01	.05	.1	.2	.3	.4	.5	.6	.7	.8	.9	x
				α = Mean								
0	.990	.951	.905	.819	.741	.670	.607	.549	.497	.449	.407	0
1	1.000	.999	.995	.982	.963	.938	.910	.878	.844	.809	.772	1
2		1.000	1.000	.999	.996	.992	.986	.977	.966	.953	.937	2
3				1.000	1.000	.999	.998	.997	.994	.991	.987	3
4						1.000	1.000	1.000	.999	.999	.998	4
5									1.000	1.000	1.000	5

x	1.0	1.1	1.2	1.3	1.4	1.5	1.6	1.7	1.8	1.9	2.0	x
					α = Mean							
0	.368	.333	.301	.273	.247	.223	.202	.183	.165	.150	.135	0
1	.736	.699	.663	.627	.592	.558	.525	.493	.463	.434	.406	1
2	.920	.900	.879	.857	.833	.809	.783	.757	.731	.704	.677	2
3	.981	.974	.966	.957	.946	.934	.921	.907	.891	.875	.857	3
4	.996	.995	.992	.989	.986	.981	.976	.970	.964	.956	.947	4
5	.999	.999	.998	.998	.997	.996	.994	.992	.990	.987	.983	5
6	1.000	1.000	1.000	1.000	.999	.999	.999	.998	.997	.997	.995	6
7					1.000	1.000	1.000	1.000	.999	.999	.999	7
8									1.000	1.000	1.000	8

x	2.2	2.4	2.6	2.8	3.0	3.5	4.0	4.5	5.0	5.5	6.0	x
					α = Mean							
0	.111	.091	.074	.061	.050	.030	.018	.011	.007	.004	.002	0
1	.355	.308	.267	.231	.199	.136	.092	.061	.040	.027	.017	1
2	.623	.570	.518	.469	.423	.321	.238	.174	.125	.088	.062	2
3	.819	.779	.736	.692	.647	.537	.433	.342	.265	.202	.151	3
4	.928	.904	.877	.848	.815	.725	.629	.532	.440	.358	.285	4
5	.975	.964	.951	.935	.916	.858	.785	.703	.616	.529	.446	5
6	.993	.988	.983	.976	.966	.935	.889	.831	.762	.686	.606	6
7	.998	.997	.995	.992	.988	.973	.949	.913	.867	.809	.744	7
8	1.000	.999	.999	.998	.996	.990	.979	.960	.932	.894	.847	8
9		1.000	1.000	.999	.999	.997	.992	.983	.968	.946	.916	9
10				1.000	1.000	.999	.997	.993	.986	.975	.957	10
11						1.000	.999	.998	.995	.989	.980	11
12							1.000	.999	.998	.996	.991	12
13								1.000	.999	.998	.996	13
14									1.000	.999	.999	14
15										1.000	.999	15
16											1.000	16

continues...

Table A.4 Continued

x	\multicolumn{11}{c	}{$\alpha = \text{Mean}$}	x
	6.5	7.0	7.5	8.0	9.0	10.0	12.0	14.0	16.0	18.0	20.0	
0	.002	.001	.001									0
1	.011	.007	.005	.003	.001							1
2	.043	.030	.020	.014	.006	.003	.001					2
3	.112	.082	.059	.042	.021	.010	.002					3
4	.224	.173	.132	.100	.055	.029	.008	.002				4
5	.369	.301	.241	.191	.116	.067	.020	.006	.001			5
6	.527	.450	.378	.313	.207	.130	.046	.014	.004	.001		6
7	.673	.599	.525	.453	.324	.220	.090	.032	.010	.003	.001	7
8	.792	.729	.662	.593	.456	.333	.155	.062	.022	.007	.002	8
9	.877	.830	.776	.717	.587	.458	.242	.109	.043	.015	.005	9
10	.933	.901	.862	.816	.706	.583	.347	.176	.077	.030	.011	10
11	.966	.947	.921	.888	.803	.697	.462	.260	.127	.055	.021	11
12	.984	.973	.957	.936	.876	.792	.576	.358	.193	.092	.039	12
13	.993	.987	.978	.966	.926	.864	.682	.464	.275	.143	.066	13
14	.997	.994	.990	.983	.959	.917	.772	.570	.368	.208	.105	14
15	.999	.998	.995	.992	.978	.951	.844	.669	.467	.287	.157	15
16	1.000	.999	.998	.996	.989	.973	.899	.756	.566	.375	.221	16
17		1.000	.999	.998	.995	.986	.937	.827	.659	.469	.297	17
18			1.000	.999	.998	.993	.963	.883	.742	.562	.381	18
19				1.000	.999	.997	.979	.923	.812	.651	.470	19
20					1.000	.998	.988	.952	.868	.731	.559	20
21						.999	.994	.971	.911	.799	.644	21
22						1.000	.997	.983	.942	.855	.721	22
23							.999	.991	.963	.899	.787	23
24							.999	.995	.978	.932	.843	24
25							1.000	.997	.987	.955	.888	25
26								.999	.993	.972	.922	26
27								.999	.996	.983	.948	27
28								1.000	.998	.990	.966	28
29									.999	.994	.978	29
30									.999	.997	.987	30
31									1.000	.998	.992	31
32										.999	.995	32
33										1.000	.997	33
34											.999	34
35											.999	35
36											1.000	36

Source: J. Banks and R. G. Heikes, *Handbook of Tables and Graphs for the Industrial Engineer and Manager*, © 1984, pp. 34–35. Reprinted by permission of John Wiley and Sons, Inc., New York.

Table A.5 Percentage Points of The Student's t Distribution with v Degrees of Freedom

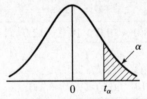

v	$t_{0.005}$	$t_{0.01}$	$t_{0.025}$	$t_{0.05}$	$t_{0.10}$
1	63.66	31.82	12.71	6.31	3.08
2	9.92	6.92	4.30	2.92	1.89
3	5.84	4.54	3.18	2.35	1.64
4	4.60	3.75	2.78	2.13	1.53
5	4.03	3.36	2.57	2.02	1.48
6	3.71	3.14	2.45	1.94	1.44
7	3.50	3.00	2.36	1.90	1.42
8	3.36	2.90	2.31	1.86	1.40
9	3.25	2.82	2.26	1.83	1.38
10	3.17	2.76	2.23	1.81	1.37
11	3.11	2.72	2.20	1.80	1.36
12	3.06	2.68	2.18	1.78	1.36
13	3.01	2.65	2.16	1.77	1.35
14	2.98	2.62	2.14	1.76	1.34
15	2.95	2.60	2.13	1.75	1.34
16	2.92	2.58	2.12	1.75	1.34
17	2.90	2.57	2.11	1.74	1.33
18	2.88	2.55	2.10	1.73	1.33
19	2.86	2.54	2.09	1.73	1.33
20	2.84	2.53	2.09	1.72	1.32
21	2.83	2.52	2.08	1.72	1.32
22	2.82	2.51	2.07	1.72	1.32
23	2.81	2.50	2.07	1.71	1.32
24	2.80	2.49	2.06	1.71	1.32
25	2.79	2.48	2.06	1.71	1.32
26	2.78	2.48	2.06	1.71	1.32
27	2.77	2.47	2.05	1.70	1.31
28	2.76	2.47	2.05	1.70	1.31
29	2.76	2.46	2.04	1.70	1.31
30	2.75	2.46	2.04	1.70	1.31
40	2.70	2.42	2.02	1.68	1.30
60	2.66	2.39	2.00	1.67	1.30
120	2.62	2.36	1.98	1.66	1.29
∞	2.58	2.33	1.96	1.645	1.28

Source: Robert E. Shannon, *Systems Simulation: The Art and Science*, © 1975, p. 372.
Reprinted by permission of Prentice Hall, Upper Saddle River, NJ.

Table A.6 Percentage Points of The Chi-Square Distribution with v Degrees of Freedom

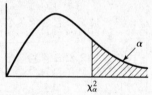

v	$\chi^2_{0.005}$	$\chi^2_{0.01}$	$\chi^2_{0.025}$	$\chi^2_{0.05}$	$\chi^2_{0.10}$
1	7.88	6.63	5.02	3.84	2.71
2	10.60	9.21	7.38	5.99	4.61
3	12.84	11.34	9.35	7.81	6.25
4	14.96	13.28	11.14	9.49	7.78
5	16.7	15.1	12.8	11.1	9.2
6	18.5	16.8	14.4	12.6	10.6
7	20.3	18.5	16.0	14.1	12.0
8	22.0	20.1	17.5	15.5	13.4
9	23.6	21.7	19.0	16.9	14.7
10	25.2	23.2	20.5	18.3	16.0
11	26.8	24.7	21.9	19.7	17.3
12	28.3	26.2	23.3	21.0	18.5
13	29.8	27.7	24.7	22.4	19.8
14	31.3	29.1	26.1	23.7	21.1
15	32.8	30.6	27.5	25.0	22.3
16	34.3	32.0	28.8	26.3	23.5
17	35.7	33.4	30.2	27.6	24.8
18	37.2	34.8	31.5	28.9	26.0
19	38.6	36.2	32.9	30.1	27.2
20	40.0	37.6	34.2	31.4	28.4
21	41.4	38.9	35.5	32.7	29.6
22	42.8	40.3	36.8	33.9	30.8
23	44.2	41.6	38.1	35.2	32.0
24	45.6	43.0	39.4	36.4	33.2
25	49.6	44.3	40.6	37.7	34.4
26	48.3	45.6	41.9	38.9	35.6
27	49.6	47.0	43.2	40.1	36.7
28	51.0	48.3	44.5	41.3	37.9
29	52.3	49.6	45.7	42.6	39.1
30	53.7	50.9	47.0	43.8	40.3
40	66.8	63.7	59.3	55.8	51.8
50	79.5	76.2	71.4	67.5	63.2
60	92.0	88.4	83.3	79.1	74.4
70	104.2	100.4	95.0	90.5	85.5
80	116.3	112.3	106.6	101.9	96.6
90	128.3	124.1	118.1	113.1	107.6
100	140.2	135.8	129.6	124.3	118.5

Source: Robert E. Shannon, *Systems Simulation: The Art and Science*, © 1975, p. 372.
Reprinted by permission of Prentice Hall, Upper Saddle River, NJ.

Table A.7 Percentage Points of The F Distribution with $\alpha = 0.05$

Degrees of Freedom for the Numerator (v_1)

v_2 \ v_1	1	2	3	4	5	6	7	8	9	10	12	15	20	24	30	40	60	120	∞
1	161.4	199.5	215.7	224.6	230.2	234.0	236.8	238.9	240.5	241.9	243.9	245.9	248.0	249.1	250.1	251.1	252.2	253.3	254.3
2	18.51	19.00	19.16	19.25	19.30	19.33	19.35	19.37	19.38	19.40	19.41	19.43	19.45	19.45	19.46	19.47	19.48	19.49	19.50
3	10.13	9.55	9.28	9.12	9.01	8.94	8.89	8.85	8.81	8.79	8.74	8.70	8.66	8.64	8.62	8.59	8.57	8.55	8.53
4	7.71	6.94	6.59	6.39	6.26	6.16	6.09	6.04	6.00	5.96	5.91	5.86	5.80	5.77	5.75	5.72	5.69	5.66	5.63
5	6.61	5.79	5.41	5.19	5.05	4.95	4.88	4.82	4.77	4.74	4.68	4.62	4.56	4.53	4.50	4.46	4.43	4.40	4.36
6	5.99	5.14	4.76	4.53	4.39	4.28	4.21	4.15	4.10	4.06	4.00	3.94	3.87	3.84	3.81	3.77	3.74	3.70	3.67
7	5.59	4.74	4.35	4.12	3.97	3.87	3.79	3.73	3.68	3.64	3.57	3.51	3.44	3.41	3.38	3.34	3.30	3.27	3.23
8	5.32	4.46	4.07	3.84	3.69	3.58	3.50	3.44	3.39	3.35	3.28	3.22	3.15	3.12	3.08	3.04	3.01	2.97	2.93
9	5.12	4.26	3.86	3.63	3.48	3.37	3.29	3.23	3.18	3.14	3.07	3.01	2.94	2.90	2.86	2.83	2.79	2.75	2.71
10	4.96	4.10	3.71	3.48	3.33	3.22	3.14	3.07	3.02	2.98	2.91	2.85	2.77	2.74	2.70	2.66	2.62	2.58	2.54
11	4.84	3.98	3.59	3.36	3.20	3.09	3.01	2.95	2.90	2.85	2.79	2.72	2.65	2.61	2.57	2.53	2.49	2.45	2.40
12	4.75	3.89	3.49	3.26	3.11	3.00	2.91	2.85	2.80	2.75	2.69	2.62	2.54	2.51	2.47	2.43	2.38	2.34	2.30
13	4.67	3.81	3.41	3.18	3.03	2.92	2.83	2.77	2.71	2.67	2.60	2.53	2.46	2.42	2.38	2.34	2.30	2.25	2.21
14	4.60	3.74	3.34	3.11	2.96	2.85	2.76	2.70	2.65	2.60	2.53	2.46	2.39	2.35	2.31	2.27	2.22	2.18	2.13
15	4.54	3.68	3.29	3.06	2.90	2.79	2.71	2.64	2.59	2.54	2.48	2.40	2.33	2.29	2.25	2.20	2.16	2.11	2.07
16	4.49	3.63	3.24	3.01	2.85	2.74	2.66	2.59	2.54	2.49	2.42	2.35	2.28	2.24	2.19	2.15	2.11	2.06	2.01
17	4.45	3.59	3.20	2.96	2.81	2.70	2.61	2.55	2.49	2.45	2.38	2.31	2.23	2.19	2.15	2.10	2.06	2.01	1.96
18	4.41	3.55	3.16	2.93	2.77	2.66	2.58	2.51	2.46	2.41	2.34	2.27	2.19	2.15	2.11	2.06	2.02	1.97	1.92
19	4.38	3.52	3.13	2.90	2.74	2.63	2.54	2.48	2.42	2.38	2.31	2.23	2.16	2.11	2.07	2.03	1.98	1.93	1.88
20	4.35	3.49	3.10	2.87	2.71	2.60	2.51	2.45	2.39	2.35	2.28	2.20	2.12	2.08	2.04	1.99	1.95	1.90	1.84
21	4.32	3.47	3.07	2.84	2.68	2.57	2.49	2.42	2.37	2.32	2.25	2.18	2.10	2.05	2.01	1.96	1.92	1.87	1.81
22	4.30	3.44	3.05	2.82	2.66	2.55	2.46	2.40	2.34	2.30	2.23	2.15	2.07	2.03	1.98	1.94	1.89	1.84	1.78
23	4.28	3.42	3.03	2.80	2.64	2.53	2.44	2.37	2.32	2.27	2.20	2.13	2.05	2.01	1.96	1.91	1.86	1.81	1.76
24	4.26	3.40	3.01	2.78	2.62	2.51	2.42	2.36	2.30	2.25	2.18	2.11	2.03	1.98	1.94	1.89	1.84	1.79	1.73
25	4.24	3.39	2.99	2.76	2.60	2.49	2.40	2.34	2.28	2.24	2.16	2.09	2.01	1.96	1.92	1.87	1.82	1.77	1.71
26	4.23	3.37	2.98	2.74	2.59	2.47	2.39	2.32	2.27	2.22	2.15	2.07	1.99	1.95	1.90	1.85	1.80	1.75	1.69
27	4.21	3.35	2.96	2.73	2.57	2.46	2.37	2.31	2.25	2.20	2.13	2.06	1.97	1.93	1.88	1.84	1.79	1.73	1.67
28	4.20	3.34	2.95	2.71	2.56	2.45	2.36	2.29	2.24	2.19	2.12	2.04	1.96	1.91	1.87	1.82	1.77	1.71	1.65
29	4.18	3.33	2.93	2.70	2.55	2.43	2.35	2.28	2.22	2.18	2.10	2.03	1.94	1.90	1.85	1.81	1.75	1.70	1.64
30	4.17	3.32	2.92	2.69	2.53	2.42	2.33	2.27	2.21	2.16	2.09	2.01	1.93	1.89	1.84	1.79	1.74	1.68	1.62
40	4.08	3.23	2.84	2.61	2.45	2.34	2.25	2.18	2.12	2.08	2.00	1.92	1.84	1.79	1.74	1.69	1.64	1.58	1.51
60	4.00	3.15	2.76	2.53	2.37	2.25	2.17	2.10	2.04	1.99	1.92	1.84	1.75	1.70	1.65	1.59	1.53	1.47	1.39
120	3.92	3.07	2.68	2.45	2.29	2.17	2.09	2.02	1.96	1.91	1.83	1.75	1.66	1.61	1.55	1.50	1.43	1.35	1.25
∞	3.84	3.00	2.60	2.37	2.21	2.10	2.01	1.94	1.88	1.83	1.75	1.67	1.57	1.52	1.46	1.39	1.32	1.22	1.00

Degrees of Freedom for the Denominator (v_2)

Source: W. W. Hines and D. C. Montgomery, *Probability and Statistics in Engineering and Management Science*, 2d ed., © 1980, p. 599.

Reprinted by permission of John Wiley & Sons, Inc., New York.

Table A.8   Kolmogorov–Smirnov Critical Values

Degrees of Freedom (N)	$D_{0.10}$	$D_{0.05}$	$D_{0.01}$
1	0.950	0.975	0.995
2	0.776	0.842	0.929
3	0.642	0.708	0.828
4	0.564	0.624	0.733
5	0.510	0.565	0.669
6	0.470	0.521	0.618
7	0.438	0.486	0.577
8	0.411	0.457	0.543
9	0.388	0.432	0.514
10	0.368	0.410	0.490
11	0.352	0.391	0.468
12	0.338	0.375	0.450
13	0.325	0.361	0.433
14	0.314	0.349	0.418
15	0.304	0.338	0.404
16	0.295	0.328	0.392
17	0.286	0.318	0.381
18	0.278	0.309	0.371
19	0.272	0.301	0.363
20	0.264	0.294	0.356
25	0.24	0.27	0.32
30	0.22	0.24	0.29
35	0.21	0.23	0.27
Over 35	$\dfrac{1.22}{\sqrt{N}}$	$\dfrac{1.36}{\sqrt{N}}$	$\dfrac{1.63}{\sqrt{N}}$

Source: F. J. Massey, "The Kolmogorov–Smirnov Test for Goodness of Fit," *The Journal of the American Statistical Association*, Vol. 46. © 1951, p. 70. Adapted with permission of the American Statistical Association.

Table A.9 Maximum Likelihood Estimates of the Gamma Distribution

$1/M$	β	$1/M$	β	$1/M$	β
0.020	0.0187	2.700	1.494	10.300	5.311
0.030	0.0275	2.800	1.545	10.600	5.461
0.040	0.0360	2.900	1.596	10.900	5.611
0.050	0.0442	3.000	1.646	11.200	5.761
0.060	0.0523	3.200	1.748	11.500	5.911
0.070	0.0602	3.400	1.849	11.800	6.061
0.080	0.0679	3.600	1.950	12.100	6.211
0.090	0.0756	3.800	2.051	12.400	6.362
0.100	0.0831	4.000	2.151	12.700	6.512
0.200	0.1532	4.200	2.252	13.000	6.662
0.300	0.2178	4.400	2.353	13.300	6.812
0.400	0.2790	4.600	2.453	13.600	6.962
0.500	0.3381	4.800	2.554	13.900	7.112
0.600	0.3955	5.000	2.654	14.200	7.262
0.700	0.4517	5.200	2.755	14.500	7.412
0.800	0.5070	5.400	2.855	14.800	7.562
0.900	0.5615	5.600	2.956	15.100	7.712
1.000	0.6155	5.800	3.056	15.400	7.862
1.100	0.6690	6.000	3.156	15.700	8.013
1.200	0.7220	6.200	3.257	16.000	8.163
1.300	0.7748	6.400	3.357	16.300	8.313
1.400	0.8272	6.600	3.457	16.600	8.463
1.500	0.8794	6.800	3.558	16.900	8.613
1.600	0.9314	7.000	3.658	17.200	8.763
1.700	0.9832	7.300	3.808	17.500	8.913
1.800	1.034	7.600	3.958	17.800	9.063
1.900	1.086	7.900	4.109	18.100	9.213
2.000	1.137	8.200	4.259	18.400	9.363
2.100	1.188	8.500	4.409	18.700	9.513
2.200	1.240	8.800	4.560	19.000	9.663
2.300	1.291	9.100	4.710	19.300	9.813
2.400	1.342	9.400	4.860	19.600	9.963
2.500	1.393	9.700	5.010	20.000	10.16
2.600	1.444	10.000	5.160		

Source: S.C. Choi and R. Wette, "Maximum Likelihood Estimates of the Gamma Distribution and Their Bias," *Technometrics*, Vol. 11, No. 4, Nov. © 1969, pp. 688–9. Adapted with permission of the American Statistical Association.

Table A.10 Operating Characteristic Curves for The Two-Sided t Test for Different Values of Sample Size n

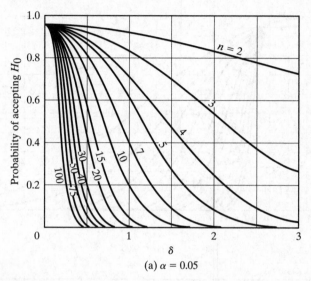

(a) $\alpha = 0.05$

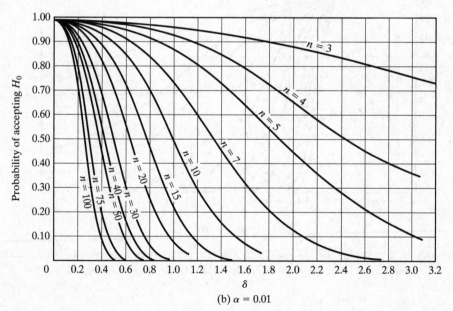

(b) $\alpha = 0.01$

Source: C. L. Ferris, F. E. Grubbs, and C. L. Weaver, "Operating Characteristics for the Common Statistical Tests of Significance," *Annals of Mathematical Statistics*, June 1946. Reproduced with permission of The Institute of Mathematical Statistics.

Table A.11 Operating Characteristic Curves for the One-Sided t Test for Different Values of Sample Size n

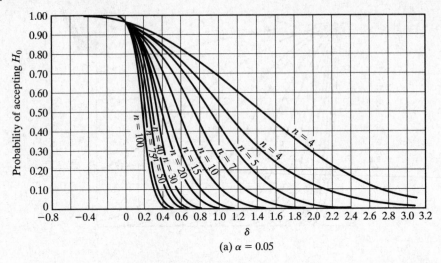

(a) $\alpha = 0.05$

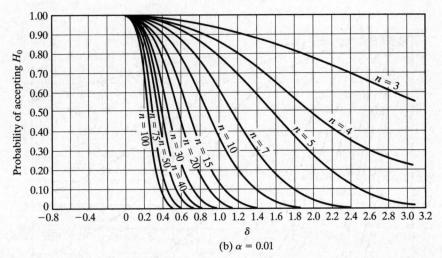

(b) $\alpha = 0.01$

Source: A. H. Bowker and G. J. Lieberman, *Engineering Statistics*, 2d ed., © 1972, p. 203.
Reprinted by permission of Prentice Hall, Upper Saddle River, NJ.

Table A.12 Values of Rinott's Constant h

$1 - \alpha$	R_0	\multicolumn{9}{c}{K}								
		2	3	4	5	6	7	8	9	10
	5	2.291	3.058	3.511	3.837	4.093	4.305	4.486	4.644	4.786
	6	2.177	2.871	3.270	3.552	3.771	3.951	4.103	4.235	4.352
	7	2.107	2.758	3.126	3.384	3.582	3.744	3.881	3.999	4.103
	8	2.059	2.682	3.031	3.273	3.459	3.609	3.736	3.845	3.941
	9	2.025	2.628	2.963	3.195	3.372	3.515	3.635	3.738	3.829
	10	1.999	2.587	2.913	3.137	3.307	3.445	3.560	3.659	3.746
	11	1.978	2.556	2.874	3.092	3.258	3.391	3.503	3.598	3.682
	12	1.962	2.531	2.843	3.056	3.218	3.349	3.457	3.551	3.632
	13	1.948	2.510	2.817	3.027	3.186	3.314	3.420	3.512	3.592
0.90	14	1.937	2.493	2.796	3.003	3.160	3.285	3.390	3.480	3.558
	15	1.928	2.479	2.779	2.983	3.138	3.261	3.364	3.453	3.530
	16	1.919	2.467	2.764	2.966	3.119	3.241	3.343	3.430	3.506
	17	1.912	2.456	2.751	2.951	3.102	3.223	3.324	3.410	3.485
	18	1.906	2.447	2.739	2.938	3.088	3.208	3.308	3.393	3.467
	19	1.901	2.438	2.729	2.926	3.075	3.194	3.293	3.378	3.451
	20	1.896	2.431	2.720	2.916	3.064	3.182	3.280	3.364	3.437
	30	1.866	2.387	2.666	2.855	2.997	3.110	3.204	3.284	3.354
	40	1.852	2.366	2.641	2.827	2.966	3.077	3.169	3.247	3.315
	50	1.844	2.354	2.627	2.810	2.948	3.057	3.148	3.225	3.292
	5	3.107	3.905	4.390	4.744	5.025	5.259	5.461	5.638	5.797
	6	2.910	3.602	4.010	4.303	4.533	4.722	4.884	5.025	5.150
	7	2.791	3.424	3.791	4.051	4.253	4.419	4.559	4.681	4.789
	8	2.712	3.308	3.649	3.889	4.074	4.225	4.353	4.463	4.561
	9	2.656	3.226	3.550	3.776	3.950	4.091	4.210	4.313	4.404
	10	2.614	3.166	3.476	3.693	3.859	3.993	4.106	4.204	4.290
	11	2.582	3.119	3.420	3.629	3.789	3.918	4.027	4.121	4.203
	12	2.556	3.082	3.376	3.579	3.734	3.860	3.965	4.055	4.135
	13	2.534	3.052	3.340	3.539	3.690	3.812	3.915	4.003	4.080
0.95	14	2.517	3.027	3.310	3.505	3.654	3.773	3.874	3.960	4.035
	15	2.502	3.006	3.285	3.477	3.623	3.741	3.839	3.924	3.998
	16	2.489	2.988	3.264	3.453	3.597	3.713	3.810	3.893	3.966
	17	2.478	2.973	3.246	3.433	3.575	3.689	3.785	3.867	3.938
	18	2.468	2.959	3.230	3.415	3.556	3.669	3.763	3.844	3.914
	19	2.460	2.948	3.216	3.399	3.539	3.650	3.744	3.824	3.894
	20	2.452	2.937	3.203	3.385	3.523	3.634	3.727	3.806	3.875
	30	2.407	2.874	3.129	3.303	3.434	3.539	3.626	3.701	3.766
	40	2.386	2.845	3.094	3.264	3.392	3.495	3.580	3.652	3.716
	50	2.373	2.828	3.074	3.242	3.368	3.469	3.553	3.624	3.687

Index